ANNUAL REVIEW OF ECOLOGY AND SYSTEMATICS

ANNUAL REVIEW OF ECOLOGY AND SYSTEMATICS

RICHARD F. JOHNSTON, *Editor*
University of Kansas

PETER W. FRANK, *Associate Editor*
University of Oregon

CHARLES D. MICHENER, *Associate Editor*
University of Kansas

VOLUME 10

1979

ANNUAL REVIEWS INC.   4139 EL CAMINO WAY   PALO ALTO, CALIFORNIA 94306

ANNUAL REVIEWS INC.
Palo Alto, California, USA

REPRINTS The conspicuous number aligned in the margin with the title of each article in this volume is a key for use in ordering reprints. Available reprints are priced at the uniform rate of $1.00 each postpaid. The minimum acceptable reprint order is 5 reprints and/or $5.00 prepaid. A quantity discount is available.

International Standard Serial Number: 0066-4162
International Standard Book Number: 0-8243-1410-7
Library of Congress Catalog Card Number: 71-135616

Annual Reviews Inc. and the Editors of its publications assume no responsibility for the statements expressed by the contributors to this Review.

PRINTED AND BOUND IN THE UNITED STATES OF AMERICA

PREFACE

This volume marks the tenth anniversary of the *Annual Review of Ecology and Systematics*. Each year for a decade we have published an average of 20 authors, 17 reviews, 480 pages. Since disciplines tend to proceed independently of review volumes, which must follow more than lead, our ambition is to affect not so much the course of science as the perspectives of individuals.

Have we often enough reached our goal, which has been to make the hard work of scholarly investigation a bit easier or more rapid—especially for young people and others new to ecology and systematics?

That the series has survived ten years heartens us, as does the variety of positive comment received from those who read and use *ARES*. And it is remarkable in view of the short half-life of a review paper that several reviews from our earlier volumes are still cited in research articles. We hope our future volumes will likewise make enduring contributions.

Volume 10 contains a greater than average number of topics in systematics. S. Conway Morris writes on the fauna of the Burgess Shale, G. R. Coope on Cenozoic coleoptera, S. Hopper on speciation in Australian floras, J. T. Wiebes on insect pollinators of figs, C. Heiser on the evolution of New World cultivated plants, and E. Nevo on subterranean mammals.

Population ecology is treated by Hamrick, Linhart & Mitton (life history and allozyme variation), J. White (the plant as a metapopulation), D. Janzen (biology of figs), and Parsons & Bock (population biology of Australian drosophilids).

Community or ecosystem-level approaches are characteristic of F. A. Bazzaz (plant succession), Brown, Reichman & Davidson (granivorous rodents), and Gorham, Reiners & Vitousek (ecosystem succession and regulation of chemical budgets).

Evolutionary ecology is the focus of reviews by Cummins & Klug (feeding ecology) and R. Silberglied (communication in the ultraviolet).

We much appreciate assistance by Stephen J. Gould, Ernst Mayr, and Edward O. Wilson in planning the contents of Volume 10. Thanks also go to our authors, who continue to do these reviews without stipend. As always, we encourage readers to tell us their thinking on possible new topics for review.

THE EDITORS AND THE EDITORIAL COMMITTEE

ERRATUM

Volume 9 (1978)

In "Group Selection, Altruism, and the Levels of Organization of Life" by Richard D. Alexander and Gerald Borgia:

Apropos Haldane's discussion (in *Causes of Evolution*, 1932) of the effects of selection on the alleles R and r in *Oenothera,* the faster growing R pollen produces a plant that is *more* (not less) cold-resistant. Haldane presumed that "the gene R cannot establish itself in the species because RR homozygotes are inviable, probably owing to linkage with a lethal." Fortunately this error does not affect the point of the discussion (both in the review and in Haldane): Selection at gametic and individual levels may conflict; as Haldane put it (p. 123), "a gene which greatly accelerates pollen tube growth will spread through a species even if it causes moderately disadvantageous changes in the adult plant."

Annual Review of Ecology and Systematics
Volume 10, 1979

CONTENTS

SOME RELATED ARTICLES IN OTHER *ANNUAL REVIEWS*

From the *Annual Review of Anthropology,* Volume 8 (1979):

Hominid Evolution in Eastern Africa During the Pliocene and Early Pleistocene, Noel T. Boaz

Demography and Archaeology, Fekri A. Hassan

Anthropological Genetics of Small Populations, Alan G. Fix

Recent Finds and Interpretations on Miocene Hominoids, David Pilbean

From the *Annual Review of Genetics,* Volume 12 (1978):

Environmental Mutagens and Carcinogens, Minako Nagao, Takashi Sugimura, and Taijiro Matsushima

Chromosomal Phylogeny of the Primates, Jean de Grouchy, Catherine Turleau, and Catherine Finaz

The Population Structure of an Amerindian Tribe, the Yanomama, James V. Neel

From the *Annual Review of Entomology,* Volume 24 (1979):

Biogeographic Dynamics of Insect–Host Plant Communities, Donald R. Strong, Jr.

Evolution of Phytophagous Mites (Acari), G. W. Krantz and E. E. Lindquist

Ecological Diversity in Trichoptera, Rosemary J. Mackay and Glenn B. Wiggins

Integrated Pest Control in the Developing World, L. Brader

Population Dynamics of Bark Beetles, Robert N. Coulson

Cytotaxonomy of Black Flies (Simuliidae), K. H. Rothfels

Annual Reviews are published in the following sciences: Anthropology, Astronomy and Astrophysics, Biochemistry, Biophysics and Bioengineering, Earth and Planetary Sciences, Ecology and Systematics, Energy, Entomology, Fluid Mechanics, Genetics, Materials Science, Medicine, Microbiology, Neuroscience, Nuclear and Particle Science, Pharmacology and Toxicology, Physical Chemistry, Physiology, Phytopathology, Plant Physiology, Psychology, and Sociology. The *Annual Review of Public Health* will begin publication in 1980. In addition, four special volumes have been published by Annual Reviews Inc.: *History of Entomology* (1973), *The Excitement and Fascination of Science* (1965), *The Excitement and Fascination of Science, Volume Two* (1978), and *Annual Reviews Reprints: Cell Membranes, 1975–1977* (published 1978). For the convenience of readers, a detachable order form/ envelope is bound into the back of this volume.

Ann. Rev. Ecol. Syst. 1979. 10:1–12

CO-EVOLUTION OF FIGS AND THEIR INSECT POLLINATORS

♦4153

J. T. Wiebes

Department of Systematics and Evolutionary Biology,
Leiden University, The Netherlands

INTRODUCTION

For the pollination of their flowers, figs (*Ficus* spp., Moraceae) are dependent upon Hymenoptera Chalcidoidea of the family Agaonidae (fig wasps). For the propagation of their kind, the fig wasps are dependent upon the ovaries of the figs, in which their larvae develop. Thus, the figs and fig wasps are interdependent; generally, their relation is strictly specific—i.e. every species of fig has its own species of pollinator wasp (21, 25, 28). For historical data on the development of our knowledge of fig–wasp relations and for a supplement to an earlier bibliography (17), general reference is made to my short history of fig wasp research (30), in which the literature on figs and fig wasps was reviewed.

POLLINATION OF FIGS

Fig flowers are of three kinds: male, female, and neuter or gall flowers (a form of female flowers, often modified, having a short style). Fig sycones, hollow receptacles bearing the flowers on the inner surface, may contain all three kinds of flowers (monoecism), or they may have male and gall flowers in the one "male" fig and female flowers only in another "female" fig, on different plants (dioecism, or better, gynodioecism). In the Edible Fig *Ficus carica* L., the situation is somewhat complicated in that several generations of wild (*caprificus*) figs are formed in the course of one year, with various numbers of male flowers, a few female flowers, and gall flowers in which the wasps, *Blastophaga psenes* (L.), develop (2). The pollen transfer and pollination in *Ficus carica* are rather primitive (11). The female wasps load pollen in the polleniferous fig into intersegmental and pleural invaginations

1

0066-4162/79/1120-0001$01.00

that form in the shrunken body of the wasp following water loss in the drying sycone. They unload in the more humid receptive fig as a result of partial swelling and contortion of the wasp's body during its futile oviposition attempts. This process, recently discovered, is quite contrary to traditional ideas. Another primitive pollination process was briefly mentioned by Ramírez (20) for wasps of the genus *Tetrapus* Mayr (pollinators of American figs of section *Pharmacosycea* Miq.), which carry pollen in the digestive tract.

Galil and Ramírez have found more intricate pollination syndromes (reminding one of the famous situation in *Yucca* with its moths) in most other figs. In many cases the pollen is loaded and transported by the female wasps in special containers—i.e. "corbiculae" on the fore coxae and "pockets" in the mesosternum (9, 20). After the wasps have entered the fig sycone (and lost their wings and the greater part of the antennae in the process), they stand on the synstigma formed by the contiguous stigmas of the fig flowers and start oviposition through the styles. Toward the end of an oviposition act the forelegs of *Pegoscapus estherae* (Grandi), an agaonid from Costa Rica associated with *Ficus costaricana* (Liebm.) Miq. (12), shovel the pollen from the thoracic pockets into the corbiculae of the fore coxae. From these corbiculae they extract some pollen grains with the (fore) tarsal arolium; pollination is effected by striking the tarsi against each other, thus shaking the pollen onto the stigmas below. In another instance [Malayan *Ceratosolen hewitti* Waterston associated with *Ficus fistulosa* Reinw. (8)] the pollen is taken directly from the thoracic pockets and smeared on the stigmatic surface with the tarsal arolia.

The pollination behavior in young fig sycones and the process of loading the pockets in ripe sycones have been described for a few species only, but much can be inferred from the morphology of wasps and figs (22). The pollen pockets may be relatively open, shallow impressions; they may be more closed, only to open when the wasp curves the thorax so that the covering membranes of the pockets stand out, as described for *Ceratosolen arabicus* (Mayr) from the African *Ficus sycomorus* (L.) (10); or they may open by an as yet unknown mechanism, as in *Platyscapa quadraticeps* (Mayr) from Israeli *Ficus religiosa* L. (13). The fore coxae may bear combs, used to shovel pollen into the pockets when loading; moreover, they may have regular corbiculae that are used as containers for the pollen, as mentioned above. These various states of the coxal apparatus have not been distinguished properly in all instances recorded (e.g. 31).

The accompanying figures (1–3) illustrate the life cycle of monoecious or dioecious figs and their wasps and indicate the apparatus the female wasps use for pollination. The male wasps are entirely different from the females. For some details on their biology see (15).

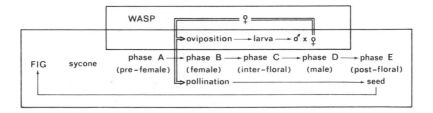

Figure 1 Generalized cycle of a monoecious fig and its pollinator wasp. The developmental phases of the syconium (A–E) are indicated in the lower rectangle, those of the wasp in the upper.

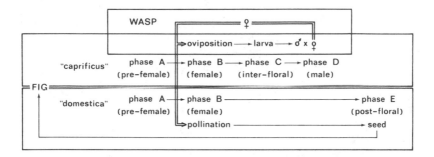

Figure 2 Generalized cycle of a (gyno)dioecious fig and its pollinator wasp. The female cycle (*domestica*-form of *Ficus carica*) runs from phase A, over the pollinated flowers in phase B, directly to the seeds of phase E; the male cycle (*caprificus*-form) from phase A, over the galled flowers in phase B and the inter-floral phase C, to the male phase D, from which the female and male wasps emerge.

SPECIFIC RELATIONS OF FIGS AND WASPS

The basis for speculations on the specificity of the relations of figs and wasps is the observation that the insects reared from a given species of fig are always of the same species. This seems fairly obvious, but actually it is where the difficulties begin. In any instance of supposed (host-) specificity one is inclined to look for similarities when the species of host are supposed to be identical and for differences when they are supposed to be distinct. Here is a potential source of serious bias.

The observation that related figs have related pollinator wasps seems to strengthen the suggestion of host-specificity. It should be borne in mind, however, that specificity may be ecologically as well as phylogenetically determined. On comparing the classifications of figs and wasps one should independently consider the arguments for both. If (and to the extent that) the two can be reconciled, the veracity of phylogenetic specificity is enhanced.

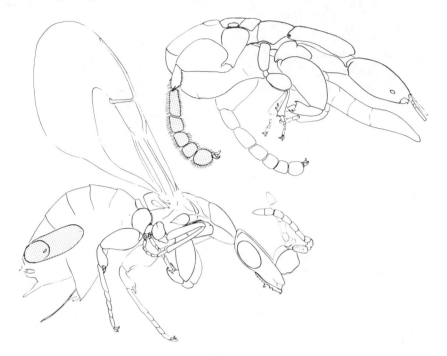

Figure 3   A fig wasp (*Ceratosolen dentifer* Wiebes), female (left) and male (right). Corbiculae are situated in the fore coxae of the female, pockets in the mesosternum of the thorax. In this species there are peculiarly large peritremes of the abdominal stigmata in the female, while the male has large hirsute hind legs.

Species-specific Relations of Figs and Wasps

As of 1963, most records of the preferences of the Agaonidae confirmed host-specificity (25). [Some did not, but either these were not documented (and therefore have no evidential value) or there was reason to doubt the taxonomic evidence.] However, soon after a list of all species of *Ficus* mentioned as hosts of fig wasps was published in 1966 (28), well-documented examples became known where strict specificity appeared to break down. These instances can be classified as follows.

ASSOCIATIONS OF ONE SPECIES OF FIG AND MORE THAN ONE SPECIES OF AGAONID   *Ficus sycomorus* is widely distributed over Africa (14), where it is often associated with two species of agaonid—*Ceratosolen arabicus* and *C. galili* Wiebes (29). On closer inspection (9), the former proved to be the pollinator, while the latter acted as a cuckoo, ovipositing in the flowers without pollinating the stigmas. A similar association may exist in several other figs [e.g. *Ficus obliqua* Forst.f. from Fiji harbors *Blastophaga greenwoodi* Grandi, while *F. obliqua* var. *petiolaris* (Benth.) Corner in Australia has *Pleistodontes imperialis* Saunders for a pollinator —as has the related *Ficus rubiginosa* Desf. ex Vent. (26)]. In several other instances a cuckoo relation is suspected, but it has only been verified in *Ficus sycomorus.*

Ramírez observed "strange" wasps to enter the sycones of the Venezuelan *Ficus turbinata* Pitt., where they oviposited and gave rise to a new generation of wasps. They may also have pollinated the fig flowers, but no viable seeds developed (21). Quite a number of double occurrences concern geographical variation of the wasp species while the host fig remains uniform over the whole range—e.g. *Ficus fistulosa* Reinw. with *Ceratosolen constrictus* (Mayr) in Java and Sumatra and *C. hewitti* Waterston in Borneo and Java (25).

ASSOCIATIONS OF ONE SPECIES OF AGAONID WITH MORE THAN ONE SPECIES OF FICUS   Here again the apparent association of one wasp and two figs may be due to the geographical variation of the one and the uniformity of the other—e.g. *Ficus soroceoides* Baker from Madagascar and *F. laterifolia* Vahl from Réunion, both associated with *Kradibia cowani* Saunders (32). In some instances, however, the figs are sympatric and there is good reason to consider them biological species, while no morphological differences can be found between their wasps. In Hong Kong, Hill (16) found the same pollinating insect in *Ficus pyriformis* Hook. et Arn., *F. variolosa* Lindl. ex Benth., and *F. erecta* Thunb. var. *beecheyana* (Hook. et Arn.) King. He argued that the insects, although morphologically indistinguishable, constituted three biological species, each pollinating its own (morphologically recognizable) species of fig. It should be noted that the figs mentioned belong to one subseries, *Podosyceae* Corner. In another instance (32), African insects identified as *Kradibia gestroi* (Grandi) were recorded from three species of fig belonging to three different series of *Ficus,* indicated in Table 1 with some related figs and their pollinators. Here, the discrepancy is not only in the rank of the taxon concerned (whether variety, subspecies, or species) but also in its affinity. Much as some identifications want confirmation, it is clear that, for instance, *Ficus asperifolia* is considered closer to *F. stortophylla* (to the extent of being almost indistinguishable) and *F. leptogramma,* than to *F. exasperata;* yet their wasps indicate different relationships.

Related Figs Have Related Wasps

When stating that related figs generally have related pollinating wasps one should indicate the level of comparison. The statement proves true for most

Table 1 Figs of the subsection *Varinga*, subdivided into three series, and their pollinating wasps (Agaonidae)

Figs	Wasps
Heterophylleae Corner	Agaonidae
Ficus heterophylla Linn. f.	*Kradibia brownii* sensu Grandi
Ficus capreaefolia Del.	⎧ *Kradibia* cf. *gestroi* (Grandi)
Exasperatae Corner	⎪
Ficus exasperata Vahl	⎨ *Kradibia gestroi* (Grandi)
Cyrtophylleae Corner	⎪
Ficus asperifolia Miq. in Hook.	⎩ *Kradibia gestroi* (Grandi)
Ficus stortophylla Warb.	*Kradibia hilli* Wiebes
⌈*Ficus asperiuscula* K. & B.	*Ceratosolen internatus* Wiebes⌉
Ficus leptogramma Corner	*Kradibia setigera* Wiebes

Ficus sections and subsections and their agaonid genera, but there are some notable exceptions on the level of the *Ficus* series—e.g. the instance of *Kradibia gestroi,* shown above. On the level of *Ficus* subgenera the statement is obviously false, as may be seen from Table 2. In the first column the subgenera, sections, and subsections of *Ficus* are listed as classified by Corner (4). Corner had some knowledge of the classification of the pollinators but in general (6) did not allow this to influence his botanical classification. The second column contains the genera of pollinator wasps; while listing (and revising) these, I had much information on *Ficus* but tried to argue the classification on entomological grounds only (32–33). Ramírez (24) proposed a new classification of *Ficus,* highly influenced by data on the pollination system in the various groups of wasps. Some of the groups of *Blastophaga* sensu lato, not revised as yet, are indicated by A–D and G, as used by Ramírez; in *Ceratosolen* I distinguished several species-groups (25), some of which are here taken together as 1–4. These groups are here treated just as are the genera. The apparent exceptions to the expected strict phylogenetic specificity fall into one of the following cases.

MORE THAN ONE GENUS OF AGAONID FOR A (SUB)SECTION OF FICUS   In some of the subsections of *Ficus* (i.e. *Conosycea, Benjamina, Ficus, Sycidium,* and *Varinga*), we find agaonid symbionts belonging to different genera. In most, these agaonid genera are peculiar to the various series or subseries contained in the section—e.g. *Deilagaon* to the fig series *Validae, Eupristina* to the *Drupaceae* and *Indicae,* and *Waterstoniella* to the *Crassirameae* (Table 3). In one instance, however, a fig belonging to the subseries *Crassirameae* (i.e. *Ficus forstenii* Miq.) harbors an agaonid of the genus *Eupristina* (*E. bakeri* Grandi) instead of a *Waterstoniella.* Similar exceptions occur in the subsections of *Sycidium,* where some pollinator species of *Ceratosolen* are classified between a series of *Blastophaga* and species of *Kradibia.* One example may be seen in Table 1.

Table 2 Subgenera,[a] sections,[b] and subsections[c] of the genus *Ficus* L. and the genera of their pollinating wasps (Agaonidae)

Figs	Wasps
UROSTIGMA	
Urostigma	*Platyscapa*
Leucogyne	*Maniella*
Conosycea	
Conosycea	*Deilagaon, Eupristina, Waterstoniella*
Dictyoneuron	*Waterstoniella*
Benjamina	*Eupristina, Parapristina*
Stilpnophyllum	*Blastophaga* G
Americana	*Pegoscapus*
Malvanthera	*Pleistodontes* ⎫
Galoglychia	*Agaon* etc. ⎪
PHARMACOSYCEA	⎬ subfamily Agaoninae
Pharmacosycea	*Tetrapus* ⎪
Oreosycea	*Dolichoris* ⎭
FICUS	
Ficus	
Ficus	*Blastophaga* A, *Ceratosolen* 2
Eriosycea	*Blastophaga* A
Rhizocladus	*Blastophaga* A
Kalosyce	*Blastophaga* A
Sinosycidium	?
Sycidium	
Sycidium	*Blastophaga* C, D, *Ceratosolen* 3, *Kradibia*
Varinga	*Ceratosolen* 3, *Kradibia*
Palaeomorphe	*Liporrhopalum*
Adenosperma	*Ceratosolen* 2
Neomorphe	*Ceratosolen* 1
Sycocarpus	
Auriculisperma	*Ceratosolen* 2
Dammaropsis	*Ceratosolen* 2
Papuasyce	*Ceratosolen* 2, 3
Lepidotus	?
Macrostyla	?
Sycocarpus	*Ceratosolen* 3, 4
SYCOMORUS	*Ceratosolen* 1

[a] Capital letters
[b] Roman letters
[c] Italic letters

Table 3 Subsections,[a] series,[b] and subseries[c] with some of the species of the *Ficus* section *Conosycea* and their pollinating wasps (Agaonidae)

Figs	Wasps
CONOSYCEA	Agaonidae
Validae Miq.	genus *Deilagaon* Wiebes
Drupaceae Corner	
Drupaceae	genus *Eupristina* Saunders
Indicae Corner	genus *Eupristina* Saunders
Zygotricheae Corner	genus *Waterstoniella* Grandi
Crassirameae Corner	
Ficus stupenda Miq.	*Waterstoniella masii* (Grandi)
Ficus crassiramea Miq.	*Waterstoniella jacobsoni* (Grandi)
[*Ficus forstenii* Miq.	*Eupristina bakeri* (Grandi)]
DICTYONEURON	genus *Waterstoniella* Grandi
BENJAMINA	
Benjamina (Miq.) Corner	genus *Eupristina* Saunders
Calophylleae Corner	genus *Parapristina* Hill

[a] Capital letters
[b] Roman letters
[c] Italic letters

ONE GENUS OF AGAONID ASSOCIATED WITH FIGS BELONGING TO DIFFERENT (SUB)SECTIONS The associations of species of *Waterstoniella* with figs of the subseries *Crassirameae,* mentioned above, may indicate a relationship of these figs with the subsection *Dictyoneuron* rather than with the other series and subseries of the subsection *Conosycea* (Table 3). Similar instances were discussed by Corner (5, 6) for *Ficus pseudopalma* Blanco (this is the species of subsection *Ficus* with a *Ceratosolen* of group 2) and *F. pritchardii* Seem. (with a species of *Ceratosolen* group 3). *F. pritchardii* was first put with doubt at the end of the section *Oreosycea* but was later classified with the subsection *Papuasyce*. Its entomological relationships are neither with *Dolichoris* (from *Oreosycea*) nor with *Ceratosolen* group 2 (from *Papuasyce*) but with *Ceratosolen* group 3 (from some subsections of the section *Sycidium;* but one is known from the subsection *Sycocarpus*!).

The Classification of Figs and Wasps

One basic discrepancy in the present classifications of figs and wasps may stem from the fact that the two are in principle incomparable. In several instances the coherence of the groups constructed has been based on similarities in primitive characters—i.e. combining taxa in what is considered an ancestral group. Examples are *Blastophaga* A in Ramírez's classification (24), *Ceratosolen* group 3 in mine (25), and the subgenus *Pharmacosycea* in Corner's (6). Recently an attempt was made to classify the pollinators of the genus *Pharmacosycea* (33). It appeared that the resemblance noted earlier (16) was based on primitive character states only, and it was shown that a more significant resemblance in derived characters existed as follows: *Tetrapus* with *Pleistodontes* and *Agaon,* etc (hence the subfamily Agaoninae) and *Dolichoris* with *Blastophaga,* etc (subfamily Blastophaginae). The basic division into the agaonid subfamilies should be a suggestion for a reappraisal of the classification of their host figs.

For the classification of the African *Galoglychia* (or *Bibracteatae,* as they were named previously) there are two suggestions (18, 19); in broad outlines these are reproduced in Figure 4, along with an indication of their agaonid genera (25 associations known out of about a hundred species of fig). It may be concluded that the recognition of the *Stipulares* as a group characterized by its pollinators (*Agaon* group ii) appears justified; the combination of several groups into the *Caducae,* however, obscures the relationships as suggested by the pollinator wasps. Actually, a finer subdivision of the *Platyphyllae* and *Chlamydorae* seems warranted.

THE ORIGIN OF BLASTOPHAGY

An accumulation of primitive characters may not serve to identify a primitive group, but it can help us to visualize the primitive association of figs and wasps. Corner (3) discussed the characters of the Moraceae, to which *Ficus* belongs, and noted the simultaneous occurrence of pachycauly and several primitive traits in the inflorescence and flowers, but this may be related to disperal rather than to pollination. The inflorescence of *Ficus* is a pseudocarp, which Berg (1) derives directly from a hypothetical ancestor showing a cymose capitulum with the terminal female flowers surrounded

BIBRACTEATAE MILDBRAED & BURRET, 1911	AGAON I	AGAON II	ALLOTRIOZOON	ELISABETHIELLA	NIGERIELLA	ALFONSIELLA	BIBRACTEATAE HUTCHINSON, 1916	
CAULOCARPAE	•							
FASCICULATAE	•						FASCICULATAE	
ELEGANTES								
CRASSICOSTAE	•							
CYATHISTIPULAE		•					STIPULARES	AXILLARES
			•					
PLATYPHYLLAE				•	•		CADUCAE	
CHLAMYDORAE				•	•	•		

Figure 4   Classifications (in broad outlines) of the *Bibracteatae* by Mildbraed & Burret (1911) and Hutchinson (1916) compared, with the genera of the agaonid pollinators indicated.

by male flowers. A remarkable aspect in the structure of the fig is the formation of urceolate receptacles with the flowers inside. This excludes wind pollination. I conclude that the symbiosis of figs and wasps made possible (and thus antedated) the special form of the syconium. Again, as for pachycauly, syconial development may have been related in the first place to disperal rather than to pollination.

Several researchers have made suggestions about the ancestral way of life of the pre-agaonid. Wiebes (25) suggested the fig wasps could have descended from gall-forming Chalcidoidea or from parasitic Chalcidoidea living on other insects in the flowers or seeds of the pre-*Ficus*. On the basis of the observation that most agaonids without pollen pockets are found to have pollen in the digestive tract, Ramírez (23) suggested that the agaonids evolved from pollen-feeding gall-makers. Why, then, the formation of the pollen pockets? It is easier to imagine that the carrying of pollen in the digestive tract was from the beginning one of the strategies followed—the alternative being to carry the pollen in pockets. This also seems to have been what Galil (7) had in mind when he opposed topocentric (i.e. passive) and ethodynamic (active) pollination. In Galil's opinion the pre-agaonid must have been a gall-producing parasite.

An obvious comparison is with certain species of *Megastigmus* Dalman (Chalcidoidea, Torymidae), which are injurious to the seeds of fir cones, juniper berries, and the like (but have also been reared from Hymenopterous gall-makers). If these plants showed the remarkable protogyny of the fig they could be pollinated by these chalcids. Probably the relationships of the Agaonidae lie with these torymids and some pteromalids, several of which [e.g. the tribes to which the fig wasp parasites belong (27)] must have evolved with the *Ficus*–agaonid symbiosis. Some (e.g. *Sycophaga* Westwood and the Otitesellini) may have remained parasites (upon the symbiosis); others "learned" to oviposit through the wall of the then closed fig. The taxonomic situation corroborates this explanation in that the Agaonidae and Sycophaginae (most parasitic tribes taken together) are related at a family level, while *Sycophaga* mentioned above and some of the parasites ovipositing from the outside are related at a tribal level—each tribe having overcome in its own peculiar way the difficulty of piercing the wall of the fig (30).

CONCLUSIONS

Pollinating fig wasps are species-specific to their host, although in some instances the fig or the wasp may have developed into distinct subspecies. In a few cases the evidence is not conclusive; it has been suggested that the wasps, although morphologically indistinguishable, behave as good species.

There may exist an isolation between fig species because of the incompatibility of foreign pollen.

Phylogenetic specificity is apparent at the level of agaonid genera (or species groups) and (sub)sections (or series) of *Ficus*. Some discrepancies are expected to disappear once the classifications of figs and wasps have been reevaluated. It may well be that a basic difference remains on the level of the subgenera of *Ficus* and the subfamilies of the Agaonidae as a consequence of a possible nonspecificity in the relation between the pre-agaonid and the pre-*Ficus*. The pre-agaonid is suggested to have been a gall-producing parasite of the pre-*Ficus* flower; the evolution of the fig syconium may be related to dispersal of the seeds rather than to the pollination of the flowers.

Literature Cited

1. Berg, C. C. 1977. Urticales, their differentiation and systematic position. *Plant. Syst. Evol.* Suppl. 1:349–74
2. Buscalioni, L., Grandi, G. 1938. Il *Ficus carica* L., la sua biologica, la sua coltivazione e i suoi rapporti con l'insetto pronubo (*Blastophaga psenes* L.). *Boll. Ist. Entomol. Univ. Bologna* 10:223–80
3. Corner, E. J. H. 1962. The classification of Moraceae. *Gard. Bull. Singapore* 19:187–252
4. Corner, E. J. H. 1965. Check-list of *Ficus* in Asia and Australasia with keys to identification. *Gard. Bull. Singapore* 21:1–186
5. Corner, E. J. H. 1969. *Ficus* sect. *Adenosperma*. *Philos. Trans. R. Soc. Ser. B* 256:319–55
6. Corner, E. J. H. 1970. *Ficus* subgen. *Pharmacosycea* with reference to the species of New Caledonia. *Philos. Trans. R. Soc. Ser. B* 259:383–433
7. Galil, J. 1973. Topocentric and ethodynamic pollination. In *Pollination and Dispersal*, ed. N. M. B. Brantjes, pp. 85–100. Nijmegen: Univ. Nijmegen Dept. Bot.
8. Galil, J. 1973. Pollination in dioecious figs. Pollination of *Ficus fistulosa* by *Ceratosolen hewitti*. *Gard. Bull. Singapore* 26:303–11
9. Galil, J., Eisikowitch, D. 1969. Further studies on the pollination ecology of *Ficus sycomorus* L. *Tijdschr. Entomol.* 112:1–13
10. Galil, J., Eisikowitch, D. 1974. Further studies on the pollination ecology in *Ficus sycomorus.* ii. Pocket filling and emptying by *Ceratosolen arabicus* Mayr. *New Phytol.* 73:515–28
11. Galil, J., Neeman, G. 1977. Pollen transfer and pollination in the common fig (*Ficus carica* L.). *New Phytol.* 79:163–71
12. Galil, J., Ramírez B., W., Eisikowitch, D. 1973. Pollination of *Ficus costaricana* and *F. hemsleyana* by *Blastophaga estherae* and *B. tonduzi* (Agaonidae) in Costa Rica. *Tijdschr. Entomol.* 116:175–83
13. Galil, J., Snitzer-Pasternak, Y. 1970. Pollination in *Ficus religiosa* L. as connected with the structure and mode of action of the pollen pockets of *Blastophaga quadraticeps* Mayr. *New Phytol.* 69:775–84
14. Galil, J., Stein, M., Horovitz, A. 1977. On the origin of the sycomore fig (*Ficus sycomorus* L.) in the Middle East. *Gard. Bull. Singapore* 29:191–205
15. Galil, J., Zeroni, M., Bar Shalom, D. 1973. Carbon dioxide and ethylene effects in the co-ordination between the pollinator *Blastophaga quadraticeps* and the syconium in *Ficus religiosa*. *New Phytol.* 72:1113–27
16. Hill, D. S. 1967. Fig-wasps (Chalcidoidea) of Hong Kong. i. Agaonidae. *Zool. Verh. Leiden* 89:1–55
17. Hill, D. S. 1967. *Figs* (Ficus spp.) *of Hong Kong*. Hong Kong: Univ. Press; London: Oxford Univ. Press. viii + 130 pp.
18. Hutchinson, J. 1916–17. *Ficus* Linn. In *Flora of Tropical Africa*, ed. D. Prain, 6 (2):78–192, 193–215. London: Lovell Reeve
19. Mildbraed, J., Burret, M. 1911. Die Afrikanische Arten der Gattung *Ficus* L. *Bot. Jahrb.* 64:163–269

20. Ramírez B., W. 1969. Fig wasps: mechanism of pollen transfer. *Science* 163:580–81
21. Ramírez B., W. 1970. Host specificity of fig wasps (Agaonidae). *Evolution* 24: 681–91
22. Ramírez B., W. 1974. Coevolution of *Ficus* and Agaonidae. *Ann. Mo. Bot. Gard.* 61:770–80
23. Ramírez B., W. 1976. Evolution of blastophagy. *Brenesia* 9:1–14
24. Ramírez B., W. 1977. A new classification of *Ficus. Ann. Mo. Bot. Gard.* 64:296–310
25. Wiebes, J. T. 1963. Taxonomy and host preferences of Indo-Australian fig wasps of the genus *Ceratosolen* (Agaonidae). *Tijdschr. Entomol.* 106:1–112
26. Wiebes, J. T. 1963. Indo-Malayan and Papuan fig wasps (Hymenoptera, Chalcidoidea). 2. The genus *Pleistodontes* Saunders (Agaonidae). *Zoöl. Meded. Leiden* 38:303–21
27. Wiebes, J. T. 1966. The structure of the ovipositing organs as a tribal character in the Indo-Australian Sycophagine

Torymidae. *Zoöl. Meded. Leiden* 41: 151–59
28. Wiebes, J. T. 1966. Provisional host catalogue of fig wasps (Hymenoptera Chalcidoidea). *Zool. Verh. Leiden* 83:1–44
29. Wiebes, J. T. 1968. Fig wasps from Israeli *Ficus sycomorus* and related East African species (Hymenoptera, Chalcidoidea). 2. Agaonidae (concluded) and Sycophagini. *Zoöl. Meded. Leiden* 42:307–20
30. Wiebes, J. T. 1977. A short history of fig wasp research. *Gard. Bull. Singapore* 29:207–32
31. Wiebes, J. T. 1977. Indo-Malayan and Papuan fig wasps (Hymenoptera, Chalcidoidea). 7. Agaonidae, mainly caught at light. *Zoöl. Meded. Leiden* 52:137–59
32. Wiebes, J. T. 1978. The genus *Kradibia* Saunders and an addition to *Ceratosolen* Mayr (Hymenoptera Chalcidoidea, Agaonidae). *Zoöl. Meded. Leiden* 53:165–84
33. Wiebes, J. T. 1979. The fig wasp genus *Dolichoris* Hill (Hymenoptera Chalcidoidea, Agaonidae). *Proc. K. Ned. Akad. Wet. C* 82:181–96

Ann. Rev. Ecol. Syst. 1979. 10:13–51

HOW TO BE A FIG   ✦4154

Daniel H. Janzen

Department of Biology, University of Pennsylvania, Philadelphia, PA 19104

The 900-odd species of *Ficus* (20, 21) constitute the most distinctive of the widespread genera of tropical plants. Figs have (*a*) a complex obligatory mutualism with their pollinating agaonid fig wasps, yet are found in almost all tropical habitat types and geographic locations [this sets them apart from ant-acacias (63–65), euglossine-orchids (24, 26), moth-yuccas (45, 88, 89,), ant-epiphytes (66, 98) and ant-fungi (109)]; (*b*) fruits eaten by a large variety of vertebrates, most of which appear to be fig seed dispersers rather than seed predators; (*c*) minute seeds despite the adults' long-lived woody life form; (*d*) exceptionally numerous congeners in almost any mainland tropical forest habitat; (*e*) every woody life-form (deciduous, evergreen; tree, strangler, epiphyte, vine, scandent shrub, bush); (*f*) intra-population inter-tree asynchronous flowering and fruiting in many habitats, yet strong intra-tree synchronous flowering and fruiting; (*g*) heavy outcrossed pollination even when the density of flowering conspecifics is extremely low; (*h*) no inter-specific competition for pollinators within a habitat irrespective of the number of *Ficus* species present and the timing of sexual reproduction; (*i*) heavy visitation of fruiting crowns by seed dispersers even when the density of conspecifics is extremely low; and (*j*) over 50% predispersal seed predation of all seed crops.

There is a voluminous literature on the taxonomy and biology of fig flowers, fig fruits, and fig wasps (see reviews in 1, 9, 16, 20–22, 34–36, 41, 47, 52, 86, 90–96, 101, 112, 113), but each author focused on particular aspects of the system. Here I stress interactions among many parts of the system.

Wiebes' chapter in this volume is the most recent review of the details of the interaction of fig wasps with figs. However, a very brief overview of fig biology is useful here. First, ignore the atypical commercial fig, *Ficus carica*, since it occupies extra-tropical to sub-tropical habitats, has decidu-

0066-4162/79/1120-0013$01.00

ous lobed leaves and large fruits in many parthenocarpic varieties, is gynodi-oecious, and has not been studied as a wild plant (11–13, 15, 17). A fig species much more representative of *Ficus* lives in tropical woody vegeta-tion below 2000 m elevation, is a fast-growing woody plant, and has me-dium to large stiff oval leaves that are shed synchronously and then immediately replaced once a year (giving the illusion of being "evergreen" even in deciduous forest habitats). The bark is smooth, gray, and epiphyte-poor, and the trunk is fluted or otherwise contorted. Buttresses and surface roots are prominent. All above-ground parts (including immature fruits) are permeated near the surface with vessels (lactifers) containing white latex that in turn contains the powerful protease, ficin, among other defensive compounds (107). Fig trees have a small fauna of leaf-eating insects but are not generally subject to massive defoliation by leafcutter ants or caterpillars. As often as twice a year, a tree produces a crop of 500–1,000,000 1–5 cm diameter fruits (called "syconia" in other literature), each containing 100–1000 seeds. The small green fruits are borne on short peduncles in the leaf axils. They are entered through a scale-occluded pore (ostiole) by one or a few minute pollen-bearing wasps of the family Agaonidae. These wasps pollinate the hundreds of single-ovuled florets inside the monoecious fruit and lay one egg in each of many of the ovaries, reaching the ovary by inserting the ovipositor down the style. A wasp larva eats the developing seed and lives inside the seed coat. The plant therefore pays 50% or more of its offspring for outcrossing services. A month or so later, the wingless male wasps emerge and then mate with the females through holes they have cut in the sides of the ovaries. Since only one or a few pollinating wasps oviposit in one fruit, many matings are brother with sister. The winged females then emerge from the seed coat and pack pollen from newly de-hisced anthers into recesses of the body to be carried off to another con-specific fig tree that bears fruit in the small green receptive stage. Females leave through a tunnel cut by the males in the wall of the nearly full-sized fig, or (rarely) through an expanded ostiole. Other minute wasps (parasitic Hymenoptera in the Agaonidae and Torymidae) parasitize fig seeds, fig wasps, or the combination, and oviposit either the same way the pollinators do or through the fig fruit wall. In gynodioecious species of *Ficus* (all Old World), some individuals or seasonal morphs bear fruits containing only florets with styles too long to allow the wasp to oviposit in the ovaries. The female wasps therefore pollinate the florets, but no seeds are lost to the wasps and no wasps are produced. Monoecious trees of these species tend to have a very high rate of fig-wasp seed predation and therefore act func-tionally as males. Various moth (Pyralidae) and weevil (Curculionidae) larvae prey on the developing seeds in maturing figs. The figs newly vacated by the pollinating wasps ripen rapidly and are avidly eaten by many species of vertebrates. These digest the fruit wall and the florets from which the wasps have exited, but the seeds undamaged by the wasps usually survive the trip through the gut. Apparently only *Treron* fruit pigeons and small parrots intensely prey on, as well as disperse, intact fig seeds. Mature seeds are also heavily preyed on by lygaeid bugs. Mature fig trees are most common in moderately disturbed sites such as riparian edges, tree crowns (as epiphytes), tree falls, secondary agricultural regeneration, and old landslides. Their lifespans are unknown, but probably do not exceed several hundred years in the wild state.

Such a natural history brings to mind many questions in population biology. This review is an effort to locate manageable questions, rather than to summarize an extensive body of information. I believe that at this stage in tropical biology it is more appropriate to try to make a few well-known systems better understood than to concentrate on the discovery of new systems (68).

TERMINOLOGY

I have deliberately avoided the word *"syconium."* If we are to have a special word for every type of inflorescence or infrutescence, the proliferation of nouns would be astronomical. The fig syconium is nothing more than a globular inflorescence internally lined with several hundred female florets and fewer male florets. Ecologically, this structure is a fruit; *Ficus* has merely reinvented the multiple-ovuled ovary and many-seeded fruit. I call the cavity inside the fig a *pseudolocule;* the ostiole with its occluding scales is functionally analogous to the stigma and style of an ordinary flower. In fact, I suspect that the stigma of the fig floret has lost much of its physiological discriminatory ability. "No family has such small standarized flowers, yet such an astonishing array of infrutescences" (21). The plant probably relies on pollinator choice and behavior, ostiolar screening, and attractant allomones to get the correct pollen tubes to the ovules.

"Condit & Flanders (18) indicated that the term *gall flower* is a misnomer since this floral type represents nothing more than short-styled female flowers" (49) in which one of the occupants of the fig has oviposited [as also concluded by Rixford (101) and other early writers]. Gordh concluded that "the term will be retained, however, because it is convenient and well established" (49). I prefer to delete it. This is not a simple semantic problem. Certainly it is false that there are specific florets destined physiologically to be no more than food for the wasps, despite the impression given by earlier writers (e.g. 19). By cutting off the ovipositors of pollen-laden wasps, Galil & Eisikowitch (38) showed quite elegantly that pollination of any floret, long- or short-styled, results in normal-appearing seed if no egg is laid in the floret. As such, then, a "gall flower" can be identified only after the fact of seed predation, and should not be called a gall flower any more than should a *Hymenaea* flower bud with an *Anthonomus* weevil larva in it. If attached florets need a special name, *wasp flowers* would be more appropriate. If a pollinating agaonid is doing the ovipositing, it is the short-styled florets that usually get the egg simply because the animal can only reach the ovaries of short-styled florets (cf 37, 38). That about half of the florets in monoecious figs have short styles [e.g. about 1 : 1 in *Ficus religiosa* (78), and see (56)] could even be a product of the mechanics of floret packing in a small space. However, optimal ovary packing could also be achieved by differential pedicel growth after pollination as well as by differential style length before pollination.

Style length, the defining trait of gall flowers, is only relative. A long ovipositor with the associated ability to oviposit in a floret with a style of any length is certainly possible, as demonstrated by the nonpollinator *Sycophaga sycomori* in *Ficus sycomorus* (39, 44). I suspect that whenever intraspecific competition begins to select for longer ovipositors in pollinating agaonids, counterselection occurs in the fig for traits that reduce the number of competing wasps that enter the fig. Such traits might be a tighter ostiole, shorter period of receptiveness by the fig, more florets per fig, etc. The wasp might counter these traits by the production of more but smaller eggs. However, smaller eggs would probably slow the development time of the wasp larvae, thereby raising the probability that they will be eaten by a dispersal agent before they can complete development and emerge. The fig can exacerbate this problem for the wasp by evolutionarily shortening the time to ripening of the fruit. Furthermore, the evolution of more seeds per fig would change the parameters of the disperser interaction.

The world is even more complicated. A gall is a proliferation of tissue abnormal for a site and generated by physiological manipulation of the plant through release of chemicals by the insect. A fruitlet containing a pollinating agaonid rather than a seed is therefore a gall. If the entering wasps are nonpollinators, such as *Sycophagus sycomori* in *Ficus sycomorus* (39, 44), the mere act of oviposition in the fruitlets leads to sclerification of the pericarp, to development of the endosperm and nucellus (which the larvae eat) (see 78), and to retention and ripening of the seed-free fig (and florets). Furthermore, the male florets do not develop in such a seed-free fig unless there are wasps developing in some female florets (44). The evolution of the ability to cause fig retention is expected in *S. sycomorus* since it never pollinates a fig and is derived from a pollinating agaonid ancestor (J. T. Wiebes, personal communication). However, in *Ficus religiosa,* if the incoming agaonids (*Blastophaga quadraticeps*) are deprived of pollen and thus can only oviposit, the tree aborts the seed-sterile figs (37). However, abortion may not occur in seed-free figs that were entered by the pollinating agaonid in Florida *Ficus aurea* and *F. laevigata* (see later discussion of host-specificity). The mere application of single hormone-like chemicals (e.g. naphthaleneacetic acid) to unpollinated *Ficus carica* figs will prevent their abortion (23). I suspect that the wasp or its larva manipulates the fig tree in the same manner. Since the normal course of events is to abort unpollinated figs, the entire fig has to be viewed as a gall when occupied by *S. sycomorus*, and each fruitlet containing a wasp larva should also be viewed this way. Unfortunately, this is not the context in which "gall" was applied originally to the system. In fact, it is not at all clear to me why the word "gall" was ever applied to the fruitlets with wasps in them since they usually look like normal fruitlets with a contained seed predator rather than like galled plant tissue. We do not speak of "gall leaves" or "gall stems," so why "gall flowers"?

The larva of a pollinating agaonid is clearly a seed parasitoid, and the adult wasp is a seed predator in analogy to the bruchid beetles and other insects whose larvae develop inside seeds and fruits (67). However, the large suite of other wasps that oviposit in the floret ovary from both inside and outside the fig may be either parasitoids of the pollinating wasps or seed predators or both. Recent studies of the biology of these animals (52, 113) make it obvious that it would be premature to conclude anything other than that they kill seeds and pollinating agaonids.

FIG SEED PARASITOIDS

Besides the pollinating agaonids, the two types of parasitic Hymenoptera found in fig seeds are competitors and parasitoids of the pollinating agaonid and are seed predators of the fig tree. Those that oviposit through the wall of the green fig and into an ovary in which a pollinating agaonid has already oviposited [Torymidae—e.g. *Apocrypta longitarsus* in *Ficus sycomorus* in Israel (38)] have an ovipositor many times the body length. The larva eats the endosperm and other seed tissues directly, and the agaonid larva starves or is eaten directly. The success ratios are unknown in this competition; they matter to the parent fig only in that the parasitoid is not a pollinator nor does it carry pollen when it leaves the fig. These wasps tend to have normal-appearing males, depend on the male pollinating wasps for an exit hole in the fig, and kill something less than 20% of the agaonids in a fig. There is strong selection for a low rate of oviposition per fig by these wasps: If too many emerge, there will not be enough agaonid males to cut the exit hole. I suspect these parasitoids to be the least fig-species–specific of the fig wasps; but, as in those that enter the fig to oviposit, strong host specificity will be selected for because the development times (which vary among fig species) must closely match those of the wasps that cut exit holes in the fig. Furthermore, ovipositor length must be appropriate for the species-specific distance between the green fig epidermis and ovaries containing agaonids. It is rumored that the brown spots on fig surfaces mark the sites of torymid oviposition punctures (52). If so, the number of oviposition attempts on a fig is much greater than the number of torymid wasps it produces.

Parasitoids that enter the pseudolocule to oviposit are much more numerous in individuals than are those that oviposit from the outside. *Sycophagus sycomori* is the best known (37–39, 44, 47) owing to its introduction to Israel along with *Ficus sycomorus* but without the pollinating agaonid. Incidentally, its survival in unpollinated figs calls into question the dogma that pollinating fig wasps pollinate the florets in which they oviposit as well as others. The ovipositor of *S. sycomorus* is long enough to oviposit in both long- and short-styled florets. In East Africa, where the pollinator is present, it oviposits in florets irrespective of whether they have pollinator eggs in them (the outcome of the ensuing competition is unknown).

S. sycomorus has no pollen pockets and does not carry pollen (J. T. Wiebes, personal communication), and the wasps leave the fig even before the anthers open. Parasitoids of this type may constitute as many as half of the wasps to emerge from a fig (J. T. Wiebes, D. Janzen, unpublished data). A continuous strong evolutionary conflict must exist between them and the fig. Any adaptation of the fig that lowers the intensity of seed predation by these parasitoids will also lower the number of pollinating agaonids incoming or outgoing with pollen. Since their males can cut their own exit from the fig, it is not obvious what keeps these parasitoids from eliminating figs from a habitat in contemporary and evolutionary time.

Both kinds of parasitoids discussed above seem to be slightly less host-specific than are the pollinating agaonids (111). Hill (56, 57) found 16 of 51 species of Hong Kong fig parasitoids to emerge from 2 or 3 species of figs. However, such comparisons are premature until more taxonomic work has been done with the wasps.

FIG PSEUDOLOCULE STERILITY

The pseudolocule of a developing fig, like the locule of any other young fruit, must be protected internally against pathogens. The ostiole constitutes a selective filter; like the stigma and style, it must admit appropriate gametes but retard passage of detrimental organisms.

The most external mechanism of pseudolocule sterility is ostiolar tightness. It is well known that ostiolar scales are so tightly appressed that they strip off wings and antennal segments as female fig wasps force their way through the ostiole. I hypothesize that this amputation is an incidental by-product of a major ostiolar scale function: wiping the wasp clean of fungal spores, pollen grains, bacteria-rich clumps of detritus, yeast, and other microorganisms. The fitness of the wasp (parasitoids as well as pollinators) should be raised if she is clean of microbes as well as wings when she arrives in the pseudolocule of the green fig. There should be strong selection, therefore, for an external morphology that is easily wiped clean. Break-away wings and antennal segments may be viewed as an adaptation allowing passage through the very tightly appressed wipers as well as an adaptation to the cramped quarters in which she will have to work; the sequestration of pollen in the corbiculae, pockets, or integumental folds (40, 42, 43, 45, 46, 90) may be viewed as a device for getting it past the wipers as well as a way of carrying it. Fig pollen is exceptional in not being exposed to airborne pathogens as it moves from anther to stigma: Pollen is acquired by the wasp before the fig is opened, and the young fig pseudolocule should contain nearly sterile air since the ostiolar scales serve as a series of air locks. This hypothesis is not falsified by the observation that pollinating agaonids have specialist mites riding on them that get past the wipers (J. T. Wiebes, personal communication) any more than beaver ear mites negate the hypothesis that the split toe-nail of a beaver functions in fur cleaning. There should also be selection against free-flying wasp females' obtaining food at contaminant-rich sites such as the accumulation of fermenting figs beneath a fruiting fig tree. Other functions of ostiolar tightness will be discussed later.

A second line of defense is needed against pathogens that gain entry to the immature fig pseudolocule. Some are bound to enter on the wasp, and in some cases so many wasps enter the ostiole that a tunnel is worn through it. This tunnel must allow entry of both dirty wasps and contaminated air. In the case of *Ficus sycomorus* in Egypt, "as soon as the eggs are laid, the fig commences to secrete a watery fluid which eventually fills the cavity to about one-fourth its capacity. Before the time arrives for the young insects to emerge from their cells, the fluid is again absorbed" (8). I have noted the same phenomenon for *Ficus insipida* and *Ficus ovalis* in Costa Rica. This fluid without doubt contains antibiotic compounds. It is analogous to the phytoalexin-rich fluid that is secreted into the bean locule when this cavity is invaded by microbes or fungi. A protective role for this fluid is further suggested by the observation that microbial or fungal clones are never found growing in the pseudolocule of undamaged developing figs; the corpses of the female wasps remain relatively intact for many weeks unless mashed by the enlarging florets.

Once the new generation of wasps has emerged from the fig [either through the self-opening ostiole (few species) or through a tunnel in the fruit wall made by male agaonids (most species)], the pseudolocule is easily accessible from the outside. However, by this time the seeds are hard and mature, and the fig will usually ripen within a few days.

POLLINATOR SPECIFICITY

There is general agreement that in most cases there is only one species of pollinating agaonid for each species of fig and that this wasp pollinates only one species of fig (2, 3, 56, 91, 111). (Instructive exceptions will be discussed later.) Both the significance for the plant and the mechanism of this extreme pollinator specificity are terra incognita.

Assume that a fig wasp can live for one week when moving between figs (but see later section). If the trees in a fig population are truly random with respect to intra-population flowering times but are highly synchronous within the tree (as they are generally believed to be), a breeding fig population is only actually 1/52 of the actual population of adult figs with respect to flowering conspecifics—if each fig tree bears receptive figs for a week and liberates wasps for a week once a year. This puts even the more common species of figs among the rarest of trees when it comes to pollination dynamics. The rare species are being pollinated at a phenomenally low density (and probably great inter-tree distances). This implies that fig wasps are extremely competent at locating their fig trees. The wasps are able to search far for a plant in which they can develop rather than fail in any nearer allospecific fig tree with figs of receptive age. On the other hand, their high specificity requires the evolution of a great ability to locate figs of the correct species. The fig tree should also be strongly selective in admitting wasps to its figs. Great ability to locate a receptive fig could easily result in a bombardment of the young figs with wasps bearing allospecific pollen. Even if these wasps develop, they would not be likely to carry the pollen to a conspecific if they were so sloppy as to enter the wrong species of fig in the first place.

There are many potential mechanisms for reinforcement of fig-wasp specificity. Most simply, the wasps are probably cued to a unique mix of chemical signals produced by each species of fig. However, incoming wasps have never been censused at a receptive fig. All specificity records are based on emergences from figs. Therefore no information exists on the species purity in the cloud of pollinators that must arrive. Mistakes do occur. In an isolated Venezuelan *Ficus turbinata* tree, Ramírez (91) found that 5% of 121 figs (syconia) produced the wrong species of agaonid. The florets in these fruits did not, however, produce viable seed. In a second case, he found a single mature fig to contain 209 individuals of the wrong species of agaonid. Again, the seeds did not develop but the wasps did.

If the wrong wasp appears at a receptive fig, the ostiole constitutes the next barrier. If the ostiole served only as a wasp-wiper and excluder of foreign macro-organisms, it and its scales should be adjusted just to the size of the wasp. Figs with identical-sized seeds and florets should have identical ostioles. Actually, however, ostiolar size, scale tightness, scale surface sculpture, and thickness of the scale pile vary greatly (e.g. 93). This variation probably reflects in part the selective exclusion of all but the correct pollinating wasp species and a morphology to minimize the number of individuals of parasitoids that enter.

Once inside the wrong fig, I would guess that a pollinating agaonid (or other fig wasp) can oviposit and develop normally if its dimensional morphology allows it. Ramírez's example cited above supports this hypothesis, as does the apparently nontoxic nature of fig seeds. No special detoxification chemistry is likely to be needed to eat fig endosperm and associated tissues since there is no reason to suspect that mature fig seeds escape from seed predators by chemical defenses. However, escape from the wrong fig by the next generation of pollen-laden adults will be complicated if the pollen-presentation behavior is not that of the usual host, if the fig wall is too tough for the wrong males to penetrate, or if the fig development time does not match that of the wrong wasp.

A fig pollinated by the wrong wasp is a bad investment for the parent tree on two counts: (a) The seeds will not develop, and therefore the wasps are acting solely as seed predators; (b) the wasp will not carry pollen to a conspecific even if it can emerge from the fig. I therefore expect the parent tree to abort fruits pollinated by the wrong wasps (though perhaps some outcrossing with congenerics is a valuable source of novel genetic information). This again means that the purity of the wasps that emerge from a fig seed crop is not evidence of fig-wasp specificity. It would be of great interest to examine the founding wasps for a set of aborted figs from a tree in a habitat rich in species of figs.

In addition to fig pollination at a very low density of flowering individuals, extreme pollinator-specificity among fig wasps also has the consequence that as new fig species are stacked into a habitat, there is no danger of exclusion through competition for pollinators. There is the possibility of a newcomer fig producing allomone messages that overlap with those of the resident fig species, but this overlap should quickly select for character displacement of this trait. I doubt that the upper limit of fig intra-habitat species richness is set by filling of this communication channel, since the vocabulary of pheromone communication is enormous.

Since there are some 900 species of *Ficus* (21), it is probably safe to guess that there are as many species of pollinating agaonids. The most parsimonious hypothesis for the generation of these wasp species is probably the classical process of speciation in allopatry followed by later reinvasion of siblings' habitats. Since there is no pool of more generalist pollinators to service a mutant fig and its offspring, sympatric speciation processes seem unlikely. Furthermore, a rather large population of conspecific fig trees would be required to sustain a mutant population unless there are numerous simultaneous behavioral changes in the fig–wasp interaction as presently understood.

There are two places in the world where I expect agaonid pollinator specificity to break down or the fig species to stop acting like the fig described in the introduction: on islands and in very harsh environments. "There is hardly any tropical island of any size but possesses one or more species of *Ficus*. . . . In Fernando de Noronha was an endemic fig, *Ficus noronhae*. In Christmas Island *F. retusa* was abundant, and reproduced itself. Both of these, as I found, possessed abundance of gall wasps. . . . The genus is absent from the Hawaii Islands [(and see 80)], but there is a species in Fanning Island, 900 miles south, and it is absent too, from Cocos-Keeling Island, 700 miles from Java. Most of the Polynesian islands, however, possess one or more species" (99). "On the small San Andres Island (Colombia) *F. aurea* and an unknown species of fig were each represented only by a few mature trees. Very few trees showed synchronized development of syconia; usually each tree had syconia in all phases of development. Thus wasps emerging from ripe figs could find figs in the receptive stage in the same tree. Apparently in small populations selection has favored a breakdown in synchrony" (91). Ramírez did not examine enough figs to know if the San Andres fig wasps are as rigidly fig-specific as mainland wasps are reputed to be.

The colonization of islands by figs and their wasps contains a relevant paradox. A single seed from a mainland cannot start a fig population because the single tree it produces cannot sustain a pollinating wasp population. Likewise, a pollinating wasp population cannot survive until there is a population of fig trees. A very unlikely solution is for a mutant with intra-crown asynchronous flowering to be the colonist seed, followed by wasp colonization (note that the colonizing wasp will bring mainland fig pollen with it). More likely is the extension of a mainland fig-seed shadow to an island by means of fruit pigeons or bats, followed by wasp colonization of the resultant island population of the mainland fig genome. This could then easily be followed by selection favoring individuals with asynchronous intra-crown fruiting (better to be self-pollinated than not at all). Once the population had begun to act like that described above on San Andres Island, it would be island-adapted and perhaps hop from island to island through even very rare seed dispersal events. In the unfolding of such a scenario, the presence of a second species of island fig could easily favor indiscriminate use of both fig species by one species of wasp and even pollination of both figs by that one wasp with attendant convergence of fig flower and fruit traits.

Very harsh mainland environments should be similar to islands: Unpredictable weather events, as well as predictable extreme ones, could occasionally reduce severely the population either of wasps or of fig trees bearing receptive figs. Florida is such a place, being the northern limit for two fig species. Without supporting data, it has been stated that "the Florida fig wasp [*Secundeisenia mexicana*] occurs abundantly in the fruits of our two native fig trees, *Ficus aurea* and *F. laevigata*. It has not been observed in the fruits of other common [introduced] *Ficus* species, including *F. altissima, benjamina, glomerata, religiosa,* and *retusa* (=*nitida*)" (10). Assuming that *S. mexicanus* in Florida is really only one species, it can pollinate both figs. It is likely that each time a new crop of pollinating agaonids is produced in a fig tree the wasps spread out over the habitat. Those that find a fig tree of the species that did not produced them complete their life cycle but do not pollinate their fig, though they will carry pollen when they leave. Ramírez's Venezuelan case mentioned above is probably a potential intermediate step in the evolution of wasp survival without pollination, and the evolution of *S. sycomorus* in Israel (47) is another. A one-wasp-two-figs system should be facilitated by three environmental traits. First, Florida is frequently subject to weather severe enough to reduce greatly the crop of receptive figs for one of the species. This should select strongly for latitude in fig choice by wasps, latitide in ability to develop in a fig for which no conspecific pollen has been brought, latitude of retention of figs pollinated by the wrong pollen (mixed sets of wasps in single fruits may facilitate this), and intra-crown asynchrony of fig flowering. Even if the wrong pollen is brought into the fig, the next generation of wasps may carry pollen off to conspecific trees and therefore render the fruit at least a pollen donor. Second, the wasp is not endemic to Florida; it occurs in *Ficus laevigata* figs in Puerto Rico (117) and probably elsewhere in the Caribbean. Florida is therefore constantly bombarded with new agaonid colonists. If the local agaonid population were eliminated, the hosts would soon be reinvaded. The opportunity for founder effects is obvious. The colonists themselves, coming from islands, may behave as they do in Florida. Third, both of these species of *Ficus* occur on Caribbean islands and undoubtedly continually bombard Florida with seed (and vice versa). If these island populations have the same pollinator overlap as do the Florida populations, then the system did not even have to evolve in Florida since it should be functional on small islands as well as at the margins of *Ficus* distributions.

Severe conditions for figs do not occur only at the margins of *Ficus* distributions. Tropical sites with severe dry seasons may select for the use of incorrect hosts by agaonids during drought-induced loss of fig crops, especially with those species that live on dry hillsides adjacent to riparian populations that are less likely to lose their fig crops during the long dry season or during rainy season droughts. No systematic search for such an example has been initiated, but there is at least one case where such circumstances seem to have led to selection for asynchrony in the fig crown (see the following section).

There is still room for skepticism about the one-on-one relationship in some cases. In Hill's (57) discussion of the situation he presents the curious anomalous definition that

> in many cases the host species of *Ficus* occur as well-defined, and genetically distinct varieties, and in some cases also subspecies; the varieties of some species may be allopatric although it is more usual for them to be more or less sympatric. As the varieties are genetically distinct, it follows that the agaonids which pollinate these varieties must either be distinct species (or sometimes subspecies) themselves or else quite separate populations of the same species, in view of the absence of natural hybridization in *Ficus* species. Clearly, it is to be expected that the agaonids inhabiting different varieties of the same fig species will be different species themselves. The present work has shown, what has been suspected for some while, namely that the agaonids inhabiting the different varieties of the same fig species are often morphologically indistinguishable, although it is felt that usually they must be biologically distinct species.

If Hill is correct about the wasps, what a fig taxonomist calls a variety of subspecies of fig would be called a species by any contemporary zoologist. Second, it is not at all obvious how one is to know that the "varieties are genetically distinct" when no artificial crosses have been made with any of these *Ficus* species.

As an example of how these taxonomic concepts can confuse the issue, Hill's (58) revision of the agaonid genus *Liporrhopalum* contains the following case. A wasp was described as *L. rutherfordi* from a single specimen from an unknown host in Sri Lanka (Peradeniya). Hill found a wasp that was morphologically similar to this one to be common in *Ficus tinctoria gibbosa* on Hong Kong, but described it as a new species, *L. gibbosae* (which has 5 or 6 lamellae on the mandibular appendage, as opposed to 4 in the one specimen of *L. rutherfordi*). Since *F. tinctoria gibbosae* occurs only as far west as Malaysia, and since *F. tinctoria parasitica* occurs all over India and Sri Lanka, Hill concludes that *L. rutherfordi* and *L. gibbosae* are separate species because they are on different varieties of host!

The most vexing problem encountered by Hill (56) was that the Hong Kong native figs *F. pyriformis, F. variolosa,* and *F. erecta* var. *beecheyana* were pollinated by morphologically indistinguishable agaonids. In his taxonomic treatment of the Hong Kong wasps (57) he rationalized the situation by concluding that they must be three species that cannot be distinguished morphologically.

PARENTAGE OF FIG SEEDS

The prevailing opinion that mainland fig seeds are obligatorily outcrossed owning to the behavior of the wasps and the intra-crown flowering synchrony is probably reasonable. The exceptions would occur when adult fig wasps survive for 3–6 months between successive crops on one tree (doubtful) and in habitats where there is some intra-crown asynchrony. A *Ficus ovalis* tree in the deciduous forest of Santa Rosa National Park, Costa Rica, bears receptive figs at the same time it is releasing pollen-bearing wasps and therefore may self-pollinate, providing that the wasps do not have a long pre-fig-entry flight requirement (72).

Figs differ from other plants in the location of the plants they mate with. While in most plant species it is likely that plants mate most often with their nearest conspecifics in space (barring incompatability of neighbors owing to excessive genetic relatedness), figs should mate most often with plants that are their nearest neighbors in past (pollen reception) or future (pollen donation) time. The longer the wasps live, the less true this generalization. This mating pattern means that if each fig tree waits a very regular time between flowerings, each would repeatedly mate with the same individuals in the population (W. Hallwachs, personal communication) and the pollen flow would never be reciprocal. Selection should thus cause figs to wait a varying number of months between flowerings, which is what Morrison (86) found to be the case with the two commonest species of rainforest figs on Barro Colorado Island (BCI). Incidentally, the greater the asynchrony of fig-wasp reception and production within a crown, the greater the number of other fig trees with which a given tree is likely to mate.

The number of pollen-donors represented by the cloud of fig wasps arriving at a receptive tree will depend very strongly on the longevity of the wasps, the density of fig trees within their flight range, and their flight behavior. If they are mixed by air over a large area and then settle out on a fig tree in response to an attractant, a fig population would be the most panmictic of any tree species in a tropical forest, yet have the most feeble of pollinators. The large and diverse nature of the seed-dispersing coterie for many fig trees will render this panmixis even more thorough. The parent fig can strongly influence the number of parents it has for its seed crop by modifying the amount of time it bears receptive figs and then aborting figs pollinated by wasps that bore pollen that was "wrong" in some sense. An increase in the period of fig receptivity to wasps would increase the number of crops to which pollen is contributed, since it would increase the period over which pollen-bearing wasps are released from the tree.

The parent fig can also increase the number of crops to which it contributes pollen through evolution of traits that lead to a larger number of smaller wasps per fruit and an increase in the amount of pollen per fruit. For example, figs of *F. religiosa* and *F. sycomorus* produce 9–19 and 60–80 anthers per fig, respectively, while the summer caprifigs of *Ficus carica* may contain 200 male florets, each with several anthers (41). With the exception of *Ficus pumila,* which may have as many as 1000 male florets to 5000–6000 female florets in a caprifig, Hill (56) found the Hong Kong figs to have ratios of about 20–170 male florets to 150–600 female florets. In commercial varieties of *Ficus carica* caprifigs, the proportion of male to female florets varies from 2:1 to 10:1, with 7 or 8:1 as the normal [for 39–1350 female florets per fig (11)]. However, these ratios mean little in the absence of information on variation in numbers of pollen grains per wasp, anther or male floret. It is of interest that the agaonid that pollinates *F. pumila* is also one of the largest of its family (56).

The seed parentage within a fig fruit is somewhat problematical. With *Ficus sycomorus* in Kenya, the number of adult agaonids gaining entry to a fig ranged from 0–13, with averages of 1–4 for different small samples. In 88 figs from a single Costa Rican tree, Ramírez (92) found that a range of 1–4 and a mean of 2 agaonids had gained entry per fig. Despite large numbers of *Pleistodontes imperialis* arriving at the fig, Froggatt (33) states that only 2 or 3 females gained entry and the remainder died stuck in the ostiole. I found (73) in 3 Costa Rican deciduous forest fig crops (3 species) that 93, 53, and 52% of the figs had only one mother for the contained wasps (average of 1.07, 2.97, and 1.72 wasps per fig, respectively). A fourth crop (*Ficus insipida*) had a mean of 7.2 wasps per fig. This crop had some figs with 20–30 wasps that had gotten past the ostiole. In general, however, the number of parents available to sire the hundreds of seeds in a fig fruit may be only 1 and usually less than 4. Furthermore, I suspect that the first and second wasps to enter do most if not all of the pollinating, and therefore seed parentage is even more monotonous than a mean of, for example, 4 wasps per fig suggests. The low number of parents for the seeds in a fig means that the decision to keep or reject a given fruit after pollination may involve very few choices of parentage. If any seeds have the wrong parentage, then likely a large number do. However, there are no records of abortion frequencies in wild figs. In doing such a study, care must be taken to distinguish between fruits aborted because no agaonids arrived and fruits aborted because the wrong species of pollen-bearers arrived. Since the wasp has made an irreversible decision when it enters the fruit, there should be strong selection for the ability to choose trees that will not reject it or its pollen load. No information is available on abortion rates of fertilized fig florets within the developing fig, but aborted (unpollinated?) florets within ripe figs certainly exist.

No data have been gathered that would allow a guess about how much inter-specific gene flow occurs in figs. Ramírez (91) agrees with Corner (20) that hybrids are rare, but that statement needs quantification. Experimental crosses are needed, as are attempts to introduce wasps into the "wrong" receptive figs. Wild agaonids do make mistakes, as mentioned in the section on Fig Wasp Specificity. The presence of the wrong agaonid wasp in a fig not only labels that fig as containing potentially hybrid seed, but should also identify the putative male parent. An extensive (and laborious) examination of the specific indentity of the remains of the fig wasps in wild developing figs (before the next generation has emerged) would be of great interest.

Ira Condit (e.g. 13, 15–17) noted that "perusal of the literature relating to the genus *Ficus* reveals few if any records of natural hybridization among the various species" (14). He then puffed the pollen of *Ficus carica* into female *Ficus pumila* syconia and got viable F_1 seeds that produced two saplings large enough to make figs. "Syconia of various common figs [*Ficus carica*] pollinated with the pollen produced by the hybrid appear to be producing fertile seeds." Before we can hope to understand the importance of intra-generic pollen flow in *Ficus,* a large number of such experiments should be attempted. When matched against the flimsy data on the frequency of mistakes by agaonids, a possible cause of the apparent lack of *Ficus* hybrids becomes more obvious. It may simply be that frequent hybridization makes little impact on the phenotype represented on herbarium sheets. Fig trees are extremely similar with respect to more than fig macromorphology. I doubt that interspecific hybrids would be noticed in most cases.

LONGEVITY AND MOVEMENTS OF AGAONIDS

The assumption that female fig wasps are "short-lived" [unquantified—e.g. (91)] is based on the observation that they live only a few days in captivity. However, no biological law dictates that a small insect must have a short adult life span. The females of one of the fig-wasp parasitoids, *Philotrypesis caricae,* have a longevity of 30–35 days (79). There is no reason that pollinating fig wasps should not live as long between figs (and see the section on Phenology of Flower and Fruit Production). However, Condit (11) says that in hot, dry, and windy California fig orchards, female *Blastophaga* live only 4 or 5 hours. If fed nectar or other liquid nutrients, parasitic Hymenoptera of similar size have much longer life spans than if not fed.

If pollinating fig wasps can live for many weeks before finally locating a receptive fig, not only will their selection for maximum sexual asynchrony among individual fig trees be somewhat relaxed, but also the chances for self-pollination by a somewhat asynchronous crown should arise. Further, the longer fig wasps live, the larger becomes the breeding population of fig trees at any given moment. At the limit, if fig wasps could live the average duration of the inter-crop period for a tree, the entire population would be in flower simultaneously.

Free-flying fig wasps must be subject to the same sorts of predation and other mortalities experienced by other small flying insects. I have seen hundreds of large dragonflies (Odonata) darting in and out of the leaves of a large Costa Rican *Ficus* just as the wasps were emerging from the figs (Playa Coco, Guanacaste Province, 1969). They were presumably preying on the wasps just as they normally prey on mosquitoes.

As indicated earlier in describing Hill's work at Hong Kong, fig wasps must move long distances at times to re-colonize areas vacated by natural catastrophes. Their apparently indigenous presence on Pacific oceanic islands [e.g. Okinawa (62)] suggests that winds may on occasion carry them very great distances. Ramírez (91) gives some indirect evidence that agaonids may sometimes move distances of many kilometers between individual trees. Condit (11) feels that a wind in a California fig orchard may carry females for several miles.

While it seems that fig wasps would easily be dispersed passively by wind, the rarity of fig trees bearing receptive figs at any given moment suggests that if wasps were dispersed only passively most would die without finding such a tree. I hypothesize that the fig tree releases a species-specific allomone at the time the fig crop comes of receptive age, and I assume that the parasitoids respond to it as do the pollinating agaonids.

SEED PREDATION BY FIG WASPS

The literature contains no information on the intensity of seed predation by pollinating agaonids and the various parasites, yet this intensity is a rather direct measure of one of the prices the fig pays for pollination (the amount of resources moved into the wasp-containing floret is one of the other direct prices). Galil & Eisikowitch (37) noted that Israeli *Ficus religiosa* had *Blastophaga quadraticeps* in 94% of the short-styled fruitlets in July; in November the figure was 92%. Of the long-styled fruitlets, 5% contained the agaonid in July, and by November this number had increased to 25% (probably more wasps enter the fig as the population builds up during the summer). The Israeli *F. religiosa* figs were losing about half of their seeds to agaonids. Indian *F. religiosa* figs contain 105–113 long-styled florets amd 81–107 short-styled florets (78). However, my unpublished records for Costa Rican figs show clearly that many long-styled as well as short-styled florets produce wasps, and therefore this ratio does not aid in determining percent seed mortality. In 5 species of Costa Rican deciduous forest figs, I found the average seed mortality due to pollinators and parasites to be 41–77% (72). In one tree of *Ficus ovalis,* the percent mortality rose from 44 to 61% per fig from the beginning to the end of two presumably overlapping generations of fig wasps on the same tree with relatively overlapping asynchronous flowering.

Over a two year period, percent seed mortality per fig per crop in four crops on this tree has ranged from 44 to 77% (D. Janzen, unpublished), which suggests that percent seed mortality in mature figs may not show much variation. There are two reasons to expect this narrow range of mortality, as well as only minor inter-specific variation in intensity of seed predation by wasps. First, figs that do not receive any wasps are aborted by the parent. Those that get at least one wasp will have 200–400 of the florets pollinated, and many of these florets will receive an egg. This means that there will be few if any figs with less than 20–25% seed predation. In the survey of deciduous forest fig-seed predation mentioned above, I counted seeds in many hundreds of figs froms tens of crops; well over 95% of the figs have greater than 25% seed predation by wasps. Second, the percent seed mortality optimal for the mutualism will be a value that generates (*a*) at least enough wasp males to cut an exit hole out of the fig and (*b*) some optimum number of pollen-laden females leaving that exit. This number is unknown but likely to be large. To keep this number high while lowering the percent seed mortality would require both a strong restriction of the number of entering pollinating wasps and an increase in the number of florets. But the increased number of florets would require an increased number of wasps for pollination. It is difficult to postulate the survival of a mutant wasp that pollinates many flowers but produces few offspring. Entrapment of pollen-carrying wasps by gynodioecious figs with purely long-styled florets is probably the only possible solution, and here the gynodioecious figs parasitize the monoecious figs (i.e. their frequency probably has the usual upper limits found in parasite-host relationships).

Within a single fig, control of intensity of seed predation is achieved in part by the ostiole. I suspect it has traits that control the average number of females entering. If the wasp were purely a seed predator, I would expect the tree to abort figs with seed predation above a certain level. However, even a fig in which all seeds are killed is of value for pollen production. In fact, depending on the relative value of seed flow versus pollen donation, it may be more valuable than one with many viable seeds. The often-noted floret dimorphism is clearly a mechanism that holds the percent seed predation in a general area. Most short-styled florets produce wasps. However, as the number of female wasps per fig rises, the unoviposited floret becomes a resource in short supply and the wasps oviposit in many of the long-styled florets. This is especially conspicuous in fig species bearing very small figs. Here, the absolute difference in style length between short-styled and long-styled florets is less than in large figs and it is likely that, with effort, more of the ovaries of long-styled florets can be reached. Furthermore, many seed and wasp parasitoids can oviposit in ovaries with styles of any length, and therefore the ratio of long- to short-styled florets is again not useful in determining percent seed predation.

SEED PREDATION BY OTHER ANIMALS

In addition to the fig wasps, some other insects prey on the fig seeds. There are at least 17 species of weevils in the neotropical genus *Ceratopus* (D. Whitehead, personal communication) whose larvae feed in nearly mature figs. In Costa Rica I have found these weevils only in rainforest species with large figs. They appear to be absent from fig species with small figs and fig species in deciduous forest or riparian habitats. The larvae mine through the fig, consuming seeds, fruit wall, intact florets, and florets from which the wasps have emerged. In some samples they occupy nearly 100% of the newly fallen figs, but since I do not know whether trees abort attacked fruits I cannot determine whether this is a high percent fig attack. The larvae pupate in the soil and adults emerge within 2–3 weeks, presumably flying off in search of new crops of ripe figs. Occasional moth larvae are encountered in maturing figs. *Boetarcha stigmosalis* (Pyralidae), for example, grazes florets and seeds and may destroy as much as 30% of a crop of *Ficus ovalis* fruits in Santa Rosa National Park, Costa Rica (D. Janzen, unpublished).

Lygaeid bugs appear to be the major nonwasp seed predators of mature and maturing fig seeds. They are small and cryptic species and have been studied best in South Africa and the West Indies. From Slater's (104, 105) study of a complex of some 46 species I have extracted a list of their traits most relevant to fig biology.

1. The fig-seed lygaeids are divisible into four groups—arboreal seed predators (mostly Heterogastrinae: *Dinomachus, Eranchiellus, Trinithignus*), obligatory terrestrial seed predators, facultative terrestrial seed predators, and accidental terrestrial seed predators. That there are four groups rather than simply a set of species that forage in the fig tree and its environment suggests specialization to different stages of development of the seed crop, driven at least in part by interspecific competition.

2. The arboreal species have exceptionally long mouth parts (as long as the abdomen or longer), presumably for penetration through the fruit wall. They may be found clustered on the fruit or, when not feeding, clustered under loose slabs of bark. These are very active insects; they fly or run

quickly when disturbed. Apparently they do not go to the ground to feed on fallen figs. Not only is this fraction of the fig seed-predator guild made up largely of one lygaeid subfamily, but the tropical Heterogastrinae are almost exclusively fig seed predators [the extra-tropical species feed on mint and nettle seed (104)]. Heterogastrinae do not occur in the neotropics, but Slater (105) suspects that the widespread rhyparochromine genus *Cholula* may act in the same manner.

3. Obligatory terrestrial seed predators "appear to feed only on the fallen seeds of figs. Several of the species swarm in great numbers under fig trees [(in Jamaica, there were estimated to be over a quarter of a million *Ozophora* under a single tree)] but have never been taken elsewhere. Generally they are found directly under the tree, and their numbers decrease drastically with increased distance from the trunk. Most are extremely active insects and when disturbed, fly readily. This latter habit is very uncommon in terrestrial Lygaeidae, but in the fig fauna it occurs in members of quite distinct tribes" (105). I suspect that this skittishness prevents the bugs from being stepped on by large mammals foraging for fallen figs.

One of the most prominent obligatory terrestrial seed predators in Africa is *Stilbicoris,* a bug that not only offers a seed to a female bug as copulatory bait, but also can fly off carrying a seed, which makes it a potential seed disperser.

A second set of obligatory terrestrial seed predators are small (seldom over 2–3 mm long), do not fly readily, and appear to prey on seeds deep in the litter. They may still be present long after the fig crop is finished and the above-mentioned species have moved on. While less abundant than the above-mentioned species, they may be especially important in eliminating the remnants of the seed crop left by the bugs that concentrate at sites where seeds are very abundant.

4. Facultative terrestrial seed predators "tend to be present in relatively small numbers and to be distributed near the periphery of the seed crop" (105). They also can develop and produce eggs while feeding on other species of seeds.

5. Accidental terrestrial seed predators are rarely taken in the litter below fig trees but are found in large numbers elsewhere. They are not taken as nymphs below fig trees and probably do not breed there. Their presence is probably due to the ability of the adults to feed on seeds that are inadequate for nymphal development or facultatively to take the fig seeds while in search of seeds of higher quality for themselves or their nymphs.

6. It is common for litter-inhabiting lygaeids to be polymorphic for brachypterous/macropterous morphs. Sweet (105a) stressed that brachyptery is most strongly correlated with permanence of the habitat. "The entire fig fauna of Lygaeidae is totally macropterous" (104). The reason seems

straightforward. Fig trees produce enormous numbers of seeds in a very short time; it may then be many months before more are available at that tree. For the arboreal species, the seeds will be present for the shortest period; for species that can find seeds deep in the litter, food will be present for the longest time. By flying readily, having well-developed wings, and being ovoviparous (*Stilbicoris* does this, presumably as a way of shortening the development time), the fig-lygaeids appear well adapted to seeking out new fig crops and rapidly migrating to them in large numbers.

7. Slater (105) has noted in Africa and I have noticed in Costa Rica that *Ficus* seedlings are extremely scarce below and near the parent, despite the very heavy seed flow into the litter below fig trees.

> The absence of juvenile plants below the trees may be due, of course, to many factors (shade, toxicity, etc.) but may be largely related to the seed predation of lygaeid bugs. To judge by the size and ubiquity of the lygaeid populations, it seems possible that they are capable of destroying nearly 100% of the seed crop that falls beneath the trees. . . . The predominance of *Ficus sycomorus* along [East African] water courses is presumably largely due to this habitat being optimum for establishment and growth of this species. It must also be realized that seeds which fall in flowing water will escape the heavy seed predation that occurs below the trees. . . . When it is realized that many birds and monkeys are concentrated along water courses, and that these vertebrates are presumably an important means of seed dispersal, it is probable that the concentration of sycomore figs along such water courses is due at least as much to greater survival of seeds in such areas as to a more favorable physiological habitat (105).

I should add, however, that no one has ever recorded what fraction of the seeds in a fig seed crop are already dead owing to fig wasps and lygaeids by the time they enter a vertebrate or fall to the ground.

In addition to direct seed predators there are insects that attack ripe and nearly ripe figs to feed on some combination of seeds and fruit wall. These have been studied only in commercial situations (4, 5, 60, 84, 110).[1] In the wild, such animals probably cause the death of many seeds by rendering the fruit unattractive to dispersal agents (and see 69). *Ficus carica* figs are attacked by the dried-fruit beetle *Carpophilus hemipterus* (Nitidulidae) in the field and in storage in California and the Mediterranean region. The adults lay their eggs in breaks in the fig epidermis or enter through the opening ostiole, and both adults and larvae feed on the mixture of figs and microbes (28, 51). The adults can be trapped with baits of ripe fig mash innoculated with a variety of spoilage fungi and yeasts (114) and are probably coevolved with these microorganisms as are figs and other fruit eaters. An anthicid beetle, *Formicomus ionicus,* treats Turkish figs in the same manner (51). The larvae of moths of the genera *Ephestia* and *Plodia* attack dry figs (as well as other fruits) in storage and after they have fallen from

[1]See (79a), encountered while this article was in press.

the tree (25, 51, 103). In Israel the Mediterranean fruit fly *Ceratitis capitata* (Tephritidae) oviposits in *F. carica* figs on the tree (50, 100). Of apple, pear, peach, and fig fruit, the fly has the shortest development time in figs (100). The larvae of the fly *Lonchaea aristella* (Lonchaeidae) also develop in Old World fig fruits (6, 50). There is a South African weevil, *Cyllophorus rubrosignatus* (Curculionidae) whose larvae apparently develop in fig fruits (81).

Ripe figs are also attacked directly by adult frugivorous insects. *Allorhina mutablis,* the large diurnal cetoniine scarab called the "green June beetle" in the southwestern United States, chews directly into ripe figs on the tree (85). The adults of the cotton leaf worm, *Alabama agrillacea* (Noctuidae), puncture Texas ripe figs with their proboscis to feed and indirectly cause premature souring of figs (61).

PHENOLOGY OF FLOWER AND FRUIT PRODUCTION

By now it should be obvious that *Ficus* sexual phenology is very different from that of other tropical trees. "Species of the genus *Ficus,* as a result of dependence for pollination on specific, short-lived symbionts (the agaonid wasps), have evolved several features to favor the continuous development of these symbionts the year around in the tropics. . . . [There is] year-around production of figs, so that in any particular area fig trees of the same species may usually be found with syconia in all phases of development, although any one tree has all syconia in the same stage. . . . Synchronization of development of all the syconia of a tree [occurs] so that usually every syconium of a particular tree is pollinated on the same day" (91). This description is probably accurate except for the population-level–selection flavor of the beginning of the first sentence. However, no detailed study of the sexual phenology of even one individual native wild fig tree, of a population of fig trees, or of an array of fig species in one habitat has been published. Furthermore, I would explain it differently: If the goal of a flowering fig is to attract a maximum number of pollen-bearing wasps that did not originate from its own figs, it should be receptive (produce new young figs) well after its wasps have left the tree, it should flower in one burst so as to maximize the amount of attractant cue while minimizing its cost per fig, and it should flower at random in the yearly cycle so as to minimize the chance of flowering at the same time as other fig trees (it cannot flower uniformly out of phase because it cannot know when other fig trees are not flowering). In addition, one cannot use ripe-fruit phenology to explain flowering phenology, or vice-versa. There is no physiological reason why the intra-crop timing of ripe fruit production has to mirror the intra-crop timing of flowering, and indeed it does not in a number of tropical tree species.

By far the best information on fig sexual phenology is contained in Hill's (56) taxonomically oriented 3 year study of Hong Kong *Ficus* and Morrison's (86, 87) 2 year study on Barro Colorado Island of wild figs as bat food. In the somewhat seasonal rainforest on BCI, Canal Zone, Morrison carefully located all the 142 *Ficus* individuals large enough to bear fruit in a 25 ha area (*F. yoponensis,* 71; *F. insipida,* 48; *F. tonduzii,* 15; *F. obtusifolia,* 5; *F. turbinata,* 1; *F. trigonata,* 1; *F.* sp. 1), counted their fruit weekly from March to November 1973, and then had them counted every other week through February 1975. Since the majority of the fig individuals were *F. yoponensis* and *F. insipida,* his generalizations were derived mainly from observations of these two species. While the published part of the study was largely based on ripe figs produced by the tree (and thus does not tell us the pattern of figs available to wasps), numerous relevant traits of the system are evident.

1. For both *F. yoponensis* and *F. insipida* a frequency distribution of the distance to the nearest conspecific neighbor of reproductive size has a peak at 20–29 m (*F. yoponensis,* n=56; *F. insipida,* n=36) with a range of 0 to 50–70 m [Figure 5 in (86)]. If the conspecific members of either of these two fig populations were in synchrony, or if a tree fruited continuously within its crown, the wasps would have to move only a short distance from their parent tree to find new receptive figs. I should add, however, that researchers on BCI regard Morrison's study area as having an exceptionally high density of fig trees.

However, "synchrony in fruiting either within or between species was not apparent. There is some suggestion of peak times for the initiation of fruiting by *F. yoponensis* in June and December. Local synchronies in the fruiting times of trees in the same area were not detected." Considering that the study area is only 25 ha in area, that for *F. insipida* there was at least one tree in ripe fruit in 22 of the 23 months of data collection, and that for *F. yoponensis* there was at least one tree in ripe fruit in 22 of the 22 months of data collection, it is evident that in some absolute sense the wasps need not live more than a month nor travel more than 700 m even when there is extreme intra-crown synchrony of new fig production.

2. The time from first appearance of the receptive figs to that of the ripe figs was about 5 weeks, and thus the generation time of the wasps is probably about 4 weeks. Ramírez (93) reports that this period for different fig species is 15–100 days but does not give the source of the information (and see 48).

3. "As many as 50 live, winged females [of agaonids?] per fig were found in several fallen, sixth week figs in 40% of the 76 fruitings from which there

were ground samples taken (86)." Where wasps are still present in a fallen fig it is likely that the fig has many fewer pollinating agaonids than normal since their presence suggests that there were not enough males to cut an exit hole. This estimate of the number of wasps produced per fig is therefore very low. Until we have a numerical study of the number of female pollinating wasps produced per fig by a wild monoecious fig species, a reasonable round number is probably about 250 per fig (based on J. T. Wiebes, personal communication, and my own unpublished observations of Costa Rican figs).

4. "The number of trees in the study area bearing prime, fifth week fruit varied from 0 to 8 per week, averaging 2.4 ± 2.0 per week. The total number of trees which came into prime fruit in any given month varies from month to month, but there is a significant correlation between the two sample years in the number of trees in fruit in any given month (Spearman rank correlation coefficient, r_s = .675, p < .05). There appears to be a particularly invariable low in figs in the months of August, September and October of both years" (86). M. Estribi (86) offers the following highly reasonable hypothesis for this seasonal low. His hypothesis is based on the assumption that the seasonal low in ripe fig production is the result of an increase in abortion of young figs owing to a failure of pollination, rather than simple nonproduction. "The incidence of aborted fruitings is correlated with the presence of insects other than fig wasps whose larvae develop inside the figs. These larvae are primarily those of two species of snout beetles (Coleoptera; Curculionidae) and several different species of Diptera. At some times of the year, these larvae become so numerous that it is possible to open fig after fig and find it eaten hollow and filled to capacity with larvae. Light trap data by N. Smythe shows that the adults of these fig parasites are most numerous from June through September. The high density of fig parasites causes a substantial reduction in the number of fig wasps available for pollinating, which in turn accounts for the abortive fig production in these months." It is not obvious why the density of non-hymenopterous fig parasites should decline after September. Fig wasps may be short-lived, but adult weevils generally are not.

5. "There was no correlation between the month an individual tree was in fruit in the two sample years (r_s = 0.165, p > .05). Further, there does not appear to be any consistent endogenous periodicity in fruiting of individual trees. The interval between fruiting varies both among individuals and within individuals. The interval between first and second fruiting is not significantly correlated with the interval between second and third fruiting for either *F. yoponensis* (Pearson correlation coefficient, r = .057, F-test, p > .25) or *F. insipida* (r = .114, F-test, .05 < p < .10). The number of crops borne by a tree per year was greater for *F. yoponensis* (1.13 ± .61) than for

F. insipida (.93 ± .46) (t-test, p < .005)" (86). For most *F. yoponensis* trees the interval between the first and second fruiting is 20–50 weeks; for most *F. insipida* it is 25–60 weeks. These figures are probably much closer to the real inter-fruit periods for wild figs (and see 83) than are statements such as "most species of figs yield 3 or 4 crops per year" (49); they reflect a situation quite different from the orderly progression of synchronized crops produced by *Ficus carica* and *Ficus sycomorus* in the Mediterranean region. The reader must also be careful to distinguish between statements about the population as a whole and statements about individual trees (i.e. the word "figs" is ambiguous in much of the literature).

6. "The number of figs in a full crop of figs varied from 5,000 to over 50,000 per tree, depending on crown size." It was estimated that this amounted to 114,000 and 78,000 figs per ha per year for *F. yoponensis* and *F. insipida,* respectively. If a figure of 250 female agaonids per fig is representative (see #3 above), the airspace for a hectare of this forest should contain about 20 million pollinating agaonid females per month or about 62,500 per day of the species specific to each of these two fig species. This is about 8 wasps per m^3 of the approximately 0.5 ha^3 occupied by 1 ha of forest. If a fig tree needs about 2 wasps per fig for normal pollination, and has 5000–50,000 figs in a crop, the degree of superfluity of wasp production (and competition among wasps) will depend on the unknown life span of the wasps. It would be of great interest to match these figures with the arrival rate of pollinating agaonids at each fig (to say nothing of their parasites).

While Morrison (86, 87) was interested in figs as bat food, Hill (56) tried (*a*) to document the phenology of the 17 species of indigenous figs and 4 species of introduced ones in Hong Kong from 1962 to 1964, (*b*) to collect their contained wasps for taxonomic purposes, and (*c*) to determine the wasps' host-specificity. Since Hong Kong experiences a cool winter and is on the edge of the tropics, *Ficus* sexual phenology there should be quite different from that in lowland tropical BCI. Hill checked his plants every two weeks; wherever possible he had a sample size of 10 individuals (20 if gynodioecious). Several of his findings are relevant:

1. "The female agaonids leave their respective figs and fly in search of young figs of the same species in which to oviposit; sometimes they will be able to find such figs on the same tree that bore their figs, but more often they will have to find other trees" (56). Allusions to a few figs being out of phase with the main crop on a single *Ficus* appear in places besides Hill's (56) monograph (e.g. 82). It would be very nice to have both a quantification of this phenomenon and information on the entrance rate of the wasps to these out-of-phase figs. Such figs could be extremely important for self-pollination and for wasp survival in seasonal habitats or in habitats where

the main crop is heavily damaged by other animals. On the other hand, newly emerged fig wasps may have the behavioral trait of having to fly or otherwise delay before they can (will) attempt to enter a receptive fig (much as some newly molted alate aphids refuse to feed until they have flown a certain length of time).

2. "All the banyans except *F. superba* v. *japonica* bear their fruit in the summer (including spring and autumn) and the trees are either bare, or have only small retarded crops, during the winter. *F. superba* v. *japonica,* however, has the main crop of figs during the winter, although some fruit (sufficient for insect propagation) is borne during the summer. Even in this case, during the coldest period of winter development is very retarded. *F. pyriformis* also bears the vast majority of its figs over the winter (from November to April) as does *F. pumila* and probably also *F. sarmentosa* v. *impressa.* Some species (*F. variolosa* and *F. hederacea*) tend to have a very large crop of figs in the spring, but for the remainder of the year only odd plants have fruit. And then usually quite small crops. The other species of *Ficus* bear their figs during the warmer months (from May to October) and just have sufficient fruit production over the winter to allow insect propagation" (56).

Hill's sexual phenological records imply that one can speak of successive synchronized crops within a mono-specific population of figs, and indeed he often says things like "The species had five crops per year [(*Ficus superba* v. *japonica*)], although no individual tree had more than three of these per year (16 trees observed), and even then one crop was small. More than 60% of the trees under observation had only one or two crops of figs per year; but then each tree always had crop 5 [(the winter crop)]." However, only 30% had the March-April crop, 10% had the June-July crop, 20% had the July-September crop, and 10% had the September-November crop. Whether these should be called "crops", with the ensuing implication of population-level synchrony, is not at all clear to me.

3. It is clear from Hill's data, albeit derived from small sample sizes, that most species have periods of 1–2 months of each year when there are no figs on the trees. Either the fig wasps live longer than is generally thought or they are repeatedly extinguished locally and have to reinvade from outside the study area.

During the years of observation it was frequently found that certain species were without agaonid wasps at some times of the year, and successive crops of figs fell prematurely after only partial development. This happened twice with trees of *F. variegata* v. *chlorocarpa* and *F. hispida* on Hong Kong Island in the spring, when the last major crop ripening had coincided with a strong typhoon the previous autumn; in 1964 a group of trees under observation in the University Compound did not get any figs infested until the third crop was well developed, and then less than 5% of this crop became lightly infested. In the case of *F. pyriformis* so few figs were produced during the summer that when the first large crops developed in November, the vast majority of the figs fell uninfested. After continuous fig production during November and December eventually some of the figs on the observed bushes began to become infested, and after further continuous production, by March most of the bushes were carrying ripe infested figs. This pattern of events was observed over a two year period (56).

Hill also notes that the figs seem to remain receptive to agaonids for several weeks.

Hill's comments suggest that the biomass of fig production by an individual tree in a particular crop is likely to be determined by the degree of pollination of previous crops. It seems reasonable that a tree that aborts its figs at an early age will have expended many fewer resources (especially of the kind needed for maturing seeds) than a tree that carries a crop through to maturity. This may even explain why unpollinated commercial figs (*Ficus carica* and *F. sycomorus*) can have numerous successive large crops of parthenocarpic figs during the year. If they are not making seeds, they have many more resources for the production of later figs.

Aside from the multiple-species studies mentioned above, one can extract a few interesting data on fig sexual phenology from studies of individual species. However, generalized comments such as "most species of figs yield 3 or 4 crops per year" (49) are not useful here because of their ambiguity. Apropos *Ficus sycomorus* as a native tree in East Africa, Galil & Eisikowitch (39) report that a single developmental cycle lasts 6–7 weeks, but did not tell how often a single tree fruits. They imply, as do many authors, that there are plants in fruit somewhere in the population throughout the year. McClure (82) found an individual of *Ficus glabella* (Ulu Gombak, West Malaysia) to fruit once in 1960, twice in 1961, once in 1962, 1963, and 1964, and not at all in 1965. Nearby, two individuals of *Ficus ruginerva* pooled produced 2 fruit crops each year for 4 consecutive years. Another nearby tree of *Ficus sumatrana* produced 3 fruit crops per year for 5 consecutive years. Considering that each mature fruit crop lasted 2 months, the latter tree had figs on it for 50% of the year. Medway (83) gives further data on these individual trees. From 1966–1969, *F. glabella* had only 2 fruit crops, one plant of *F. ruginerva* produced 2 crops per year from 1963–1969 (except for 1968 when it had one), and the *F. sumatrana* tree continued to fruit 3 times a year until 1969.

As was discussed under "Pollinator specificity", fig trees on islands (91) and in extremely seasonal warm climates may have intra-crown asynchrony of flowering.

As an introduced plant in the Egypt-Israel region, *Ficus sycomorus* seems to be synchronized by the winter. In Egypt, the first fig crop is in April, the next in May, and the third in the first half of June. "After this there is more or less continuous production of new syconia until autumn and even throughout the winter" (8). *F. sycomorus* can have up to 6 generations of figs per year in Israel (38, 44). Does this mean that each tree fruits 6 times (as is probably possible since no seeds are produced) or that the population has 6 peaks?

DIOECIOUS FIGS

The subgenus *Ficus* contains what have commonly been called dioecious species of figs (e.g. 56). Described more accurately they are gynodioecious trees-to-shrubs and creepers. While there are trees in a gynodioecious species that bear figs with no male florets (e.g. the commercial fig of *Ficus carica*), the other morph has figs containing both male and female florets (e.g. the caprifig of *F. carica*). Apparently all caprifig florets are short-styled (11, 15, 56, 101). They may have very high levels of seed predation and may even become effectively male plants if all the seeds are destroyed.

In short, the pollinating agaonids have a normal life cycle in the figs of the trees containing both male and female florets. However, the pollen-bearing wasps also enter the figs containing nothing but female florets. They pollinate them and probe them with the ovipositor. They lay no eggs, apparently because the styles are all too long. They then die in the fig. While this process has been studied best in *F. carica* (11), it occurs in a number of wild species of *Ficus* and is thus not the result of selection by humans. Also, the large difference in edibility between Smyrna and caprifigs applies to wild gynodioecious species as well (e.g. 19). Since monoecious figs of gynodioecious species in nature commonly lose all their seeds to pollinating wasps, I expect them not to attract animals. Galil (35) gives the best account of pollination of a wild gynodioecious fig.

I expect strong selection for agaonids that are able to distinguish between the female figs and the monoecious ones. On the other hand, there should be strong selection for chemical mimicry between the two fig morphs. It seems to me that this situation is most likely to evolve where it is common-place for many more female wasps to arrive at a receptive monoecious fig than that fig needs and where the plants are at their peak of production of figs. A mutant plant that deleted its male florets and had only long styles could take advantage of such a wasp surplus by avoiding both the male floret costs and the loss of seeds to pollinating agaonids. However, such a cheater has the disadvantage that it is reproducing only by seed and not by pollen. As the proportion of such plants rises in the population, the possibility of pollination failure for both morphs arises. Just as with ordinary sex ratios then, there should be some ideal ratio of female and monoecious plants (or figs) from the viewpoint of the fitness of the parent producing them. This ratio has never been recorded for any wild population of *Ficus*. In a planting of seed of *F. carica* in California, of 139 seedlings, 74 were caprifigs (i.e. monoecious) and 65 were Smyrna figs (female) (11). In commercial orchards, it is recommended that 3–5 caprifigs be planted for each 100 Smyrna figs (11), but here the caprifigs are harvested and hung in the Smyrna fig tree, so the wasp's search problem is eliminated. Condit also noted that in *F. carica* caprifigs it is the spring fruit crop that has the high number of male florets; in later crops there are fewer florets, and even some (short-styled) florets set seed. There is also inter-variety variation in the amount of pollen produced per fig.

DISPERSAL OF FIG SEEDS

Who eats figs? Everybody. Wild figs are famous for being consumed by a very large number of species of vertebrates. They constitute a large part of the diet for more species of animals than any other genus of wild tropical perennial fruit. Ridley (99) records 44 tropical species of birds, bats, and nonvolant mammals feeding on figs. McClure (82) lists 32 species of vertebrates feeding on the figs of a single tree of *Ficus sumatrana* in West Malaysia. Freeland (32) found that mangabey monkeys (*Cercocebus albigena*) eat the figs of five species of *Ficus* in the Ngogo Reserve, Uganda. Figs comprised 16–17% of their diet; they ate the fruit of an average of 2.4 fig species each month for 8 months of the year. In a mainland lowland tropical forest with a normal complement of 5–10 *Ficus* species, all terrestrial species of herbivorous, frugivorous, and omnivorous vertebrates eat some species of fig at some time during the year. As a working generalization, nearly all of these animals (except small parrots) disperse fig seeds rather than intensely preying on them or spitting them out directly beneath the parent tree. They therefore probably generate the most thorough and extensive seed shadows found in any vertebrate-dispersed tropical perennial. This is not to say that fig seed shadows lack high intensity peaks (see e.g. 71, 86, 87), but rather that a fig seed probably lands occasionally on every square meter of the habitat. On the other hand, the user of this generalization must keep the following qualifiers in mind: Some molars and gizzards (as in *Treron* fruit pigeons) may grind up seeds as small and hard as fig seeds. Small parrots and very small rodents extract seeds directly from the fig and crack them. I found approximately 5000 cracked *Ficus ovalis* seeds inside an adult *Brotogeris gularis* in Costa Rican deciduous forest (D. Janzen, unpublished). Some primates eat immature figs [e.g. howler monkeys (86, 87)] and therefore act as seed predators. Finally, since figs are generically easy to identify and to lump into one ecological category, and since they often grow in gardens near forest, many naturalists have not had the time or inclination to record what visits different species of figs in different parts of the forest, or even to keep the observations separate for different species of wild figs. A *fig* is not a FIG is not a **fig**.

There are several different but not evolutionarily independent reasons why figs are eaten by so many kinds of animals. They have a high nutrient value per fruit fresh weight, and much of their weight is edible flesh. The seeds are apparently not toxic even if ground up during consumption. Figs occur in very large numbers and total weights when a crop ripens. In most tropical habitats ripe figs are available at any time of year. Most species of ripe figs do not appear to contain secondary compounds that would make them available only to very specialized frugivores (dispersal agents). They occur in a variety of sizes. Given these traits, it is not surprising that certain vertebrates seem to be heavily specialized at feeding on figs—bats being the most conspicuous of these.

Nutrient Value

Figs are no exception to the general rule that it is almost impossible to find relevant analyses of nutrient content of wild fruits (e.g. 33, 102). There are three indirect measures of their high nutrient value (and see 97). First, Mediterranean fruit fly larvae develop roughly twice as fast in fresh figs as in apples, pears, and peaches (100). Second, their obvious popularity among a very wide variety of frugivores suggests that they are either exceptionally free of secondary compounds, very rich in nutrients, or both. I suspect the latter to be the case. Third, if it is really true that a number of species of bats eat a diet of almost pure figs (7, 86, 87), then they must provide a moderately balanced diet.

Of a more direct but not necessarily more biologically meaningful nature, Hladik et al (59) found that *F. yoponensis* and *F. insipida* figs were, respectively, 4.5% and 6.1% (dry weight) protein, a percentage 2–3 times higher than that of the fleshy fruit pulp of *Spondias mombin,* an anacardiaceous fruit commonly eaten by bats. However, these protein percentages are difficult to interpret. Part of the protein is derived from wasp and seed fragments still in the fig. What fraction of this protein is obtained by the bat or other animal depends on the stage of ripening at which the animal eats the fig as compared with when the investigator harvests the fruit; it also depends upon whether the animal eats the trash in the pseudolocule. The wasps may be viewed as a legitimate part of the fig, since whatever nitrogen they contain came largely from the seeds they ate and thus from the parent plant. A second and much more serious complication is that the protein analyses were done on the fig wall with the good seed included. By grinding and otherwise digesting the seeds to different degrees, different animals will get variable amounts of the total protein in the fig. I suspect that bats, for example, rarely if ever actually chew up fig seeds, and thus 4.5% is much too high a measure of the protein they can get from the fruit. On the other hand, for an animal that thoroughly grinds its food much of the seed nitrogen might well be available.

While commercial figs (*F. carica*) have undoubtedly been bred for increased sugar content, and perhaps other nutrients, their nutrient content is at the high end of the range for commercial fruits. Winton & Winton (116) tried to summarize the literature on fig-nutrient analysis. The results are reported in too garbled a manner to be of use in a study of wild fig biology.

Seed Toxicity

I have found no suggestion that fig seeds contain any toxic secondary compounds. If this is the case, it is unusual for a tropical tree seed but not surprising for figs since many dispersal agents must grind some seeds. Seed toxicity need not be incompatible with the wasps, since many insects specialize on toxic seeds (e.g. 74). However, there is the problem that a mutant with a toxic seed would have to encounter simultaneously a resistant pollinating agaonid. Fig seeds are similar to the other tiny seeds embedded in tropical fruits eaten by tropical animals (e.g. *Cecropia, Piper, Miconia, Trema, Guazuma, Macaranga,* etc) and none of the seeds of these are known to be toxic.

Size of Fig Crops

When a fig tree comes into fruit, its branches are laden heavily with figs. No comparative data are available, but fig trees are certainly in the upper end of the frequency distribution of kilos of fruit per crop per tree for all tree species. Figs are generally spheres 1–4 cm in diameter. Many species of Hong Kong figs may have tens of thousands in a crop (56). On Barro Colorado Island, fresh ripe *F. insipida, F. obtusifolia,* and *F. yoponensis* figs weigh about 9, 17, and 3 g, respectively (7). [However, Morrison (86) gives the fresh weight of *F. insipida* figs as 5.6 g and that of *F. yoponensis* as 1.8 g.] This range is representative of the majority of tropical figs. Other trees may produce many large fruits in a crop, but figs are exceptional because most of the fruit is edible pulp. Among the Hong Kong figs an inverse relationship seems to exist between the number of figs in a crop and the size of the individual figs. For example, *F. microcarpa* var. *microcarpa* has one of the smallest figs (only 150 seeds per fig) and may have up to 100,000 figs in a large crop. *F. pumila* var. *pumila* has the largest fig (up to 6000 seeds in a fig), and a large crop is 200 figs (56). Morrison (86) presents the only data that can be used for a population estimate for wild-fig crop sizes. Calculating backwards from his figures for fig trees per hectare gives an average fruit crop size of 33,731 and 42,791 for *F. yoponensis* (n=71) and *F. insipida* (n=48), respectively. This works out to 61 and 240 kg of fresh figs (with seeds) per tree per crop. Morrison (86) calculated that all the species of figs (142 individuals of 7 species in 25 ha) in the BCI rainforest site were producing about 200,000 ± 75,000 figs per ha per year (low estimate is 650 kg fresh weight figs per ha per year, or 195 kg dry weight).

In determining the amount of food available to animals, it is not clear what measure is best, if any is. In theory dry weight is closer to the "nutrient" content. However, animals do not eat nutrients, they eat figs or fig parts. The water in the fig may be an important dietary item. Furthermore, the water-solid mix may be important for digestive processes and passage rates. Finally, the resources in the fig are not of some fixed and intrinsic value. Their particular value arises in the context of all the other foods that the animal eats. In order to substantiate the subjective impression that a fig tree in fruit represents a large, high-quality food resource one must compare the animals' use of or dependence on it with their use of or dependence on other fruit crops in the forest. The ideal measure of figs' importance is what happens to the animal population if figs are removed.

A. jamaicensis *as a Fig Specialist*

Bonaccorso (7) and Morrison (86) both did their Ph.D. dissertation research on the frugivorous bats of BCI. *Artibeus jamaicensis* was the most extensively studied and appears to be a specialist on fig fruits. Space is not available to discuss these studies, but they should be read by those dealing with fig biology (and see 27, 29–31, 53–55, 76, 77, 95, 106, 115).

Size and Other Traits of Ripe Figs

Different-sized bats on Barro Colorado Island appear to collect (carry away) different-sized fruits of *Ficus insipida* (presumably on occasion from the same tree) (7). This strongly suggests the possibility that not only may a tree choose its dispersers from among the total array of vertebrates through manipulation of the average size of its ripe figs, but it may do the same by the generation of array of fig sizes within its crop. In other words, variation in the sizes of ripe figs may be due to more than just the number of figs that happen to be produced on a branch, the number of florets fertilized in a fig, the amount of water the tree has, etc. Selection may lead to a given distribution of fruit sizes. Fruit-size distributions have never been recorded for any wild fruit tree.

Hill (56) noted a large variety of fig sizes, shapes, colors, and textures among 14 Hong Kong figs that bore fruit in his 1962–1964 study. I assume that this variety is adaptive in molding the disperser coterie and that it is a product of the forces generating intra-crown variation mentioned in the previous paragraph. Hill (56) noted that the subsection *Urostigma* of *Ficus* are all banyans, large monoecious trees "usually bearing large numbers of small figs." Banyans are generally strangler figs. Their disperser coterie should be molded to generate a seed shadow that spreads the seeds among the cracks and crevices of certain trees in the habitat. Perhaps one way to do this is to produce a very large number of very small figs that will be taken by the small vertebrates (small bats and small birds, since they should be the most arboreal frugivores). The big vertebrates can also eat many small figs. If the figs are large, the reverse is not necessarily true.

Ridley (99) has rather liberally interpreted the sizes, location, and color of figs. He interprets the Old World large subterranean and ground-level figs [e.g. *Ficus geocarpa, F. cunea, F. auriculata, F. capensis* (see 16)] as probably eaten by pheasants, pigs, rodents, and large mammals. He notes that fruits of smaller shrub species (e.g. *F. urophylla, F. diversifolia,* and *F. alba*) are small and yellow to red; on low climbers (e.g. *F. punctata, F. apiocarpa*) they are large and orange or red. Most large trees have brown, purple, or green figs. He feels these are usually taken by bats. Malaysian "fruit bats do not fly low; the smaller ones (*Cynopterus*) often feed on trees 15 to 20 feet tall, but not shorter; *Pteropus* only on trees 30 feet tall and higher. . . . The larger green or brownish inconspicuous figs borne on the trunks and boughs . . . of *F. polysyce* are eaten by bats only. In that species the figs are . . . about 1 inch long. . . . I watched for some time one fine tree in Singapore Gardens when the figs were ripe, but did not see a bird ever touch them. . . . A smaller tree . . . was regularly visited by *Cynopterus marginatus,* which flew up to the tree and carried off a fig to a distant point, ate it, and flew back for another till all were gone." McClure (82) noted that the mahogany-red, golf-ball-sized figs of *Ficus ruginervia* were ignored by birds but eaten by simiang whitehanded gibbons and 2 species of squirrel (one of which appeared to select them carefully by odor).

While I have been unable to locate a detailed study of the array of animals that might visit a single fig tree and thus constitute its disperser coterie, there is ample field evidence to suggest that for some individuals and species the array of species may be quite broad. In Singapore, "When a tree of *Ficus benjamina* or *F. retusa* is covered with the small, inconspicuous, purple-black figs, myriads of bulbuls (*Pycnonotus*) and often glossy starlings (*Calornis chalybaea*) and green pigeons (*Treron vernans*) and many other fruit-eating birds appear and devour the fruit all day. . . . During the night these birds are replaced by fruit-bats (*Pteropus* and *Cynopterus*). . . ." (99).

As mentioned earlier, McClure (82) listed 24 birds ("and many others") and 7 mammals that took the figs of a single *Ficus sumatrana* at Ulu Gombak, West Malaysia. On the other hand, a nearby *Ficus ruginervia* had its larger figs taken by only 4 species of mammals and no birds.

While Morrison (86) and Bonaccorso (7) stressed the highly synchronized intra-crown ripening period for figs on Barro Colorado Island, McClure's (82) account of *Ficus sumatrana* suggests a more complex picture. He states that "all of the feeding by birds and mammals was selective, for the fruit were examined carefully and only the ripe ones eaten. Since ripening was progressive along the twigs, the tree provided a continuous food supply for almost two months during each fruiting period."

It is particularly interesting that the members of the subgenus *Ficus* (all Old World) are all dioecious "but with no external differences morphologically between gall [(monoecious figs)] and female plants, except for the differences in the shape and size of the figs, and sometimes in their colour and seasonal occurrence" (56). The gynodioecious habit might be related in some manner to subtle patterning of the dispersal coterie. What would be the consequences if the disperser coteries that visited the female trees and the monoecious ones were somewhat different? In this context it may be relevant that all species with only monoecious figs are trees. Shrubs (epiphytic) and climbers are largely restricted to the dioecious subgenus *Ficus*.

In closing this section, it should be noted that the "size" of a fig is particularly difficult to interpret. First, it is my impression that figs contain an exceptionally large pulp/seed volume ratio. The pulp nutrient/seed volume ratio is impossible to guess, but I suspect it is at the high end of the scale as well. Second, a 17 g *Ficus obtusifolia* fig may weigh the same as an *Astrocaryum standleyum* fruit (7), but chewing the thin pulp off an enormous, hard *Astrocaryum* nut may be a much more difficult (and slippery) operation than mushing up a fig containing many seeds. Third, it is possible that fig size may be evolutionarily altered by changing the size of the pseudolocule without changing the dry weight of the rest of the fig. This would be particularly important for volume-responsive dispersers.

SPECIES PACKING IN *FICUS*

When the plant genera of a mainland tropical habitat are ranked according to the number of species they contain, *Ficus* is almost always at or near the top of the list. For example, in the deciduous forest of Santa Rosa National Park, Costa Rica, there are at least 7 species of *Ficus* trees; only a couple of shrub genera (*Mimosa, Cassia*) surpass this in species richness (75). *Ficus* is one of the largest genera of woody plants in regional tropical floras. Yet in seeming contradiction to the concept of limiting similarity, *Ficus* is notorious for its similarity of flowers, fruits, seeds, leaves, and branching patterns. Once one has learned to recognize a few species of *Ficus,* one can recognize them all easily anywhere from fruit or foliage.

I hypothesize that *Ficus* has such high intra-habitat species richness because the species have no pollinators in common, can be pollinated at very low population density, fruit very asynchronously, have seeds dispersed by many vertebrates, and have many vegetative life forms.

Having no pollinator species in common, *Ficus* species can be stacked into a habitat without competing for pollinators just as can wind-pollinated plants such as grasses, conifers, oaks, etc that can also have many congeners in small areas. This should be especially important for trees that have flowering individuals at low density at the time of flowering. Because *Ficus* pollinators seek out particular *Ficus* and are not seduced by other trees as the density of a particular species of *Ficus* declines, pollination will occur at a very low density of individuals, thereby allowing both extreme intra-population asynchrony of flowering and great inter-individual distances between reproducers. The ability to be pollinated at a low density means that the tree can exist on a very scarce resource type or with a very low probability of seedling survival on a common resource type. Fig pollination differs from other pollination by animals in a way very important to successful pollination of scarce individuals in time or space. The wasp has to make only one trip between conspecifics and no trips to a nest site, to nectar hosts, or to pollen hosts. (However, fig wasps may well visit calorie-rich sources such as extra-floral nectaries or flowers.) The fig tree is therefore the most animal-like of all trees in the forest in that its mating is highly active.

By fruiting synchronously within the crown but asynchronously at the population level, and by being a relatively rare tree when in fruit, figs have a minimum chance of competitively excluding each other over the services of dispersal agents. When the fruits are edible to most of the vertebrates in the habitat, this affect is accentuated. Chances are that when one tree is in fruit, few others will be in fruit in the vicinity. The dispersal agents will therefore be divided among a minimum number of fruit-bearing trees (i.e. there will be a maximum number of dispersal agents per tree). At the opposite extreme, if all fig trees in the forest were to fruit simultaneously, the vast majority of the figs would rot beneath their parents. Incidentally, I am certain that the large crops of fruits produced by figs are at least in part possible because figs put no resources (*a*) into chemical protection of ripe seed, (*b*) into the fruit in the form of protective woody tissue, fibrous tissue, or chemicals designed to cause all but a certain small set of animals to ignore figs, and (*c*) into large flowers with nectar flow and pollen as pollinator food. To the degree that vertebrate populations are in fact maintained by fig crops, fig species even synergistically augment each other's densities. If a common species doubles the density of monkeys by its presence, for example, then several rare species may have their seeds much better dispersed if their individuals on average happen to fruit between local individuals of common species than if they were the only fig trees in the habitat. An upper limit to this process should occur when the density of any one fig species rises to a point that occasions strong competitive interactions between individual trees in fruit. The feedback should be direct. A fig tree not visited by aerial or arboreal dispersers will drop its figs directly below. If they are not eaten by terrestrial vertebrates the seeds will probably die, owing to lygaeid bug seed predation and the inadequacy of heavy shade as a site for seedling growth.

Figs are noted for many adult vegetative life forms. This implies that the seeds get dropped in many kinds of safe sites. This can be done in two ways in plants. The fruits of some plants have chemical or nutritional qualities that cause them to be dispersed by a specific part of the animal disperser guild—animals that will put the seeds in very specific places. Figs, on the other hand, appear to generate a thin sheet, with occasional peaks, of very small seeds over much of the habitat (a diffuse but thorough seed shadow). The peaks may not even be associated with particularly safe sites for a given species of *Ficus*. In this manner, fig seeds can hit very small safe sites—a resource type that may be likened to what a dandelion can hit when it subdivides apomictically each year into many small pieces (70). The locations of the small safe sites of the fig and the dandelion change from year to year, and their exact positions are unpredictable with respect to other major traits of the habitat such as the perches of certain kinds of birds.

Figs turn out to be very different from other plants. They deserve careful study for reasons besides the details of their peculiar pollination or their direct value as bat food. They are almost everywhere in the tropics and are often left standing even when the forest is cut. They should quickly provide that animal-plant interaction in the tropics about which we know the most.

CONCLUSION

A fig tree is a specialist at producing a large crop of highly edible fruits rich in small edible seeds that are dispersed into a large and thorough seed shadow. The large crops of outcrossed seeds are produced by massive pollination by minute wasps, for which there appears to be no competition among the sympatric fig species. This pollination is achieved at a very low density of flowering trees, probably by chemical attraction of the wasps. Substantial seed predation by the wasps is the price paid for the pollination service. Despite the large literature on the interactions of the wasp and fig, many major questions of natural history remain unapproached, and the system is long overdue for analyses at the level of the populations of wasps, figs, and the complex of their species to be found in any tropical habitat.

ACKNOWLEDGMENTS

This study was supported by NSF DEB 77-04889 and the Servicio de Parques Nacionales de Costa Rica. D. E. Gladstone, D. W. Morrison, J. T. Wiebes, W. Hallwachs, and G. C. Stevens offered constructive commentary on the manuscript.

Literature Cited

1. Askew, R. R. 1971. *Parasitic Insects.* London: Heinemann. 316 pp.
2. Baker, H. G. 1961. *Ficus* and *Blastophaga. Evolution* 18:378–79
3. Baker, H. G., Hurd, P. D. 1968. Interfloral ecology. *Ann. Rev. Entomol.* 13:385–414
4. Barnes, D. F. 1949. Information on beetles infesting figs. *Calif. Fig Inst. Ann. Res. Conf. Proc.* 3:21–23
5. Barnes, D. F. 1952. Observations on the spring food habits of nitidulid beetles which attack figs. *Calif. Fig Inst. Ann. Res. Conf. Proc.* 6:7–8
6. Bezzi, M. 1978. Two new Ethiopian Lonchaeidae with notes on other species. *Bull. Entomol. Res.* 9:241–54
7. Bonaccorso, F. 1975. *Foraging and reproductive ecology in a community of bats in Panama.* PhD thesis. Univ. Florida, Gainesville. 122 pp.
8. Brown, T. W., Walsingham, F. G. 1917. The sycomore fig in Egypt. *J. Hered.* 8:3–12
9. Burger, W. 1977. Moraceae. *Fieldiana Bot.* 40:94–215
10. Butcher, F. G. 1964. The Florida fig wasp, *Secundeisenia mexicana* (Ashm.), and some of its hymenopterous symbionts. *Fla. Entomol.* 47:235–38
11. Condit, I. J. 1920. Caprifigs and caprification. *Calif. Agr. Exp. Sta. Bull.* 319:341–77
12. Condit, I. J. 1926. Fruit-bud and flower development in *Ficus carica. Proc. Am. Soc. Hort. Sci.* 23:259–63
13. Condit, I. J. 1941. Fig characteristics useful in the identification of varieties. *Hilgardia* 14:1–68
14. Condit, I. J. 1950. An interspecific hybrid in *Ficus. J. Hered.* 41:165–68
15. Condit, I. J. 1955. Fig varieties: a monograph. *Hilgardia* 23:323–538
16. Condit, I. J. 1969. *Ficus,* the exotic species. Berkeley, Calif: Univ. Calif. Div. Agr. Sci. 363 pp.
17. Condit, I. J., Enderud, J. 1956. A bibliography of the fig. *Hilgardia* 25:1–663
18. Condit, I. J., Flanders, S. E. 1945. "Gall-flower" of the fig, a misnomer. *Science* 102:129–30
19. Corner, E. J. H. 1939. A revision of *Ficus,* subgenus *Synoecia. Gard. Bull. Straits Settlement* 10:82–161
20. Corner, E. J. H. 1958. An introduction to the distribution of *Ficus. Reinwardtia* 4:325–55
21. Corner, E. J. H. 1962. The classification of Moraceae. *Gard. Bull. Straits Settlement* 19:187–252
22. Corner, E. J. H. 1976. The climbing species of *Ficus:* derivation and evolution. *Philos. Trans. R. Soc. London Ser. B* 273:359–86
23. Crane, J. C., Blondeau, R. 1949. Use of hormones as a substitute for caprification in the production of calimyrna figs *Proc. Ann. Res. Conf. Calif. Fig Inst.* 3:11–3
24. Dodson, C. H., Dressler, R. L., Hills, H. G., Adams, R. M. 1969. Biologically active compounds in orchid fragrances. *Science* 164:1243–49
25. Donohoe, H. C., Barnes, D. F. 1934. Notes on host materials of *Ephestia figulilella* Cregson. *J. Econ. Entomol.* 27:1075–77
26. Dressler, R. L. 1968. Pollination by euglossine bees. *Evolution* 22:202–10
27. Eisenberg, J. F., Wilson, D. E. 1979. Relative brain size and feeding strategies in the Chiroptera. *Evolution* 32:740–51
28. Essig, E. O. 1915. The dried fruit beetle, *Carpophilus hemipterus. J. Econ. Entomol.* 8:396–400
29. Fleming, T. H. 1971. *Artibeus jamaicensis:* delayed embryonic development in a neotropical bat. *Science* 171:402–4
30. Fleming, T. H. 1973. The number of mammal species in several North and Central American communities. *Ecology* 54:555–63
31. Fleming, T. H., Hooper, E. T., Wilson, D. E. 1972. Three Central American bat communities: structure, reproductive cycles and movement patterns. *Ecology* 53:555–69
32. Freeland, W. J. 1977. *Dynamics of primate parasite guilds.* PhD thesis, Univ. Michigan, Ann Arbor. 202 pp.

33. French, M. H. 1938. The composition and nutritive value of *Ficus sycomorus*. *Tanganyika Dept. Vet. Sci. Anim. Husb. Ann. Rep.* 2:51–52

34. Galil, J. 1973. Topocentric and ethodynamic pollination. In *Pollination and Dispersal*, ed. N. B. M. Brantjes, H. F. Liskens, pp. 85–100. Nijmegen, Netherlands: Dept. Bot., Univ. Nijmegen

35. Galil, J. 1973. Pollination in dioecious figs. Pollination of *Ficus fistulosa* by *Ceratosolen hewitti*. *Gard. Bull. Straits Settlement* 26:303–11

36. Galil, J. 1977. Fig biology. *Endeavor* 1:52–56

37. Galil, J., Eisikowitch, D. 1968. On the pollination ecology of *Ficus religiosa* L. in Israel. *Phytomorphology* 18:356–63

38. Galil, J., Eisikowitch, D. 1968. Flowering cycles and fruit types of *Ficus sycomorus* in Israel. *New Phytol.* 67:745–58

39. Galil, J., Eisikowitch, D. 1968. On the pollination ecology of *Ficus sycomorus* in East Africa. *Ecology* 49:259–69

40. Galil, J., Eisikowitch, D. 1969. Further studies on the pollination ecology of *Ficus sycomorus* L. (Hymenoptera, Chalcidoidea, Agaonidae). *Tijdschr. Entomol.* 112:1–13

41. Galil, J., Eisikowitch, D. 1971. Studies on mutualistic symbiosis between syconia and sycophilous wasps in monoecious figs. *New Phytol.* 70:773–87

42. Galil, J., Eisikowitch, D. 1974. Further studies on pollination ecology in *Ficus sycomorus*. II. Pocket filling and emptying by *Ceratosolen arabicus* Mayr. *New Phytol.* 73:515–28

43. Galil, J., Neeman, G. 1977. Pollen transfer and pollination in the common fig (*Ficus carica* L.). *New Phytol.* 79:163–71

44. Galil, J., Dulberger, R., Rosen, D. 1970. The effects of *Sycophaga sycomori* L. on the structure and development of the syconia in *Ficus sycomorus* L. *New Phytol.* 69:103–11

45. Galil, J., Ramírez, W., Eisikowitch, D. 1973. Pollination of *Ficus costaricana* and *F. hemsleyana* by *Blastophaga esterae* and *B. tonduzi* in Costa Rica (IIymenoptera: Chalcidoidea, Agaonidae). *Tijdschr. Entomol.* 116:175–83

46. Galil, J., Snitzer-Pasternak, Y. 1970. Pollination in *Ficus religiosa* L. as connected with the structure and mode of action of the pollen pockets of *Blastophaga quadriticeps* Mayr. *New Phytol.* 69:775–84

47. Galil, J., Stein, M., Horovitz, A. 1976. On the origin of the sycomore fig (*Ficus sycomorus* L.) in the Middle East. *Gard. Bull. Straits Settlement* 29:191–205

48. Galil, J., Zeroni, M., Bar Shalom, D. 1973. Carbon dioxide and ethylene effects in the coordination between the pollinator *Blastophaga quadraticeps* and the syconium in *Ficus religiosa*. *New Phytol.* 72:1113–27

49. Gordh, G. 1975. The comparative external morphology and systematics of the Neotropical parasitic fig wasp genus *Idarnes* (Hymenoptera: Torymidae). *Univ. Kans. Sci. Bull.* 50:389–455

50. Grünberg, A. 1938. The Mediterranean fruit-fly (*Ceratitis capitata*) in the Jordan Valley. *Bull. Entomol. Res.* 29:63–76

51. Hagan, H. R. 1929. The fig-insect situation in the Smyrna district. *J. Econ. Entomol.* 22:900–9

52. Hamilton, W. D. 1979. Wingless and fighting males in fig wasps and other insects. In *Reproduction, Competition and Selection in Insects*, ed. M. S. Blum, N. A. Blum. New York: Academic. In press

53. Heithaus, E. R., Fleming, T. H. 1978. Foraging movements of a frugivorous bat, *Carollia perspicilata* (Phyllostomatidae). *Ecol. Monogr.* 48:127–43

54. Heithaus, E. R., Fleming, T. H., Opler, P. A. 1975. Foraging patterns and resource utilization by eight species of bats in a seasonal tropical forest. *Ecology* 56:841–54

55. Heithaus, E. R., Opler, P. A., Baker, H. G. 1974. Bat activity and pollination of *Bauhinia pauletia*: plant-pollinator coevolution. *Ecology* 55:412–19

56. Hill, D. S. 1967. Figs (*Ficus* spp.) of Hong Kong. Hong Kong: Hong Kong Univ. Press. 130 pp.

57. Hill, D. S. 1967. Fig-wasps (Chalcidoidea) of Hong Kong. I. Agaonidae. *Zool. Verhand.* 89:1–55

58. Hill, D. S. 1969. Revision of the genus *Liporrhopalum* Waterston, 1920 (Hymenoptera, Chalcidoidea, Agaonidae). *Zool. Verhandl.* 110:1–36

59. Hladik, C. M., Hladik, A., Bousset, J., Valdebouze, P., Viroben, G., Delort-Laval, J. 1971. Le regime alimentaire de primates de l'ile de Barro Colorado (Panama). *Folia Primatol.* 16:85–122

60. Howard, B. J. 1933. The influence of insects in the souring of figs. *J. Econ. Entomol.* 26:917–18

61. Hull, F. M. 1929. Some possible means of control of the damage caused by the cotton leaf worm moth to the fig. *J. Econ. Entomol.* 22:792–96

62. Ishii, T. 1934. Fig chalcidoids of Japan. *Kontyû* 8:84–100
63. Janzen, D. H. 1966. Coevolution of mutualism between ants and acacias in Central America. *Evolution* 20:249–75
64. Janzen, D. H. 1967. Interaction of the bull's horn acacia (*Acacia cornigera* L.) with an ant inhabitant (*Pseudomyrmex ferruginea* F. Smith) in eastern Mexico. *Univ. Kans. Sci. Bull.* 47:315–558
65. Janzen, D. H. 1974. Swollen-thorn acacias of Central America. *Smithson. Contrib. Bot. No. 13.* 131 pp.
66. Janzen, D. H. 1974. Epiphytic myrmecophytes in Sarawak: mutualism through the feeding of plants by ants. *Biotropica* 6:237–59
67. Janzen, D. H. 1975. Interactions of seeds and their insect predators/parasitoids in a tropical deciduous forest. In *Evolutionary Strategies of Parasitic Insects and Mites*, ed. P. W. Price, pp. 154–86. NY: Plenum
68. Janzen, D. H. 1977. Promising directions in tropical animal-plant interactions. *Ann. Mo. Bot. Gard.* 64:706–36
69. Janzen, D. H. 1977. Why fruits rot, seeds mold, and meat spoils. *Am. Nat.* 111:691–713
70. Janzen, D. H. 1977. What are dandelions and aphids? *Am. Nat.* 111:586–89
71. Janzen, D. H. 1978. A bat-generated fig seed shadow in rainforest. *Biotropica* 10:121
72. Janzen, D. H. 1979. How many babies do figs pay for babies? *Biotropica.* In press
73. Janzen, D. H. 1979. How many parents do the wasps from a fig have? *Biotropica.* In press
74. Janzen, D. H. 1979. Prey-specificity of coleopteran seed-predators in the deciduous forests of Guanacaste Province, Costa Rica. *J. Ecol.* Submitted
75. Janzen, D. H., Liesner, R. 1979. Annotated check-list of plants of lowland Guanacaste Province, Costa Rica, exclusive of grasses, sedges, ferns and lower plants. *Brenesia.* In press
76. Janzen, D. H., Miller, G. A., Hackforth-Jones, J., Pond, C. M., Hooper, K., Janos, D. P. 1976. Two Costa Rican bat-generated seed shadows of *Andira inermis* (Leguminosae). *Ecology* 56:1068–75
77. Jimbo, S., Schwassman, H. O. 1967. Feeding behavior and daily emergence pattern of *Artibeus jamaicansis. Atas Simp. Biota Amazon.* 5:239–53
78. Johri, B. M., Konar, R. N. 1956. The floral morphology and embryology of *Ficus religiosa* Linn. *Phytomorphology* 6:97–111
79. Joseph, K. J. 1966. Taxonomy, biology and adaptations in fig insects (Chalcidoidea). In *Second All-India Congr. Zool. Proc., Varanasi* (*1962*), 2:400–3
79a. Lachaise, D. 1977. Niche separation of African *Lissocephala* within the *Ficus* drosophilid community. *Oecologia* 31:201–14
80. Lyon, H. L. 1929. Figs in Hawaiian forestry. *Hawaii. Plant. Bull.* 33:83–96
81. Marshall, G. A. K. 1914. Four new injurious weevils from Africa. *Bull. Entomol. Res.* 5:235–39
82. McClure, H. E. 1966. Flowering, fruiting and animals in the canopy of a tropical rain forest. *Malay. For.* 29:182–203
83. Medway, L. 1972. Phenology of a tropical rainforest in Malaya. *Biol. J. Linn. Soc.* 4:117–46
84. Miller, M. W. 1952. Yeast associated with the dried fruit beetle in figs. *Calif. Fig Inst. Ann. Res. Conf. Proc.* 6:8–9
85. Morrill, A. W. 1913. Entomological pioneering in Arizona. *J. Econ. Entomol.* 6:185–95
86. Morrison, D. W. 1975. *The foraging behavior and feeding ecology of a neotropical fruit bat,* Artibeus jamaicensis. PhD thesis. Cornell Univ., Ithaca. 94 pp.
87. Morrison, D. W. 1978. Foraging ecology and energetics of the frugivorous bat *Artibeus jamaicensis. Ecology* 59:716–23
88. Powell, J. A. 1974. Biological interrelationships of moths and *Yucca schottii. Am. Philos. Soc. Yearb.* 1973:342–43
89. Powell, J. A., Mackie, R. A. 1966. Biological interrelationships of moths and *Yucca whipplei* (Lepidoptera: Gelechiidae, Blastobasidae, Prodoxidae). *Univ. Calif. Publ. Entomol.* 42:1–46
90. Ramírez, W. 1969. Fig wasps: mechanism of pollen transfer. *Science* 163:580–81
91. Ramírez, W. 1970. Host specificity of fig wasps (Agaonidae). *Evolution* 24:680–91
92. Ramírez, W. 1970. Taxonomic and biological studies of Neotropical fig wasps (Hymenoptera: Agaonidae). *Univ. Kans. Sci. Bull.* 49:1–44
93. Ramírez, W. 1974. Coevolution of *Ficus* and Agaonidae. *Ann. Mo. Bot. Gard.* 61:770–80
94. Ramírez, W. 1976. Evolution of blastophagy. *Brenesia* 9:1–13
95. Ramírez, W. 1976. Germination of seeds of New World *Urostigma* (*Ficus*) and of *Morus rubra* L. (Moraceae). *Rev. Biol. Trop.* 24:1–6

96. Ramírez, W. 1977. Evolution of the strangling habit in *Ficus* L., subgenus *Urostigma* (Moraceae). *Brenesia* 12/13:11–19

97. Reinherz, O. 1904. Note on the chemical composition of the fruits of *Ficus* spp. *Agr. Ledg.* 4:387–94

98. Rickson, F. R. 1979. Absorption of animal tissue breakdown products into a plant stem—the feeding of a plant by ants. *Am. J. Bot.* 66:87–90

99. Ridley, H. N. 1930. *The Dispersal of Plants Throughout the World.* Kent, England: L. Reeve & Co. 744 pp.

100. Rivnay, E. 1950. The Mediterranean fruit fly in Israel. *Bull. Entomol. Res.* 41:321–41

101. Rixford, G. P. 1912. Fructification of the fig by *Blastophaga. J. Econ. Entomol.* 5:349–55

102. Saldova, A. K. 1938. The biochemical value of fig fruit. *Sov. Subtrop.* 10:48–49

103. Simmons, P., Reed, W. D. 1929. An outbreak of the fig moth in California. *J. Econ. Entomol.* 22:595–96

104. Slater, J. A. 1971. The biology and immature stages of South African Heterogastrinae, with the description of two new species (Hemiptera: Lygaeidae). *Ann. Natal Mus.* 20:443–65

105. Slater, J. A. 1972. Lygaeid bugs (Hemiptera: Lygaeidae) as seed predators of figs. *Biotropica* 4:145–51

105a. Sweet, M. H. 1964. The biology and ecology of the Rhyparochrominae of New England (Heteroptera: Lygaeidae). Pts. I, II. *Entomol. Am.* 43:1–124; 44:1–201

106. Thomas, S. P. 1975. Metabolism during flight in two species of bats, *Phyllostomus hastatus* and *Pteropus gouldii. J. Exp. Biol.* 63:273–93

107. Thomen, L. F. 1939. The latex of *Ficus* trees and derivatives as anthelmenthics. *Am. J. Trop. Med.* 19:409–18

108. Turner, D. C. 1975. *The Vampire Bat.* Baltimore: Johns Hopkins Univ. Press. 145 pp.

109. Weber, N. A. 1972. Gardening ants, the attines. Philadelphia: *Am. Philos. Soc.* 146 pp.

110. Wheeler, G. B. 1953. Molds and souring in relation to infestation. *Calif. Fig Inst. Ann. Res. Conf. Proc.* 7:26–29

111. Wiebes, J. T. 1966. Provisional host catalog of fig wasps (Hymenoptera: Chalcidoidea). *Zool. Verh.* 83:1–44

112. Wiebes, J. T. 1976. A short history of fig wasp research. *Gard. Bull. Straits Settlement* 29:207–36

113. Wiebes, J. T. 1979. Figs and their insect pollinators. *Ann. Rev. Ecol. Syst.* 10:1–12

114. Wildman, J. D. 1933. Notes on the use of micro-organisms for the production of odors attractive to the dried fruit beetle. *J. Econ. Entomol.* 26:516–17

115. Wilson, J. W. 1974. Analytical zoogeography of North American mammals. *Evolution* 28:124–40

116. Winton, A. L., Winton, K. B. 1935. *The Structure and Composition of Foods,* Vol. II. New York: Wiley. 904 pp.

117. Wolcott, G. N. 1951. The insects of Puerto Rico. *J. Agr. Univ. Puerto Rico* 32:1–771

Ann. Rev. Ecol. Syst. 1979. 10:53–84

THE REGULATION OF CHEMICAL BUDGETS OVER THE COURSE OF TERRESTRIAL ECOSYSTEM SUCCESSION

❖4155

Eville Gorham

Department of Ecology and Behavioral Biology, University of Minnesota, Minneapolis, Minnesota 55455

Peter M. Vitousek

Department of Biology, Indiana University, Bloomington, Indiana 47401

William A. Reiners

Department of Biological Sciences, Dartmouth College, Hanover, New Hampshire 03755

INTRODUCTION

Vitousek & Reiners (158) have suggested that change in net ecosystem production is a major determinant of the balance between inputs and outputs of elements in terrestrial ecosystems. They argued that in the course of primary succession element outputs are initially relatively high (approximating inputs), that they then drop to a minimum because of element accumulation in biomass and detritus when net ecosystem production is highest, and that eventually output rates rise again approximately to equal inputs in late succession when net ecosystem production approaches zero (Figure 1). In most cases of secondary succession, net ecosystem production is negative immediately following disturbance, and in such cases output rates can exceed input rates. The probable importance of change in net ecosystem production is supported in particular by observations of higher element outputs from later successional ecosystems (22, 90, 157–159) and

53

0066–4162/79/1120–0053$01.00

also by the recorded increases in element outputs following disturbance (17, 86, 91, 150, 170).

Other processes can also affect chemical budgets systematically in the course of terrestrial ecosystem succession. These processes may lessen or obscure the importance of net ecosystem production in particular instances, or they may cause the level of inputs and outputs, though still in balance, to be higher or lower at the end of succession than at the beginning. The objective of this paper is to identify and discuss the diverse processes that can influence inputs and outputs in the course of terrestrial succession. The processes examined are those that affect element inputs to systems, including rock and soil weathering, nitrogen fixation, particle impaction, and gas

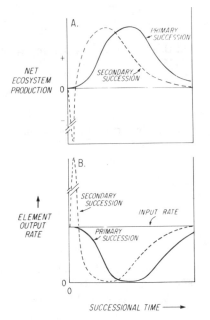

Figure 1 A. Patterns of change in net ecosystem production (NEP) during primary and secondary succession of terrestrial ecosystems. The starting time for primary succession is immediately following exposure of a new site; for secondary succession it is immediately following a destructive disturbance. Note the scale break for the pulse of negative NEP following disturbance. This negative pulse may be accentuated by combustion of biomass, accelerated decomposition, removal of forest products, erosion, and other factors. The integrated area under a properly scaled negative pulse should approximate the integrated area under the positive NEP part of the curve. See Vitousek & Reiners (158) for further explanation.

B.   Patterns of output rates for a limiting element in primary and secondary succession, assuming that changes in storage are controlled principally by NEP and that *input rates are constant*. Note the break in scale for the pulse of high output rates following disturbance in secondary succession. The integrated area under this pulse above the line for input rate should approximate the integrated area for outputs below the line for input rate.

absorption; those that change with alterations in hydrology, including losses of dissolved substances, erosion, and oxidation-reduction reactions; and those biological processes that directly or indirectly affect the balance between inputs and outputs, including net ecosystem production, element mobilization or immobilization, cation/anion balance, the production of allelochemic substances, and changes in element utilization by the biota. Each of these processes varies systematically in the course of succession, though the magnitude of each change may be difficult to predict in the course of any particular succession.

We seek in this review to establish an organizational framework that will stimulate research upon the mechanisms controlling ecosystem inputs and outputs, and thus to facilitate the development of general mechanistic models for biogeochemical cycling in terrestrial ecosystems. We cannot reduce the complexity of nature to simplicity, but we can provide a clearer appreciation and understanding of the diverse interacting mechanisms at work over the course of terrestrial succession. The new insights that will lead to a fuller understanding of the chemical budgets of ecosystems can come only from further detailed examination of these mechanisms.

Before we analyze in detail the processes controlling chemical budgets in ecosystems, we define precisely what types of ecosystems we are considering and explain what we mean by succession and by element inputs and outputs.

Succession

Succession is a venerable and central concept in ecological thought (21, 26, 27, 50), yet today it is undergoing intense reconsideration (24, 36, 38, 39, 75, 76, 167). Much of this reconsideration addresses the regularity and directionality of processes and the causal mechanisms underlying change in community structure. In this reexamination new generalizations, paradigms, and terminology are being tested as replacements for the old. We wish to examine how some ecosystem properties vary as the biota changes in composition and mass with time. Until a better term is introduced, we shall use the term "succession" to describe ecosystem change following exposure of new sites or disturbance of previously occupied sites.

The Watershed-Ecosystem Approach

For convenience and clarity, we consider as our conceptual model a watershed ecosystem occupying a self-contained basin not subject to appreciable groundwater input. In such a situation, the only hydrologic transfers of elements into an ecosystem are in precipitation, and the only hydrologic transfers of elements out are loss to groundwater and stream water (10). Under these conditions, hydrologic fluxes of elements can be measured quite rigorously, and the effects of any perturbations of these fluxes can be assessed (10). However, in this paper we are also concerned with other kinds of fluxes, and for some systems and some elements the watershed model is not particularly germane (171).

Definition of Inputs

By convention, chemical budgets for watersheds or catchment areas are usually recorded as inputs in bulk precipitation minus outputs in stream water (46, 72, 86, 95). Inputs by gaseous absorption and particulate impaction are sometimes included, but their treatment is inconsistent. In some cases they are calculated by difference and included as inputs [(98) for sulfur.] In other cases they are calculated by difference but not included in inputs [(11, 98) for nitrogen]. Gaseous outputs (e.g. of nitrogen and sulfur) may also occur and add to hydrologic outputs. Weathering is not usually included as an input because it is generally calculated by difference (85). Consequently, elements derived primarily from weathering will most often have negative budgetary balances.

Although it is convenient to calculate element budgets for ecosystems in terms of hydrologic transfers alone, and there is some conceptual justification for considering weathering as an internal transformation, we believe that this approach can lead to some conceptual and practical problems. First, ions derived from weathering of primary minerals in the solum can represent inputs in the same sense as ions derived from the atmosphere. Weathering of primary minerals is not a form of recycling—by definition (100) there is no chance for elements to return to primary forms within the system. Second, regarding weathering as an input can aid in understanding the processes of element retention and loss. For example, calcium in control watersheds of the Hubbard Brook Experimental Forest has inputs in bulk precipitation of 2.2 kg ha^{-1}yr^{-1}, and hydrologic outputs of 13.7 kg ha^{-1} yr^{-1} (98). One might conclude that the system does not "conserve" calcium, that the calcium cycle is relatively "leaky." Yet Likens et al (98) estimate that approximately 3 kg ha^{-1}yr^{-1} of calcium is sequestered in biomass each year—an amount equal to 20% of total inputs (bulk precipitation plus weathering). This view of a system losing less calcium than it receives from all sources each year better reflects the actual dynamics of calcium in this ecosystem.

Practical considerations suggest a third reason for treating weathering as an input. New forestry methods have made harvesting of whole trees on a short rotation possible over large areas. In calculating the costs of such practices to long-term productivity, weathering must be considered as a nutrient source (15, 118, 161, 163). According to some of these calculations, weathering becomes essential for maintenance of a sufficient supply of metallic cations.

This relatively simple approach of defining the soluble products of weathering as inputs is complicated by the fact that weathering may take place below the zone in which the elements released could be incorporated into biota (71). Conceptually, material weathered in such locations is neither an input to nor an output from a terrestrial ecosystem (though it certainly represents an input to a downstream ecosystem). Such material is excluded from lysimetric studies (to their detriment as techniques for the examination of land/water interactions) but is included in watershed studies (to the detriment of their ability to focus on terrestrial biogeochemical processes). The formation of secondary soil minerals (e.g. calcium carbonate, calcium sulfate, and the oxides of iron, manganese, aluminum, and silica) is a further complicating factor but may be regarded as an internal transfer unless the secondary minerals are themselves weathered to soluble products.

Throughout, when we use the word "input" without qualification we intend it to include inputs from bulk precipitation, aerosol impaction, gaseous absorption and transformation, and weathering of primary minerals to soluble products. This approach is logical and consistent for conceptual purposes. We recognize, however, that for practical reasons involving the difficulty of measuring other inputs, the input term in chemical budgets for watersheds will generally continue to be calculated in terms of bulk precipitation and perhaps gaseous inputs and impacted particles. But because rock weathering is such an extremely important budgetary item, its calculation by difference is undesirable. Considerable efforts should be made to develop methods of direct measurement.

Finally, our perspective comes from the biogeochemistry of forested ecosystems, which at present is better understood than that of other ecosystems. We believe that most of the processes we discuss are also important in other types of ecosystems, but their applicability remains to be tested. Woodmansee (171) has explained some of the important differences between the components of ecosystem budgets in forests and dry grasslands.

PROCESSES AFFECTING CHEMICAL BUDGETS OF ECOSYSTEMS

For the remainder of this paper we shall examine a number of ecosystem processes that can vary because of successional development and can therefore alter chemical budgets. We shall restrict the discussion to processes intrinsic to ecosystems, ignoring external factors (such as changes in the amount or composition of bulk precipitation) that might change input rates. For the most part we shall consider changes beginning with the exposure of a new site (primary succession) or following destruction of a former ecosystem on a previously occupied site (secondary succession). Our discussion of processes extends to the time at which the rate of change is small. This is not to be interpreted as a conviction on our part that single-point steady states are likely phenomena now or that they were in the past (cf 13, 36, 172). Our focus is mainly upon characteristic directions of change.

The relevant ecosystem processes are as follows:

1. Processes affecting inputs
    a. Rock and soil weathering
    b. Nitrogen fixation
    c. Particle impaction and gas absorption
2. Hydrologic processes affecting outputs
    a. Loss of dissolved substances
    b. Erosion
    c. Regulation of redox potential
3. Biological processes affecting the balance of inputs and outputs
    a. Net ecosystem production
    b. Decomposition and element mobilization
    c. Regulation of soil solution chemistry
    d. Production of allelochemics
    e. Variability in utilization of elements

We make no pretense that this list of processes is complete, but these are the most general and quantitatively the most important processes affecting biogeochemical cycling in developing ecosystems. Furthermore, we realize that a linear series of mechanisms in such a list is an inherently misleading portrayal of relationships. The mechanisms themselves are interrelated and interact through time in complex ways.

Processes Affecting Inputs

As the major processes capable of altering element inputs over the course of terrestrial succession we shall examine weathering, nitrogen fixation, particle impaction, and gas absorption.

WEATHERING   Weathering entails both physical and chemical processes, but we focus mainly on chemical processes. Chemical weathering includes the processes by which environmental agents acting within the zone of influence of the atmosphere produce relatively stable, new mineral phases (100). In these processes dissolved substances are produced and removed. It is convenient to view chemical weathering as the conversion of primary mineral X to secondary mineral Y plus dissolved product Z. Our principal attention will be directed to Z, although Y may weather further to produce additional dissolved products.

Relative patterns and rates of chemical weathering for broad climatic regions and general rock types are reported in the geochemical literature. The chief determinants of weathering rates are temperature, water flux, parent material, relief, and biota (8, 100, 148). There is a distinct parallel between the analysis of weathering rates and the analysis of processes of soil formation as presented by Jenny (77). Chemical weathering, like primary productivity and decomposition, is greatly enhanced by heat and water, and so may correlate well with actual evapotranspiration (105, 133). Strakhov's widely duplicated figure illustrating the extent and nature of weathering along a latitudinal gradient of varying temperature and precipitation [(148), p. 6] is a forceful representation of the role of climate.

Within climatic regions the patterns and rates of chemical weathering are highly dependent upon parent material. Its texture, structure, and composition will have an influence on rate of percolation, surface area affected, and chemical reactions. Loughnan (100), who dealt only with silicate rocks, carefully differentiated weathering characteristics of four classes: basic crystalline, acid crystalline, alkaline igneous, and argillaceous sedimentary rocks. Carbonate rocks might provide even more dramatic contrasts in comparison with igneous, metamorphic, or even other sedimentary rocks. The effect of major contrasts between rock types was demonstrated by a study of the chemical degradation of opposite flanks of the Wind River Range, Wyoming (70). Paleozoic and Mesozoic rocks on the northeast flank yielded nearly twice as much dissolved material in runoff as the more resistant rocks on the southwest flank, in spite of a higher runoff volume on the southwest side.

Parent material will have a powerful effect on the change in rate of weathering through time. If a rock is composed wholly of easily weathered minerals, weathering may accelerate to an asymptotic maximum in parallel with the development of the acid-generating capability of plant roots and detritus decomposers. If the rock is composed of differentially weatherable minerals, the change in weathering rate will be more complex. It will be a compound function of the relative proportions of easily weathered and resistant minerals, and the change in biotic influence over time. In such a case, chemical weathering may accelerate and then decline asymptotically as resistates accumulate at the expense of more soluble minerals. William Graustein (personal communication) suggests that in the Tesuque River watersheds of the Sangre de Cristo Range of New Mexico, the initial granite contained both biotite and potassium feldspar as sources of potassium. Initially, potassium was liberated rapidly from the easily weathered biotite but now it is made more slowly available from the more resistant feldspar.

Loughnan [(100), p. 74] gives another example of how rock type can influence the rate of change. Rain striking fresh lava is rapidly lost through surface runoff or evaporation. It is only after a thick crust forms that water is held and chemical weathering is promoted. In this situation weathering provides a positive feedback leading to acceleration of chemical breakdown. Such an acceleration could, for a time, be represented by a geometrically ascending curve.

During secondary succession the degree of prior weathering will influence subsequent weathering rates. If the disturbed soil represents a mature body of minerals more or less in equilibrium with the climatic regime, succession will have relatively little effect on weathering rates.

The influence of relief on chemical weathering is not easily predictable. Strakhov (148) stressed that increased relief accelerates physical weathering and suppresses chemical weathering (cf 101). Clearly the relative importance of these two components shifts in that fashion, but it is not so clear that the absolute rate of chemical weathering must be diminished. Reynolds & Johnson (126) observed very high rates of chemical weathering in the temperate, glacial environment of the Cascade Mountains. Similarly, Gibbs (48) reported high rates of chemical weathering in the mountainous headwaters of the Amazon. If the high rainfall associated with regions of high relief is taken into account, it seems likely that removal of weathered minerals by erosion takes place under conditions that favor chemical weathering of primary minerals. The ability of a local ecosystem to sequester the dissolved substances liberated under such conditions is questionable, however. In the extreme, these conditions of strong relief and severe erosion may not permit the development of a vegetative cover or solum adequate to delay the escape of ions from the weathering region by recycling them within the local ecosystem.

The biota can affect chemical weathering in several ways. Roots can have a physical action on parent material by fracturing rocks and enlarging fractures through wedging. Plants and detritus will retard runoff and promote slow percolation. These effects will enhance chemical weathering. Increased transpiration will decrease the water available for weathering reactions to some extent and will decrease the potential of erosion (12), thereby preventing the stripping of weathered materials and the exposure of unweathered material.

The most important effect of biota is probably through its generation of acidity. Biological respiration by roots and soil biota adds carbon dioxide to the soil atmosphere, thus tending to maintain or even increase the supply of hydrogen ions to the parent material. There may also be "contact exchange" with hydrogen ions released from root surfaces or from the surfaces of bacteria, fungi, or mycorrhizae in contact with mineral surfaces (78). Recent evidence, however, suggests that ionic uptake by plant roots often involves excretion of more bicarbonate than hydrogen ions (113a). Organic acids (30) resulting from decomposition or root exudation (142) may also play a significant role in soil and rock weathering. The stoichiometric importance of organic acids as compared with carbonic acid seems to increase in cooler, humid environments (83). More importantly, some of these organic acids may be powerful chelators and play leading roles in the alteration of primary minerals to secondary minerals and in the cheluviation of dissolved solids downward (3, 18, 19, 33, 63, 100, 141). Not all organic acids are effective in this way (107), and careful field studies are needed to document the role of organic chelators in chemical weathering (31, 62). It is possible that plants of particular successional stages produce unusually large amounts of organic acids, either directly through their roots or indirectly through decomposition of their litter. They may thus change the rates of mineral decomposition and ion release. If the chemistry of soil solutions generated under major species of particular seres were known, changes in biologically engendered weathering rates might be hypothesized and tested. Changes at Glacier Bay, Alaska, are especially rapid during the older stage of primary succession dominated by alder (156). It has not, however, been determined whether the accelerated weathering is due to organic acids characteristic of alder, or to nitric acid produced in the humus layer. Alder litter is unusually rich in nitrogen because of the activities of nitrogen-fixing microorganisms associated with alder roots.

Variability is to be expected in chemical weathering rates, and such variability makes prediction of changes in rate over the course of terrestrial succession most difficult. Nevertheless, we offer the following tentative generalizations.

To a large extent the rate of chemical weathering, and its changes during succession, will be determined by climate, parent material, and relief. The biological effects associated with succession will be modifiers of the fundamental trajectories set by the physical factors. From this perspective, three points can be made. First, changes in weathering rates are likely to be greater during primary succession than during secondary succession—given the same climate, parent material, and relief (Figure 2a). This fundamental contrast between primary and secondary succession is due to the chemically unstable nature of the fresh parent material in a new site. Second, other things being equal, weathering will take longer to reach a steady state during primary than during secondary succession (Figure 2a). This difference could be due to initial lags (to be discussed later) and to the fact that chemical weathering will go on longer where large amounts of fresh minerals are exposed near the surface. Third, a good deal of weathering may be independent of the biota, especially in primary succession, so that changes in weathering rates through succession are not wholly or necessarily biotically induced effects of succession itself. Ecologists should beware of attributing too much change in weathering rates to the biotic influences of succession.

Both primary and secondary succession probably exhibit an initial increase in weathering rate over time, because some factors change together in ways that promote chemical weathering. These factors may include physical weathering, increased water percolation into physically weathered material, and increasing acid-generating capacity through the accumulation of biomass. This acid potential could be especially important in those cases of secondary succession where acidity is increased by nitrification following

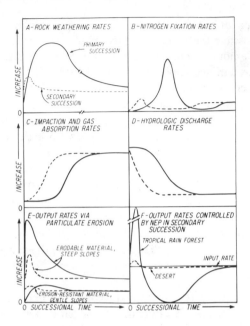

Figure 2   Patterns of change in some biogeochemical processes with primary (solid line) and secondary (dashed line) successional development of terrestrial ecosystems. These graphs represent generalized expressions of changes in processes only. With the exception of weathering in primary succession (A), these graphs are scaled to match rates of change illustrated in Figure 1. A: Note hash marks denoting a break in the time scale for weathering in primary succession. B: The two initial peaks represent pulses of N-fixation characterizing many seres in early- to mid-succession. The rate increases denoted later represent fixation in epiphytes, decaying logs, and other components characteristic of more mature systems. C: Impaction and gas absorption rates are scaled to parallel physical extension of plant canopies. D: Hydrologic discharge (surface runoff and deep seepage) rates are graphed as inversely mirroring the development of full utilization of soil water. Full utilization is believed to be attained in advance of maximum biomass (see text). E: Two conditions of susceptibility to erosion are illustrated. The most erodable site will have higher output rates than the less erodable site even if a steady state is reached. A slight lag in peak outputs is shown for secondary succession on erodable material, in consideration of residual biotic resistance produced by old roots, etc. The peak output rate for secondary succession precedes that of primary succession on erosion-resistant sites, because in primary succession a period of rock weathering and organic accumulation is probably necessary before erosion of resistant materials can become significant. Output rates from primary successional sites on rock substrates would logically start from zero, whereas output rates from other primary successional sites would initiate at higher values. Output rates from secondary successional states would initiate at values approximating the steady-state rates. F: This graph differs from the others in this set in that both curves portray secondary successional behavior. It is essentially a version of Figure 1B, illustrating variation in the degree of control exerted by NEP on output rates depending on the biomass of the system. Note the break in scale of the peak output pulse. The integrated areas under the curves above the line for input rate should approximate the integrated areas under the line for input rate.

cutting (94). Based on the changes in silica outputs from the devegetated watershed at Hubbard Brook (97), we estimate rock and soil weathering nearly doubled under those extreme conditions. Because secondary usually proceeds faster than primary succession, we expect that this acceleration will begin sooner in secondary succession (Figure 2a).

Because by definition secondary succession takes place on previously occupied sites, the amplitude of response should be small or nonexistent unless there is extensive stripping of weathered materials by erosion or accelerated weathering caused by fire (9, 105). We also anticipate that rates will rather quickly resume their original levels (Figure 2a). If erosion is extensive, initial conditions more closely resemble those of primary succession; and the generalizations for change in weathering rates likewise shift towards those for primary succession.

In the course of primary succession, various changes in weathering rate may be postulated following an initial acceleration. Where physical weathering is not very important, as in unconsolidated deposits, and where resistates such as quartz do not accumulate, chemical weathering may rise to an asymptotic level, remaining at the maximum rate until geological changes intervene. The solution weathering of limestone and andesite may be examples of this. We think the more general case for primary succession is a large initial increase followed by a slow logarithmic decline (Figure 2a). Initial lags in chemical weathering may or may not occur. For very rocky terrain, such as polished bedrock or basalt in cool or dry climates, both water contact with mineral surfaces and colonization by plants would be limited, and the lag in weathering very significant. In other materials—such as glacial till, alluvial deposits, or calcareous sands—the lag would be less significant. Lack of nitrogen may impose a lag on all primary successions (see next section).

With or without a lag, there are many examples of an initial surge in the weathering of minerals. Stevens & Walker (146) reviewed the voluminous literature on soil development, from which we can infer generalizations about changes in chemical weathering rates throughout succession. The most consistent rapid change is loss of carbonates (and to a large degree, associated calcium and magnesium). This is documented in studies of soil genesis, experimental research, and stratigraphic investigations of lake sediments (20, 99, 135, 136, 165). The initial high rate of weathering is seen especially in glacial materials where grinding has disrupted crystal structure and increased surface area (cf 20, 88). Slower changes are observed in the forms of iron, aluminum, and phosphorus. Other weathering phenomena such as clay formation seem to take place at still slower rates throughout chronosequences, but there is evidence in a few cases of ultimate degradation of the soil (loss of fertility) linked to completion of weathering to ultimate resistates (see also 79). A chronosequence of podsol development

(45) is particularly graphic. In this sequence, ranging from 2,250 to 10,000 years of age, the rate of change of phosphorus peaked at 4,000 years and decreased thereafter. Rates of change in iron and aluminum peaked at 8,000 years and then decreased; the rates of change in clay and silt continued to accelerate through 10,000 years. The review of these chronosequences points toward another consideration: The shape of the weathering curve will be different for different elements. Nonlithophilic elements such as nitrogen will be extreme in their independence of weathering rates, except insofar as indirect influences such as pH or redox potential affect their transformations.

The gradual decline of chemical weathering rate seems inevitable in primary succession where resistates accumulate. Reactive minerals will become restricted to ever greater depths in the solum, or within weathering rinds of rock fragments (8, 116). However, the concept of weathering rinds at the crystal level is questionable (7). The expected pattern of decline is represented in Figure 2a, but it is highly dependent on the nature of the parent material. Some might doubt that it is the most typical case, or that it is possible to generalize any curve for this process. Nevertheless, we offer these generalizations as reference points at which consideration of patterns in particular systems can begin.

We are concerned not only with rate of production of dissolved products, but also with whether these products are liberated at a position in which they could become involved in local chemical cycles. Dissolved weathering products tend to move downhill, and often they will not be available to the ecosystem in which they were produced. In such a case chemical weathering, regardless of its rate, will not represent an input to the system in which the weathering took place, although it may be an input to another system downslope or downstream. The availability of weathering products to the local system will depend on the depth of unstable minerals in the soil profile in relation to the distribution of roots therein. A great range of possibilities can be imagined [for one contrasting pair see (41, 106)].

A first approximation to the importance of mineral weathering as an input to a particular system would most likely begin with a bulk analysis of soil mineral composition versus depth. The weatherability of those minerals in that particular physical-chemical milieu, and their distribution relative to the roots, would put some upper limits on minerals as a source of biogenic elements to that system. A finer assessment would require field estimates of weathering rates and of the proportions of dissolved products that escape the rooting zone with percolating water.

NITROGEN FIXATION   In most studies of primary succession to date, nitrogen inputs from symbiotic nitrogen-fixers are high at some point early in succession. Stevens & Walker (146) state: "Nearly all chronosequence studies report that very early in the vegetation succession (though not necessarily at the pioneer stage) plants capable of fixing atmospheric nitrogen are present. Afterwards, during a 'transition period', the N-fixing plants are eliminated, leaving other plants to utilize the accumulated N."

Walker (160) further proposed that nitrogen fixation in an ecosystem will be greatly reduced when forms of phosphorus available to plants become limiting through conversion to unavailable forms and diminished weathering of primary phosphate minerals. This hypothesis suggests an interesting control of the cycle of one element by that of another. It is similar to Redfield's (125) hypothesis for the control of fixed nitrogen in marine waters by the solubility of phosphorus (see also 16, 138), and it deserves further study.

Considerably less definitive information is available on variations in nitrogen fixation during the course of secondary succession. *Ceanothus* and *Alnus* are important components of secondary seres in parts of western North America, and they may add significant amounts of fixed nitrogen to an ecosystem (111, 174). Becker & Crockett (6) suggest that fixation associated with prairie legumes declines with succession. In other regions, relatively few nitrogen-fixers are present. Nonsymbiotic fixers may be favored in such areas (cf 87), but few data are available. It may be useful to extend Walker's (160) hypothesis and speculate that in systems with a high fire frequency, nitrogen fixers will be more prevalent in secondary succession, since fire removes nitrogen by volatilization (cf 64) and releases phosphorus. In systems disturbed by other mechanisms (windthrow, etc), nitrogen is not volatilized and changes in nitrogen fixation will be less important.

Figure 2b represents the predicted pattern for nitrogen fixation in primary and secondary succession. The peak in primary succession is well documented, but the one in secondary succession is by no means certain. The secondary rise later in succession is hypothesized on the basis of a few data that demonstrate nitrogen fixation in epiphytic lichens [(44, 122); W. A. Reiners, unpublished] and decaying wood (25), both of which are more prevalent late in succession.

PARTICLE IMPACTION AND GAS ABSORPTION   The presence of a plant canopy increases inputs of elements owing to impaction of particles and aerosols of various kinds and absorption of gases by plant surfaces. The importance of impaction and absorption is primarily a function of the concentrations and kinds of particulates and gases in the local environment, and secondarily a function of the wind speed at a local site. After these factors, the nature of vegetation is probably the most important variable controlling this kind of input. But because of the role of vegetation, impaction and gas absorption rates could well change with successional development. The sizes of particles and their chemical behavior will also be important.

Unfortunately, impaction and subsequent resuspension phenomena are poorly understood (140); but according to Droppo (35) the following intrinsic vegetation characteristics are important variables in this process: canopy height, foliage density, canopy roughness, leaf area index, stomatal openings, and surface characteristics. Obviously these characteristics can change in complex ways in succession, but we know of no studies relating inputs via impaction to successional status.

Estimates of the magnitude of impaction vary widely, depending upon ecosystem location and the methodology used, but significant quantities of elements can be added to systems through this pathway. Gorham (55) reviewed early estimates of the importance of impaction in supplying elements to streams and lakes. Since then, indirect estimates of "dry deposition" (including impaction) have been made for a few terrestrial ecosystems of northwestern Europe, where it has been found that large amounts (sometimes more than half) of some atmospherically supplied ions are delivered by this process (49, 112, 162). In the White Mountains of New Hampshire, deposition of cations (calcium, magnesium, sodium, potassium, and lead) in buckets beneath artificial foliage was 5.5–8 times greater than in buckets placed in the open, largely as a result of a 4.5-fold increase in water collected (139).

Element inputs by impaction would probably be lower in moderately moist mid-continental locations but relatively high in drier but well vegetated regions where dust blown from adjacent cultivated or otherwise disturbed soils may be important (57, 59). Ecologists have generally tended to measure impaction directly in systems where they have had reason to suspect its importance, for instance in coastal zones (1, 2, 164). Impaction supplied more than 60% of total atmospheric inputs of sodium, magnesium, and potassium, and more than 25% of atmospheric calcium and phosphorus, to an English forest (164). It supplied more potassium, sodium, calcium, and magnesium to a coastal forest ecosystem than did either rain or dry fallout (1). For sodium and magnesium, supplied by sea spray, inputs by impaction substantially exceeded the sum of inputs by rain and dry fallout. Estimates of annual inputs based on direct measurements of impaction in nonmaritime environments are scarce (139).

Even less is known about the dynamics of gas capture by plant canopies. In general, capture increases nonlinearly with wind velocity, canopy height, light intensity, and gas concentration (74). The mechanisms for capture of gases are quite different from those of particulates. Nevertheless, changes in canopy height and geometry caused by succession would seem likely to affect particulate and gas capture similarly.

Figure 2c summarizes the way in which we expect the rates of impaction and gas absorption to change in the course of succession. It is drawn on the assumption that such changes reflect primarily changes in canopy height; other effects could be included given familiarity with a particular successional pattern. The curve suggested for secondary succession reflects a typically faster rate of canopy extension.

Atmospheric outputs of these same types can occur. Evidence is accumulating that aerosol particles are generated by plants under normal conditions (cf 42) and that those aerosols can contain elements of biogeochemical importance (5). The magnitude of this process is unknown, especially in the field, but it clearly will vary with succession, probably with the same general shape (but opposite effect) as that of aerosol capture and gas absorption (Figure 2c). Gaseous outputs are well appreciated (171). In general such outputs emanate from soils rather than canopies, however, so the dynamics of change follow processes other than canopy growth.

Hydrologic Processes Affecting Outputs

Because unvegetated systems transpire no water, runoff is higher than from vegetated systems. Increased runoff following devegetation has been demonstrated a number of times (67, 96, 97, 149). Vegetation decreases runoff in two ways: (a) through transpiration and (b) by increasing watershed storage capacity and impeding flow, thus decreasing storm flows. Whereas flow impedance may be proportional to accumulated biomass and detritus, transpiration usually reaches an upper limit more rapidly (120).

Evapotranspiration is fundamentally limited by supplies of water and evaporative energy. In broad outline, ecosystems can influence one or the other of these supplies by variations in oasis effects, canopy roughness, storage capacity for soil water, effective rooting volume, leaf area, seasonality of leaf exposure, and interception capacity (120). Once these variables reach certain limits, evapotranspiration usually becomes limited by the timing and amount of precipitation, and by energy. The biologically mediated effects usually reach their highest level long before an ecosystem reaches maximum biomass (120). In secondary succession, the limits for evaporative surface area and rooting volume can be reached in less than two decades (151). Second-order differences can occur through succession if there are shifts from deciduous to evergreen species, insofar as evergreen species prolong the effective transpiration season and can increase evaporative loss by higher levels of interception (149).

These variables must be taken into account for each particular case. In general, however, (120, 151) evapotranspiration will increase rapidly in succession, leveling off well before rates of change in physiognomy or composition have slowed significantly (120, 151). The shapes of the curves portraying evapotranspirational changes with time in either primary or secondary succession might be similar to, but steeper than, the curves for impaction rates in Figure 2c. As evapotranspiration increases, the precipitation available for runoff decreases. Low levels of transpiration in early succession can lead to high runoff and thus affect element outputs from ecosystems by increasing the transport of material in solution, by increasing erosion, and by favoring the waterlogging of certain sites—thereby increasing the possibility of manganese and iron mobilization under reducing conditions. Waterlogging can have significant effects upon redox potentials at later stages of succession as well, particularly where paludification or swamping of upland soils by peat accumulation has been extensive.

LOSS OF DISSOLVED SUBSTANCES   When element concentrations are plotted against water flux from terrestrial ecosystems, the resultant slope can be positive, negative, or zero. A slope of −1 indicates that a change in water flux alone will not change total element output. A more negative slope would indicate that an increase in water flux would cause a decrease in element output, but to our knowledge such a response has not been observed. When the slope is greater than −1, total outputs increase with increasing flow. Studies of changes in element concentration versus changes in stream flow within a watershed have reported slopes of greater than −1 for all elements, whether the comparison was within one flow event (61, 73), within one or a few years (84, 86), or among years (98).

Consequently, low evapotranspiration in early successional ecosystems can be responsible for high total outputs of elements because of high water flux through the system. A conservative prediction for changes in output of dissolved substances would be one that followed changes in runoff and deep seepage with no change in concentration (Figure 2d). The actual impact of increased water flux would depend on the slope of the relationship between water flux and element concentrations. Elements with positive slopes would increase more than is suggested in Figure 2d and elements with negative slopes would increase less. Many other processes could complicate this simple prediction by affecting the availability and/or susceptibility to loss of ions early in succession. For example, devegetation at the Hubbard Brook Experimental Forest resulted in increased water yields of 28 and 39% in the first two years, whereas concentrations of calcium increased 4-fold, and of nitrate 57-fold (97). In less humid regions, changes in water flow might exert greater effects on outputs of dissolved substances.

The importance of greater water flux in increasing element outputs could differ in primary and secondary succession. If plant cover increases more rapidly after disturbance, transpiration will also increase more rapidly. Therefore water flux to streamwater and groundwater will decrease more rapidly, as will outputs of dissolved substances through discharge alone (Figure 2d).

EROSION   Soil scientists and geomorphologists have formalized the regulation of fluvial erosion in the "universal soil loss equation," $A = f(R, K, L, S, C, P)$ where A is soil eroded, R is rainfall intensity, K is a soil erodability factor, L and S are slope length and gradient factors, C is cover, and P represents erosion control practices (134, 169). The factors C and K are of greatest interest here, because R is independent of successional status, L and S are generally so (except in rare cases such as the accretion and stabilization of sand dunes), and P is applicable only to managed systems.

Cover clearly varies in the course of succession. Numerous studies, both paleolimnological (cf 165) and current (cf 12, 34a, 89, 92, 137) demonstrate that bare ground loses far more soil than vegetated ground and that forest and grass best reduce erosion. Because erosion removes smaller particles preferentially, and because clay particles are richer in plant nutrients than are other soil fractions, significant amounts of nutrients can be removed during episodes of erosion.

The shape of a curve depicting fluvial erosional outputs of particulates from ecosystems during the course of succession may be generalized (Figure 2e). Erosion will be inversely related to cover and will thus be greatest early in a primary or secondary succession. It will remain high longer in a primary succession because cover usually takes longer to develop. A lag preceding the maximum erosion rate may or may not occur in primary succession, depending on the role of weathering in consolidated material, or the change from dispersed to concentrated runoff across unconsolidated materials. A lag in secondary succession is more likely because of residual resistance from humus, roots, and debris (12, 152). As cover develops erosion will decline, and once the site is completely occupied (which may be very early in some secondary successions) erosion is unlikely to vary systematically.

The magnitude of erosional outputs cannot be so easily generalized. Soil erodability (K) is substrate-dependent, and a system developing on bare rock (to choose one extreme example) will erode considerably less than one developing on unconsolidated silt. Moreover, K itself is a variable in the course of primary succession, with added organic matter decreasing the erodability of most soils (168). In primary successions developing on rock, however, K will increase during long-term system development as a consequence of soil development. Because more erodable material can be available following disturbance of mature systems there may be cases with very high initial rates in secondary succession in spite of residual erosional resistance.

Erosion and chemical weathering are interrelated processes because relatively unstable terrestrial soils are continually losing little-weathered materials. As fresh unweathered surfaces are exposed, chemical weathering can liberate ions at a rapid rate. In contrast, stable soils can undergo relatively long periods of weathering with low rates of ion liberation. Mackereth (101) has suggested that in the lake sediments of the English Lake District sodium and potassium derived largely from clastic minerals can be useful indexes to erosion and weathering. High ratios of these elements to clastic material indicate that the sediment comes from a relatively unstable, unweathered upland soil, whereas low ratios indicate more stable land surfaces on which the soil has undergone a longer period of weathering. Whether such an interpretation can be widely applied remains to be seen.

Wind erosion must be a very important output mechanism in semi-arid and arid landscapes—environments where hydrological outputs become less important and the catchment approach less useful. We are not familiar with quantitative studies on the pattern of change in this output through succession. On the basis of "dust-bowl" experiences and intuition, it seems likely that such outputs will change in a fashion parallel to that depicted in Figure 2e for fluvial erosion.

One minor but interesting exception to these general patterns is the fact that plant and litter biomass on steep slopes can precipitate mass movements of soil and thus in the long run possibly increase erosion (43). Peat flows (50, 52) provide an extreme example of this phenomenon. Such exceptions highlight the value of gaining a fundamental geomorphological understanding of a particular situation (cf 152) before attempting to make predictions of the relationships between succession and erosion.

REGULATION OF REDOX POTENTIAL   The accumulation of organic matter in wet or waterlogged soils can affect element mobility in ecosystems, as has been demonstrated in lake sediments in the English Lake District (54, 56, 58, 101). Metals such as iron and manganese are precipitated in aerobic soils, but may be rapidly reduced, mobilized, and lost from anaerobic soils along with trace elements scavenged by their oxidized precipitates. The paludification or swamping of upland soils through succession has been recognized for a long time (50, 52). In North America, Crocker & Major (28) and Heinselman (68) have described seres in which organic matter accumulation and *Sphagnum* invasion have led to waterlogging and eventual acidification of upland soils and presumably to increased metal mobility on both accounts.

Waterlogging can also lead to elevated losses of nitrogen and sulfur through denitrification and sulfate reduction. Denitrification will be particularly marked in systems where nitrification occurs in aerobic surface soils, supplying nitrate to underlying anaerobic subsoils (119). Additionally, gaseous outputs that are not dependent on soil oxygen content can be important. For example, ammonia volatilization is likely to be significant in alkaline soils (132).

These processes will not be readily apparent in every ecosystem or every successional stage, but they can be quantitatively very important at certain stages of development in some ecosystems

Biological Processes Affecting the Balance of Inputs and Outputs

The processes described to this point affect chemical inputs and outputs. Biological changes in element utilization or processing within an ecosystem may also cause changes in the balance between inputs and outputs. The processes discussed below involve changes in storage patterns within ecosystems or controls on output rates. As biological processes they tend to be of more intrinsic interest to ecologists. Probably they are also more complex and less susceptible to generalization.

NET ECOSYSTEM PRODUCTION   One important way in which intrasystem processes can alter element budgets is through net storage or release of elements in living organisms or detritus, which we shall combine in the term "biomass" (158). The basic exposition of this idea was given in the introduction of this paper (Figure 1). We wish to make three points to clarify and extend the original hypothesis.

First, this process extends to short- as well as long-term changes in biomass (and thus in element storage or release). An extreme example of a short-term change in net ecosystem production is seasonality in temperate and boreal ecosystems. There may be an excess of element uptake over element release during the growing season, yielding a seasonal cycle with a minimum in ecosystem outputs of important plant nutrients during the growing season (84, 157). Changes in biomass storage on an intermediate time-scale are exhibited by ecosystems subject to frequent destructive disturbances. For example, severe fires cause a pulse in element outputs during (64), and in many cases immediately after (104, 173), the fire. The pulse is then followed by variable periods, depending on the ecosystem, during which nutrients are accumulated as biomass increases. This development is likely to be terminated by another fire (69), causing another pulse in nutrient outputs and initiating another cycle of regrowth and element accumulation. If the disturbance (which may also be caused by insects, wind, or other processes) is repeated frequently, the system may be termed pulse-stable (115) even though net ecosystem production, and thus element storage, is never zero in a particular patch (171). Finally, there is the possibility of a long-term steady state, which may be more a theoretical construct than a reality (cf 13, 36, 172).

The second point is that this hypothesis must be considered in spatial as well as temporal scales. A landscape may be composed of a series of patches of differing ages since disturbance, each of which will be undergoing the mass dynamics described above. If the unit of study is large relative to the unit of natural disturbance, the unit of study as a whole may be in a "mosaic steady state" (143) with respect to net ecosystem production and element dynamics. If the size of the disturbed patch is large relative to the unit of study, a steady state of any kind will rarely be observed (143, 172).

The third point is that the importance of this process is proportional to the ratio of element storage in biomass to that in the inorganic fractions of the soil. This is true both for comparisons among elements within an ecosystem [e.g. of nitrogen, calcium, and sodium in (157)] and for the same elements in different ecosystems. Thus, we predict that net ecosystem production will have the greatest effect on elemental budgets in tropical rain forests and the least effect in deserts (Figure 2f). Even within biomes we predict differing responses to net ecosystem production. For example, changes in element storage in seres on shallow, coarse soils will reflect changing net ecosystem production better than in seres on deep, fine soils. In some cases the other processes described in this paper will override the effect of net ecosystem production, especially in ecosystems with low proportions of elements bound in biomass. Such processes may well be important in explaining those cases in which disturbed watershed ecosystems have not shown significant increases in element outputs following disturbance (4, 32, 131).

DECOMPOSITION AND ELEMENT MOBILIZATION   A large fraction of available nutrients is bound in the detritus of many terrestrial ecosystems. Consequently, any change in storage by the detritus pool can cause a significant change in the balance of inputs and outputs. Change in the detritus pool is controlled by litter inputs and by rates of decomposition and element release (117). This section focuses chiefly on decomposition.

Two factors can cause changes in rates of decomposition in successional sequences: changes in physical conditions and changes in litter quality. Early in succession physical conditions in the soil change rapidly. Litter temperatures may be 9–10°C higher in exposed soils than under plant canopies, which could approximately double the rate of decomposition and presumably of element release (147). Stone (147) argued that the high losses of nitrate from New England hardwood forests following cutting are in part a consequence of elevated nitrogen release from the litter in clearcut forest soils. Although this process cannot be the only reason for elevated losses (157), it could contribute significantly to the high losses observed (121). Changes in litter moisture due to microclimatic or hydrological alterations may also contribute to significantly decreased decomposition rates. Harcombe (66) suggested that the physical effects of revegetation on soil moisture and temperature were more important than the reestablishment of plant uptake in reducing nutrient losses from an early successional ecosystem in Costa Rica. Paludification can also lead to a pronounced decline in the rate of decomposition and to a major long-term sequestration of elements in peat deposits

Changes in litter quality can also affect element outputs from ecosystems by affecting the rate and the amount of organic matter decomposition and of element release (60). Litter with a high C:N or C:P ratio, or litter with a high content of lignin or other resistant chemicals (29, 105), will decompose more slowly than other litter, thus effectively retaining elements in detrital storage pools. These factors were apparently the underlying causes for a complex change in decay rate with succession in Douglas fir stands (155). Pioneers in secondary succession may be selected for rapid dispersal and growth in nutrient-rich environments, and their litter may consequently be more rapidly decomposable than that of species occurring later in succession (29). Rapid mobilization of nutrients from pioneer litter may not result in high element outputs from whole ecosystems, however, if net uptake by the increasing biomass of pioneer plants exceeds element mobilization from their litter.

A related process may explain why some forests do not show high rates of element output following severe disturbance. Sometimes a large amount of woody biomass is left behind after logging or fires in the form of stumps, large logs, roots, and slash. The ratio of carbon to mineral nutrients in woody material is usually high, and immobilization of nutrients by microflora in and around this rich carbon source could significantly decrease the rate of element escape to ground-water and streams.

Decomposition may exert another effect in ecosystems whose cation exchange sites are largely organic. If disturbance is such that inputs of organic matter decline or cease for a period of time (or if a severe fire destroys soil organic matter), the exchange sites themselves may decompose. Any cations held by them would then be released into the soil solution, from which losses are likely to be rapid (34). Organic exchange sites decompose in all systems, of course, but there may be little or no replacement following some kinds of disturbance. This process may have been important in controlling element outputs from a devegetated watershed at the Hubbard Brook Experimental Forest (34).

REGULATION OF SOIL SOLUTION CHEMISTRY Changes in the chemistry of soil solutions can change the mobility of elements, thereby changing storage levels of these nutrients in the soil. For the most part, the important changes involve the formation of acids with mobile anions. Extrinsic changes in acid concentration of bulk precipitation have been demonstrated for some areas (23, 53, 93), but of more interest here are acids produced by endogenous, biological processes—processes that might change in rate because of succession itself. Of particular concern are acids produced from uptake of carbon, sulfur, and nitrogen and their incorporation into organic matter. The organic matter is decomposed in turn and transformed to yield carbonic, sulfuric, and nitric acids as well as other decomposition products. In the soil solution these acids dissociate, and the hydrogen ions can substitute for metallic cations on exchange surfaces. Metallic cations (balanced by anions) can then be leached down the soil profile and even out of the system with the percolate. Thus, a faster rate of acid production during a particular state of succession can accelerate outputs of cations.

Such acids can also accelerate weathering rates, so that in this sense the acids contribute to inputs via rock weathering.

The introduction of living or dead organic matter to soil in primary succession increases the pCO_2 of the soil solution. As discussed under weathering, this directly raises the concentrations of carbonic acid and its dissociation products. Bicarbonate, the derivative anion of this process, is the dominant anion of many soil solutions (80, 81, 102, 103) over the pH range from about 5.5 to 10. The production of carbonic acid can vary in several ways throughout succession. Accumulated organic matter and biological respiration tend to increase early in primary succession. This increase is probably paralleled by carbonic acid production. With disturbance, accumulated organic matter could decompose more rapidly because of higher temperatures and greater moisture supply. Consequently, the carbonic acid flux could be increased. Once the accumulated labile organic matter becomes exhausted, the carbonic acid flux would presumably return to much lower levels.

Organic acids, also produced by plant roots and soil microorganisms, are implicated in the regulation of soil chemistry, especially in cool humid environments (19, 31, 63, 81). The mobility of these acids, their dissociation characteristics, buffering capacities, chelation behavior, and stability are poorly understood. The production of organic acids, including the complex humic acids, probably increases during succession.

The behavior of nitrogen is more complex. It enters ecosystems not only as a gas through the process of nitrogen fixation, but in precipitation. Thus, it can accumulate in organic matter through time from either exogenously or endogenously controlled inputs. Inorganic nitrogen can exist as a cation, ammonium, or as an anion, chiefly nitrate. As ammonium it is usually retained in soils, but as nitrate it is easily leached away. Consequently, a shift from forms of organic nitrogen or ammonium in the soil to nitrate (owing to nitrification) generates nitric acid, which can accelerate cation loss and rock weathering (94). Nitrate content of soil is a function of rates of nitrogen mobilization, nitrification, denitrification, plant uptake, immobilization, and leaching. If plant uptake is slowed or stopped by a disturbance, and especially if the decomposition of organic matter is accelerated by the disturbance, uptake may not match the rate of nitrification, in which case nitric acid can dominate or at least play a stronger role in the soil solution. In humid areas this can lead to very rapid losses of nitrate and accompanying cations (97). If plant uptake matches nitrogen input and mobilization rates, plants can compete successfully with nitrifying bacteria for ammonium, leading to a reduction in the rate of nitrification (37, 82). In such a situation root demand for ammonium and nitrate will reduce losses of nitrate in the soil percolate. Either of these cases may occur in typical seres, and an intermediate case of plant uptake being slightly less than nitrate production and supply rates may be more common over long periods of succession. Low rates of nitric acid formation and loss may then prevail (22).

Sulfur can function in a way similar to that of nitrogen. Organic sulfur is accumulated in the biomass of growing successional systems, amounting to about 10% of accumulated nitrogen. Following some kinds of disturbance, mobilization of organic sulfur may yield enough sulfate ion to cause a net flux of sulfate and associated, positively charged ions out of the ecosystem. In many cases hydrogen will be the associated cation, leading to increased weathering. In waterlogged habitats, dry weather can lead to the aeration of surface soils and the oxidation of sulfur compounds to sulfuric acid. The acid can then attack the lattices of soil minerals as well as exchange hydrogen ions for adsorbed metallic cations (55).

We wish to stress that the production of acids *with mobile anions* can control cation movement from terrestrial ecosystems. Plants that take up an excess of cations over anions achieve electrical neutrality by releasing hydrogen ions (40). Such hydrogen ion production can acidify a soil, but it will not lead to net cation transport from a system (or to the acidification of downstream aquatic systems). Instead, the supply of mobile anions limits cation transport (83, 103), so that the addition of hydrogen ions without a mobile anion could even decrease cation transport from a system by shifting the equilibrium of carbonic acid and some organic acids to the associated form. Likens et al (98) discussed the sources of hydrogen ions responsible for net cation outputs from the Hubbard Brook watersheds, including hydrogen-ion production by roots within the system. Although root-derived hydrogen ions may be important in driving weathering, the mobile sulfate added to the Hubbard Brook system by absorption of sulfur dioxide and impaction of sulfate aerosol accounts for most of the net transport of cations from this rather acid system.

PRODUCTION OF ALLELOCHEMICS   Rice (127) has argued strongly for the importance of allelochemic substances in controlling the early stages of succession. He suggested (128–130) that plants characteristic of "mature" ecosystems produce allelochemic substances that suppress nitrification, thus leading to the retention of nitrogen and other elements within those systems. Early successional systems produce smaller amounts of inhibitory substances, and have higher rates of nitrification and element losses. It is further suggested that these differences are adaptive on the ecosystem level (128).

Secondary plant substances certainly can reduce or prevent nitrification, and the production of such substances can vary in the course of succession. Although the process is thus potentially very important, it is premature to generalize about monotonic increases in inhibition during the course of succession. Alternative mechanisms, including phosphorus deficiency as well as competition for ammonium among roots, mycorrhizae, and nitrifying bacteria, can explain most of the evidence previously adduced for inhibition of nitrification (123, 124).

Allelopathy may influence succession, and thus element outputs, in other ways. By affecting rates and directions of vegetation change, biogeochemical processes may be altered indirectly (109, 127, 166). Allelopathy may also affect rates of nitrogen fixation in the rhizosphere and soil (87), thereby influencing the input term of the nitrogen budget.

VARIABILITY IN UTILIZATION OF ELEMENTS   Changes in the efficiency of uptake and/or utilization may affect element budgets of ecosystems in several ways. For instance, if plants of a particular successional stage utilize nutrients more efficiently through evergreenness (108) or internal cycling (153, 154), external nutrient requirements will be less and element losses will be greater than where similar net ecosystem production is accompanied by less efficient nutrient utilization. However, evergreenness and efficient internal recycling are likely to develop as a result of nutrient-poor conditions, so that high losses as a consequence of efficient nutrient utilization would not be expected. The same may be said of Stark's (144, 145) hypothesized successional development of highly effective recycling in tropical forests by means of increasing surface concentrations of mycorrhizal feeder roots.

If early successional plants accumulate higher concentrations of nutrients in net ecosystem production (12), this may operate against high outputs of plant nutrients early in succession. Conversely, if net ecosystem production is composed primarily of nutrient-poor wood biomass rather than nutrient-rich leaf biomass (as it must become at some point in forest succession), nutrient accumulation rates could be lessened, thereby favoring increased element outputs later in succession.

Another factor in nutrient utilization is the affinity of roots or mycorrhizae for nutrients. Certain plants characteristic of a particular stage of succession may have a higher affinity (lower Michaelis constant) for nutrients than those at some other stage. The plants with the higher affinity might reduce levels of nutrients in the soil to lower concentrations, and thus possibly decrease losses. However, diffusion rates through soil are more likely to limit plant nutrient uptake (113). A similar mechanism may come into play if early successional plants have a high affinity for nitrate while later successional plants preferentially absorb ammonium (65).

A successional stage with a high diversity of species or growth forms may obtain nutrients (perhaps at different times, or at different levels or sites within the soil) that another, less diverse system does not (e.g. 110). Such differences among seral stages are as yet undocumented, and their significance to element losses from ecosystems is unknown.

Finally, the tendency for certain kinds of organisms to accumulate specific elements in high concentration (14) must be taken into account. Some plants are also inherently poor in certain elements, the low nitrogen content of *Sphagnum* being a good example (51).

A SUMMARY OF ITEMS IN THE CHEMICAL BUDGET

Chemical budgets, if they are to be complete, must include a considerable number of terms even for a tightly sealed watershed without groundwater seepage. Atmospheric inputs will include precipitation (P), dry fallout (F), impacted particles (I) and absorbed gases (G). We have treated weathering (W) as an additional, lithospheric input. Output terms will include materials dissolved in outflow streams (S), gaseous emissions (E), particles released to the air from vegetation (R), and particulate material eroded into streams (D). Storage terms will include sequestration into secondary minerals (M) and biomass (B), as well as by adsorption (A) and chelation (C). Under certain circumstances these storage terms may change their sign, as for instance if disturbance leads to loss of biomass and to destruction of organic materials responsible for adsorption and chelation, or if at some stage of succession secondary minerals are themselves weathered faster than they are formed.

An overall description of a chemical budget might be written as follows:

$$P + F + I + G + W \pm \Delta M \pm \Delta B \pm \Delta A \pm \Delta C = S + E + R + D$$

$$|\!\longleftarrow \text{input} \longrightarrow\!|\!\longleftarrow \text{storage} \longrightarrow\!|\!\longleftarrow \text{output} \longrightarrow\!|$$

The large number of terms, of which only P, F, and S are commonly measured, makes the construction of an accurate short-term budget, even for a self-contained basin not subject to groundwater seepage, far from easy.

CONCLUSION

Chemical budgets can be altered by successional processes in terrestrial ecosystems in a number of ways. The degree of biotic control of these processes varies among environments and over time. Although inputs and outputs must be in balance both at time zero and in any eventual steady state, the absolute levels of input and output may vary considerably from the start to the end of ecosystem succession. Whether they will be greater or less in steady state than at time zero cannot be predicted from current generalizations about "nutrient conservation" (114) or "leakage" (170) but will depend upon the element under consideration, the nature of the parent material, and the interrelation among the processes listed on page 58.

Considerable empirical work, together with modeling of individual processes, is needed to increase our understanding of the successional aspects of ecosystem budgets, and we hope that this review will stimulate and organize research in a number of directions. A tremendous amount of work is clearly required before we can attain a reasonable understanding of these processes, and other processes we have not discussed may be important in local situations. However, no problem appears more insoluble than when it is considered as a whole, and we believe that this dissection of element budgets in ecosystems into a large number of component processes will make the eventual mechanistic understanding of land/water interactions more attainable. An analysis of these processes will at least aid in interpreting, and even predicting, the budgetary behavior of particular ecosystems. More optimistically, we hope that with more experience in focusing on these and other mechanisms, ecologists will forge general models for all terrestrial ecosystems.

We thank J. M. Melillo and G. P. Robertson for their helpful comments on this manuscript, and particularly W. C. Graustein for his penetrating contribution to our consideration of weathering.

Literature Cited

1. Art, H. W. 1976. Ecological Studies of the Sunken Forest, Fire Island National Seashore, New York. *Natl. Park Serv. Sci. Monogr. Ser. No. 7.* 237 pp.
2. Art, H. W., Bormann, F. H., Voigt, G. K., Woodwell, G. M. 1974. Barrier island forest ecosystem: role of meteorological nutrient inputs. *Science* 184:60–62
3. Ascaso, C., Galvan, J. 1976. Studies on the pedogenetic action of lichen acids. *Pedobiologia* 16:321–31
4. Aubertin, G. M., Patric, J. H. 1974. Water quality after clearcutting a small watershed in West Virginia. *J. Environ. Qual.* 3:243–49
5. Beauford, W., Barber, J., Barringer, A. R. 1977. Release of particles containing metals from vegetation into the atmosphere. *Science* 195:571–73
6. Becker, D. A., Crockett, J. S. 1976. Nitrogen fixation in some prairie legumes. *Am. Midl. Nat.* 96:113–43

7. Berner, R. A., Holdren, G. R. Jr. 1977. Mechanism of feldspar weathering: some observational evidence. *Geology* 5:369–72

8. Birkeland, P. W. 1974. *Pedology, Weathering, and Geomorphological Research.* NY: Oxford Univ. Press. 285 pp.

9. Blackwelder, E. B. 1927. Fire as an agent in rock weathering. *J. Geol.* 35:134–40

10. Bormann, F. H., Likens, G. E. 1967. Nutrient cycling. *Science* 155:424–29

11. Bormann, F. H., Likens, G. E., Melillo, J. M. 1977. Nitrogen budget for an aggrading northern hardwoods forest ecosystem. *Science* 197:981–83

12. Bormann, F. H., Likens, G. E., Siccama, T. G., Pierce, R. S., Eaton, J. S. 1974. The export of nutrients and recovery of stable conditions following deforestation at Hubbard Brook. *Ecol. Monogr.* 44:255–77

13. Botkin, D. B., Sobel, M. J. 1975. Stability in time-varying ecosystems. *Am. Nat.* 109:625–46

14. Bowen, H. J. M. 1966. *Trace Elements in Biochemistry.* London: Academic. 241 pp.

15. Boyle, J. R., Phillips, J. J., Ek, A. R. 1973. "Whole tree" harvesting: nutrient budget evaluation. *J. For.* 71:760–62

16. Broecker, W. S. 1974. *Chemical Oceanography.* NY: Harcourt Brace Jovanich. 214 pp.

17. Brown, G. W., Gahler, A. R., Marston, R. B. 1973. Nutrient losses after clearcut logging and slash burning in the Oregon Coast Range. *Water Resour. Res.* 9:1450–53

18. Bruckert, S., Jacquin, F. 1966. Relation entre l'évolution des acides hydrosolubles de deux litières forestières et les processus pedogenétiques. *Bull. Ecole Nat. Sup. Agron., Nancy* 8:95–111

19. Bruckert, S., Jacquin, F. 1969. Interaction entre la mobilité de plusieurs acides organiques et de divers cations dans un sol à mull et dans un sol à mor. *Soil Biol. Biochem.* 1:275–94

20. Burger, D. 1969. Calcium release from 11 minerals of fine-sand size by dilute sulfuric acid. *Can. J. Soil Sci.* 49:11–20

21. Clements, F. E. 1916. Plant succession. Washington DC: Carnegie Inst. Washington Publ. 242. 512 pp.

22. Coats, R. N., Leonard, R. L., Goldman, C. R. 1976. Nitrogen uptake and release in a forested watershed, Lake Tahoe Basin, California. *Ecology* 57:995–1004

23. Cogbill, C. V., Likens, G. E. 1974. Acid precipitation in the northeastern United States. *Water Resour. Res.* 19:1133–37

24. Connell, J. H., Slatyer, R. O. 1977. Mechanisms of succession in natural communities and their role in community stability and organization. *Am. Nat.* 111:1119–44

25. Cornaby, B. W., Waide, J. B. 1973. Nitrogen fixation in decaying chestnut logs. *Plant Soil* 39:445–48

26. Cowles, H. C. 1899. The ecological relations of the vegetation on the sand dunes of Lake Michigan. *Bot. Gaz.* 27:95–117, 167–202, 281–308, 361–91

27. Cowles, H. C. 1901. The physiographic ecology of Chicago and vicinity. *Bot. Gaz.* 31:73–108, 145–82

28. Crocker, R. L., Major, J. 1955. Soil development in relation to vegetation and surface age at Glacier Bay, Alaska. *J. Ecol.* 43:427–48

29. Cromack, K., Monk, C. D. 1975. Litter production, decomposition, and nutrient cycling in a mixed hardwood watershed and a white pine watershed. In *Mineral Cycling in Southeastern Ecosystems,* ed. F. G. Howell, J. B. Gentry, M. H. Smith, pp. 609–24. US/ERDA Publ. CONF-740513.

30. Cronan, C. S., Reiners, W. A., Reynolds, R. C. Jr., Lang, G. E. 1978. Forest floor leaching: contributions from mineral, organic, and carbonic acids in New Hampshire subalpine forests. *Science* 200:309–11

31. Dawson, H. J., Ugolini, F. C., Hrutfiord, B. F., Zachara, J. 1978. Role of soluble organics in the soil processes of a podzol, Central Cascades, Washington. *Soil Sci.* 126:290–96

32. Debyle, N. V., Packer, P. E. 1972. Plant nutrient and soil losses in overland flow from burned forest clearcuts. In *Watersheds in Transition, Am. Water Resour. Assoc. Proc. Ser. 14,* pp. 296–307. Urbana, Ill: AWRA

33. DeKimpe, C. R., Martel, Y. A. 1976. Effects of vegetation on the distribution of carbon, iron, and aluminum in the B horizons on northern Appalachian spodosols. *Soil. Sci. Soc. Am. Proc.* 40:70–80

34. Dominski, A. S. 1971. *Nitrogen transformations in a northern hardwood podzol on cutover and forested sites.* PhD thesis. Yale Univ., New Haven, Conn. 165 pp.

34a. Douglas, I. 1967. Man, vegetation, and the sediment yields of rivers. *Nature* 215:925–28

35. Droppo, J. G. 1976. Dry deposition processes on vegetation canopies. In *Atmospheric-Surface Exchange of Par-*

ticulate and Gaseous Pollutants, coord. R. J. Engelmann, G. A. Sehmel, pp. 104–11. US/ERDA Ser., CONF-740921 (1974). Springfield, Va: Natl. Tech. Info. Serv.

36. Drury, W. H., Nisbet, I. C. T. 1973. Succession. *J. Arnold Arboretum Harv. Univ.* 54:331–68

37. Edwards, N. T. 1978. The effects of stem girdling on biogeochemical cycles within a mixed deciduous forest in eastern Tennessee. I. Soil solution chemistry, soil respiration, litterfall, and root biomass studies. *Bull. Ecol. Soc. Am.* 59:63 (Abstr.)

38. Egler, F. E. 1947. Arid southeast Oahu vegetation, Hawaii. *Ecol. Monogr.* 17: 383–435

39. Egler, F. E. 1954. Vegetation science concepts. 1. Initial floristic composition —a factor in old field vegetation development. *Vegetatio* 4:412–17

40. Epstein, E. 1972. *Mineral Nutrition of Plants: Principles and Perspectives.* NY: Wiley. 412 pp.

41. Feth, J. H., Roberson, C. E., Polzer, W. L. 1964. Sources of mineral constituents in water from granitic rocks Sierra Nevada California and Nevada. *US Geol. Surv. Water-Supply Pap. 1535–I.* Washington DC: GPO

42. Fish, B. R. 1972. Electrical generation of natural aerosols from vegetation. *Science* 175:1239–40

43. Flaccus, E. 1959. Revegetation of landslides in the White Mountains of New Hampshire. *Ecology* 40:692–703

44. Forman, R. T. T. 1975. Canopy lichens with blue-green algae: a nitrogen source in a Columbian rain forest. *Ecology* 56:1176–84

45. Franzmeier, D. P., Whiteside, E. P., Mortland, M. M. 1963. A chronosequence of podzols in northern Michigan. III. Mineralogy, micromorphology, and net changes occurring during soil formation. *Q. Bull. Mich. State Univ. Agric. Exp. Stn.* 46:37–57

46. Fredriksen, R. L. 1972. Nutrient budget of a Douglas-fir forest on an experimental watershed in western Oregon. In *Proc. Res. Coniferous Forest Ecosystems —A Symp.,* pp. 115–31. Portland, Ore: USDA For. Serv., Pac. NW For. Range Exp. Stn.

47. Gibbs, R. J. 1967. Amazon River: Environmental factors that control the dissolved and suspended load. *Science* 156:1734–37

48. Gibbs, R. J. 1967. Geochemistry of the Amazon River system. *Geol. Soc. Am. Bull.* 78:1203–32

49. Gjessing, E. T., Henriksen, A., Johannessen, M., Wright, R. F. 1976. Effects of acid precipitation on freshwater chemistry. In *Impact of Acid Precipitation on Forest and Freshwater Ecosystems in Norway,* ed. F. H. Braekke, pp. 64–85. Oslo: SNSF

50. Gorham, E. 1953. Some early ideas concerning the nature, origin and development of peat lands. *J. Ecol.* 41:257–74

51. Gorham, E. 1953. Chemical studies on the soils and vegetation of some waterlogged habitats in the English Lake District. *J. Ecol.* 41:345–60

52. Gorham, E. 1957. The development of peat lands. *Q. Rev. Biol.* 32:145–66

53. Gorham, E. 1958. The influence and importance of daily weather conditions in the supply of chloride, sulphate and other ions to fresh waters from atmospheric precipitation. *Philos. Trans. R. Soc. London, Ser. B,* 241:147–78

54. Gorham, E. 1958. Observations on the formation and breakdown of the oxidized microzone at the mud surface in lakes. *Limnol. Oceanogr.* 3:291–98

55. Gorham, E. 1961. Factors influencing supply of major ions to island waters with special reference to the atmosphere. *Geol. Soc. Am. Bull.* 72:795–840

56. Gorham, E. 1964. Molybdenum, manganese, and iron in lake muds. *Int. Ver. Theor. Angew. Limnol. Verh.* 15:330–32

57. Gorham, E. 1976. Acid precipitation and its influence upon aquatic ecosystems—an overview. *Water, Air Soil Poll.* 6:457–81

58. Gorham, E., Swaine, D. J. 1965. The influence of oxidizing and reducing conditions upon the distribution of some elements in lake sediments. *Limnol. Oceanogr.* 10:268–79

59. Gorham, E., Tilton, D. L. 1978. The mineral content of *Sphagnum fuscum* as affected by human settlement. *Can. J. Bot.* 56:2755–59

60. Gosz, J. R., Likens, G. E., Bormann, F. H. 1973. Nutrient release from decomposing leaf and branch litter in the Hubbard Brook Forest, New Hampshire. *Ecol. Monogr.* 43:173–91

61. Gosz, J. R. 1975. Nutrient budgets for undisturbed ecosystems along an elevational gradient in New Mexico. See Ref. 29, pp. 780–99

62. Graustein, W. C. 1975. On chemical weathering and forests. *Geol. Soc. Am. Abstr. Programs* 7(7):1090–91

63. Graustein, W. C., Cromack, K. Jr., Sollins, P. 1977. Calcium oxalate: occurrence in soils and effect on nutrient

and geochemical cycles. *Science* 198: 1252–54

64. Grier, C. C. 1975. Wildfire effects on nutrient distribution and leaching in a coniferous ecosystem. *Can. J. For. Res.* 5:559–607

65. Haines, B. L. 1977. Nitrogen uptake. Apparent pattern during old field succession in southeastern U.S. *Oecologia* 26:295–303

66. Harcombe, P. A. 1977. Nutrient accumulation by vegetation during the first year of recovery of a tropical forest ecosystem. In *Recovery and Restoration of Damaged Ecosystems*, ed. J. Cairns, K. L. Dickson, E. E. Herricks, pp. 347–78. Charlottesville, Va: Univ. Virginia Press

67. Harr, R. D. 1976. Forest practices and streamflow in western Oregon. *USDA For. Serv. Gen. Tech. Rep. PNW-49.* 18 pp.

68. Heinselman, M. L. 1970. Landscape evolution, peatland types, and the environment in the Lake Agassiz Peatlands Natural Area, Minnesota. *Ecol. Monogr.* 40:235–61

69. Heinselman, M. L. 1973. Fire in the virgin forests of the Boundary Waters Canoe Area, Minnesota *Quat. Res.* 8:329–82

70. Hembree, C. H., Rainwater, F. H. 1961. Chemical degradation on opposite flanks of the Wind River Range Wyoming. *US Geol. Surv. Water-Supply Pap. 1535-E.* Washington DC: GPO 9 pp.

71. Henderson, G. S. 1975. Element retention and conservation. *BioScience* 25:770

72. Henderson, G. S., Harris, W. F. 1975. An ecosystem approach to the characterization of the nitrogen cycle in the deciduous forest watershed. In *Forest Soils and Land Management*, ed. B. Bernier, C. H. Winget, pp. 179–82. Quebec: Les Presses de L'Université Laval.

73. Henderson, G. S., Hunley, A., Selvidge, W. 1977. Nutrient discharge from Walker Branch Watershed. In *Watershed Research in Eastern North America*, ed. D. L. Correll, pp. 307–22. Washington DC: Smithsonian Inst.

74. Hill, A. C., Chamberlain, E. M. Jr. 1976. The removal of water-soluble gases from the atmosphere by vegetation. See Ref. 35, pp. 153–70

75. Horn, H. S. 1974. The ecology of secondary succession. *Ann. Rev. Ecol. Syst.* 5:25–37

76. Horn, H. S. 1976. Succession. In *Theoretical Ecology: Principles and Applications*, ed. R. M. May, pp. 187–204. Oxford: Blackwell

77. Jenny, H. 1941. *Factors of Soil Formation.* NY: McGraw-Hill. 281 pp.

78. Jenny, H. 1951. Contact phenomena between adsorbents and their significance in plant nutrition. In *Mineral Nutrition of Plants*, ed. E. Truog, pp. 108–32. Madison, Wis: Univ. Wisconsin Press

79. Jenny, H., Arkley, R. J., Schultz, A. M. 1969. The pygmy forest–podzol ecosystem and its dune associates of the Mendocino coast. *Madrono* 20:60–74

80. Johnson, D. W. 1975. *Processes of elemental transfer in some tropical, temperate, alpine, and northern forest soils: factors influencing the availability and mobility of major leaching agents.* PhD thesis. Univ. Washington, Seattle. 169 pp.

81. Johnson, D. W., Cole, D. W. 1977. Anion mobility in soils: relevance to nutrient transfer from terrestrial to aquatic ecosystems. *US/EPA Report EPA-6001 3–77–068.* Corvallis, Ore: USEPA. 28 pp.

82. Johnson, D. W., Edwards, N. T. 1978. The effects of stem girdling on biogeochemical cycles within a mixed deciduous forest in eastern Tennessee. II. Soil nitrogen mineralization and nitrification rates. *Bull. Ecol. Soc. Am.* 59:63

83. Johnson, D. W., Cole, D. W., Gessel, S. P., Singer, M. J., Minden, R. B. 1977. Carbonic acid leaching in a tropical, temperate, subalpine, and northern forest soil. *Arctic Alp. Res.* 9:329–43

84. Johnson, N. M., Likens, G. E., Bormann, F. H., Fisher, D. W., Pierce, R. S. 1969. A working model for variations in streamwater chemistry at the Hubbard Brook Experimental Forest, New Hampshire. *Water Resour. Res.* 5:1353–63

85. Johnson, N. M., Likens, G. E., Bormann, F. H., Pierce, R. S. 1968. Rate of chemical weathering of silicate minerals in New Hampshire. *Geochim. Cosmochim. Acta* 32:531–45

86. Johnson, P. L., Swank, W. T. 1973. Studies of cation budgets in the southern Appalachians in four experimental watersheds with contrasting vegetation. *Ecology* 54:70–80

87. Kapustka, L. A., Rice, E. L. 1976. Acetylene reduction (N_2-fixation) in soil and old-field succession in central Oklahoma. *Soil Biol. Biochem.* 8:497–503

88. Keller, W. D., Reesman, A. L. 1963. Glacial milks and their laboratory-simulated counterparts. *Geol. Soc. Am. Bull.* 74:61–79

89. Langbein, W. B., Schumm, S. A. 1958. Yield of sediment in relation to mean annual precipitation. *Am. Geophys. Union Trans.* 39:1076–84

90. Leak, W. B., Martin, C. W. 1975. Relationship of stand age to stream-water nitrate in New Hampshire. *USDA For. Serv. Res. Note NE-211.* Upper Darby, Pa: USDA For. Serv. 4 pp.

91. Likens, G. E., Bormann, F. H. 1972. Nutrient cycling in ecosystems. In *Ecosystem Structure and Function,* ed. J. A. Wiens, pp. 26–67.

92. Likens, G. E., Bormann, F. H. 1974. Linkages between terrestrial and aquatic ecosystems. *BioScience* 24: 447–56

93. Likens, G. E., Bormann, F. H. 1974. Acid rain: a serious regional environmental problem. *Science* 184:1176–79

94. Likens, G. E., Bormann, F. H., Johnson, N. M. 1969. Nitrification: importance to nutrient losses from a cutover forested ecosystem. *Science* 163:1205–6

95. Likens, G. E., Bormann, F. H., Johnson, N. M., Pierce, R. S. 1967. The calcium, magnesium, potassium, and sodium budgets for a small forested ecosystem. *Ecology* 48:772–85

96. Likens, G. E., Bormann, F. H., Pierce, R. S., Reiners, W. A. 1978. Recovery of a deforested ecosystem. *Science* 199: 492–96

97. Likens, G. E., Bormann, F. H., Johnson, N. M., Fisher, D. W., Pierce, R. S. 1970. Effects of forest cutting and herbicide treatment on nutrient budgets in the Hubbard Brook watershed-ecosystem. *Ecol. Monogr.* 40:23–47

98. Likens, G. E., Bormann, F. H., Pierce, R. S., Eaton, J. S., Johnson, N. M. 1977. *Biogeochemistry of a Forested Ecosystem.* NY: Springer. 146 pp.

99. Lindeman, R. L. 1941. The developmental history of Cedar Bog Lake, Minnesota. *Am. Midl. Nat.* 25:101–12

100. Loughnan, F. C. 1969. *Chemical Weathering of the Silicate Minerals.* NY: Elsevier. 154 pp.

101. Mackereth, F. J. H. 1966. Some chemical observations on post-glacial lake sediments. *Philos. Trans. R. Soc. London, Ser. B* 250:165–213

102. McColl, J. G. 1973. A model of ion transport during moisture flow from a Douglas-fir forest floor. *Ecology* 54: 181–87

103. McColl, J. G., Cole, D. W. 1968. A mechanism of cation transport in a forest soil. *Northwest. Sci.* 42:134–40

104. McColl, J. G., Grigal, D. F. 1975. Forest fire: effects on phosphorus movement to lakes. *Science* 188:1109–11

105. Meentemeyer, V. 1978. Macroclimate and lignin control of litter decomposition rates. *Ecology* 59:465–72

106. Miller, W. R., Drever, J. I. 1977. Chemical weathering and related controls on surface water chemistry in the Absaroka Mountains, Wyoming. *Geochim. Cosmochim. Acta* 41:1693–1702

107. Moghimi, A., Tate, M. E. 1978. Does 2-ketogluconate chelate calcium in the pH range 2.4 to 6.4? *Soil Biol. Biochem.* 10:289–92

108. Monk, C. D. 1966. An ecological significance of evergreenness. *Ecology* 47:504–5

109. Muller, C. H. 1966. The role of chemical inhibition (allelopathy) in vegetational composition. *Bull. Torrey Bot. Club* 93:332–52

110. Muller, R. N., Bormann, F. H. 1976. The role of *Erythronium americanum* Ker. in energy flow and nutrient dynamics of a northern hardwood forest ecosystem. *Science* 193:1126–28

111. Newton, M., El-Hassan, B., Zavitkovski, J. 1968. Role of red alder in western Oregon forest succession. In *Biology of Alder, Proc. NW Sci. Assoc. Meet. 1967,* ed. J. M. Trappe, pp. 73–84, Portland, Ore: USDA For Serv. Pac. NW For. Range Exp. Stn.

112. Nihlgård, B. 1970. Precipitation, its chemical composition and effect on soil water in a beech and a spruce forest in south Sweden. *Oikos* 21:208–17

113. Nye, P. H. 1977. The rate-limiting step in plant nutrient absorption from soil. *Soil Sci.* 123:292–97

113a. Nye, P. H., Tinker, P. B. 1977. *Solute Movement in the Soil-Root System.* Berkeley & Los Angeles: Univ. California Press. 342 pp.

114. Odum, E. P. 1969. The strategy of ecosystem development. *Science* 164: 262–70

115. Odum, H. T. 1968. Work circuits and system stress. In *Primary Production and Mineral Cycling in Natural Ecosystems,* ed. H. E. Young, pp. 81–138. Orono, Me: Univ. Maine Press.

116. Ollier, C. D. 1969. *Weathering.* NY: Elsevier. 304 pp.

117. Olson, J. S. 1963. Energy storage and the balance of producers and consumers in ecological systems. *Ecology* 44: 322–21

118. Patric, J. H., Smith, D. W. 1975. Forest management and nutrient cycling in eastern hardwoods. *USDA For. Serv.*

Res. Pap. NE-324. Upper Darby, Pa: USDA For Serv. 12 pp.

119. Patrick, W. H., Tusneem, M. E. 1972. Nitrogen loss from flooded soil. *Ecology* 53:735–37

120. Penman, H. L. 1963. *Vegetation and Hydrology.* Farnham Royal, Bucks: Commonwealth Agric. Bur. Tech. Commun. 53.

121. Pierce, R. S., Martin, C. W., Reeves, C. C., Likens, G. E., Bormann, F. H. 1972. Nutrient losses from clearcutting in New Hampshire. See Ref. 32, pp. 285–95

122. Pike, L. H., Tracy, D. M., Sherwood, M. A., Nielsen, D. 1972. Estimates of biomass and fixed nitrogen of epiphytes from old-growth Douglas fir. See Ref. 46, pp. 177–87

123. Purchase, B. S. 1974. Evaluation of the claim that grass root exudates inhibit nitrification. *Plant Soil* 41:527–39

124. Purchase, B. S. 1974. The influence of phosphate deficiency on nitrification. *Plant Soil* 41:541–47

125. Redfield, A. C. 1958. The biological control of chemical factors in the environment. *Am. Sci.* 46:205–21

126. Reynolds, R. C., Johnson, N. M. 1972. Chemical weathering in the temperate glacial environment of the northern Cascade Mountains. *Geochim. Cosmochim. Acta* 36:537–54

127. Rice, E. L. 1974. *Allelopathy.* NY: Academic. 353 pp.

128. Rice, E. L., Pancholy, S. K. 1972. Inhibition of nitrification by climax ecosystems. *Am. J. Bot.* 59:1033–40

129. Rice, E. L., Pancholy, S. K. 1973. Inhibition of nitrification by climax ecosystems. II. Additional evidence and possible role of tannins. *Am. J. Bot.* 60:691–702

130. Rice, E. L., Pancholy, S. K. 1974. Inhibition of nitrification by climax ecosystems. III. Inhibitors other than tannins. *Am. J. Bot.* 61:1095–1103

131. Richardson, C. J., Lund, J. A. 1975. Effects of clearcutting on nutrient losses in aspen forests on three soil types in Michigan. See Ref. 29, pp. 673–86

132. Rolston, D. E., Nielsen, D. R., Biggar, J. W. 1972. Desorption of ammonia from soil during displacement. *Soil Sci. Soc. Am. Proc.* 36:905–11

133. Rosenzweig, M. L. 1968. Net primary productivity of terrestrial communities: prediction from climatological data. *Am. Nat.* 102:67–74

134. Ruhe, R. V. 1975. *Geomorphology.* Boston: Houghton-Mifflin. 246 pp.

135. Salisbury, E. J. 1922. The soils of Blakeney Point: a study of soil reaction and succession in relation to the plant covering. *Ann. Bot.* 36:391–431

136. Salisbury, E. J. 1925. Note on the edaphic succession in some dune soils with special reference to the time factor. *J. Ecol.* 13:322–28

137. Saxton, K. E., Spomer, R. G., Kramer, L. A. 1971. Hydrology and erosion of a loessial watershed. *Hydraul. Div., Am. Soc. Civ. Engr., Proc.* 97:1835–51

138. Schindler, D. W. 1977. Evolution of phosphorus limitation in lakes. *Science* 195:260–62

139. Schlesinger, W. H., Reiners, W. A. 1974. Deposition of water and nutrients on artificial foliar collectors in fir Krummholz of New England mountains. *Ecology* 55:378–86

140. Slinn, W. G. 1976. Dry deposition and resuspension of aerosol particles: a new look at some old problems. See Ref. 35, pp. 1–40

141. Smith, W. H. 1970. Root exudates of seedling and mature sugar maple. *Phytopathology* 60:701–3

142. Smith, W. H. 1976. Character and significance of forest tree root exudates. *Ecology* 57:324–31

143. Sprugel, D. G. 1976. Dynamic structure of wave-regenerated *Abies balsamea* forests in the north-eastern United States. *J. Ecol.* 64:889–911

144. Stark, N. 1971. Nutrient cycling I. Nutrient distribution in some Amazonian soils. *Trop. Ecol.* 12:24–50

145. Stark, N. 1971. Nutrient cycling II. Nutrient distribution in Amazonian vegetation. *Trop. Ecol.* 12:177–201

146. Stevens, P. R., Walker, T. W. 1970. The chronosequence concept and soil formation. *Q. Rev. Biol.* 45:333–50

147. Stone, E. 1973. The impact of timber harvest on soil and water. In *Report of the President's Advisory Panel on Timber and the Environment,* pp. 427–67. Washington DC: GPO

148. Strakhov, N. M. 1967. *Principles of Lithogenesis,* Vol. 1. Edinburgh: Oliver and Boyd. 245 pp. (Engl. transl.)

149. Swank, W. T., Douglass, J. E. 1974. Streamflow greatly reduced by converting deciduous forest to pine. *Science* 185:857–59

150. Swank, W. T., Douglass, J. E. 1977. Nutrient budgets for undisturbed and manipulated hardwood forest ecosystems in the mountains of North Carolina. See Ref. 73, pp. 343–64

151. Swank, W. T., Schreuder, H. T. 1973. Temporal changes in biomass, surface

area, and net production for a *Pinus strobus* L. forest. In *IUFRO Biomass Studies*, ed. H. E. Young, pp. 173–82. Orono, Maine: Univ. Maine

152. Swanston, D. N., Swanson, F. J. 1976. Timber harvesting, mass erosion, and steepland forest geomorphology in the Pacific Northwest. In *Geomorphology and Engineering*, ed. D. R. Coats, pp. 199–221. Stroudsburg, Pa: Dowden, Hutchinson, & Ross

153. Turner, J. 1975. *Nutrient cycling in a Douglas-fir ecosystem with respect to age and nutrient status*. PhD thesis. Univ. Washington, Seattle 151 pp.

154. Turner, J. 1977. Effect of nitrogen availability on nitrogen cycling in a Douglas-fir stand. *For. Sci.* 23:307–16

155. Turner, J., Long, J. N. 1975. Accumulation of organic matter in a series of Douglas-fir stands. *Can. J. For. Res.* 5:681–90

156. Ugolini, F. C. 1966. Soils. In *Soil Development and Ecological Succession in a Deglaciated Area of Muir Inlet, Southeast Alaska*, ed. A. Mirsky, pp. 29–72. Inst. Polar Studies Rep. No. 20. Columbus, Ohio: Ohio State Res. Found.

157. Vitousek, P. M. 1977. The regulation of element concentrations in mountain streams in the northeastern United States. *Ecol. Monogr.* 47:65–87

158. Vitousek, P. M., Reiners, W. A. 1975. Ecosystem succession and nutrient retention: a hypothesis. *BioScience* 25:376–81

159. Vitousek, P. M., Reiners, W. A. 1976. Ecosystem development and the biological control of streamwater chemistry. In *Environmental Biogeochemistry*, ed. J. O. Nriagu, pp. 665–80. *Ann Arbor, Mich*: Ann Arbor Science

160. Walker, T. W. 1965. The significance of phosphorus in pedogenesis. In *Experimental Pedology*, ed. E. G. Hallsworth, D. V. Crawford, pp. 295–315. London: Butterworth's

161. Weetman, G. F., Webber, B. 1972. The influence of wood harvesting on the nutrient status of two spruce stands. *Can. J. For. Res.* 2:351–69

162. White, E., Starkey, R. S., Saunders, M. J. 1971. An assessment of the relative importance of several chemical sources to the waters of a small upland catchment. *J. Appl. Ecol.* 8:743–49

163. White, E. H. 1974. Whole-tree harvesting depletes soil nutrients. *Can. J. For. Res.* 4:530–35

164. White, E. J., Turner, F. 1970. A method of estimating income of nutrients in catch of airborne particles by a woodland canopy. *J. Appl. Ecol.* 7:441–61

165. Whitehead, D. R., Rochester, H., Rissing, S. W., Douglass, C. D., Sheehan, M. C. 1973. Late glacial and postglacial productivity changes in a New England pond. *Science* 181:744–47

166. Whittaker, R. H., Feeny, P. O. 1971. Allelochemics: chemical interactions between species. *Science* 171:757–70

167. Whittaker, R. H., Levin, S. A. 1977. The role of mosaic phenomena in natural communities. *Theor. Pop. Biol.* 12:117–39

168. Wischmeier, W. H., Mannering, J. V. 1965. Effect of organic matter content of the soil on infiltration. *J. Soil Water Conserv.* 20:150–52

169. Wischmeier, W. H., Smith, D. D. 1965. Predicting rainfall erosion losses from cropland east of the Rocky Mountains. *USDA Handbook No.* 282. Washington DC: GPO. 47 pp.

170. Woodwell, G. M. 1974. Success, succession, and Adam Smith. *BioScience* 24:81–87

171. Woodmansee, R. G. 1978. Additions and losses of nitrogen in grassland ecosystems. *Bioscience* 28:448–53

172. Wright, H. E. Jr. 1974. Landscape development, forest fires, and wilderness management. *Science* 186:487–95

173. Wright, R. F. 1976. The impact of forest fire on the nutrient influxes to small lakes in northeastern Minnesota. *Ecology* 57:649–63

174. Zavitkovski, J., Newton, M. 1968. Ecological importance of snowbrush *Ceanothus velutinus* in the Oregon Cascades. *Ecology* 49:1134–45

Ann. Rev. Ecol. Syst. 1979. 10:85–107

SYSTEMATICS AND ECOLOGY OF THE PALMAE

♦4156

P. B. Tomlinson

Harvard Forest, Harvard University, Petersham, MA 01366

INTRODUCTION

Although Linnaeus named only 10 species of true palms (31) his view of the family is still accepted. Its 2800 species are still regarded as a natural and isolated group (Palmae or Arecaceae) within its own order (Arecales). No problems exist in either ascribing a plant to the family or excluding it despite superficial similarity (to *Pandanus,* cyclanths, cycads, aroids, or Agavaceae). This review attempts to show that although they constitute one of the taxonomically best-known tropical families of flowering plants, our knowledge of the true palms is still imperfect and is not yet balanced by an equal appreciation of their ecology and physiology. In view of the obvious difficulty in collecting and studying them (1) and the small number of systematists who have investigated them it is perhaps surprising that the palms are well known taxonomically.

The palms are economically important because they include major plantation crops—e.g. oil-palm, coconut, and date-palm. Numerous species have minor economic importance as sources of oil, wax, starch, fiber, sugar, and alcohol but are of tremendous importance to local commerce as sources of food, thatch, fiber, wax, oil, timber, sugar, salt, alcoholic beverages, masticatories, and stimulants. Some are symbolic in several cultures. They have considerable aesthetic value, are used in magic and folk-medicine, and are an essential ecological associate of many primitive tribes. They have become increasingly important in commercial horticulture because of their elegant and predictable shapes.

The Botany of Palms

Most recent work on the family has been directly or indirectly influenced by H. E. Moore, Jr., of the L. H. Bailey Hortorium at Cornell University,

0066-4162/79/1120-0085$01.00

who has supervised the collaborative effort to apply evidence from a diversity of disciplines to the classification of the Palmae. Systematically studied topics have been elaborated and extended into related fields. Examples include (a) systematic vegetative anatomy (68) continued to studies of vascular construction (96-98), development, and physiology (95, 97), on the one hand, and to studies of structural and eventually ultrastructural development of the phloem, on the other hand (46, 48, 49); (b) comparative floral anatomy (e.g. 76–79, 82) extended to a study of floral biology (60, 80, 81); (c) monographic study extended to ecology (55); (d) comparative inflorescence morphology extended to a study of plant-animal interaction (72, 80, 81); and (e) study of geographic distribution applied to problems of conservation (39). All of this work must be elaborated further; future directions include systematic studies of the anatomy of palm fruits (18) and extension of the foundational work done on palm pollen (63, 65, 66), work that has only begun to use scanning-electron-microscopic (SEM) techniques extensively. A survey of tracheary elements (30) provides a background for future studies of xylem transport.

Much of this recent literature has been summarized by Moore and his associates (e.g. 37, 38, 44, 80, 81). Palms are well known morphologically even though they are difficult to study by standard methods. They are usually bulky organisms, not designed for routine herbarium procedures. One has to know the palms well in order to collect representative diagnostic samples. The few modern specialists have all been field workers, appreciative of the fact that palms can be studied best as natural populations or at least in cultivation. Moore has studied living representatives of all but 18 known genera of palms, mostly in their native environments. This has led to the appreciation of these "unruly monsters" as whole organisms. Systematically oriented field work has brought back a wealth of biological information unparalleled for any other tropical group of plants. Botanic gardens with large collections of cultivated palms have played an important role because close proximity of a diversity of living specimens facilitates the extensive comparative morphological analysis necessary in systematic work. My own interest in palms dates from an opportunity to study cultivated specimens in glasshouses at the Royal Botanic Gardens, Kew, followed by periods in the Singapore Botanic Garden, the University Botanic Garden, Legon, Ghana, and finally in Fairchild Tropical Garden, Miami, Florida, where one of the largest collections of cultivated palms has been assembled. The extensive cytological work of Read (e.g. 50–56) is based on this collection, with some counts from samples air-mailed from other sources (51, 56).

The botany of palms can certainly be said to be well understood. However, there are regrettable gaps in our knowledge, and especially between pure and applied science. Many disease problems of economically important palms remain insoluble because of a deficiency in our understanding of the genetics, physiology, nutrition, and development of palms. Knowledge of the reproductive biology of palms is still too deficient to allow predictions about the future of the many rare species of palms threatened by destruction of their habitat (39). Knowledge of the metabolic processes in palms that lead to the accumulation of starch or oil in commercially usable quantities is very small, even though such knowledge is crucial to selective crop improvement. There is, for example, no single study of starch accumulation in the stem of *Metroxylon,* an important food source (58). We have no extensive information about translocation processes in palms, even though a recent major epidemic of lethal yellowing in coconuts (see below) is the presumed result of a phloem-borne pathogen, and even though tapping a palm inflorescence is one of the most direct sources of sugar and alcohol in the tropics (e.g. 34, 75). We have insufficient knowledge of the physical properties of coconut stems to make them an economic resource (9).

Palm Literature

The taxonomic literature on the palms is very extensive and begins with the work of C. F. P. von Martius (87a), set out sumptuously in the pages of his *Historia Naturalis Palmarum* (1849–53). This was followed by the monographic series of the Italian botanist Beccari (e.g. 4–7). The recent and excellent *Natural History of Palms* by Corner (10) is much colored by his evolutionary interpretations. Regional studies may be found in (10, 43, 90, 92). The palms include many relatively small genera that lend themselves to monographic treatment. In addition to the steady stream from the pen of H. E. Moore (e.g. 40), we have treatments of *Copernicia* (11), *Johannesteijsmannia* (12), *Maxburretia* (13), *Pseudophoenix* (54), *Ptychosperma* (17), *Thrinax* (55), and studies of the geonomoid palms (91). Moore's foundational taxonomic overview has been concisely summarized (38). We await his definitive "Genera Palmarum," which is in preparation.

However, these statements should not generate complacency. Most regions in the tropics have no specialized palm literature; in most tropical countries it is impossible to get an up-to-date statement about the systematics and nomenclature of native palms. The older taxonomic literature needs bringing up-to-date. Most genera of palms have not been monographed; this includes such familiar ones as *Arenga, Caryota, Eugeissona, Euterpe, Metroxylon.* Letouzey (32) has recently called for a monographic study of the important palm *Raphia* and has pointed out some of the existing taxonomic problems. This monographic study, he emphasizes, would require a small team of botanists working throughout Africa and making field studies of morphological and ecological details. Moore (36) indicates that 20 years of experience went into his monograph of *Synechanthus,* a genus of 2 species!

RELATIONSHIPS OF THE PALMAE

The true palms have been related traditionally to certain other groups of woody monocotyledons, notably the Pandanaceae, Cyclanthaceae, and Araceae, largely on the basis of superficial similarity which is seen to be spurious on detailed examination. In some instances the putative similarity is based on quite elementary misconception of morphological features, as that which attempts a comparison of inflorescences in Araceae and Palmae simply because the terms "spadix" and "spathe" have been used *descriptively* for the two families.

These four groups of monocotyledons are easily distinguished by a combination of gross morphological features, substantiated by evidence from other disciplines, of which that from anatomy is still incomplete.

A Conspectus of the Palmae

Plants typically with woody self-supporting aerially unbranched trunks (rarely plagiotropic), rarely aerial branching dichotomous; if scandent, never root climbers; stem branching, if present, almost always basal and never branching distally below a terminal inflorescence. Leaves 2-, 3-, or usually many-ranked; leaf base (at least initially) a closed tubular sheath, blade well-developed, on a longer or shorter petiole (or pseudopetiole) with a single midrib or rachis; the blade plicately folded with a marginal nonplicate strip, usually split partly or completely into leaflets (pinnate leaves) or leaf segments (palmate leaves). Shoots usually pleonanthic, but hapaxanthy (sometimes leading to monocarpy) occasional. Flowers sometimes perfect, but usually diclinous by abortion and plants monoecious or dioecious. Lateral inflorescences (or first-order branches of terminal inflorescences) typically much-branched with a basal prophyll and one or more enlarged basal (but never petaloid) bracts, the distal bracts reduced. Axis rarely unbranched as a true spike (= spadix). Single inflated (sometimes woody) enveloping bract (= spathe) if present never on the trunk axis. Flowers either solitary or more usually aggregated in 2s, 3s, or more, the aggregations commonly representing condensed cincinni. Flowers usually trimerous (rarely dimerous or polymerous) with a well-developed floral envelope (the envelope rarely vestigial or absent). Flowers typically with 3, 6, or sometimes more numerous stamens, gynoecium apocarpous with 1–3 carpels or more commonly syncarpous with 3 (sometimes more) locules each with 1 functional ovule, 2 locules and ovules sometimes aborted. Fruit almost always indehiscent, baccate, or drupaceous 1–3– several-seeded, the pericarp woody, fleshy or fibrous, the endocarp sometimes thick and woody. Endosperm abundant (usually of hemicellulose), embryo small, germination hypogeal.

Stem vascular bundles always simple, collateral with 1–2–numerous wide metaxylem vessels. Stegmata (unequally thick-walled isodiametric cells), including a spherical or hat-shaped silica body common next to fibers in all parts except roots. Latex, mucilage, trichoscClereids, lysigenous air-lacunae or secretory cavities not developed.

Araceae, Cyclanthaceae, Pandanaceae

These families differ in fundamental ways, each showing a different set of diagnostic characters. The following are suggested major differences between them and palms:

Plants typically branched aerially, either as linear sympodia below terminal inflorescences (Chamberlain's model, as in many Araceae and Cyclanthaceae) or with branched sympodia (Leeuwenberg's model, Pandanaceae) or with sympodial branches on a monopodial trunk (Scarrone's and Stone's model, some Pandanaceae). Root climbers common (e.g. Araceae, Cyclanthaceae, *Freycinetia* in Pandanaceae). Leaves usually few-ranked, or spirally 3-ranked in Pandanaceae; 4-ranked in *Sararanga*. Leaf sheath apparently always an *open* tube, the blade either lanceolate, folded, and little differentiated (Pandanaceae) or if with a differentiated blade, the blade rolled in bud (Araceae), or if multi-plicate without conspicuous marginal strips and usually with 2 or more major ribs (Cyclanthaceae). Shoots most typically hapaxanthic, the terminal inflorescence unbranched and usually a true spike, with a dense aggregation of flowers (spadix) or least commonly a branched system terminating in such axes (Pandanaceae), with one or more, often petaloid, bracts on the main axis below the flower, or in the Araceae with a single often conspicuous or enveloping bract (spathe) in this position. Flowers without protective envelopes, or perianth reduced (some Araceae, Pandanaceae); flowers dimerous, trimerous, or tetramerous (as in Cyclanthaceae and some Araceae) usually with few stamens (sometimes synandrous in Araceae) or the stamens terminating branching structures (Pandanaceae) or numerous in Cyclanthaceae. Carpels various, uni- or multi-ovulate, gynoecium commonly syncarpous (e.g. Pandanaceae), the fruit either fleshy in various ways or apparently secondarily drupaceous in Pandanaceae, sometimes dehiscent. Seed reserve various, starchy, oily or hemicellulosic.

Stem with either simple vascular bundles (and then either collateral, with 1 or more wide metaxylem elements, or amphivasal) or frequently with compound vascular bundles resulting from a complex regular or irregular association of simple bundles. Crystals, latex, mucilaginous or secretory lysigenous or schizogenous cavities and trichosclereids of great diversity, but siliceous stegmata not observed.

The morphological nature of the reproductive organs in the Cyclanthaceae and especially Pandanaceae has not been fully resolved so that some of my statements may be controversial. However, the principles of construction in Palmae seem different. The flowers and inflorescence of the Araceae seem relatively easy to interpret in typological terms; they differ from those of palms in never being aggregated in complex ways, and for aroids the terms "spathe" and "spadix" have the most consistent descriptive applications.

These data must be elaborated more fully because considerable morphological variation exists within each family; but the basic conclusion, that the Palmae stand apart from the other 3 families, is not likely to be challenged.

The palms thus stand in no close relation to any one group of monocotyledons, and accepting them as a monotypic order best expresses this in conventional systematics. Certain Cyclanthaceae probably approach the palms closest in details of leaf morphology but still can be distinguished (93); the similarity may reflect parallelism in view of the numerous other differences between palms and cyclanths.

With the exception of Hutchinson (29), no one has emphasized a possible systematic relation between palms and woody Liliflorae. Although there are frequent similarities (or parallels) the Agavaceae are certainly unlike the palms in their simple unplicate leaves, secondary thickening, petaloid and often tubular flowers, frequent inferior ovary, and multi-ovulate carpels. Perhaps the closest similarity could be drawn between palms and certain members of the Australian assemblage, the Xanthorrhoeaceae, an unnatural family based primarily on geographical juxtaposition of its species. *Kingia* and *Dasypogon* are quite palm-like in their anatomy but have linear or lanceolate simple leaves. The anomalous rush *Prionium* (Juncaceae) has the appearance of a diminutive palm. The Strelitziaceae have an appreciable range of habit that parallels that in palms but are very different in floral morphology. I do not wish to emphasize arborescence as a basis for phyletic reasoning but do point out that there are possible candidates for palm relatives among other groups of monocotyledons as well as the traditionally maintained palm associates like pandans, aroids, and cyclanths. It seems that these alternatives are simply less familiar.

SUBDIVISION OF THE PALMAE

Since the time of von Martius, clearly circumscribed subgroups within the Palmae have been recognized with continuous and progressive refinement by successive specialists. Hierarchical rankings have varied, but the usual principle has been to recognize subfamilies and tribes. Currently Moore (38) recognizes 15 unranked "major groups" that are readily comparable to the taxa formally treated in earlier accounts. The essential features of these groups are set out in Table 1.

The assemblages vary in size from the monotypic nypoid group (*Nypa fruticans*), podococcoid group [*Podococcus barteri* incl. *P. acaulis* (32)], and oligotypic pseudophoenicoid to the arecoid alliance with over 700 species; the larger groups are capable of further subdivision.

The classification relies initially on gross morphological features—e.g. the orientation of the leaf plication that distinguishes induplicate ($\smile$) from reduplicate ($\frown$) groups. Small as this difference may appear, it seems to reflect the two alternative fundamental pathways in leaf development, a dichotomy even more basic than the degree of rachis extension, which produces either fan or feather leaves based on a single plan. Beyond this, characters of inflorescence, flower association, sex distribution, and flower and fruit structure are variously used.

The subdivision based on morphological evidence continues to be substantiated by evidence from other disciplines. Vegetative and floral anatomy has been particularly supportive (e.g. 38, 46, 68). Diagnostic features of leaf anatomy have been pursued even to the specific level (25, 40). Now that cumbersome and often misleading terminology of older authors has been replaced by a neutral, descriptive, and minimal set of terms (72), information of taxonomic value has been derived from a study of inflorescence morphology. Frequently the major groups have a distinctive and fairly constant kind of inflorescence morphology, an aspect revealed only by dissection of entire palms. The palm inflorescence provides material for comparative study that reveals probable phyletic trends of inflorescence elaboration, frequently paralleled in different groups (e.g. 22). Among the trends that can be recognized are: progressive reduction in number of branch orders, so that the spike or spike-like axis is a derived condition (not fundamental, as in aroids); progressive reduction in the number of large, proximal bracts, leading to inflorescences with few, 2, or even a single basal enveloping bract; and progressive sexual specialization of different parts of one inflorescence, and of different inflorescences of a single plant, and ultimately (in the dioecious state) of different plants. Aggregations of flowers on ultimate inflorescence axes are diverse and variable. Detailed studies of ultimate flower aggregations and of floral anatomy and development have provided information of taxonomic value and have led ultimately to a greater appreciation of features of pollination ecology (see below).

At the generic level there are many homogeneous and distinct genera (e. g. *Borassus, Copernicia, Metroxylon, Roystonea, Sabal*) or aggregations of genera (e.g. within the coryphoid palms), but generic limits are still uncertain in some groups (e.g. *Syagrus, Attalea,* certain iriarteoid taxa). Future field work and more intensive morphological analysis will continue to reveal undiscovered genera (41, 42) and so will undoubtedly refine the classification of palms; the outline is certainly substantial and clear but needs extensive infilling.

Table 1 The major groups of palms and their characteristics[a, b]

Group	Geographical distribution	Sex distribution	Number of taxa (Genera/species)	Leaf morphology	Distinguishing features
Induplicate — Leaved Palms					
Coryphoid (n = 18)	Pantropical	Hermaphrodite, (monoecious) (dioecious)	32/322	Palmate (Costapalmate)	Many-bracted branched inflorescences
Phoenicoid (n = 18)	Old World (Africa–Indo-China)	Dioecious	1/17	Pinnate	Basal leaflet spines
Borassoid (n = 14, 17, 18)	Old World (Africa–New Guinea)	Dioecious	6/56	Palmate	Thick inflorescence axes
Caryotoid (n = 16, 17)	Old World; eastern tropics	Monoecious	3/35	Pinnate (Bipinnate)	Hapaxanthic basipetal flowering; toothed leaflets
Reduplicate — Leaved Palms					
Nypoid (n = 17)	Ceylon to New Guinea and Ryukyu 1st	Monoecious	1/1	Pinnate	Rhizomatous, saline estuarine swamps
Lepidocaryoid (n = 14)	Pantropical; mainly eastern tropics	Monoecious	22/664	Pinnate (rarely palmate)	Scaly fruits
Pseudophoenicoid (n = 17)	New World (Caribbean)	Hermaphrodite (polygamous)	1/4	Pinnate	Single bract pseudopedicel
Ceroxyloid (n = ?)	Disjunct (S. America, Indian Ocean)	Dioecious	4/30	Pinnate	Diverse
Chamaedoroid (n = 13, 14, 16)	Mainly New World; plus Mascarenes	Monoecious Dioecious	6/146	Pinnate	Diverse
Iriarteoid (n = ?)	New World; Tropical America	Monoecious	8/52	Pinnate	Frequently stilt-rooted
Podococcoid (n = ?)	West Africa	Monoecious	1/2	Pinnate	Leaf morphology Elongated fruit
Arecoid (n = 16, 18)	Pantropical	Monoecious	88/760	Pinnate	Crown shaft frequent
Cocosoid (n = 15, 16)	Tropical America (South Africa)[c]	Monoecious	28/583	Pinnate	Bony endocarp with 3 pores
Geonomoid (n = 14)	Tropical America	Monoecious	6/92	Pinnate	Flowers in pits
Phytelephantoid (n = 16)	Tropical America	Dioecious	4/15	Pinnate	Numerous stamens per flower, fruits in heads

[a] After (38)

[b] The characters listed are generalized and there may be exceptions. Horizontal lines suggest major discontinuities.

[c] The natural range of *Cocos* into the Asian tropics is a disputed topic.

GEOGRAPHICAL DISTRIBUTION OF PALMS

Information about the geographic distribution of palms has been summarized by Moore (37). There are a few widely distributed or large genera of palms. The average number of species per genus is only 13 [2779 spp., 212 genera (38)]; 73 genera are monotypic. More than half the genera of palms have 5 or fewer species. Thus it is not surprising to find a high degree of endemism; in the South Pacific, each island or island group has its endemic species and often genera. Continental Africa is poor in palms, with about 50 species, a number much exceeded by adjacent Madagascar (115 spp.), which has 12 endemic genera. New Caledonia (30 species) represents an extraordinarily rich center of endemism; no less than 17 genera of palms are limited to the island (42).

The large size and wide distribution of a few palm genera, notably *Calamus* (370 spp.) and *Daemonorops* (115 spp.), suggests recent adaptive radiation. *Chamaedorea* (133 spp.), *Licuala* (108 spp.), and *Pinanga* (120 spp.) are also large genera, but they are exceeded in geographical range by a number of moderate-sized or even small genera like *Borassus* (7 spp.), *Hyphaene* (41 spp.), *Phoenix* (17 spp.), and *Raphia* (28 spp.). Dioecious species tend to be wide-ranging.

The otherwise generally restricted range of palms makes it relatively easy to define floristic regions and is further reflected even in the range of the major groups. There is a marked dissimilarity between the palm floras of the Old and New World. The iriarteoid, geonomoid, pseudophoenicoid, phytelephantoid, and cocosoid palms (with 3 exceptional genera in the last group) are entirely New World. The phoenicoid, caryotoid, borassoid, nypoid, and podococcoid palms are all Old World. The lepidocaryoid palms are also Old World except for *Lepidocaryum, Mauritia,* and one species of *Raphia* in the New World. Ceroxyloid and chamaedoroid palms both have a disjunct distribution in New and Old Worlds. This leaves only the coryphoid and arecoid with an essentially pan-tropical distribution. At the generic level only two palms have ranges that span the Atlantic (*Elaeis* and *Raphia*).

THE HABIT OF PALMS

Growth Limitations

Current knowledge of the growth-form of palms can be set against our recently increasing knowledge of the architectural diversity of tropical trees (27). This knowledge is often relevant to an understanding of the ecological role of palms.

Palms are growth limited (28) because they have no secondary vascular cambium and so lack any mechanism for secondary increase of vascular tissue; they do have a limited capacity for diffuse secondary growth (68) not dependent on meristematic activity but sometimes so localized at the base of the trunk as to be mechanically useful (88). The fixed primary conducting and mechanical ability of the trunk accounts for the fixed crown size. Palms normally remain unbranched above ground simply because they either lack completely any lateral vegetative meristems or such meristems are restricted to the base of the stem. *Serenoa* is exceptional in that there are vegetative lateral meristems interspersed among the inflorescence axes along the stem, but the axes are usually creeping (23). Other irregularities of branching in palms have been described and discussed by Fisher (19–22).

In the terminology of Hallé et al (27), palms are thus precisely model-conforming. This precise growth programming suggests that they might be restricted to climatically and microclimatically predictable environments. They seem to be restricted to tropical and subtropical environments primarily because vegetative growth is essentially continuous and they lack dormancy mechanisms. Few palms can withstand extended freezing temperatures. The hardiest palm is probably *Rhapidophyllum hystrix,* ranging from central Florida to Alabama and Georgia.

Establishment Growth

The development of a massive trunk and crown from an initially narrow embryonic axis is dependent on a preliminary phase of "establishment growth" (74) in which successive internodes are progressively wider as the primary thickening meristem becomes progressively more massive. The seedling axis in palms (indeed in all monocotyledons) is thus obconical. This is evident only in stilt-palms (e.g. many iriarteoid palms) in which the seedling internodes are elongated and above ground, but supported by a series of progressively wider stilt-roots. This phase of development is evidently critical in the successful establishment of the palm, and several developmental modifications have evolved to render it more efficient. This usually results in a burying of the plumular axis, with the axis developing to its maximum diameter underground. In several genera (e.g. *Diplothemium, Rhopalostylis, Sabal*) the seedling is saxophone-shaped. The plumule initially grows obliquely downward but is reorientated quite abruptly into an erect position. The distribution of this juvenile morphology in palms is not known. The diversity of establishment processes in woody monocotyledons generally is even greater than in palms (71).

Palm Architecture

Within the considerable limits imposed by these constraints, palms achieve an appreciable diversity of growth habit. Nevertheless, only 4 of the 23 models recognized in the Hallé-Oldeman system (27) can be identified (i.e. the models of Holttum, Corner, Tomlinson, and Schoute). Holttum's model is uncommon and refers to palms like *Corypha* and most *Metroxylon* spp., which are vegetatively unbranched and consist of one hapaxanthic axis (i.e. the vegetative axis ends in a terminal inflorescence). Corner's model is common and represents single-stemmed palms with lateral inflorescences. Both these models are distinctive among trees because the whole tree is programmed by a single vegetative shoot meristem, and reiteration does not occur. Tomlinson's model refers to multiple-stemmed palms—i.e. palms that circumvent growth restrictions by branching basally, each new trunk developing an adventitious root system. Tomlinson's model is the most versatile as well as commonest among palms. Species vary much in overall stature; axes may be partially plagiotropic or rhizomatous at the base, or scandent distally, and may or may not flower terminally. Schoute's model is rare since it includes trees with apical dichotomy of the vegetative axis, which may be erect as in species of *Hyphaene* or horizontal as in *Nypa* (70). Dichotomy here seems derived and not primitive; in *Hyphaene* the size of crowns is progressively reduced at each bifurcation so that the transport and mechanical limitations of the initial trunk axis are not exceeded; in *Nypa,* dichotomy permits proliferation of lateral-flowering plagiotropic axes that have presumably lost their ability to produce axillary vegetative meristems.

Scandent Palms

Lianescent palms have usually slender axes, with very long (up to 2 m) internodes, supported by grapnels that may be hooked extensions of the leaf rachis or inflorescence axis, the hooks either modified and backwardly directed terminal leaflets, as in *Desmoncus,* or clusters of spiny appendages like cat's claws. In most *Calamus* species, the inflorescence is reduced to a flowerless unbranched flagellum, partly adnate to distal organs (21). The scandent habit has evolved independently in the New and Old World and in different groups, chamaedoroid, cocosoid, and lepidocaryoid. These lianes are usually rhizomatous, although individual axes may be hapaxanthic. *Plectocomia* is exceptional since it is monocarpic (Holttum's model). Scandent palms represent a very successful elaboration of the palm habit; *Calamus* and *Daemonorops* are the two largest genera of palms and are associated with a number of satellite rattan genera. However, this group of palms is in need of taxonomic revision. Juvenile and adult phases are often markedly contrasted in morphology and the plants may be dioecious. Collections are often mixed, and specimens are so spiny that they are awkward to handle. This group is an important source of canes, the basis of a minor industry in Malaysia.

Ecologically rattans are weedy, often characteristic of wet and disturbed sites in the forest but readily persisting into communities with closed canopies. Do tropical vines "grow up" or are they "carried up" into the canopy?

A single rattan may well do both, developing as a juvenile plant in a gap and exploiting the enlarging canopy that closes the gap, subsequently persisting in the closed canopy by throwing up new axes from a permanent, branched rhizome system. Whether we view rattans as the curse of the forester, the pride of the cane merchant, or the puzzle of the ecologists, they certainly represent a biologically fascinating group of palms.

ECOLOGY OF PALMS

Palms occupy a diversity of habitats at different altitudes (especially in South America), are absent from truly xeric environments, and show a strong predilection for wet habitats (where they may be dominant) (37). Because some seem especially characteristic of disturbed sites, the term "weedy" may be applied—notably to *Euigeissona tristis* in Malaya, since its development inhibits forest regeneration (14). Dransfield (14) estimates that 75% of palms are rainforest species, but their great diversity of habit in the forest understorey suggests an appreciable diversity of ecological roles. The association among habit, physiology, and edaphic preference still remains virtually unexplored for palms. However, even though palms are a relatively minor component of the total forest biomass, they can be an important determining factor in forest composition because of their likely competitive interaction with canopy components when these are at sapling stages (84, 86) and can have a special influence on the soil profile (24). Bannister (2) concludes that *Euterpe globosa* (correctly *Prestoea montana*) is a normal component of "climax" forest vegetation in Puerto Rico, but the mosaic nature of succession in tropical forests suggests that palm species may have particular light-demanding attributes that make a linear "successional" concept seem too simplistic. Vandermeer et al (86) suggested that the populations of *Welfia, Socratea,* and *Iriartea* that they studied in Costa Rica have a cyclic interaction with the physical factors of their understorey forest environment. The physical environment (chiefly light patterns determined by gap size and regeneration phase) dictates the survival potential of these populations, which in turn determines the pattern of light within the environment. Somewhat similar conclusions are suggested on a more intuitive foundation in the study by de Granville (26) of the rôle of palms and other monocotyledons in the forest vegetation of French Guiana. In certain forest types he suggests that the forest understorey may become dominated by a single species—e.g. "astrocaryosed" by the dominant tendency of *Astrocaryum paramaca,* as also with *A. sciaphilum* in Suriname. Similarly, *Astrocaryum mexicanum* provides the characteristic feature of the forest understorey in Veracruz, Mexico (59). Palms thus seem a particularly representative group of organisms to study in relation to canopy-understorey interactions in tropical forests.

Palms in open sites tend to compensate for lack of species by large numbers of individuals. They frequently dominate the vegetation particularly of wet edaphically limited sites. The marked tendency of palms to make aerial root pneumatophores seems adaptive in this respect (26). Genera that characterize swamps include *Manicaria, Mauritia,* and *Raphia* in South America; *Phoenix* and *Raphia* in Africa; and *Metroxylon, Nypa,* and *Salacca* in Asia. The list of palm taxa in wetter tropical habitats is quite long (37). Palm habit in relation to habitat has been discussed most recently by Dransfield (14), and the contrasted biotope exploration of single-stemmed versus multiple-stemmed palms has been illustrated by de Granville (26). The difference in architecture between Holttum's, Corner's, and Tomlinson's model (27) in terms of overall size, root volume, detritus recycling, floral phenology, and breeding mechanism still remains unexplained. The difference between palms that do or do not sucker is usually quite clear, but the relative benefits of the two life-styles are not obvious.

Palm savannah is a common vegetation type in which the palm is the only tall tree. Examples in all three major tropical regions variously have species of such genera as *Borassus, Copernicia, Hyphaene,* and *Sabal* as the conspicuous element; but again the palms more frequently inhabit wet sites or those that are seasonally flooded. *Elaeis* is considered an aboriginal inhabitant of gallery or fringing forest in West Africa, but its range and habitat have been much broadened by human distribution (32). The most obvious feature of palm construction (the absence of secondary thickening) promotes fire-resistance and accounts for their frequent abundance in fire-climaxes (e.g. *Serenoa* associated with pines in Florida). However, the association between palms and fire has never been scrutinized very fully. Dransfield (14) cites the suggested example of *Hyphaene compressa* in East Africa; because of its branched crown (Schoute's model) it provides a nuclear site for forest regeneration in areas of savannah maintained by burning.

These brief comments about palm ecology show that the physiological basis for the ecological preferences of palms is not understood. The needed research might well study physiological tolerances of individual palm species first in artificially controlled environments and then in field circumstances.

Demography of Palms

Palms provide ideal subjects for demographic studies on woody plants because they are easily recognized and counted. Age determination is considered easy as compared with tropical trees in general because leaf scars are obvious and more or less permanent; rates of leaf production can be determined over a limited period and extrapolated to the total life span of the tree (59, 83). An estimate for the period of establishment growth is also necessary. This can be provided by an examination of a seedling population, which in time also provides the initial entry into life tables and survivorship curves. Care is needed in applying extrapolated values since growth rates of individual palms vary enormously. And in the enthusiasm for extended analyses one must not forget the assumptions made. The extended observations of Waterhouse & Quinn (88) on *Archontophoenix* are unique in the literature on palms and provide a demonstration of the way in which diffuse secondary growth of palms can be long continued, bringing about age-dependent changes inexplicable in terms of a simple analysis of a size table (e.g. that tall trees are wider at the base, even though they have no vascular cambium, and that taller trees have longer internodes). The close dependence of morphological, demographic, and phenological analysis is well demonstrated in this study.

Sarukhán (59) has provided a population flux model for *Astrocaryum* into which data can be progressively inserted as they are accumulated. His preliminary survivorship plots show the expected concave curve, and it is suggested that older trees in a population make the greatest reproductive contribution to a population. The life span is apparently determined by accident rather than senescence. Van Valen's reconstruction of a life table for *Prestoea montana* (83) based on Bannister's observations in Puerto Rico (2) adds further assumptions but seems comparable to the results obtained for *Astrocaryum* in a number of respects. It is also interesting that the reproductive potential of *Prestoea* (Corner's model) in terms of seeds produced per individual in relation to life expectancy (a tree entering the canopy has a future life expectancy of 70 years or 350,000 seeds) is quite comparable to that of *Corypha* (Holttum's model), in which an adult palm 44 years old was estimated to have produced 250,000 seeds in its single flowering (73).

The suitability of palms for biomass measurement is also demonstrated in Van Valen's analysis. In most palms either the individual parts are few and discrete (trunks and leaves) or they are produced in conveniently sized and easily harvested units (fairly large seeds on lateral inflorescences) (87). It is estimated for *Prestoea* that reproduction uses about 5% of net photosynthesis, compared with 15–22% in *Corypha*; these values, however, are based on a population and an individual, respectively. The values, crude as they are, still suggest that Holttum's model is the more efficient producer of seed meristems. The possibility for elaborating this kind of analysis is extensive, and palms may well play a major role in developing demographic theories about tropical trees.

Palms and Predators

Because palms have little regenerative ability (in the simplest situation one vegetative meristem functions throughout the life span) they are peculiarly vulnerable to predator attack; destruction of the one apical meristem destroys the whole tree. Uhl & Moore (80) have discussed this problem with reference to mechanisms that, directly or indirectly, protect pollen and ovules. The discussion may be extended to the ways in which the survival and reproductive potential of palms in the vegetative state may be affected by predators.

Wilson & Janzen (94) have provided information about seed mortality in *Scheelea rostrata,* where 80% of the seeds produced may be destroyed by a single species of bruchid beetle larva. This is an extreme example of the predispersal seed predation to which palms in general seem highly susceptible because of their phenology, seed size, and method of dispersal (85).

The apical meristem is protected primarily by its enclosure within the terminal crown of leaves; the youngest parts of the developing leaves are protected in the same way because they grow from basal meristems. Outer protection is provided by an enveloping series of leaf sheaths, which may become massive and woody (e.g. many coryphoid, cocosoid, and borassoid genera). The crownshaft of many arecoid palms results because most expansion is accommodated by growth; the tubular leaf base dehisces along precisely determined separation regions. In many palms the mechanical protection of the crown is frequently supplemented by prickles or similar sharp appendages.

The inflorescence is initially protected within the crown, especially by its own subtending leaf if it is lateral. It may not be exposed until this falls, as in most palms with a crownshaft. The expanding inflorescence axes, whether they protrude through the mouth of the subtending leaf, pierce its dorsal surface via a dorsal suture, or wait until the leaf falls, are themselves protected by one or more sometimes woody bracts. In the simpler situation enveloping bracts are many, the inflorescence has several orders of branching, each with associated bracts, and ultimate flower-bearing axes are unspecialized. In presumably derived inflorescences the number of bracts become few, or even one; basal bracts become large and assume the major protective function; ultimate bracts become vestigial; and the flowers are often aggregated and variously protected, as by overlapping distal bracts or by the margins of the pits into which the flowers are sunken.

Direct protection of the sex organs is provided by the floral envelope, which is rarely petaloid. Thick, hard imbricate or tightly valvate perianth segments are very important in palms, which, because of functional dioecism, retain exposed flowers of at least one sex for a considerable time. Uhl & Moore have demonstrated a diversity of structural and biochemical features likely to discourage chewing insects. There are trichomes, fibers, tannins, sclereids, raphides, and silica bodies. Carpels are protected because they mature basipetally, with many of the unpalatable structures in the apical parts (80). The same features also apply in the developing fruit, which at maturity may be attractive to animals via a colored pericarp. A stony endocarp or bony endosperm frequently protects the small embryo.

Poisonous secondary metabolites are not usually developed in palms, wherein lies one reason for their great value to man.

POLLINATION ECOLOGY AND BREEDING MECHANISMS

The early and entirely theoretical view that palms are exclusively wind-pollinated because they have "reduced" flowers has been supplanted by field observation, which shows that although wind-pollination does occur in a diversity of palms, methods of pollen transfer involving animals are not only frequent, but diverse (60). However, few examples have yet been studied in detail, and generalizations about whether palms are primitively anemophilous or zoophilous seem premature. Uhl & Moore (81) have presented a series of case histories based on a summation of field observation and an intimate knowledge of flower structure. This sets a standard for the much needed field and laboratory examination of flower function in palms. Flower distribution in palms is diverse; dicliny is the most common arrangement, with frequent marked size differences between male and female flowers.

Dioecy

Dioecy with obligate outbreeding is found throughout the borassoid, phoenicoid, and phytelephantoid palms, in some chamaedoroid and lepidocaryoid palms, but rarely elsewhere. It may be associated with wind-pollination, as in the phoenicoid palms (in *Phoenix dactylifera,* the date palm, it has been known since pre-history). The syndrome of characters related to wind-pollination includes abundant powdery pollen and synchronous and short-term flowering of numerous and relatively well-exposed flowers. The correlation between wind-pollination and dioecy is not fixed, since *Thrinax,* with perfect flowers, is wind-pollinated (55). *Cocos* is a good example of a predominantly wind-pollinated monoecious palm; some borassoid palms, on the other hand, have septal nectaries (79).

Monoecy

Monoecism is much the commonest condition in palms, being almost universal in the arecoid line, but the distribution of male and female flowers is variable. There may be separate male and female inflorescences in a single tree, as in *Elaeis* and some other cocosoid palms; or male flowers distally and female flowers proximally on a single inflorescence, as in *Cocos;* but most commonly the two kinds of flowers are closely aggregated, as in diads, triads, cincinni, or in linear series (acervuli). Such aggregations may still function in outbreeding because the different sexes mature at different times, often without overlap. In most coconut varieties outbreeding is virtually assured since there is usually only one inflorescence with functioning male or female flowers at one time. In caryotoid palms the separate sexual phases of one inflorescence are widely separated in time, but there may be overlap between different phases of one trunk. The existence of this chronological sexuality has to be appreciated by the collector. In multiple-stemmed palms, the advantages of sexual phases may be lost since the behavior of different trunks is nonsynchronous.

Perfect Flowers

Palms with perfect flowers are in a minority, and this condition may be associated with polygamy (e.g. *Pseudophoenix,* some coryphoid palms). Selfing is possible and indeed may be an essential feature of reproductive strategy. Thus *Corypha* has perfect flowers, but is evidently self-compatible, as indicated by the abundant fruit set of isolated cultivated individuals (73). Abundant seeding would seem to be important in this monocarpic species. Outcrossing in natural populations could then only occur between synchronously flowering individuals, but even this is minimized by the short flowering period (3 weeks). We lack detailed information about other monocarpic palms like *Raphia* and *Metroxylon.*

Breeding Mechanisms

Protogyny as an outbreeding device is known. *Nypa* is visited by a variety of insects but is said to be pollinated by drosophiloid flies that use the fleshy male axes as breeding sites (16); *Bactris* is apparently pollinated by nitidulid and curculionid beetles (15); *Hydriastele* is visited by bees, small flies, and weevils, the latter the most likely pollen visitors (16). In *Nypa* the protogyny is determined by the inflorescence structure, since the terminal aggregate of female flowers on the main axis becomes receptive before the male flowers are exposed on lateral branches. Pollen is sticky (probably related to its distinctive spinous morphology) and could not be transported by wind. *Bactris* and *Hydriastele* are protogynous because the female flowers of the triads are all receptive first. Individual flowers in *Sabal palmetto* are protogynous because stigma receptivity precedes anther dehiscence by at least 2.5 hr (8).

Protandry is recorded for *Asterogyne,* with pollination by syrphid flies (60), and is shown in *Ptychosperma,* visited by syrphid flies and *Nomia* bees. Among the insect attractants recorded for palm flowers are abundant pollen, nectar, odor (either sweet or foetid), and conspicuous aggregations of flowers, commonly against a dark background. Heat emission, which occurs in *Bactris* (61), may be an attractant. Bees are common visitors to palm flowers, as can be seen in cultivated specimens. The need to discourage the unwanted visitor seems real enough . However, the biology of the palm inflorescence may be very complex. In *Bactris gasipaes* (peach palm) a succession of visitors occurs—beetles, drosophilid flies, bees, and moths—so that it is not clear which insect is the effective pollen vector (J. Beach, personal communication). Information of this kind is needed for other palms of potential economic importance.

ECONOMIC ASPECTS

Justification for continued and even intensified study of palms resides in their economic importance. Much still needs to be done with major crops, as is shown later, but the value of minor palm products in local economy is often underestimated. Where palms of local value enter into world trade, their products become subject to world market fluctuations, they may not compete well against plantation products with highly efficient marketing procedures, and they may not provide the stable income a grower would need if he were to exploit them effectively.

Sago

The trunk starch of *Metroxylon saga,* one source of commercial sago, illustrates the problem well (58). Palm sago from this source (mainly Sarawak) has never competed well with other sources of starch. Nevertheless this palm is locally important both as a staple for local consumption and as a trade material, particularly since it grows in large natural populations in swampy habitats otherwise unsuited to cultivated crops. However, there has been no modern revision of the genus, its reproductive biology has been little investigated, and neither its ecology nor its physiology has been explored in useful detail. All these factors would be involved in selection for yield improvement, successful plantation cultivation, or management of wild stands for sustained high yield.

Pejibaye

The edible fruit of *Bactris gasipaes* (*Guilielma gasipaes*), a food crop of growing importance, has fared somewhat better at the hands of agronomists; there is an extensive literature on this palm (G. Hartshorn, personal communication). The biggest need is a careful assessment of high-yielding varieties and likely pest problems. For this, knowledge of reproductive biology and the physiology of fruit development is needed. Part of this study should certainly involve a detailed study of the other *Bactris* species, which are frequent in tropical America and evidently diverse.

Lethal Yellowing

Of the numerous diseases of tropical crops none has more spetacular effects than lethal yellowing decline of coconuts, which will destroy a healthy palm within six months of the appearance of the first visible symptoms. This disease reached epidemic proportions in the Caribbean in the 1960s. In Jamaica, where coconuts are a major factor in the economy, losses due to lethal yellowing of up to 100,000 trees a year have been regular. An outbreak in the Miami area (where the coconut is only of ornamental value) brought it to the attention of a large urban population. The establishment of a research team by the University of Florida accelerated the rate of existing research. An International Council on Lethal Yellowing now meets on a biennial basis to report progress of the effort to combat the spread of the disease and find a cure (57).

Present knowledge implicates a phloem-inhabiting mycoplasma-like organism (MLO) as the pathogen (3, 47), but the method of spread of the disease remains unknown at the time of writing. No artificial transfer of the disease has been obtained. Remission of disease symptoms has been obtained by injecting individual palms with massive doses of antibiotics; but this is not a solution to the problem on a plantation scale and is a procedure probably detrimental to the whole research effort, since it may select resistant strains of the pathogen. The future of the coconut industry is currently dependent on the existence of varieties of coconut resistant to the disease. One aspect of research effort is a breeding program that should generate high-yielding resistant coconut varieties.

Lethal yellowing impinges on our knowledge of the taxonomy of palms. The disease is known in other species; at least other palms in infected areas die with symptoms like lethal yellowing and recognizable MLO in the phloem (67). This knowledge has been obtained in South Florida where the many species of palms in cultivation have been exposed to the disease. So far putative susceptibility has been demonstrated in 23 species representing a diversity of groups. Most of the palms are only of immediate horticultural value, but the commercial date is on the list and there is a reason to suspect cultivated *Nypa fruticans* may have succumbed to the disease.

Susceptibility to the disease thus has a peculiar taxonomic distribution. The pathogen distinguishes between different cultivars of one species, between closely related genera and species, but not between groups. The genetic basis for susceptibility or resistance is thus not clear and may depend on the behavior of an as yet unknown vector.

Current quarantine precautions are intended to contain the spread of the disease, but outbreaks in other major coconut growing areas, such as India, the Philippines, and the South Pacific would constitute an international disaster, such is the dependence of large populations on this palm. The possibility that the disease could affect other commercial palms is alarming.

CONCLUSIONS

I have been able to present only a sampling of a rich, recent literature. There is a firm systematic foundation, surprisingly complete in view of the small number of scientists who have specialized in the group; this needs to be strengthened by continued extensive field work. In contrast, understanding of ecology and reproductive biology—to which may be added most aspects of development and physiology—remains deficient. This deficiency is serious in view of the dependence of applied research on basic knowledge.

The palms remain one of the most economically important groups of tropical plants, a major source of food and raw material that remains under-explored; they certainly increase the chances of survival for people in tropical developing countries.

Most of the fundamental research on palms has been provided by private institutions, botanic gardens, herbaria, and universities; in the United States there has been indirect governmental support, largely through the National Science Foundation. Nevertheless the gap between need and effort in the study of this important group of plants is still very evident. It should be closed, on an international cooperative basis, since the accumulation of knowledge about these plants can no longer be left to the dedication of a few, often isolated, individuals. Continuity of support and guaranteed access to the organisms over lengthy periods are needed.

ACKNOWLEDGMENTS

I am indebted to Drs. H. E. Moore, Jr., Natalie Uhl, and M. H. Zimmermann for critical commentary on the manuscript. However, errors of fact, interpretation, or emphasis are entirely my own. Comparative work on stem anatomy of arborescent monocotyledons, referred to in the text, was supported by grant GB 31844 - X from the National Science Foundation, Washington DC.

Literature Cited

1. Bailey, L. H. 1933. Palms, and their characteristics. *Gentes Herbarum* 3: 3–29
2. Bannister, B. A. 1970. Ecological life cycle of *Euterpe globosa* Gaertn. In *A Tropical Rain Forest: a Study of Irradiation and Ecology at El Verde, Puerto Rico,* ed. H. T. Odum, R. F. Pigeon, pp. B. 299–314. Oak Ridge, Tenn: US Atomic Energy Commission
3. Beakbane, A. B., Slater, C. H. W., Posnette, A. F. 1972. Mycoplasmas in the phloem of coconut, *Cocos nucifera* L., with lethal yellowing disease. *J. Hort. Sci.* 47:265
4. Beccari, O. 1908. Asiatic palms, Lepidocaryeae. I. The species of *Calamus. Ann. R. Bot. Gard. Calcutta* 11.
5. Beccari, O. 1918. Asiatic palms, Lepidocaryeae. *Ann. R. Bot. Gard. Calcutta* 12.
6. Beccari, O. 1924. *Palme della tribù Borasseae.* Firenze.
7. Beccari, O. 1933. Asiatic palms—Corypheae. *Ann. R. Bot. Gard. Calcutta* 13:1–356
8. Brown, K. E. 1976. Ecological studies of the cabbage palm, *Sabal palmetto. Principes* 20:3–10
9. Coconut Stem Utilization Seminar, 1977, Nuku'alota, Tonga. Wellington, New Zealand: Ministry of Foreign Affairs
10. Corner, E. J. H. 1966. *The Natural History of Palms.* London: Weidenfeld & Nicholson. 393 pp.
11. Dahlgren, B. E., Glassman, S. F. 1961–63. A revision of the genus *Copernicia.* 1. South American species. 2. West Indian species. *Gentes Herbarum* 9:1–232
12. Dransfield, J. 1972. The genus *Johannesteijsmannia* H. E. Moore, Jr. *Gard. Bull. Singapore* 26:63–83
13. Dransfield, J. 1978. The genus *Maxburretia* (Palmae). *Gentes Herbarum* 11: 197–99
14. Dransfield, J. 1978. Growth form of rain forest palms. In *Tropical Trees as Living Systems,* ed P. B. Tomlinson, M. H. Zimmermann, pp. 247–68. NY: Cambridge Univ. Press
15. Essig, F. B. 1971. Observations on pollination in *Bactris. Principes* 15:20–24,35
16. Essig, F. B. 1973. Pollination in some New Guinea palms. *Principes* 17:75–83
17. Essig, F. B. 1978. A revision of the genus *Ptychosperma* Labill. (Arecaceae). *Allertonia* 1:415–78
18. Essig, F. B. 1977. A systematic histological study of palm fruits. I. The *Ptychosperma* alliance. *Syst. Bot.* 2:151–68
19. Fisher, J. B. 1973. Unusual branch development in the palm, *Chrysalidocarpus. Bot. J. Linn. Soc.* 72:83–95
20. Fisher, J. B. 1974. Axillary and dichotomous branching in the palm *Chamaedorea. Am. J. Bot.* 61:1045–56
21. Fisher, J. B., Dransfield, J. 1977. Comparative morphology and development of inflorescence adnation in rattan palms. *Bot. J. Linn. Soc.* 75:119–40
22. Fisher, J. B., Moore, H. E. Jr. 1977. Multiple inflorescences in palms (Arecaceae): their development and significance. *Bot. Jahrb. Syst.* 98:573–611
23. Fisher, J. B., Tomlinson, P. B. 1973. Branch and inflorescence production in saw palmetto (*Serenoa repens*). *Principes* 17:10–19
24. Furley, P. A. 1975. The significance of the Cohune palm *Orbignya cohune* (Mart.) Dahlgren on the nature and in the development of the soil profile. *Biotropica* 7:32–36
25. Glassman, S. F. 1972. Systematic studies in the leaf anatomy of palm genus *Syagrus. Am. J. Bot.* 59:775–88
26. de Granville, J.-J. 1978. *Recherches sur la flore et la vegetation Guyanaises.* Thesis (Docteur des Sciences Naturelles), Univ. Sci. Tech. Languedoc, Montpellier, France.
27. Hallé, F., Oldeman, R. A. A., Tomlinson, P. B. 1978. *Tropical Trees and Forests: an Architectural Analysis.* Berlin, Heidelberg, New York: Springer. 441 pp.
28. Holttum, R. E. 1955. Growth habits of monocotyledons. Variations on a theme. *Phytomorphology* 5:399–413
29. Hutchinson, J. 1934. *The Families of Flowering Plants. Vol. II. Monocotyledons.* Oxford: Clarendon. 792 pp.
30. Klotz, L. H. 1978. Form of the perforation plates in the wide vessels of metaxylem in palms. *J. Arnold Arbor.* 59:105–28
31. Linnaeus, C. 1753. *Species Plantarum.* Stockholm. 2 vols.
32. Letouzey, R. 1979. Notes phytogéographiques sur les palmiers du Cameroun. *Adansonia Ser. 2.* 18:293–325
33. Deleted in proof
34. Miller, R. H. 1964. The versatile sugar palm. *Principes* 8:115–47
35. Deleted in proof
36. Moore, H. E. Jr. 1971. The genus *Synechanthus* (Palmae). *Principes* 15:10–19
37. Moore, H. E. Jr. 1973. Palms in the tropical forest ecosystems of Africa and South America. In *Tropical Forest Ecosystems of Africa and South America: a*

Comparative Review, ed. B. J. Meggers, E. S. Ayensu, D. D. Duckworth, pp. 63–88. Washington DC: Smithsonian Inst. Press

38. Moore, H. E. Jr. 1973. The major groups of palms and their distribution. *Gentes Herbarum* 11:27–141

39. Moore, H. E. Jr. 1977. Endangerment at the specific and generic levels in palms. In *Extinction is Forever. The Status of Threatened and Endangered Plants of the Americas,* ed. G. T. Prance, T. S. Elias, pp. 267–83. NY: N.Y. Bot. Gard.

40. Moore, H. E. Jr. 1978. The genus *Hyophorbe* (Palmae). *Gentes Herbarum* 11:212–45

41. Moore, H. E. Jr. 1978. *Tectiphiala,* a new genus of Palmae from Mauritius. *Gentes Herbarum* 11:284–90

42. Moore, H. E. Jr. 1978. New genera and species of Palmae from New Caledonia. *Gentes Herbarum* 11:291–309

43. Moore, H. E. Jr. 1979. Arecaceae (Fam. 39). In *Flora Vitiensis Nova,* Vol. 1, A. C. Smith. Lawai, Kauai, Hawaii: Pac. Trop. Bot. Gard.

44. Moore, H. E. Jr., Uhl, N. W. 1973. The monocotyledons: their evolution and comparative biology. VI. Palms and the origin and evolution of monocotyledons. *Q. Rev. Biol.* 48:414–36

45. Deleted in proof

46. Parthasarathy, M. V. 1968. Observations on metaphloem in the vegetative parts of palms. *Am. J. Bot.* 55:1140–68

47. Parthasarathy, M. V. 1974. Mycoplasmalike organisms associated with Lethal Yellowing Disease of palms. *Phytopathology* 64:667–74

48. Parthasarathy, M. V. 1974. Ultrastructure of phloem in palms. I. Immature sieve elements and parenchymatic elements. II. Structural changes and fate of the organelles in differentiation sieve elements. III. Mature phloem. *Protoplasma* 79:59–91; 93–125; 265–315

49. Parthasarathy, M. V., Klotz, L. H. 1976. Palm "Wood." I. Anatomical aspects. II. Ultrastructural aspects of sieve elements, tracheary elements and fibers. *Wood. Sci. Tech.* 10:215–29; 247–71

50. Read, R. W. 1963. Palm chromosomes. *Principes* 7:85–88

51. Read, R. W. 1965. Palm chromosomes by air mail. *Principes* 9:4–10

52. Read, R. W. 1965. Chromosome numbers in the Coryphoideae. *Cytologia* 30:385–91

53. Read, R. W. 1966. New chromosome counts in the Palmae. *Principes* 10:55–61

54. Read, R. W. 1968. A study of *Pseudophoenix* (Palmae). *Gentes Herbarum* 10:160–213

55. Read, R. W. 1975. The genus *Thrinax* (Palmae; Coryphoideae). *Smithson. Contrib. Bot.* 19:1–98

56. Read, R. W., Moore, H. E. Jr. 1967. More chromosome counts by mail. *Principes* 11:77

57. Romney, D. H. 1976. Second meeting of the International Council on Lethal Yellowing. *Principes* 20:57–79

58. Ruddle, K., Johnson, D., Townsend, P. K., Rees, J. D. 1978. *Palm Sago—a Tropical Starch from Marginal Lands.* Honolulu: Univ. Hawaii Press. 207 pp.

59. Sarukhán, J. 1978. Studies on the demography of tropical trees. See Ref. 14, pp. 163–84

60. Schmid, R. 1970. Notes on the reproductive biology of *Asterogyne martiana* (Palmae). I. Inflorescence and floral morphology—phenology. II. Pollination by syrphid flies. *Principes* 14:3–9; 39–49

61. Schroeder, C. A. 1978. Temperature elevation in palm inflorescences. *Principes* 22:26–29

62. Deleted in proof

63. Sowunmi, M. A. 1972. Pollen morphology of the Palmae and its bearing on taxonomy. *Rev. Palaeobot. Palynol.* 13:1–80

64. Deleted in proof

65. Thanikaimoni, G. 1966. Contribution à l' étude palynologique des palmiers. *Inst. Fr. Pondichery, Trav. Sec. Sci. Technol.* 5:1–91

66. Thanikaimoni, G. 1971. Les palmiers: palynologie et systématique. *Inst. Fr. Pondichery. Trav. Sec. Sci. Technol.* 11:1–286

67. Thomas, D. L. 1979. Mycoplasmalike bodies associated with lethal declines of palms in Florida. *Phytopathology* 69: In press

68. Tomlinson, P. B. 1961. *Anatomy of the Monocotyledons, Vol. II: Palmae.* Oxford: Clarendon Press. 453 pp.

69. Deleted in proof

70. Tomlinson, P. B. 1971. The shoot apex and its dichotomous branching in the *Nypa* palm. *Ann. Bot. London* 35: 865–79

71. Tomlinson, P. B., Esler, A. E. 1973. Establishment growth of woody monocotyledons native to New Zealand. *N. Z. J. Bot.* 11:627–44

72. Tomlinson, P. B., Moore, H. E. Jr.

1968. Inflorescence in *Nannorrhops ritchiana*. *J. Arnold Arbor.* 49:16–34

73. Tomlinson, P. B., Soderholm, P. K. 1975. The flowering and fruiting of *Corypha elata* in South Florida. *Principes* 19:83–99

74. Tomlinson, P. B., Zimmermann, M. H. 1966. Anatomy of the palm *Rhapis excelsa*. III. Juvenile phase. *J. Arnold Arbor.* 47:301–12

75. Tuley, P. 1965. The production of Raphia wine in Nigeria. *Exp. Agric.* 1:141–44

76. Uhl, N. W. 1966. Morphology and anatomy of the inflorescence axis and flowers of a new palm, *Aristeyera spicata*. *J. Arnold Arbor.* 47:9–22

77. Uhl, N. W. 1972. Inflorescence and flower structure in *Nypa fruticans* (Palmae). *Am. J. Bot.* 59:729–43

78. Uhl, N. W. 1976. Developmental studies in *Ptychosperma* (Palmae). I. The inflorescence and flower cluster. II. The staminate and pistillate flower. *Am. J. Bot.* 63:82–96; 97–109

79. Uhl, N. W., Moore, H. E. Jr. 1971. The palm gynoecium. *Am. J. Bot.* 58:945–92

80. Uhl, N. W., Moore, H. E. Jr. 1973. The protection of pollen and ovules in palms. *Principes* 17:111–49

81. Uhl, N. W., Moore, H. E. Jr. 1977. Correlations of inflorescence, flower structure, and floral anatomy with pollination in some palms. *Biotropica* 9:170–90

82. Uhl, N. W., Moore, H. E., Jr. 1977. Centrifugal stamen initiation in Phytelephantoid palms. *Am. J. Bot.* 64:1152–61

83. Van Valen, L. 1975. Life, death, and energy of a tree. *Biotropica* 7:259–69

84. Vandermeer, J. H. 1977. Notes on density dependence in *Welfia georgii* Wendl. ex Burret (Palmae), a lowland rainforest species from Costa Rica. *Brenesia* 10/11:9–15

85. Vandermeer, J. H. 1979. Hoarding behavior of captive *Heteromys desmarestianus* on the fruits of *Welfia georgii*, a rainforest dominant from Costa Rica. *Brenesia* 13: In press

86. Vandermeer, J. H., Stout, J., Miller, G. 1974. Growth rates of *Welfia georgii*,

Socratea durissima, and *Iriartea gigantea* under various conditions in a natural rainforest in Costa Rica. *Principes* 18:148–54

87. Vandermeer, J. H., Stout, J., Risch, S. 1979. Seed dispersal of a common Costa Rican rain forest palm (*Welfia georgii*). *Trop. Ecol.* In press

87a. von Martius, C. F. P. 1849–1853. *Historia Naturalis Palmarum* 3:307–41. Leipzig

88. Waterhouse, J. T., Quinn, C. 1978. Growth patterns in the stem of the palm *Archontophoenix cunninghamiana*. *Bot. J. Linn. Soc.* 77:73–93

89. Deleted in proof

90. Wessels-Boer, J. G. 1965. The indigenous palms of Suriname. Leiden: E. J. Brill. 172 pp.

91. Wessels-Boer, J. G. 1968. The geonomoid palms. *Verh. Kon. Ned. Akad. Wetensch. Afd. Natuurk. Tweede Seck. Ser.* 2. 58:1–202

92. Whitmore, T. C. *Palms of Malaya*. London: Oxford Univ. Press. 132 pp.

93. Wilder, G. J. 1976. Structure and development of leaves of *Carludovica palmata* (Cyclanthaceae) with reference to other Cyclanthaceae and Palmae. *Am. J. Bot.* 63:1237–56

94. Wilson, D. E., Janzen, D. H. 1972. Predation on *Scheelea* palm seeds by bruchid beetles: seed density and distance from the parent plant. *Ecology* 53:954–59

95. Zimmermann, M. H. 1973. The monocotyledons: their evolution and comparative biology. IV. Transport problems in arborescent monocotyledons. *Q. Rev. Biol.* 48:314–21

96. Zimmermann, M. H., Tomlinson, P. B. 1965. Anatomy of the palm *Rhapis excelsa*. I. Mature vegetative axis. *J. Arnold Arbor.* 4:160–78

97. Zimmermann, M. H., Tomlinson, P. B. 1971. The vascular system of monocotyledonous stems. *Bot. Gaz.* 133:141–55

98. Zimmermann, M. H., Tomlinson, P. B. 1974. Vascular patterns in palm stems: variations on the *Rhapis* principle. *J. Arnold Arbor.* 55:402–24

Ann. Rev. Ecol. Syst. 1979. 10:109–45
Copyright © 1979 by Annual Reviews Inc. All rights reserved

THE PLANT AS A METAPOPULATION

♦4157

James White

Department of Botany, University College Dublin, Dublin 4, Ireland

INTRODUCTION

"If a bud be torn from the branch of a tree and cut out and planted in the earth with a glass cup inverted over it to prevent the exhalation from being at first greater than its power of absorption; or if it be inserted into the bark of another tree, it will grow, and become a plant in every respect like its parent. This evinces that every bud of a tree is an individual vegetable being; and that a tree therefore is a family or swarm of individual plants, like the polypus, with its growing young out of its sides, or like the branching cells of the coral-insect. . . . the shoot is a succession of individual vegetable members". These observations by Erasmus Darwin occur on the first page of his *Phytologia,* published in 1800 (50). They reflect one version of a view by then current among botanists that a single plant (derived from one seed) was not a simple unit but a collection of subunitary parts. Through various refinements this concept became widespread during the nineteenth century and persisted among some plant anatomists and morphologists until modern times.

Recently a few plant ecologists have espoused the same concept (the plant as a collection of unit parts), attempting to interpret better some aspects of plant demography (104, 105, 108, 262). I have contended (262) that an understanding of the demography of subunitary parts is the starting point for some aspects of the demography of separate individuals (particularly in competition studies), and Harper (104) has drawn particular attention to the demography of plants with clonal growth. A wide overview of the population dynamics of organisms (plants and animals) whose growth can be interpreted as an accumulation of some more 'elementary' constructional

0066-4162/79/1120-0109$01.00

units has been presented by Harper & Bell (105). It is also clear that many biologists without any explicit theoretical justification have simply used the seemingly obvious subunitary structures of plants (however interpreted) to describe the dynamics of pasture grasses, the geometry of canopy development in trees, or some features of insect population behavior. And some plant morphologists still find the idea fruitful in new guises.

At the outset a few definitions and clarifications are necessary. The title of this review itself needs explanation. It seems desirable to retain the word *population* for collections of genetically distinct organisms and to coin a modified term for aggregations of parts that comprise, or are derived from, a single genetic individual. The term *metapopulation* is suggested, since the classical etymology of *meta-* expresses the notion of 'sharing, action in common' (*Oxford English Dictionary*). It also connotes 'change' (of place, order, nature) and, in modern formulations, 'beyond' or 'higher' or 'more fundamental.' But in the context of this review it is used to imply the *shared* elements that make up the morphological structure of a genetic individual, by whatever means these elements are defined. The variety of ways in which they *have* been defined is in large part the subject of the review.

Van Valen (253) has independently considered the analogies between plants and some animals and proposed new terms to describe their structures: 'individuoids' are parts of an organism that have the same general structure as whole free-living individuals but that connect to each other to form a 'colonoid,' which functions as a single individual; the 'individuoids' may function as organs of an individual. On the other hand, physiologically separate individuals of similar genetic identity (part of a clone) constitute a colony. In the terminology of this paper both 'colonoid' and colony are metapopulations since they share a common genetic identity. This use of metapopulation differs from Wilson's (267). Apart from these neologisms, several other terms are currently used for naming the parts of a plant and these are given in the Appendix. Some standardization is desirable because certain terms, such as *module,* first used in print by Harper & White (108) as a translation of the French *article* (99), have themselves begun to sprout new and less precise meanings.

This review is more an attempt to collate an extremely diffuse literature than a theoretical essay on the application of the concept of the plant as a metapopulation. The beginnings of an application to the general theory of plant demography have already been made (104, 262), and flourishing developments in plant morphology (100) will underpin the concept more firmly than heretofore. The review is restricted to gymnosperms and angiosperms and excludes reproductive structures. Historical aspects are treated first to give perspective to the modern studies reviewed subsequently.

HISTORICAL SURVEY

It is essential to review some aspects of the history of the idea of a plant as an assembly of unit parts before considering its modern applications. This is not to suggest that contemporary botanists using this notion place their work in the perspective I outline here: No more than a few are aware of it. Nowadays the plant as a metapopulation is mostly a utilitarian concept, but it was once more keenly debated on epistemological grounds. Since the history of plant demographic studies (in the widest sense, including plant competition) is still largely neglected (63, 102) it may be useful to record here a small fragment relevant to the present theme.

Buds as Individuals

Darwin (50) was by no means the first philosophical biologist to consider the multiplicity of similar parts within what we might now call a single genetic individual or genet (108). Part of this history has been documented by Guédès (95). Two students of Linnaeus whose theses are collated in *Amoenitates Academiae* (141) briefly considered the problem. Loefling's thesis *Gemmae arborum* (1749) and Dahlberg's thesis *Metamorphosis plantarum* (1755) both contain passages in which the formation of stems and branches in plants is compared to that of the polyps of coelenterates. For Dahlberg each bud initiated a new individual, which remained fixed on the parent. Loefling seems to have regarded a seed rather than a bud as a new individual. In the nineteenth century the 'bud-root theory' became influential, a notion in which roots were supposed to descend from a bud through the stem—in fact were supposed largely to constitute the stem. [The anatomical researches of von Mohl proved otherwise (231).] The polyp-plant analogy of Erasmus Darwin was used by his grandson (but now in reverse) when speaking of corals: In "these compound animals . . . surprising as this union of separate individuals in a common stock must always appear, every tree displays the same fact, for buds must be considered individual plants . . . " (47). He evidently believed (49) that the analogy "now universally accepted" (as he put it) had originated with his grandfather. Perhaps Spencer's use of the same analogy in *The Principles of Biology* (224) led Darwin to a view of 'universal acceptance'. Nor is the comparison quite moribund: Wilson (267), in his discussion of colonial organisation in coelenterates, says that "growth is plant like" in the order Stolonifera, in some species "the result is a dendritic colony"; in the order Gorgonacea "growth is treelike." Harper & Bell (105) have recently expanded these analogies.

The Shoot

From earliest times botanists have been intrigued by the problems of repetitive growth, both in vegetative and reproductive shoots (6). The opening pages of Theophrastus' *Enquiry into Plants* are concerned with the problem of defining 'the parts of the plant,' particularly in relation to the structure of animals. Among the contrasts he made between plants and animals was that "the number of parts is indeterminable" in plants. But he gave as "the primary and most important parts . . . root, stem, branch, twig . . . parts into which we might divide the plant, regarding them as members corresponding to the members of animals . . . " (117). The idea of the shoot as a plant unit has been recognized since the early days of botany (6), but not until the seventeenth century, two thousand years later, was much advance made on any of the botanical principles of Theophrastus. Malpighi and others believed that branches formed at the nodes corresponded to young plants that remained fixed to the maternal plant (95). Bradley (17) observed that "the twigs and branches of trees are really so many plants growing upon one another."

In the writings of Goethe this idea, along with so much other nature philosophy, was effectively promulgated. In his influential treatise *Versuch die Metamorphose der Pflanzen zu erklären,* published in 1790, he stated that "the lateral branches which spring from the nodes of plants may be regarded as individual plantlets which take their stand upon the body of the mother, just as the latter is fixed in the earth" (5), a simile later rendered by Arber as a "matriarchal tribe of shoots" (6). Much of Goethe's 1790 essay is concerned with the 'metamorphosis' of an ideal conceptual or abstract unit of plant structure into various actual physical expressions: For example, an ideal *Urblatt* might find expression variously as a foliage leaf, as a petal, or as a carpel. Such a conception was to prove extremely fruitful, and homology has lain at the core of plant morphology since Goethe: His treatment of the leaf as an irreducible unit to which certain other plant structures might be homologized has remained permanently influential. De Candolle (52) formulated similar ideas apparently independently of Goethe, and his treatment of the angiosperm flower has had a lasting influence (96, 209). The search for idealized plant parts seemed naturally to predicate some basic unit or other of which higher plants might be constructed. While for Goethe such a basic unit might be an intellectual abstraction, for his successors it became an object to be more firmly identified in the real form of plants. Idealistic plant morphology of the sort first made popular among botanists by Goethe had its roots in Platonic scholarship, especially in the philosophical ideas denoted as *essentialism* by Popper (189) or *typological thinking* by Mayr (153), the overthrow of which and its replacement by populational thinking has been perhaps the most significant effect of the 'Darwinian revolution' in biology (153). Typological thinking is nowadays less evident in plant morphology though it is by no means obsolete (80, 198, 244).

The topic of the individuality of plants was first thoroughly reviewed by Braun (21), who, after discussing at length the philosophical and botanical issues involved, considered that "the shoot is the morphological vegetable individual." He had earlier touched on it in a lengthy monograph redolent of the nature philosophy and vitalism characteristic of his time in Continental Europe. [Nägeli's discourse (170) is another example on the same subject]. It was "not the separated bud which we must regard as the individual but the entire shoot which ... includes several superimposed buddings" (20). He prefigured the concept of reiteration, which has now become a significant concept in modern studies of tree architecture (100, 181), when he speaks of "repetition sprouts" having an "influence on the habit of plants, on the architectural design of the stock ... " and asserts that "the peculiar forms of the crowns of trees depend on the proportion of the vigour and abundance of the repetition sprouts to the main sprout or trunk of the tree." His views (set out at length in 1853) were influential enough to merit translation and publication in two of the leading scientific journals of the time (21), perhaps through the interest of Asa Gray, who attached editorial footnotes to the translation.[1] Gray held similar opinions (90). While conceding that "individuality in plants seems as obscure and ambiguous as in animals it appears clear and simple," he embarked on an attempt to summarize opinion on the question. He was reluctant to accept the view that the bud is an individual and the shoot "a succession of individual vegetable members," still less to regard the cell as that individual, and least of all a single plant as an individual: "Our feelings aroused by the sight of the most ramified plant-stocks, especially by a tree ... excite the presentiment that this is not one single being ... comparable with the animal or human individual, but rather a world of united individuals which have sprung from each other in a succession of generations ... " (21).

Braun made a significant distinction between lateral and terminal buds. Lateral buds were the only ones from which branches originate and therefore they alone were regarded as new lines of development, i.e. as individuals. Terminal buds were, on the contrary, only the undeveloped parts of a single axis, providing continuation and augmentation of an individual already existing. Having emphasized that individuality lay in the shoot, he

[1]Braun was Professor of Botany at Berlin and a brother-in-law of Louis Agassiz. Sachs (205) has given an account of his botanical philosophy and Guédès (97) an exposition of his morphological ideas.

held further that only lateral buds on a shoot could beget new individuals: "Hence trees which produce no terminal buds, such as linden, willow, elm, develop new individuals and nothing else at each renewal of vegetation; while oak and poplar, which do produce terminal buds also, bear a mixed annual generation, partly of new individuals, partly of old ones continuing their development" (21).

Whether or not contemporary plant morphologists acknowledge or realize any influence of Braun, it seems clear that his approach to plant individuality had a commonsense basis that has since his time been mirrored among botanists. This is particularly evident in the modern literature on monocotyledons, perhaps most clearly expressed by Holttum (115), who regards the sympodial growth pattern of most monocotyledons as a necessary consequence of their (almost universal) lack of cambium. Sympodial units (falling firmly within Braun's wider conception of new individuals derived from lateral buds) are convenient structural units to enumerate and describe, although they may assume a great variety of shapes and sizes. It has become widespread practice to accept them as plant growth units (11, 13, 238).

The diversity of meristem activity complicates the structures and developmental patterns of shoot units for dicotyledons, but significant advances have been made by the architectural analyses of Hallé and his school (100), which will be mentioned later. They echo clearly Braun's original emphasis on the shoot as the unit of plant construction. Champagnat et al (34) have particularly stressed the *integrity* of the shoot.

The Plant as Phytons

The phyton theory of plant construction was first popularized by Gaudichaud (81). The theory held that the stem of a vascular plant represented the fused basal parts of leaves. Gaudichaud considered a phyton to consist of an internode and its upper node with the attached leaf: The phyton was the individual in plants. The influential American botanist Asa Gray held similar views (90): "The branch . . . is manifestly an assemblage of similar parts, placed one above another in a continuous series, developed one from another in successive generations. . . . [Each] is a plant-element or as we term it a phyton. This view . . . [is] essential to a correct philosophical understanding of the plant." The first figure in his widely read textbook *Structural Botany* gives an example of these phytons, or (a word he preferred) phytomers, "plant parts, the structures which produced in a series, make up a plant of the higher grade." [The term phytomer is still used in this sense by agronomists (119)]. Gray chose, perhaps significantly, to illustrate the concept with an annual monocotyledon; lively debate has continued over the identity of phytomers in dicotyledons. A variety of papers has since appeared to support phyton theories [reviewed by Arber (4) and Schoute (211)]. Celakowsky's (32) influential monograph revived a flagging debate. By 1915 it could be asserted by one botanist that "on the phyton theory, which is here held, the stem is really built up entirely of leaf bases" (269). Celakowsky supposed that in the shoot of monocotyledons the phytons (or *Sprossglieder*) were superposed, each internode being one morphological unit, whereas in the dicotyledons each internode was a complex of such units: In a shoot with whorled or spirally arranged leaves each stem segment was regarded as built up of many juxtaposed units, not readily distinguishable. In a comprehensive review of morphological theories Chauveaud (38) refers to Celakowsky's phytons as *articles* (Fr.), a term later to be used by Hallé & Oldeman (99) in a different context. As Chauveaud showed, phyton theories were but one manifestation of a range of reductionist morphological theories of the structure and organization of plants.

It seems that the last revival of phytonism (as the doctrine came to be called) occurred in the 1930s, largely associated with the school of J. H. Priestley in Leeds (England). A discussion on phytonic theories at the Botanical Congress at Cambridge in 1930 (23) probably marked their end as a useful conceptual or practical guide to plant morphology. Schoute's (211) and Arber's (4) criticisms at the congress were radical: "Phytonic ideas are not so much the outcome of investigation as the consequence of a philosophical tendency to simplify our conceptions" (Schoute); "the phyton theory seems to me to belong to that group of over-ingenious, academic conceptions which are difficult to discuss because they bear so little relation to reality" (Arber). Priestley's group produced a series of papers (93, 146, 147, 191–193) espousing a modified form of the phyton theory. They held that the shoot was divided into a series of articulated though closely integrated growth units, and invoked anatomical evidence of vascular differentiation and development to support their view. Some quite explicit representations of their thinking are shown in Figure 1 (192). The unit of shoot growth is "the segment of the axis which subtends a leaf initial and surrounds its leaf trace as it develops." In *Alstroemeria* each unit is considered in fact to extend through sixteen internodes. (In its lower portion it becomes narrow and encloses a vascular bundle only.) Majumdar (146) also asserted that the axis of *Heracleum* was a "composite structure built up of discrete yet continuous growth units" and supported Priestley's interpretation of the plant stem. Priestley's ideas were grounded on a combination of anatomy and morphology, but more sophisticated anatomical analyses—of monocotyledons especially (272, 273)—undermine rather than substantiate them.

Madison (145) has found the concept of phyton or phytomer useful in interpreting the gross morphology of some monocotyledons, especially

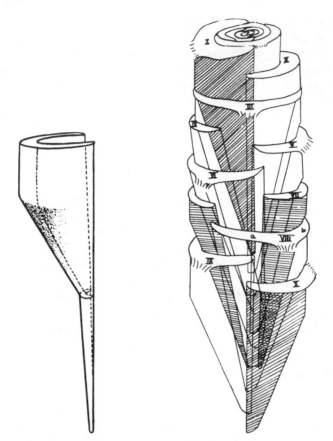

Figure 1 The conceptualized phyton or unit of shoot growth in *Alstroemeria* and their arrangement in a stem segment, according to Priestley et al (192).

grasses. However, the concept is not applicable in monocotyledons lacking a localized intercalary meristem at the base of the internode, since defined repeating units of growth do not occur (76). Many monocotyledons lack a nodal plexus and have little or no distinction between node and internode except for the actual leaf insertion. A region of meristematic activity at first expanding over the entire internode and later confined to the upper region of the developing internode is characteristic of such monocotyledons: This region has been termed an "uninterrupted meristem" (77).

It seems that the view of phytonism expressed by Arber has prevailed among plant morphologists: Such ingenious academic metamorphoses of plant structure have been almost disregarded in recent times (but see 44) and are not mentioned in standard reviews (202) or monographs (e.g. 58,

67, 69, 226) concerned with plant structure and development. "Concepts of the shoot as consisting of a series of structural units are old and have been obscured by the dominance of the stem- and leaf-theory. Anatomically units like these do not exist: the shoot is the basic unit" (61). Nonetheless plant ecologists may still find the idea fruitful, and a few modern manifestations of phytonism are mentioned below (see the section on Metameric Construction).

ARCHITECTURE OF SHOOT SYSTEMS

For nearly a hundred years plant ecologists have been concerned with the shapes and sizes of plants under the general rubric of life-forms or growth-forms. There had been earlier attempts to describe the physiognomy of plants, but Warming's monograph of 1884 was a seminal influence on subsequent attempts to classify plants into biological groups (60). To varying degrees almost all classifications of life-forms are based on the nature and organization of the shoot system. The growth-form system of du Rietz (60) is based primarily on what he calls 'shoot architecture' and owes a heavy debt to Warming. Although the use of Raunkiaer's system has become widespread among ecologists, there are cogent arguments against its appropriateness (60, 214) for describing vegetation structure, chiefly because of its emphasis on one major structural attribute of plants (the position of the resting bud). This tradition of vegetation analysis stimulated an enormous amount of research on plant morphology: Fekete & Szujko-Lacza (73) have partly updated du Rietz's review, paying particular attention to the extensive Soviet tradition. An intimate knowledge of plant morphogenesis is critical to the rather unique sort of plant demography practiced in Russia (107, 108).

Quite apart from the life-form investigations of plant ecologists, research on plant morphology *sui generis* has resulted in spectacular syntheses like Troll's *Vergleichende Morphologie der höhern Pflanzen,* published in Berlin between 1937 and 1943; or on a somewhat more modest scale (relatively) the studies of Rauh, Meusel, Champagnat, and others (e.g. 33, 156, 195). A recent flowering of this brilliant morphological tradition has been the study of tropical tree architecture by Hallé & Oldeman (99), lately expanded and refined by their collaboration with P. B. Tomlinson (100). The significance of these studies for the understanding of plant construction, both of trees and herbaceous plants (62, 121), cannot be over-emphasized. Among their signal contributions to the elucidation of the constructional principles of vascular plants is a coherent terminology to describe the sub-unitary morphology of plants. Their architectural descriptions of plants have already found wide acceptance (e.g. 242).

Architecture of Trees

As the point of departure from previous studies Hallé et al (100) define the plan of growth, the 'architectural model,' of trees using a few straightforward criteria: the life span of meristems (determinate or indeterminate) and the differentiation of meristems (sexual or vegetative; giving rise to erect or horizontal axes; functioning episodically or continuously). Only primary (extension) growth is considered. Twenty-four architectural models are defined a priori and all but one of them may be found in living trees. A great variety of tree species fall into one or another of these twenty-three models. There is generally no correspondence between systematic affinity of plants and architectural models. Since the models define patterns of morphogenesis that are not necessarily size-related, similar models have been found in many herbaceous plants and in lianas, although it is likely that further (different) models may be found among them (42, 121). For the purposes of the present review these studies are quite significant, because throughout their exposition the sub-unitary construction of plants is explored in detail and acts as a leitmotiv. Plants (especially trees) are conceived to have 'modular construction', a module being "an axis in which the entire sequence of aerial differentiation is carried out from the initiation of the meristem to the onset of sexuality which completes its development." Most commonly modules form sympodia, either linear or branched. These units are regarded as developmental units, not necessarily structural units. When the succession of sympodial modules constitutes the trunk of the tree they may be referred to as "relay axes." *Structural* units (such as the rhythmic increments of growth within the monopodium) may be referred to as *caulomers,* a term that has this connotation traditionally (204). Hallé et al (100) describe these morphologically distinct growth increments as "units of extension." A summary does little justice to the richness of their analysis, remarkable for its clear espousal of the concept of the plant as a population of shared parts (metapopulation).

Attention should be drawn to a few further points. Not all modules in a plant are equivalent. Those on plagiotropic (horizontal) axes may differ significantly from those on erect axes. Modules may differ in size from plant to plant: In *Phillodendron selloum* each sympodial unit in the axis consists of a prophyll, a foliage leaf and a pair of inflorescences ($\equiv$ module), tightly condensed to give the superficial appearance of a simple monopodial axis; this is probably an extreme example of modular condensation (100).

Reiteration

An important additional concept was introduced by Oldeman (181): reiteration, defined as "any modification of the tree's architecture not inherent in the definition of its model and which is occasioned by damage, environmental stress or supra-optimal conditions" (100). The striking thing is that reiterations usually recapitulate almost entirely the architectural model. In mature trees, consequently, the architecture may be interpreted as a population of architectural models. The basic model gives an 'initial complex' of a tree, but this is typically augmented by more of these complexes in the process of reiteration, giving a 'reiterated complex.' As the tree grows, the size of the reiterations tends to diminish and ultimately only parts of the 'initial complex' or model are recapitulated: Reiterations are described as being successively tree-like, shrub-like, and herb-like in their stature. The ultimate 'wave of herbaceous reiteration' in the top canopy of a tree has been compared to a population of herbs or a field of weeds (100). Reiterated complexes within a single tree may even behave like a population: In the tropics their phenology may be unsynchronized—some may be leaf bearing, others bare except for inflorescences, and others past flowering (100). Asynchronous flowering of meristems on a single tree is most frequent in those trees with "an above average number of lateral branches or stems" (i.e. reiterations) (43). The phenomenon may also be observed in temperate trees, especially if their internal water-conducting system is sectorial rather than spiral, leading temporarily to differential water relations within the canopy (271); but the asynchrony is measured in days rather than weeks.

Starting with the insights of Hallé et al we find the growth patterns of trees increasingly easy to interpret in terms of a plant metapopulation. At perhaps the grossest level, reiterations of model complexes are seen clearly as *populations* of complexes, as an integral part of the structure of the tree crown (where their delimitation is sometimes obscure) or more obviously as sucker sprouts (coppice) or clones. The clonal growth of *Populus* species has been particularly studied: Offshoots or ramets arise from the root system of the genet, and 'parental' connection may be maintained for up to 50 years (51). Not only are intragenet connections maintained but intergenet connections by root grafting are frequent, giving rise in a sense to a superorganism, uniting and transcending the limits of single genets (51). [Graham & Bormann (89) have comprehensively reviewed this phenomenon, which is widespread in trees. Natural root grafts may influence the survivorship, spacing, and dominance-hierarchy in tree populations, but this phenomenon has been little investigated.] The establishment and development of *Populus* clones have been reviewed by Barnes (8), with detailed maps of their geometry. A clone develops gradually as successive generations of reiterations arise from a continually expanding root system. Their life span is indeterminate, and ages of 8000 years have been suggested for some (41). Two clones of *Populus tremuloides* in Utah covered areas of 10.1 ha and 43.3 ha, containing 15,000 and 47,000 ramets, respectively (128a). Ramet age was about 100 years, but the genet age may exceed 10,000 years. *Fagus gran-*

difolia may also reiterate from root buds under certain circumstances: Ward (257) mapped and measured 81 sprouts within an 8 m radius of a 'parent' tree 90 years old. Coppice shoots, reiterations arising near the base of the initial model, are a common feature of angiosperm trees but are rare in conifers (29, 271). The capacity of *Sequoia sempervirens* to reiterate occasionally from buds just above the root crown may confer on genets of this species the greatest known longevity of any living organism. Some groves of redwood are clonal (201a, 229). Reiterations in *Populus sargentii* (cottonwood), which can be precisely aged, have been used to describe rates of channel migration and sediment transport in the river courses along which they grow. When overwhelmed by alluvial deposits they may be prostrated and buried, but clumps of shoots can rise vertically by reiteration (71).

Multiple reiterations are a conspicuous feature of trees growing at the timberline: Griggs (94) referred to them as 'cripples.' The uppermost trees surrounded by dwarf shrubs and herbs may form 'timber atolls' (94) or 'tree islands' (150). These are caused by layering and rooting of branches that then grow erect, giving the appearance of clumps of small trees. If the older parts die, the clump becomes hollow in the center; one such fir 'atoll' had a diameter of 20 m and an open interior 10 m across (94). Popovic (188) has given detailed charts of the geometry of *Pinus mugo* genets growing at timberline. Layering of basal lateral branches is a common feature of willows and conifers growing in the Arctic (40, 210), resulting in thickets of ramets. Some of these clones may be 900 years old (25).

A spectacular form of layering may be seen in *Ficus benghalensis,* a large spreading evergreen tree with massive pillar roots that extend the crown laterally. It has the largest crown of any tree, one specimen having a surface area of over 2 ha and a circumference of 530 m (illustrated in 244). Less impressive perhaps are some of the records of crown size in temperate-zone trees, achieved by layering. Near Edinburgh a beech (about 300 years old and 35 m high in 1906) had a circumference of about 130 m. Numerous branches had taken root and then formed successive circles of subsidiary upright stems around the main trunk (64). Braun (21) drew attention to a similar phenomenon in *Juniperus sabina* as a pertinent example of the plant as a metapopulation.

A peculiar manifestation of the plant as a metapopulation occurs in *Eucalyptus* ('mallee' species), where abundant shoots are reiterated from a persistent subterranean lignotuber containing numerous dormant buds (e.g. 166). More generally the distribution and ecology of these so-called geoxylic structures have been reviewed by White (260). They are a conspicuous feature of the *campos cerrados* of Brazil, but are most diversified in Africa, where there are 109 species in 31 plant families. Sometimes the aerial shoots are herbaceous and renewed annually: *Tamus communis* is a familiar European example.

The shape of trees, however genetically determined, is an outcome of demographic processes, birth and death of meristems (262). While many meristems remain dormant for long periods, growing just enough to keep pace with radial stem increments (271), others form shoots of varying sizes. In time as the canopy expands some shoots become shaded, die, and are abscised. Münch (167) expressed vividly the dynamics that underlie 'the harmony of tree shape:'

> The growing shoots behave towards the other parts in an egotistical fashion. They promote themselves and the ones that carry them, but they treat their neighbors as competitors for light and nutrients. Thus a large number of shoots cannot develop at all, others remain very small and still others reach only limited size. . . . Without this correlation, every crown would become a witchbroom. Only the never-ending struggle of all shoots against all others produces the shape of the trees which is not only physiologically harmonious, but which we consider aesthetically pleasing (translation of M. H. Zimmermann).

Certainly tree structure is highly ordered, not a witchbroom (7, 183). There have been attempts to simulate branching patterns, but until recently few have been biologically realistic no matter how mathematically complex. Some studies point the way towards more sophisticated modelling (31, 39, 55, 78). Undoubtedly the architectural analyses of Hallé et al (100) will be a seminal influence.

Architecture of Herbs

The application of the principles of Hallé et al to herbs is only beginning (121), but a huge literature otherwise clearly demonstrates their subunitary construction. Considering the distribution of herbaceous species by offshoots, Kerner von Marilaun (129) concluded that they "may be formed on all parts of the plant": Commonly these offshoots are what we may now term 'reiterations.' Möbius (162) has since reviewed vegetative propagation comprehensively (his Figure 7 being a striking example of reiteration in a spruce tree). Offshoot production enables a genet to augment its population without sacrificing immediate and local phenotypic fitness to the hazards (*sensu* chance) of sexual recombination. Some clonal herbaceous plants may persist for centuries (108) and, as Williams (263a) argued, must show "extraordinarily high fitness" for their region of occupancy against new genotypes. Genet fragmentation may also occur, enabling effective dispersal: Stolon or rhizome dispersal is common in aquatic plants, bulbil-dispersal in terrestrial plants (197a). The life expectancy of vegetative propagules is higher than that of seeds (13, 207), at least locally. However, the selective advantage of genet proliferation and fragmentation in all its variety remains uninvestigated. Harper has suggested (personal communication) that fragmentation may be selectively advantageous if it hinders the spread of viruses and other pathogens within the plant. Williams (263a) argued that genet fragmentation and dispersal would avoid intense competition within the clone.

Plant ecologists have investigated the vegetative spread of herbaceous perennials under the term 'pattern-analysis' (e.g. 130), but this work is noted more for its statistical demonstration of 'pattern' (nonrandomness) than for its contribution to plant morphology or demography. It may be a useful approach in certain circumstances (when dealing with clonal growth of plants whose genetic identity is clear) to study patterns of plant morphology in natural communities, since it reveals what one may term 'postinteractive' morphology as distinct from the morphology of isolated plants grown in artificial conditions. However, no such comparisons appear to have been systematically conducted.

Soviet plant demographers have made detailed studies of herb morphogenesis. Their diagrams commonly show the metameric or modular structure of plants (e.g. 247, 248), but this is only part of a much wider tradition of plant morphogenesis (e.g. 14, 59, 187, 212, 213). Notwithstanding this, there appear to be no accounts of the demography of plant metamers and modules in the Soviet literature I have examined.

MONOCOTYLEDONS Because of the general absence of secondary growth and the consequent prevalence of sympodial growth in monocotyledons (115, 238), their modular structure is at once apparent. There is a vast literature on the modular dynamics of monocotyledons, concerned chiefly with rhizomes, tillers, and bulbs. The architecture of rhizomatous plants has been fully reviewed recently by Bell & Tomlinson (13). They show clearly the importance of demographic thinking in interpreting the morphology of rhizomatous plants. Bell's work (10–12) is especially notable in this regard. Noble (175, 176) has conducted a detailed study of the modular dynamics of *Carex arenaria*. In rhizomatous plants that accumulate sympodial modules (or monopodial caulomers in the case of dicotyledons) in a regular annual fashion, detailed age-structure analyses may be carried out—as Rabotnov (194) showed for several species [see Harper & White (107) for graphical analyses] and Kawano (126) showed for *Polygonatum macranthum* and *Smilacina robusta*. Tomlinson (239) has argued that an understanding of the modular growth of herbs underlies a proper understanding of productivity, at least in seagrasses: His paper is especially useful in its definitions of branching as 'regenerative' when the general form of the plant is maintained and 'proliferative' when the number of indeterminate meristems is increased (leading to vegetative propagation).

Typical rhizome modules bear buds that are determinate and few in number [as in *Medeola virginana* (10) or *Alpinia speciosa* (12)]. However, the architectural model is expressed not by the proliferation of all meristems but by the growth of a few, except in special circumstances (e.g. disturbance) where normally quiescent meristems reiterate the model. The dormant axillary meristems or buds act as a reservoir that may be termed a 'bud-bank' (103) by analogy with the 'seed-bank' (107). Having acknowledged the important genetic distinction between them, we nevertheless note that buds may have functions similar to those of seeds in perennation, multiplication, and (if sundered from the genet) in local dispersal. The dynamics of bud-banks have been neglected except in some troublesome weeds, notably *Agropyron repens* (see 103), *Cyperus esculentus* (236), and *Sorghum halepense* (116). *Agropyron repens* in particular has been the subject of extensive physiological investigation (e.g. 35, 154, 199) and a few demographic studies. Tripathi & Harper (243) found that four *Agropyron* plants each produced an average of 215 buds per plant after 20 week's growth from single tillers in a 20 cm diameter pot ($\equiv 2.7 \times 10^4$ m^{-2}). A single node from the rhizome may produce 825 dormant rhizome buds in a season (103). Fail (72) found 620 buds m^{-2} in disturbed natural environments. Noble's (175) census of *Carex arenaria* rhizome modules included bud numbers that varied seasonally, ranging up to 1400 m^{-2}. He concluded that the accumulation of buried dormant buds guaranteed the ecological homeostasis of *Carex arenaria* populations. This comment seems particularly pertinent to the (not very successful) efforts to eradicate *Agropyron repens* in horticulture! The dynamics of the reservoir of dormant buds in the soil has been a neglected aspect of plant demography. They are produced by underground stems (rhizomes) and by roots and are significant in the propagation of some weedy species such as *Senecio jacobaea* (101), *Linaria vulgaris* (37), *Chamaenerion angustifolium* (250), *Convolvulus arvensis,* and others (see 186). I have found no demographic studies of root buds.

An essential difference between rhizomatous and tillering habits in monocotyledons is the disposition of the modules horizontally or vertically. Many species (especially grasses) show both habits. The modular construction of tillering monocotyledons is perhaps the most self-evident expression of the plant as a metapopulation. In the overwhelming number of studies on pasture and meadow grasses the tiller (a module) rather than the whole plant (genet) is the unit of study. Examples are so extensive that reviews have been made repeatedly (122, 139, 264). Some examples of the striking demographic content of studies on tiller dynamics are given by White (262). Clearly the sympodial unit (module) is pragmatically the most useful unit in the ecology of tillering grasses. (Many grasses are monopodial of course.) Several detailed studies have been made on the physiological interdependence of tillers within a single genet (134, 151, 180, 182). But while the genet may show physiological integration (transfer of organic materials between tillers), the tiller remains the vulnerable unit ecologically. Under crowded conditions tiller death precedes genet death (128). Harris (109) showed how the population structure (frequency distribution of individual size) of rye-grass plants (genets) in swards is a function of tiller number per plant. Detailed studies on the morphogenesis of tillers have enabled the age structure of tillers within a genet to be determined for *Poa pratensis* (see 262) and *Nardus stricta* (185).

The demography of plants (mostly monocotyledons) that perennate or propagate by bulbs or corms has been neglected. Rees (197) has reviewed many aspects of their morphology and physiology and Burns (28) has documented their occurrence in grasses. The effects of density on bulb size and numbers of 'daughter' bulbs have been investigated for species of commercial importance (197, 237). Kawano's researches (125–127) on *Allium, Erythronium, Cardiocrinum, Scilla, Fritillaria* are the most detailed demographic studies so far published on bulbs; age- and size-class structures have been determined for these species in natural communities. J. P. Barkham (personal communication) has made a detailed demographic study of *Narcissus pseudonarcissus*. *Ranunculus bulbosus* (corms) has been the subject of an outstanding demographic analysis (207).

DICOTYLEDONS   Evidently the modular structure of many monocotyledons enables one to recognize within-genet units rather easily. But dicotyledons present a stumbling-block, since they encompass a much wider range of architectures (121), and in the absence of good architectural analyses convenient sub-unitary elements are not easy to define. The models of Hallé et al (100) and their application by Jeannoda-Robinson (121) to herbs may facilitate future work. Modules in monocotyledons within a genet are relatively invariant in size (e.g. 10–12), but this is unlikely to be the case in dicotyledons with their more diverse meristematic activity. A few attempts have been made to count branch numbers in crop plants (86, 110, 113, 169, 227); but while they may show some response to experimental treatments (e.g. competition-density effects), their inherently greater plasticity than monocotyledonous modules seems to have been an intractable problem in establishing a formal demography of shoots. An analysis of Stern's (227) experiment on *Trifolium subterraneum* grown at 4, 16, and 36 plants dm^{-2} does however show that the number of *apical* meristems per unit area is asymptotic, notwithstanding the variability of branch size (Figure 2). Perhaps the most detailed morphogenetic studies on branch numbers, sizes, and arrangement have been on *Gossypium* (cotton) (168)—not surprising in view of its outstanding commercial importance. In stoloniferous or

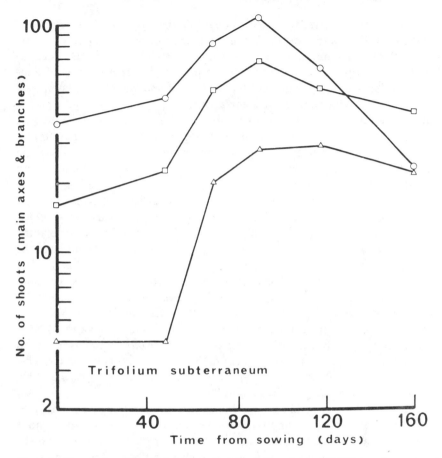

Figure 2   The number of shoots (with apical meristems) on plants of *Trifolium subterraneum* grown at starting densities of 4, 16, and 36 plants dm⁻². Although shoot size varies their number is convergent after 160 days, part of the decline at the highest density being due to genet death. Derived from data in Stern (227).

rhizomatous dicotyledons the definition of denumerable units (ramets) is relatively straightforward, as in *Ranunculus repens* (207), *Mercurialis perennis* (118), and others (104).

METAMERIC CONSTRUCTION

The plant has been conceptually fragmented in a variety of ways concerned with the shoot. The unit of shoot growth in grasses, the phytomer, has been mentioned above. Introduced by Gray (90), the concept has never lost its appeal to morphologists. The phytomer is still regarded as a basic constructional unit (68, 70, 152) or used to quantify the rhythm of shoot growth (119, 221). Etter's (68) detailed study remains a classical exposition: "The grass plant is a community of phytomers . . . [which] in various combinations unite to produce the characteristic parts of the mature plant." Rhizome, tiller, and flowering shoot (culm) of *Poa pratensis* are all composed of phytomers in his view. He recognizes that the phytomers are quite plastic in themselves depending on their location on the shoot, thereby enlarging on the traditional notion of relatively uniform units. The regulation and reduction of the number of phytomer units on the culm are among the criteria suggested by Donald (57) in his design for a high-yielding variety of wheat. Although internode length has been varied by breeding in some cereals such as rice, phytomer number is more difficult to alter (36).

In a particularly detailed study of the growth of *Helianthus annuus,* Kobayashi (136) sought to analyze the plant in terms of metameric units, each consisting of a leaf and an internodal stem segment. He simulated the growth of the whole plant as the sum of these segments. This is the first detailed application of this 'metametric approach' to dicotyledons; a comparable attempt has been made for *Pimpinella anisum* (233). The accumulation of metameric segments may be crucial in determining the 'biological age' of shoot systems: For example, Fernandez (74) has shown that before inflorescences are produced (in leaf axils) in *Phaseolus coccineus* a critical minimum number of metamers (stem-nodes plus leaves) is necessary. This number varies in an altitudinal sequence for natural populations: 7 at 3000 m, 13 at 2800 m, 17 at 2200 m, and 23 at 1800 m.

Not all metamers within a shoot system are alike, as Nozeran et al (179) have emphasized. Regarding the shoot as a succession of metamers [node + leaf + axillary bud(s) + internodal segment], they have shown that the metamers may "present an evolution in their aspect"—that is, are differentiated to produce varying types of lateral branches (orthotropic or plagiotropic vegetative axes, or inflorescences). The first-formed metamers (the oldest) "are always those which remain in the least differentiated state" (179), a conclusion that underlines the significance of a proper understanding of the metameric construction of plant stems.

The segmented or articulated nature of the vegetative axis of *Streptocarpus* spp. has long attracted attention. Jong & Burtt (123) have interpreted it as composed of repetitive structural units called phyllomorphs. While they suggest comparisons with phytons, they believe phyllomorphs are structural units peculiar to some members of the family Gesneriaceae.

Most metameric conceptions (perhaps fantasies) have usually segmented the vegetative axis transversely, as it were. One rather unique conception segments the axis vertically: the pipe model theory (215). Plant (especially tree) stems are conceived to consist of an assemblage of *unit pipes,* each supporting a unit amount of photosynthetic tissue. In this model the unit pipe is the metamer. Allometric relations between leaf area and cross-sectional area of supporting stem tissue lend weight to the analogy, but mechanically and functionally such structures do not exist. Water movement in trees, for example, is not usually restricted to straight files of xylem elements ('pipes') but is commonly spiral (271). However, a few trees mirror in their construction the abstraction of the pipe model, trees with strongly fluted or even fenestrated ('see-through') trunks. Examples are *Aspidosperma* spp., *Haematoxylon campechianum, Inocarpus edulis, Adina* spp., *Minquartia guianensis* [strikingly illustrated in Hallé et al (100)], and *Geissospermum seriaceum,* all of which occur in tropical forests (100, 155). Some temperate ring-porous trees with sectorial water movement may show slight fluting (271).

Particularly striking examples of vertical metamers are found in some desert shrubs. Starting from a single stem, the plant undergoes vertical fission, essentially by isolating strips of xylem tissue from each other (in a variety of ways). The splitting occurs both in roots and shoots, but all the 'daughter' splits remain together and are not propagated. A whole plant is thus "a bunch of separate units produced by splitting of a mother plant and isolated from each other by a layer of cork" (84). This may enable the plant to abscise a large part of its shoot system, reducing the transpiring area, without damage to the remaining part. The surviving metamers are probably those in slightly more favorable microhabitats than their neighbors (84). In a unique study Waisel et al (255) have shown how water movement in such shrubs is sectorial—i.e. along continuous files of xylem elements within a metamer. Functionally, the shrubs they examined seemed to be better examples of the pipe model theory than many trees, since the sectorial ascent ensures a direct supply of water and nutrients from one root to one of the branches. Conversely, when a single root was desiccated, only the branch connected to it was damaged; the rest of the crown was undamaged. Under conditions of water shortage the metameric structure of such plants may have important survival value. Interxylary cork and stem fission occurs in several plant families (165) and is not always related to xeric environments. Splitting of roots and rhizomes into strands isolates vascular tissues connected to dead stems: *Mertensia maritima* is especially notable since its roots and rhizomes form a 'cable of strands' (219), doubtless a significant adaptation to its shingle-beach habitat.

LEAF DEMOGRAPHY

Leaf Number

A few angiosperms may be characterized by the small number (1–3) of leaves on the shoot [*Dentaria, Jeffersonia, Listera, Maianthemum, Monophyllaea, Panax, Podophyllum, Sanguinaria, Trillium,* and some *Umbel-*

liferae (e.g. 114)], but most vascular plants bear several leaves on the shoot, often an indeterminate number. The annual shoot extension growth of trees may bear a determinate or indeterminate number of leaves, resulting either from the expansion of the fixed number in the overwintering bud or from the expansion of such leaves plus leaves newly formed during the growing season (100, 240, 271); the ecological significance of this distinction has been partially investigated (149). In neither case, however, does one attempt (usually) to characterize the trees by leaf numbers, since they may exceed 10^6 on all shoot systems combined. There is probably an inverse relation between the number of leaves on a plant and individual leaf size, but this has so far not been documented beyond an anecdotal level (see Figure 3). The leaves of some palms may be huge, e.g. those of *Corypha elata* are up to 4 m long (241); the longest determinate leaf recorded is that of *Raphia regalis,* reaching an astonishing length of 25 m. [The indeterminate leaves of some climbing ferns may exceed this (100).] *Corypha* produces about ten leaves per year (241).

Numbers of leaves on plants have been counted quite accurately in some studies, which seem to fall into two major categories: those concerned with primary production processes in canopies, and those undertaken by entomologists. Most of this literature seems to be modern. Phytonists, despite their preoccupation with the ultimate metameric structure of plants, seem to have produced no quantitative estimates of numbers of phytons per plant and nothing on their demography. The only attempts at quantification appeared after Delpino's discussion (54) of phyllotaxy, a subject that has given rise to a large literature (58, 265). The rate of leaf appearance, the plastochron index (138), and the growth rates of individual leaves (265) have been documented, but the *demography* of leaves has been singularly neglected.

Unexpected as it may appear, there seem to be more estimates of leaf numbers on trees than on herbaceous plants, notwithstanding the immensely greater numbers. Büsgen & Münch (29) review some early studies. MacDougal's work (144) on *Pinus radiata* is a rich source of undigested raw data; one tree he observed from 1919 (age 31 years) to 1934 increased its leaf number from 4×10^6 to 8×10^6; another from 1919 (age 5 years) to 1932 increased its leaf number from 4×10^4 to 3×10^6. While many of his figures are estimates based on branch sampling, he actually counted all the leaves on a few trees by defoliating them. Leaf numbers increase exponentially with age in early life (up to about 20 years) and then increase at a slower rate. Although he does not mention it, leaf area is allometrically related to trunk diameter (measured 1 m from base). About the same time Burger (27) began a series of studies on a variety of tree species (white pine, Scots pine, Douglas fir, white fir, spruce, larch, beech, and oak) in which he measured leaf numbers among a range of other attributes. His primary purpose was to estimate the amount of foliage required to produce 1 m³ of wood. His data are very extensive, very accurate, and have never been emulated. Like MacDougal, he made no attempt at an allometric formulation of a leaf-number–stem-diameter relationship. Some examples drawn from his work are shown in Figure 3. Other censuses of tree leaf numbers have been made for *Pinus monticola* (26), *Quercus* spp. (249), *Acer saccharinum* (45), *Pseudotsuga menziesii* (160), and *Picea glauca* (79) by a variety of techniques. A review of earlier studies on leaf numbers on fruit trees (apples, apricots, peaches) was made by Vyvyan & Evans (254), who stated that "no investigation had attempted to measure every leaf on any except very young trees though counts have sometimes been made." They themselves give much detail on the numbers of leaves per shoot of two apple trees—overall the trees had 10,435 and 20,006 leaves each.

By far the commonest means of estimating leaf number in the past twenty years has been by allometry. The concordance between leaf number or leaf area and stem diameter at breast height (dbh) has proved to be remarkably good. While Sinnott (218) had earlier tried to apply allometry to plants and found it worked less well for plants than animals, it remained for Kittredge (135) to show its value. Perhaps because Kittredge was a forester, the method has been applied most frequently to woody plants. Kittredge estimated the dry weight of leaves on trees of various sizes and concluded that it was closely related to stem dbh. A large literature has since developed, largely from Japanese workers (reviewed by J. White, "Allometry in Plants," manuscript). Among the few studies that estimate leaf numbers (besides weight or area) are those on *Populus davidiana* (208), *Quercus* spp. (201), and *Tectona grandis* (124). Most studies document leaf weight or area, but these are in fact good predictors of leaf numbers, as analyses of Burger's extensive data show (J. White, manuscript). Turrell (245, 246) estimated leaf numbers in *Citrus* cultivars and showed a somewhat unusual log-log relationship between leaf numbers and age; age in turn had a log-log relationship to tree size (weight). Recently the leaf-area–stem-dbh allometry has been refined by considering the cross-sectional area of sapwood or conducting tissue (92, 258), a precision prefigured in the pipe-stem model (215). Although leaf numbers or area per plant increase allometrically with stem dbh for some herbs (225) as well as trees, the leaf area per unit of ground covered (LAI) is asymptotic and rarely exceeds 10; the leaves are differentially distributed between plants of widely different sizes [(261); J. White, unpublished]. Most studies on leaf numbers of herbaceous plants have been concerned with their rate of formation or turnover and are reviewed below in the section on Leaf Dynamics.

Apart from the interest of foresters and horticulturalists in leaf numbers on trees (usually as part of wider studies on productivity) a significant entomological aspect should be mentioned. Many insects breed or feed on

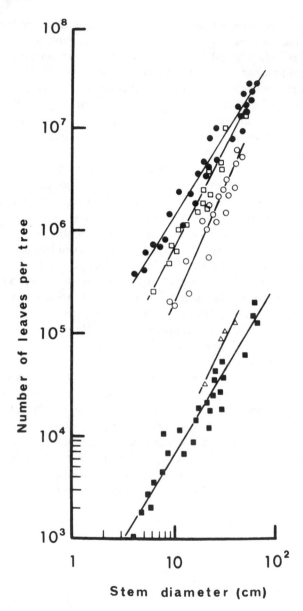

Figure 3 The allometric relationship between stem diameter at breast height (*d*) and leaf number (*N*) per tree for white pine (○), Douglas fir (□), larch (●), beech (△), and oak (■). The *N-d* relation for larch is $N = 3.5 \times 10^4\, d^{1.6}$ and for oak $N = 1.7 \times 10^2\, d^{1.6}$. Note the slope constants: for given *d*, larch has ∼200 times as many leaves as oak. Leaf number and size are inversely related in trees of given *d*. (Further details will be published elsewhere —J. White, "Allometry of Plants," ms.) Derived from data tabulated in (27).

leaves; for the entomologist the leaf is an important plant unit in the demography of insects. Leaf numbers have been determined for *Abies balsamea* (164), *Quercus rubra* (66), *Acer saccharum* (3), *Viburnus tinus* (223), *Carya* sp. (46). In some studies the 'leaf cluster' is the basic sampling unit (140, 143, 184). Lord (143) discusses the relative merits of using branch limb or foliage cluster in determining insect numbers. For Le Roux & Reimer (140) the leaf cluster was the most suitable unit since it was relatively stable for mature trees; on it are found almost all immature stages of both insect species they were investigating. The mean number of leaf clusters per tree (for ten 35-year-old apple trees) was $11,300 \pm 800$ and the mean leaf number per cluster 11.2 ± 0.2. A limited survey of the entomological literature reveals repeated reference to some sub-unitary part of the plant such as the leaf or bud when it is significant in determining potential insect numbers. Bud numbers have been accurately counted in *Abies balsamea*, for example, in studies on spruce budworm (15, 159). Nef (171) counted all the branchlets on young trees of *Pinus silvestris*. Nielsen has studied the patterns of herbivory in phyllophagous arthropods of *Fagus silvatica* (173, 174). Dixon, observing the impact of aphids on the leaf numbers and area of *Acer pseudoplatanus* (e.g. 56), noted that a 20 m high tree had 116,000 leaves supporting 2.25×10^6 aphids at one time. Such botanical details are not often the concern of botanists.

Leaf Dynamics

Discussion of this topic falls into a few discrete categories: rate of leaf appearance, mostly based on agronomic research on crops or pastures; leaf abscission and leaf fall, predominantly observed by tree biologists; and studies on leaf age structure and survivorship.

Apart from such studies on plastochron index reviewed by Dormer (58) or Lamoreaux et al (138), the development of the leaf canopy is of primary importance to crop ecologists (e.g. 217). Two generalizations emerge: The number of leaves formed per plant declines with increasing density, and individual leaf longevity also declines with increasing density. Among the earliest detailed studies that support the first conclusion is that on *Trifolium subterraneum* by Stern (227), who showed that at densities of 4, 16, and 36 plants dm^{-2} the numbers of leaves per plant at the peak of leaf production were 35, 14, and 8, respectively. [Of course, the number of leaves per unit area and LAI increases asymptotically with density (261)]. Similar conclusions were reached independently by Hodánová (112) for wheat, by Kirby & Faris (133) for barley, by Bazzaz & Harper (9) for *Linum usitatissimum*, and may be inferred for *Helianthus annuus* (111). Leaf longevity declines with increasing density; for example, the first pair of leaves in *Helianthus annuus* lives 28, 23, and 19 days at densities of 6.25, 25, and 100 plants

m^{-2}, respectively (111). With increasing density and increasing shading in the lower canopy, older leaves die more rapidly than leaves of similar age in well-lit canopies. Bazzaz & Harper (9) seemingly produced a contrary result, since they found that leaf life expectancy *increases* with increasing density, but their e_x value is calculated for the cohorts of leaves present at $x = 63$ days after germination (more than half-way through the whole life of the plants), and this may have influenced their conclusion. Among other studies on leaf dynamics of crop plants are those of Brougham (24), Greenwood et al (91), Syme (232), and Stern & Donald (228).

Most observations on leaf longevity have been on trees rather than herbs (e.g. 83). Molisch (163, also 256) conveniently summarizes early work: Apart from the singular case of *Welwitschia* (whose two leaves may live for over 100 years), few leaves live more than 10 years (those of *Araucaria* and *Abies* being possible exceptions). Species with leaves living longer than 5 years are mostly conifers. Obviously in plants that carry cohorts of leaves for more than one year an age-structure may develop if successive annual cohorts overlap. Studies on leaf age-structure exist for certain species since the physiological significance of leaves of different age classes (their relative contributions to net photosynthesis) affects tree productivity (e.g. 132, 142, 268): *Pseudotsuga menziesii* (160, 216, 222), *Tsuga heterophylla* (222), *Pinus radiata* (200), *Castinopsis cuspidata, Chamaecyparis obtusa,* and *Pinus taeda* (234). Although it should be relatively easy to document (e.g. 241), there seem to be no reports of age structures for palm leaves.

Most leaves are determinate in growth [though a few are not and retain a functioning apical meristem (100)] and have a determinate life span. Within a species this life span may be environmentally determined. For example, *Picea excelsa* may retain some needles for 4–6 years at 230 m a.s.l. (above sea level) and 10–13 years at 1750 m a.s.l. (163); similar results are reported for *Pinus montana* (163). However, such reports are somewhat misleading because, in common with all populations, leaf populations undergo mortality over a time period. Survivorship curves best represent this pattern and have been constructed for the following species: *Abies Veitchii* (131), *Actinodaphne longifolia* (270), *Tilia japonica* (161), *Ledum groenlandicum, Kalmia polifolia, Chamaedaphne calyculata* (196), *Linum usitatissimum* (9), and *Viola sororia* (O. T. Solbrig, unpublished). Without exception they all show Deevey Type 1 (53) survivorship curves, as might be expected from structures with access to 'maternal' food resources in early life. A somewhat similar pattern was noted for five species of meadow grass (266) where the percentage survival of green material of sets of leaves was followed through a twelve-month period. Mitchell (160) applied a life-table analysis to the foliage of *Pseudotsuga menziesii,* and while he does not present the usual format for a life-table (e_x values are not computed), his study represents a significant attempt to give details of leaf dynamics in a formal demographic manner. Maconochie (144a) has shown how the current leaf number on some shrub species represents a dynamic balance between leaf formation and leaf death, using a graphic technique familiar to plant demographers recording plant numbers in permanent quadrats (e.g. 207). There is a detailed but isolated study on leaf death in field beans (75), but its extensive observations cannot be simply interpreted demographically. As was emphasized elsewhere (108, 262), much potential demographic information has been gathered by applied plant scientists. It must be reanalyzed for inclusion within the framework of formal demography. However, too often key data are not collected or reported.

Rates of leaf abscission and death, mostly in trees, have been recorded frequently and reviewed comprehensively (2, 22, 137, 206). Almost all such estimates are based on total leaf weights rather than numbers. There is a climatically determined pattern in rate of leaf fall (22, 98). Van der Pijl (251, 252) has given special attention to the mechanisms of leaf abscission.

CONCLUSION: PLANT MORPHOLOGY AND PLANT DEMOGRAPHY

Plasticity in morphogenesis is almost universal in plants: Varying numbers of meristems are activated to form organs of diverse shape and form, responding to environmental conditions. Braun (21) gave an early quantitative example of plants of *Erigeron canadensis* varying in height from 5 cm (with 34 flowers in a single terminal capitulum) to 1 m (2000 capitula on a plant with nearly 100 first-order branches). This has important consequences for plant demography that are only now being clearly spelled-out (108, 262). We now realize that the traditional methods of animal demography (which has largely ignored colonial invertebrates) cannot be applied uncritically to plants. Indeed there is a wider sense in which plant population biology may be distinct from animal population biology, as Bradshaw has argued (19). The size and longevity of genets, the dispersal of ramets, the genetic composition of natural plant populations and its flux, and the stability of populations, are all topics still little investigated. While plant plasticity has been a commonplace observation, its ecological consequences are only beginning to be explored (18, 235). Many studies that record the variability of plant size in artificial and natural populations have paid no attention to its possible genetic basis. In most competition experiments plasticity is directly (and reasonably) ascribed to density stress. [Virtually all competition studies involve intergenotypic competition; intragenotypic, clonal competition experiments are extremely rare (203)]. Gottlieb (88) could find no genetic differences between large and small plants in natural populations of *Stephanomeria exigua.* But there is increasing evidence that intraspecific plasticity *per se* has a genetic basis [in *Senecio vulgaris* (1), *Bromus mollis* (120), *Collomia linearis* (263), *Poa annua* (259)—all annual or short-lived plants] and shows ecotypic differentiation. The genetics of plasticity is in need of much research. Within plants there is clearly a 'hierarchy of plasticities' (16); structures associated with reproduction (106), essential perennating organs, or vegetative spread (85) are among the least plastic. Genetic studies of plant shape (30) are still rudimentary.

In the analysis of plant plasticity, choice of which subunits to count may sometimes be straightforward: leaves or buds or (grass) tillers. But it may be very difficult to demarcate subunits that are realistic ecologically. I believe the choice must be pragmatic, depending on the purpose of the study and the nature of the plant. This review has outlined some of the choices that have been made. Oldeman has argued that one should ideally count meristems (100) and has even attempted to map their distribution in the forest canopy. Meristems may even show an age structure (256); they certainly have life expectancies (82, 83). The cell itself is the ultimate subunit [and the effects of density on root cell numbers in sugar beet have been studied (158)]. With Braun (21), who wrote the last review of this subject 126 years ago, we ask "how can one steer a middle course between indefinite subdivision and indefinite expansion . . . . [W]hich member of the series. . . . deserves pre-eminently the title of individual"? Morphological idealism has almost been displaced (209), though not quite (e.g. 157), and so it seems foolish now to suggest which member of the series deserves the title.

No matter how the plant is subdivided for pragmatic purposes, physiologically united meristems are not independent. Witchbrooms *are* abnormalities. The meristems are integrated into a galaxy of possible but not random morphologies. The French School [partially reviewed by Maresquelle (148)] has been preeminent in its advocacy of plant integrity (33, 34, 177–179) despite the qualitative and quantitative variety of meristem expression. 'Meristem dependence' (239) is as much a feature of plants as their plasticity.

A rapprochement between plant morphology and plant demography is desirable and necessary. Erasmus Darwin's grandson cannot be gainsaid in his belief that "morphology [is] the most interesting department of natural history and may be said to be its very soul" (48).

Acknowledgments

I am grateful to Adrian Bell and Barry Tomlinson for discussions on this topic over the past few years, for their constructive criticisms, for making manuscripts available in advance of publication, and for freely sharing ideas. As plant morphologists they have made clear to me the significance of morphology in plant demography and ecology. I thank John Harper and T. R. E. Southwood for information on some topics.

ideas. As plant morphologists they have made clear to me the significance of morphology in plant demography and ecology. I thank John Harper and T. R. E. Southwood for information on some topics.

APPENDIX

Many terms or phrases have been employed to name the units (often conceptual rather than tangible) recognized in plants. Indeed there has been lively controversy (6, 65, 209) about such apparently straightforward categories as root, stem, and leaf. As a contribution to a standardized vocabulary the following usages and definitions are suggested.

Genet
: the plant, of whatever size or however subdivided or propaged, derived from a single seed. An equivalent term (common in forestry) is ortet (e.g. 8, 51, 172).

Shoot
: an axis with associated appendages (leaves generally) and lateral meristems. (Colloquial usage)

Module
: a monopodial shoot terminated by an inflorescence, by a spine or tendril, or by parenchymatization of the apical meristem (100, 190); may also be referred to as a sympodial unit or a "relay axis" (100).

Caulomer
: a segment of a module showing morphologically distinct growth increments; a "unit of extension" (100).

Shoot complex
: a sequence of modules linked sympodially, together forming a coherent structure. The term is usefully applied to plants that contain distinct shoot complexes (commonly erect and lateral complexes). If the modules form a linear sequence the axis (e.g. tree trunk) may be referred to as being composed of "relay axes" (100).

Metamer
: a generic term (e.g. 179, 220) for a unitary part of a shoot, defined morphologically and/or anatomically. One of the several segments ('metameric segments'— usually of animals) that *share* in the construction of a shoot, or into which a shoot may be conceptually (at least) resolved. More concrete examples are caulomers, phytons, or in some plants modules.

Phyton
: a particular kind of metamer, defined as a node (or part of a node) and its attached single leaf, together with the internode (or part of an internode) just below it. An equivalent term is phytomer, commonly used of grasses.

Ramet	a single module or shoot complex of a genet that is conveniently enumerated. May be attached to the genet or become detached and independent. (Colloquial usage, not strictly defined morphologically)
Clone	a collection of plants derived by vegetative propagation (natural or artificial) from a single genet. Although they may remain attached to the 'parental' genet they are usually physiologically independent and commonly there is complete separation. Formerly referred to as a clon (230).
Architectural model	the normal growth program that determines the successive architectural phases of a plant derived from a seed, undisturbed by unusual or severe extrinsic forces (such as pruning, defoliation, injury), and excluding reiterations. The visible morphological expression of the genotype at any one time is the architecture —a static concept distinguished from the dynamic concept of the architectural model (99, 100).
Reiteration	recapitulation or rejuvenescence of the whole or part of the architectural model within the same genet, as a result of improved environmental conditions or damage to the initial model. It is usually initiated from dormant meristems (100).

Literature Cited

1. Abbott, R. J. 1976. Variation within common groundsel, *Senecio vulgaris* L. 1. Genetic response to spatial variations in the environment. *New Phytol.* 76:153–64
2. Addicott, F. T. 1978. Abscission strategies in the behavior of tropical trees. In *Tropical Trees as Living Systems,* ed. P. B. Tomlinson, M. H. Zimmermann, pp. 381–98. Cambridge: Cambridge Univ. Press
3. Allen, D. C. 1976. Methods for determining the number of leaf-clusters on sugar maple. *For. Sci.* 22:412–16
4. Arber, A. 1930. Root and shoot in the angiosperms: a study of morphological categories. *New Phytol.* 29:297–315
5. Arber, A. 1946. Goethe's botany. *Chron. Bot.* 10(2):63–126
6. Arber, A. 1950. *The Natural Philosophy of Plant Form.* Cambridge: Cambridge Univ. Press. 247 pp.
7. Barker, S. B., Cumming, G., Horsfield, K. 1973. Quantitative morphometry of the branching structure of trees. *J. Theor. Biol.* 40:33–43
8. Barnes, B. V. 1966. The clonal growth habit of American aspens. *Ecology* 47:439–47
9. Bazzaz, F. A., Harper, J. L. 1977. Demographic analysis of the growth of *Linum usitatissimum. New Phytol.* 78:193–208
10. Bell, A. D. 1974. Rhizome organization in relation to vegetative spread in *Medeola virginiana. J. Arnold Arbor. Harv. Univ.* 55:458–68
11. Bell, A. D. 1976. Computerized vegetative mobility in rhizomatous plants. In *Automata, Languages, Development,* ed. A. Lindenmayer, G. Rozenberg, pp. 3–14. Amsterdam: North Holland
12. Bell, A. D. 1979. The hexagonal branching pattern of *Alpinia speciosa* L. (Zingiberaceae). *Ann. Bot.* 43:209–23
13. Bell, A. D., Tomlinson, P. B. 1979. Adaptive architecture in rhizomatous plants. *Bot. J. Linn. Soc.* In press

14. Bezdeleva, T. A. 1975. The morphogenesis and life form of *Corydalis buschii* and the rhythm of its seasonal development. *Biull. Mosk. Obsch. Isp. Prir. Otd. Biol.* 80(2):56–67 (In Russian)

15. Blais, J. R. 1952. The relationship of the spruce budworm (*Choristoneura fumiferana* Clem.) to the flowering condition of balsam fir [*Abies balsamea* (L.) Mill.] *Can. J. Zool.* 30:1–29

16. Bonaparte, E. E. N. A., Brawn, R. I. 1975. The effect of intraspecific competition on the phenotypic plasticity of morphological and agronomic characters of four maize hybrids. *Ann. Bot.* 39:863–69

17. Bradley, R. 1721. *A Philosophical Account of the Works of Nature.* London: W. Mears

18. Bradshaw, A. D. 1965. Evolutionary significance of phenotypic plasticity in plants. *Adv. Genet.* 13:115–55

19. Bradshaw, A. D. 1972. Some of the evolutionary consequences of being a plant. *Evol. Biol.* 5:25–47

20. Braun, A. 1851. Betrachtungen über die Erscheinung der Verjüngung in der Natur. Leipzig. Transl. "Reflections on the phenomenon of rejuvenescence in nature especially in the life and development of plants." In *Botanical & Physiological Memoirs*, ed. A. Henfrey, pp. 1–341. London: The Ray Society

21. Braun, A. 1853. Das Individuum der Pflanze in seinem Verhältniss zur species—Generationsfolge Generationswechsel, und Generationstheilung der Pflanze. *Abh. Kgl. Preuss. Akad. Wiss. Berlin*, 1853:19–122. Transl. "The vegetable individual in its relation to species." *Am. J. Sci. Arts* 19:297–318 (1855); 20:181–201 (1855); 21:58–79 (1856). Also *Ann. Nat. Hist.* 16:233–56, 333–54 (1855); 18:363–86 (1857)

22. Bray, J. R., Gorham, E. 1964. Litter production in forests of the world. *Adv. Ecol. Res.* 2:101–57

23. Brooks, F. T., Chipp, T. F. 1931. *5th Int. Bot. Congr. Cambridge, 16-23 August 1930. Rep. Proc.* Cambridge: Cambridge Univ. Press. 680 pp.

24. Brougham, R. W. 1958. Leaf development in swards of white clover (*Trifolium repens* L.). *N.Z.J. Agric. Res.* 1:707–18

25. Bryson, R. A., Irving, W. N., Larsen, J. A. 1965. Radiocarbon and soil evidence of former forest in the southern Canadian tundra. *Science* 147:46–48

26. Buchanan, T. S. 1936. An alinement chart for estimating number of needles on western white pine reproduction. *J. For.* 34:588–93

27. Burger, H. 1929–1953. Holz, Blattmenge und Zuwachs. I–XIII. *Mitt. Schweiz. Anst. forstl. Versuchswes.* 15:243–92; 19:21–72; 20:101–14; 21:307–48; 22:10–62, 377–445; 24:7–103; 25:211–79, 435–93; 26:419–68; 27:247–86; 28:109–56; 29:38–130

28. Burns, W. 1946. Corm and bulb formation with special reference to the Gramineae. *Trans. Bot. Soc. Edinburgh* 34:316–47

29. Büsgen, M., Münch, E. 1929. *The Structure and Life of Forest Trees.* London: Chapman & Hall. 436 pp.

30. Caligari, P. D. S., Hanks, M. J. 1978. Genetical analysis of components of overall plant shape. *Theor. Appl. Genet.* 52:65–72

31. Cannell, M. G. R. 1974. Production of branches and foliage by young trees of *Pinus contorta* and *Picea sitchensis:* provenance differences and their simulation. *J. Appl. Ecol.* 11:1091–115

32. Celakowsky, L. J. 1901. Die Gliederung der Kaulome. *Bot. Z.* 59:79–114

33. Champagnat, P. 1974. Introduction a l'étude des complexes de corrélations. *Rev. Cytol. Biol. Vég.* 37:175–208

34. Champagnat, P., Barnola, P., Lavarenne, S. 1971. Premières recherches sur le détermisme de l'acrotonie des végétaux ligneux. *Ann. Sci. For.* 28:5–22

35. Chancellor, R. J. 1974. The development of dominance among shoots arising from fragments of *Agropyron repens. Weed Res.* 14:29–38

36. Chandler, R. F. Jr. 1969. Plant morphology and stand geometry in relation to nitrogen. In *Physiological Aspects of Crop Yield*, ed. J. D. Eastin, F. A. Haskins, C. Y. Sullivan, C. H. M. Van Bavel, pp. 265–85. Madison, Wis: Am. Soc. Agron.

37. Charlton, W. A. 1966. The root system of *Linaria vulgaris* Mill. I. Morphology and anatomy. *Can. J. Bot.* 44:1111–16

38. Chauveaud, G. 1921. *La Constitution des Plantes Vasculaires relévée par leur Ontogénie.* Paris: Payot. 155 pp.

39. Combe, J. C., du Plessix, C. J. 1974. Étude du développement morphologique de la couronne de *Hevea brasiliensis. Ann. Sci. For.* 31:207–28

40. Cooper, W. S. 1923. The recent ecological history of Glacier Bay, Alaska. II. The present vegatation cycle. *Ecology* 4:223–46

41. Cottam, W. P. 1954. Prevernal leafing

of aspen in Utah Mountains. *J. Arnold Arbor. Harv. Univ.* 35:239–48

42. Cremers, G. 1973, 1974. Architecture de quelques lianes d'Afrique Tropicale. *Candollea* 28:249–80; 29:57–110
43. Cruden, R. W., Hermann-Parker, S. M. 1977. Temporal dioecism: an alternative to dioecism. *Evolution* 31:863–66
44. Cuénod, A. 1951. Du rôle de la feuille dans l'édification de la tige. *Bull. Soc. Sci. Nat. Tunis* 4:3–15
45. Cummings, W. H. 1941. A method for sampling the foliage of a silver maple tree. *J. For.* 39:382–84
46. Cutler, B. L. 1976. *Instar determination and foliage consumption of the walnut caterpiller on pecan in Texas.* MS thesis. Texas A. & M. Univ.
47. Darwin, C. 1839. *Journal of Researches into the Geology and Natural History of the Various Countries Visited by H.M.S. Beagle, under the Command of Capt. Fitzroy, R.N., from 1832–36.* London: Colburn. 520 pp.
48. Darwin, C. 1859. *On the Origin of Species by Means of Natural Selection.* London: John Murray. 502 pp.
49. Darwin, C. 1879. *The Life of Erasmus Darwin (with an Essay on his Scientific Works by Ernst Krause).* London: Murray. (Reprint 1971, Farnborough, Hants: Gregg International)
50. Darwin, E. 1800. *Phytologia; or the Philosophy of Agriculture and Gardening.* London: J. Johnson. 612 pp.
51. de Byle, N. V. 1964. Detection of functional intraclonal aspen root connections by tracers and excavation. *For. Sci.* 10:386–96
52. de Candolle, A.-P. 1813. *Théorie Élémentaire de la Botanique.* Paris. 500 pp.
53. Deevey, E. S. 1947. Life tables for populations of animals. *Q. Rev. Biol.* 22:283–314
54. Delpino, F. 1883. Teoria generale della Fillotassi. *Atti R. Univ. Geneva* 4(2), 345 pp.
55. de Reffye, P., Snoeck, J. 1976. Modèle mathématique de base pour l'étude et la simulation de la croissance et de l'architecture du *Coffea robusta. Café, Cacao, Thé* 20:11–32
56. Dixon, A. F. G. 1971. The role of aphids in wood formation. 1. *J. Appl. Écol.* 8:165–79
57. Donald, C. M. 1968. The breeding of crop ideotypes. *Euphytica* 17:385–403
58. Dormer, K. J. 1972. *Shoot Organization in Vascular Plants.* London: Chapman & Hall. 240 pp.
59. Dorokhina, L. N. 1969. Life forms and evolutionary relationships in the sub-genus *Dracunculus*, genus *Artemisia. Biull. Mosk. Obsch. Isp. Prir. Otd. Biol.* 74(2): 77–89 (In Russian)
60. du Rietz, G. E. 1931. Life-forms of terrestrial flowering plants. I. *Acta Phytogeogr. Suec.* 3:1–95
61. Eames, A. J. 1961. *Morphology of the Angiosperms.* New York: McGraw-Hill. 518 pp.
62. Edelin, C. 1977. *Images de l'Architecture des Conifères.* Thèse, Université des Sciences et Techniques du Languedoc, Montpellier
63. Egerton, F. N. 1977. A bibliographic guide to the history of general ecology and population ecology. *Hist. Sci.* 15:189–215
64. Elwes, H. J., Henry, A. 1906–1913. *The Trees of Great Britain and Ireland.* Edinburgh: R & R Clark. 7 vols. 2022 pp.
65. Emberger, L. 1952. Tige, racine, feuille. *Ann. Biol.* 28:C109–28
66. Embree, D. G. 1965. The population dynamics of the winter moth in Nova Scotia, 1954–1962. *Mem. Entomol. Soc. Can.* 46:57
67. Esau, K. 1965. *Plant Anatomy.* New York: Wiley. 767 pp. 2nd ed.
68. Etter, A. G. 1951. How Kentucky bluegrass grows. *Ann. Mo. Bot. Gard.* 38:293–375
69. Evans, G. C. 1972. *The Quantitative Analysis of Plant Growth.* Oxford: Blackwell. 734 pp.
70. Evans, M. W. 1958. Growth and development in certain economic grasses. *Ohio Agric. Exp. Stn. Agron. Ser.* 147:1–123
71. Everitt, B. L. 1968. Use of the cottonwood in an investigation of the recent history of a flood plain. *Am. J. Sci.* 266:417–39
72. Fail, H. 1956. The effect of rotary cultivation on the rhizomatous weeds. *J. Agric. Eng. Res.* 1:68–80
73. Fekete, G., Szujko-Lacza, J. 1970, 1971. A survey of the plant life-form systems and the respective research approaches. II, III. *Ann. Hist.–Natur. Mus. Nat. Hung., Pars Bot.* 62:115–27; 63:37–50
74. Fernandez, P. 1978. *Patrones de ciclos de vida de poblaciones silvestres de* Phaseolus coccineus *L. en un gradiente altitudinal.* PhD thesis. UNAM, Mexico
75. Finch-Savage, W. E., Elston, J. 1977. The death of leaves in crops of field beans. *Ann. Appl. Biol.* 85:463–65
76. Fisher, J. B. 1973. The monocotyledons: their evolution and comparative

biology. II. Control of growth and development in the monocotyledons—new areas of experimental research. *Q. Rev. Biol.* 48:291–98

77. Fisher, J. B., French, J. C. 1976. The occurrence of intercalary and uninterrupted meristems in the internodes of tropical monocotyledons. *Am. J. Bot.* 63:510–25

78. Fisher, J. B., Honda, H. 1977. Computer simulation of branching pattern and geometry in *Terminalia* (Combretaceae), a tropical tree. *Bot. Gaz.* 138:377–84

79. Fraser, D. A., Belanger, L., McGuire, D., Zdrazil, Z. 1964. Total growth of the aerial part of a white spruce tree at Chalk River, Ontario, Canada. *Can. J. Bot.* 42:159–79

80. Froebe, H. A. 1971. Die Wissenschaftstheoretische Stellung der Typologie. *Ber. Dtsch. Bot. Ges.* 84:119–29

81. Gaudichaud, C. 1841. Recherches generales sur l'organographie, la physiologie et l'organogenie des vegetaux. *C. R. Acad. Sci. Paris* 12:627–37

82. Gill, A. M. 1971. The formation, growth and fate of buds of *Fraxinus americana* L. in central Massachusetts. *Harvard For. Pap. 20.* 16 pp.

83. Gill, A. M., Tomlinson, P. B. 1971. Studies on the growth of the red mangrove (*Rhizophora mangle* L.).3. Phenology of the shoot. *Biotropica* 3:109–24

84. Ginzburg, C. 1963. Some anatomic features of splitting of desert shrubs. *Phytomorphology* 13:92–97

85. Ginzo, H. D., Lovell, P. H. 1973. Aspects of the comparative physiology of *Ranunculus bulbosus* L. and *Ranunculus repens* L. 1. Response to nitrogen. *Ann. Bot.* 37:753–64

86. Godfrey-Sam-Aggrey, W. 1978. Effects of plant population on sole-crop cassava in Sierra Leone. *Exp. Agric.* 14:239–44

87. Deleted in proof

88. Gottlieb, L. D. 1977. Genotypic similarity of large and small individuals in a natural population of the annual plant *Stephanomeria exigua* ssp. *coronaria* (Compositae). *J. Ecol.* 65:127–34

89. Graham, B. F., Bormann, F. H. 1966. Natural root grafts. *Bot. Rev.* 32:255–92

90. Gray, A. 1849. On the composition of the plant by phytons, and some applications of phyllotaxis. *Proc. Am. Assoc. Adv. Sci.* 1849:438–44

91. Greenwood, E. A. N., Farrington, P., Beresford, J. D. 1975. Characteristics of the canopy, root system and grain yield

of a crop of *Lupinus angustifolius* cv. Unicrop. *Aust. J. Agric. Res.* 26:497–510

92. Grier, C. C., Waring, R. H. 1974. Conifer foliage mass related to sapwood area. *For. Sci.* 20:205–6

93. Griffiths, A. M., Malins, M. E. 1930. The unit of shoot growth in dicotyledons. *Proc. Leeds Philos. Soc.* 2:125–39

94. Griggs, R. F. 1938. Timberlines in the northern Rocky Mountains. *Ecology* 19:518–64

95. Guédès, M. 1969. La théorie de la métamorphose en morphologie végétale: Des origines à Goethe et Batsch. *Rev. Hist. Sci. Appl.* 22:323–63

96. Guédès, M. 1972. La théorie de la métamorphose en morphologie végétale. A.-P. de Candolle et P.-J.-F. Turpin, *Rev. Hist. Sci. Appl.* 25:253–70

97. Guédès, M. 1973. La théorie de la métamorphose en morphologie végétale. La métamorphose et l'idée d'évolution chez Alexandre Braun. *Episteme* 7:32–51

98. Hagihara, A., Suzuki, M., Hozumi, K. 1978. Seasonal fluctuations of litter fall in a *Chamaecyparis obtusa* plantation. *J. Jpn. For. Soc.* 60:397–404

99. Hallé, F., Oldeman, R. A. A. 1970. *Essai sur l'Architecture et la Dynamique de Croissance des Arbres Tropicaux.* Paris: Masson. 178 pp.

100. Hallé, F., Oldeman, R. A. A., Tomlinson, P. B. 1978. *Tropical Trees & Forests: an Architectural Analysis.* Berlin: Springer. 441 pp.

101. Harper, J. L. 1958. The ecology of ragwort (*Senecio jacobaea*) with especial reference to its control. *Herb. Abstr.* 28:151–57

102. Harper, J. L. 1974. A centenary in population biology. *Nature* 252:526–27

103. Harper, J. L. 1977. *Population Biology of Plants.* London: Academic. 892 pp.

104. Harper, J. L. 1978. The demography of plants with clonal growth. *Verh. Kon. Nederland. Akad. Weten., Afd. Nat.* 70:27–48

105. Harper, J. L., Bell, A. D. 1979. The population dynamics of growth form in organisms with modular construction. In '*Population Dynamics*', 20th Symp. *Brit. Ecol. Soc.*, ed. R. M. Anderson, B. D. Turner, L. R. Taylor. Oxford: Blackwell Sci. Publ. In press

106. Harper, J. L. Lovell, P. H., Moore, K. G. 1970. The shapes and sizes of seeds. *Ann. Rev. Ecol. Syst.* 1:327–56

107. Harper, J. L., White, J. 1971. The dynamics of plant populations. In *Proceedings of the Advanced Study Institute*

on 'Dynamics of Numbers in Populations', ed. P. J. den Boer, G. R. Gradwell, pp. 41–63. Wageningen: Cent. Agric. Publ. Doc.

108. Harper, J. L., White, J. 1974. The demography of plants. *Ann. Rev. Ecol. Syst.* 5:419–63

109. Harris, W. 1970. Competition effects on yield and plant and tiller density in mixtures of ryegrass cultivars. *Proc. N.Z. Ecol. Soc.* 17:10–17

110. Hinson, K., Hanson, W. D. 1962. Competition studies in soybeans. *Crop Sci.* 2:117–23

111. Hiroi, T., Monsi, M. 1966. Dry-matter economy of *Helianthus annuus* communities grown at varying densities and light intensities. *J. Fac. Sci. Univ. Tokyo* III, 9:241–85

112. Hodánová, D. 1967. Development and structure of foliage in wheat stands of different density. *Biol. Plant. (Praha)* 9:424–38

113. Holliday, R. 1960. Plant population and crop yield: Pt. II. *Field Crop Abstr.* 13:247–54

114. Holm, T. 1925. Hibernation and rejuvenation exemplified by North American herbs. *Am. Midl. Nat.* 9:439–512

115. Holttum, R. E. 1955. Growth-habits of monocotyledons—variations on a theme. *Phytomorphology* 5:399–413

116. Horowitz, M. 1973. Spatial growth of *Sorghum halepense*. *Weed Res.* 13:200–8

117. Hort, A. 1916. *Theophrastus: Enquiry into Plants*. (English transl.). London: Heinemann. 2 vols.: 475, 499 pp.

118. Hutchings, M. J., Barkham, J. P. 1976. An investigation of shoot interactions in *Mercurialis perennis* L., a rhizomatous herb. *J. Ecol.* 64:723–43

119. Hyder, D. N. 1972. Defoliation in relation to vegetative growth. In *The Biology and Utilization of Grasses*, ed. V. B. Youngner, C. M. McKell, pp. 304–17. NY: Academic

120. Jain, S. K. 1978. Inheritance of phenotypic plasticity in soft chess *Bromus mollis* L. (Gramineae). *Experientia* 34:835–36

121. Jeannoda-Robinson, V. 1977. *Contribution à l'étude de l'architecture des Herbes*. Thèse, Université des Sciences et Techniques du Languedoc, Montpellier.

122. Jewiss, O. R. 1972. Tillering in grasses —its significance and control. *J. Br. Grassl. Soc.* 27:65–82

123. Jong, K., Burtt, B. L. 1975. The evolution of morphological novelty exemplified in the growth patterns of some Gesneriaceae. *New Phytol.* 75:297–311

124. Kaewla-iad, T., Yarwudhi, C. 1975. The estimation of the total amount of teak foliage. *Fac. For. Kasetsart Univ. Bangkok, Res. Note 12.* 10 pp.

125. Kawano, S. 1970. Species problems viewed from productive and reproductive biology. 1. Ecological life histories of some representative members associated with temperate deciduous forests in Japan. *J. Coll. Lib. Arts, Toyama Univ.* 3:1181–213

126. Kawano, S. 1975. The productive and reproductive biology of flowering plants. II. The concept of life history strategy in plants. *J. Coll. Lib. Arts, Toyama Univ.* 8:51–86

127. Kawano, S., Nagai, Y. 1975. The productive and reproductive biology of flowering plants. I. Life history strategies of three *Allium* species in Japan. *Bot. Mag. Tokyo* 88:281–318

128. Kays, S., Harper, J. L. 1974. The regulation of plant and tiller density in a grass sward. *J. Ecol.* 62:97–105

128a. Kemperman, J. A., Barnes, B. V. 1976. Clone size in American aspens. *Can. J. Bot.* 54:2603–7

129. Kerner von Marilaun, A. 1904. *The Natural History of Plants*. (English transl. F. W. Oliver). London: Gresham. Vol. 2, 983 pp.

130. Kershaw, K. A. 1973. *Quantitative and Dynamic Plant Ecology*. London: Edward Arnold. 308 pp.

131. Kimura, M., Mototani, I., Hogetsu, K. 1968. Ecological and physiological studies on the vegetation of Mt. Shimagare. VI. Growth and dry matter production of young *Abies* stand. *Bot. Mag. Tokyo* 81:287–96

132. Kinerson, R. S., Higginbotham, K. O., Chapman, R. C. 1974. The dynamics of foliage distribution within a forest canopy. *J. Appl. Ecol.* 11:347–53

133. Kirby, E. J. M., Faris, D. G. 1970. Plant population induced growth correlations in the barley plant main shoot and possible hormonal mechanisms. *J. Exp. Bot.* 21:787–98

134. Kirby, E. J. M., Jones, H. G. 1977. The relations between the main shoot and tillers in barley plants. *J. Agric. Sci.* 88:381–89

135. Kittredge, J. 1944. Estimation of the amount of foliage of trees and stands. *J. For.* 42:905–12

136. Kobayashi, S. 1975. Growth analysis of plant as an assemblage of internodal segments—a case of sunflower plants in pure stands. *Jpn. J. Ecol.* 25:61–70

137. Kozlowski, T. T. 1973. Extent and significance of shedding of plant parts. In *Shedding of Plant Parts*, ed. T. T. Kozlowski, pp. 1–44. NY: Academic

138. Lamoreaux, R. J., Chaney, W. R., Brown, K. M. 1978. The plastochron index: a review after two decades of use. *Am. J. Bot.* 65:586–93

139. Langer, R. H. M. 1963. Tillering in herbage grasses. *Herb. Abst.* 33:141–48

140. Le Roux, E. J., Reimer, C. 1959. Variation between samples of immature stages and of mortalities from some factors of the eyespotted bud moth *Spilonota ocellana* and the pistol casebearer *Coleophora serratella* on apple in Quebec. *Can. Entomol.* 91:428–49

141. Linnaeus, C. 1743–1776. *Amoenitates Academiae. Seu dissertationes variae physicae, medicae, botanicae*.... Uppsala. 19 vols.

142. Linzon, S. N. 1958. The effect of artificial defoliation of various leaves upon white pine growth. *For. Chron.* 34: 50–56

143. Lord, F. T. 1968. An appraisal of methods of sampling apple trees and results of some tests using a sampling unit common to insect predators and their prey. *Can. Entomol.* 100:23–33

144. MacDougal, D. T. 1936. Studies in treegrowth by the dendrographic method. *Carnegie Inst. Wash. Publ. 462*

144a. Maconochie, J. R. 1975. Shoot and foliage production of five shrub species of *Acacia* and *Hakea* in a dry sclerophyll forest. *Trans. R. Soc. S. Aust.* 99(4): 177–81.

145. Madison, J. H. 1970. An appreciation of monocotyledons. *Notes R. Bot. Gard. Edinburgh* 30:377–90

146. Majumdar, G. P. 1947. Growth unit or the phyton in dicotyledons with special reference to Heracleum. *Bull. Bot. Soc. Bengal* 1:61–66

147. Majumdar, G. P. 1957. The shoot of higher plants: its morphology and phylogeny: a review. *J. Asiatic Soc. Lett. Sci.* 23:39–62

148. Maresquelle, H. J. 1977. Architektonische Morphologie der höheren Pflanzen als Ausdruck eines Korrelationsnetzes. *Ber. Dtsch. Bot. Ges.* 90:309–23

149. Marks, P. L. 1975. On the relation between extension growth and successional status of deciduous trees of the northeastern United States. *Bull. Torrey Bot. Club* 102:172–77

150. Marr, J. W. 1977. The development and movement of tree islands near the upper limit of tree growth in the southern Rocky Mountains. *Ecology* 58:1159–64

151. Marshall, C., Sagar, G. R. 1968. The interdependence of tillers in *Lolium multiflorum* Lam.—a quantitative assessment. *J. Exp. Bot.* 19:785–94

152. Matheis, P. J., Tieszen, L. L., Lewis, M. C. 1976. Responses of *Dupontia fischeri* to simulated lemming grazing in the Alaskan Arctic tundra. *Ann. Bot.* 40:179–97

153. Mayr, E. 1971. The nature of the Darwinian revolution. *Science* 176:981–89

154. McIntyre, G. I. 1970. Studies on bud development in the rhizome of *Agropyron repens*. I. The influence of temperature, light intensity and bud position on the pattern of development. *Can. J. Bot.* 48:1903–9

155. Menninger, E. A. 1967. *Fantastic Trees*. NY: Viking. 304 pp.

156. Meusel, H. 1952. Über Wuchsformen, Verbreitung und Phylogenie einiger mediterran–mittel europäischen Angiospermen-Gattungen. *Flora* 139:333–93

157. Meyen, S. V. 1973. Plant morphology in its nomothetical aspects. *Bot. Rev.* 39:205–60

158. Milford, G. F. J. 1976. Sugar concentration in sugar beet: varietal differences and the effects of soil type and planting density on the size of the root cells. *Ann. Appl. Biol.* 83:251–57

159. Miller, C. A. 1977. The feeding impact of spruce budworm on balsam fir. *Can. J. For. Res.* 7:76–84

160. Mitchell, R. G. 1974. Estimation of needle populations on young, open-grown Douglas fir by regression and life table analysis. *USDA For. Serv. Res. Pap. PNW-181*. 14 pp.

161. Miyaji, K.-I., Tagawa, H. 1973. A life-table of the leaves of *Tilia japonica*. *Rep. Ebino Biol. Lab. Kyushu Univ.* 1:98–108

162. Möbius, M. 1940. *Die vegetative Vermehrung der Pflanzen*. Jena: Gustav Fischer. 82 pp.

163. Molisch, H. 1938. *The Longevity of Plants*. NY: E. H. Fulling. 226 pp.

164. Morris, R. F. 1951. The effects of flowering on the foliage production and growth of balsam fir. *For. Chron.* 27:40–57

165. Moss, H. H., Gorham, A. L. 1953. Interxylary cork and fission of stems and roots. *Phytomorphology* 3:285–94

166. Mullette, K. J. 1978. Studies on the lignotubers of *Eucalyptus gummifera*. 1. The nature of the lignotuber. *Aust. J. Bot.* 26:9–13

167. Münch, E. 1938. Untersuchungen über die Harmonie der Baumgestalt. *Jahrb. Wiss. Bot.* 86:581–673

168. Munro, J. M., Farbrother, H. G. 1967. Composite plant diagrams in cotton. *Cotton Grow. Rev.* 46:261–82

169. Musgrave, D. J., Langer, R. H. M. 1977. Crown development of two diverse genotypes of lucerne. *N. Z. J. Agric. Res.* 20:453–58

170. Nägeli, C. 1856. Die Individualität in der Natur mit Vorzüglicher Berücksichtung des Pflanzenreiches. *Monatsshr. Wiss. Ver. Zürich* 1(4/5):171–212

171. Nef, L. 1959. Etude d'une population de larves de *Retinia buoliana* (Schiff.). *Z. Angew. Entomol.* 44:167–86

172. Nicholls, J. W. P., Pawsey, C. K., Brown, A. G. 1976. Further studies on the ortet-ramet relationship in wood characteristics of *Pinus radiata*. *Silvae Genet.* 25:73–79

173. Nielsen, B. O. 1978. Food resource partition in the beech leaf-feeding guild. *Ecol. Entomol.* 3:193–201

174. Nielsen, B. O., Ejlersen, A. 1977. The distribution pattern of herbivory in a beech canopy. *Ecol. Entomol.* 2:293–99

175. Noble, J. C. 1976. *The population biology of rhizomatous plants.* PhD thesis. Univ. Wales, Bangor

176. Noble, J. C., Bell, A. D., Harper, J. L. 1979. The structural demography of rhizomatous plants: *Carex arenaria* L. 1. The morphology and flux of modular growth units. *J. Ecol.* In press

177. Nozeran, R. 1978. Reflexions sur les enchaînements de fonctionnements au cours du cycle des végétaux superieurs. *Bull. Soc. Bot. Fr.* 125:263–80

178. Nozeran, R. 1978. Polymorphism des individus issus de la multiplication végétative des végétaux superieurs avec conservation du potentiel génétique. *Physiol. Vég.* 16:177–94

179. Nozeran, R., Bancilhon, L., Neville, P. 1971. Intervention of internal correlations in the morphogenesis of higher plants. *Adv. Morphol.* 9:1–66

180. Nyahoza, F., Marshall, C., Sagar, G. R. 1973. The interrelationships between tillers and rhizomes of *Poa pratensis* L. —an autoradiographic study. *Weed Res.* 13:30-9

181. Oldeman, R. A. A. 1974. L'architecture de la forêt guyanaise. *Mém. Off. R. Sci. Tech. Outre-Mer* 73. 204 pp.

182. Ong, C. K. 1978. The physiology of tiller death in grasses. I. The influence of tiller age, size and position. *J. Br. Grassl. Soc.* 33:197–203

183. Oohata, S., Shidei, T. 1971. Studies on the branching structure of trees. I. Bifurcation ratio of trees in Horton's Law. *Jpn. J. Ecol.* 21:7–14

184. Paradis, R. O., Le Roux, E. J. 1962. A sampling technique for population and mortality factors of the fruit-tree roller *Archips argyrospilus* on apple in Quebec. *Can. Entomol.* 94:561–73

185. Perkins, D. F. 1968. Ecology of *Nardus stricta* L. 1. Annual growth in relation to tiller phenology. *J. Ecol.* 56:633–46

186. Peterson, R. L. 1975. The initiation and development of root buds. In *Development and Function of Roots,* ed. J. G. Torrey, D. Clarkson, pp.125–61. NY: Academic

187. Polyntseva, N. 1972. Developmental rhythms of *Carex macroura*. *Bot. Zhur.* 57:804–9 (In Russian)

188. Popovic, M. 1976. Growth of the mountain pine (*Pine mugo* Turr.) in Yugoslavia. *J. Biogeogr.* 3:261–67

189. Popper, K. R. 1957. *The Poverty of Historicism.* London: Routledge & Kegan Paul. 166 pp.

190. Prévost, M.-F. 1978. Modular construction and its distribution in tropical woody plants. In *Tropical Trees as Living Systems,* ed. P. B. Tomlinson, M. H. Zimmermann, pp. 223–31. Cambridge: Cambridge Univ. Press

191. Priestley, J. H., Scott, L. I. 1933. Phyllotaxis in the dicotyledon from the standpoint of developmental anatomy. *Biol. Rev.* 8:241-68

192. Priestley, J. H., Scott, L. I., Gillett, E. C. 1935. The development of the shoot in Alstroemeria and the unit of shoot growth in monocotyledons. *Ann. Bot.* 49:161–79

193. Priestley, J. H., Scott, L. I., Mattinson, K. M. 1937. Dicotyledon phyllotaxis from the standpoint of development. *Proc. Leeds Philos. Soc.* 3:380–88

194. Rabotnov, T. A. 1950. Life cycles of perennial herbage plants in meadow communities. *Proc. Komarov. Bot. Inst. Acad. Sci. USSR* Ser. 3(6): 7–240 (In Russian)

195. Rauh, W. 1950. *Morphologie der Nutzpflanzen.* Heidelberg: Quelle & Meyer. 285 pp.

196. Reader, R. J. 1978. Contribution of overwintering leaves to the growth of three broad-leaved, evergreen shrubs belonging to the Ericaceae family. *Can. J. Bot.* 56:1248–61

197. Rees, A. R. 1972. *The Growth of Bulbs.* London: Academic. 311 pp.

197a. Ridley, H. N. 1930. *The Dispersal of Plants Throughout the World.* Ashford, Kent: L. Reeve. 744 pp.

198. Ritterbusch, A. 1977. Homolog- und Analog-Modell einer spermatophyten und einer terrestren Pflanzen. *Ber. Dtsch. Bot. Ges.* 90:363–68

199. Rogan, P. G., Smith, D. L. 1976. Experimental control of bud inhibition in rhizomes of *Agropyron repens* (L.) Beauv. *Z. Pflanzenphysiol.* 78:113–21

200. Rook, D. A., Corson, M. J. 1978. Temperature and irradiance and the total daily photosynthetic production of the crown of a *Pinus radiata* tree. *Oecologia* 36:371–82

201. Rothacher, J. S., Blow, F. E., Potts, S. M. 1954. Estimating the quantity of tree foliage in oak stands in the Tennessee Valley. *J. For.* 52:169–73

201a. Roy, D. F. 1966. Silvical characteristics of redwood (*Sequoia sempervirens*), *U.S. For. Serv. Res. Pap. PSW-28.* 20 pp.

202. Ruhland, W. 1965. *Encyclopedia of Plant Physiology.* XV. *Differentiation and Development.* Berlin: Springer. 1647 pp.

203. Rumbaugh, M. D. 1970. A clonal competition experiment with alfalfa, *Medicago sativa* L. *Agron J.* 62:51–55

204. Sachs, J. 1875. *Textbook of Botany, Morphological and Physiological.* Oxford: Clarendon Press. 858 pp.

205. Sachs, J. 1890. *History of Botany (1530–1860).* (English transl. H. E. F. Garnsey, I. B. Balfour). Oxford: Clarendon. 568 pp.

206. Saito, H. 1977. Litterfall. In *Primary Productivity of Japanese Forests,* ed. T. Shidei, T. Kira, pp. 65–75; 93–95. Tokyo: Univ. Tokyo Press

207. Sarukhán, J., Harper, J. L. 1973. Studies on plant demography: *Ranunculus repens* L., *R. bulbosus* L., *R. acris* L. I. Population flux and survivorship. *J. Ecol.* 61:675–716

208. Satoo, T., Kunugi, R., Kumekawa, A. 1956. Materials for the studies of growth in stands. 3. Amount of leaves and production of wood in an aspen (*Populus Davidiana*) second growth in Hokkaido. *Bull. Tokyo Univ. For.* 52:33–51

209. Sattler, R. 1974. A new conception of the shoot of higher plants. *J. Theor. Biol.* 47:367–82

210. Savile, D. B. O. 1972. Arctic adaptations in plants. *Can. Dep. Agric. Monogr. 6.* 81 pp.

211. Schoute, J. C. 1931. On phytonism. *Rec. Trav. Bot. Neerl.* 28:82–96

212. Serebryakova, T. I. 1971. *Shoot Morphogenesis and Evolution of Life Forms in Grasses.* Moscow: Nauka (In Russian)

213. Serebryakova, T. I. 1971. Types of major life cycle and structure of aerial shoots in flowering plants. *Biull. Mosk. Obsch. Isp. Prir. Otd. Biol.* 76(1): 105–19 (In Russian)

214. Shimwell, D. W. 1971. *Description and Classification of Vegetation.* London: Sidgewick & Jackson. 322 pp.

215. Shinozaki, K., Yoda, K., Hozumi, K., Kira, T. 1964. A quantitative analysis of plant form—the pipe model theory. I. Basic analyses. *Jpn. J. Ecol.* 14:97–105

216. Silver, G. T. 1962. The distribution of Douglas fir foliage by age. *For. Chron.* 38:433–38

217. Singh, B. N., Lal, K. N. 1935. Investigations of the effect of age on assimilation of leaves. *Ann. Bot.* 49:291–307

218. Sinnott, E. W. 1921. The relation between body size and organ size in plants. *Am. Nat.* 55:385–403

219. Skutch, A. F. 1930. Repeated fission of stem and root in *Mertensia maritima*—a study in ecological anatomy. *Ann. NY Acad. Sci.* 32:1–52

220. Smirnova, E. S. 1970. Morphological classification of flowering plants according to vegetative characteristics. *Dokl. Akad. Nauk. SSSR.* 190:1243–45 (English transl.)

221. Smith, D. L., Rogan, P. G. 1975. Growth in the stem of *Agropyron repens* (L.) Beauv. *Ann. Bot.* 39:871–80

222. Smith, J. H. G. 1972. Persistence, size and weight of needles of Douglas fir and western hemlock branches. *Can. J. For. Res.* 2:173–78

223. Southwood, T. R. E., Reader, P. M. 1976. Population census data and key factor analysis for the viburnum whitefly, *Aleurotrachelus jelinekii* (Frauenf.) on three bushes. *J. Anim. Ecol.* 45:313–25

224. Spencer, H. 1867. *The Principles of Biology.* London: Williams & Norgate. 2 vols. 492, 566 pp.

225. Splinter, W. E., Beeman, J. F. 1968. The relationship between plant stem diameter and total leaf area for certain plants exhibiting apical dominance. *Tobacco Sci.* 12:139–43

226. Steeves, T. A., Sussex, I. M. 1972. *Patterns in Plant Development.* Englewood Cliffs, NJ: Prentice-Hall. 302 pp.

227. Stern, W. R. 1965. The effect of density on the performance of individual plants in subterranean clover swards. *Aust. J. Agric. Res.* 16:541–55

228. Stern, W. R., Donald, C. M. 1962.

Light relations in grass-clover swards. *Aust. J. Agric. Res.* 13:599–614

229. Stone, E. C., Vesey, R. B. 1968. Preservation of coastal redwood on alluvial flats. *Science* 159:157–61

230. Stout, A. B. 1929. The clon in plant life. *J. NY Bot. Gard.* 30:25–37

231. Studhalter, R. A. 1955. Tree growth. I. Some historical chapters. *Bot. Rev.* 21:1–72

232. Syme, J. R. 1974. Leaf appearance rate and associated characters in some Mexican and Australian wheats. *Aust. J. Agric. Res.* 25:1–7

233. Szujko–Lacza, J., Szucs, Z. 1975. The architecture and the quantitative investigation of some characteristics of anise, *Pimpinella anisum* L. *Acta Bot. Acad. Sci. Hung.* 21:443–50

234. Tadaki, Y., Kagawa, T. 1968. Studies on the production structure of forest. 13. Seasonal change in litter-fall in some evergreen stands. *J. Jpn. For. Soc.* 50:7–13

235. Thompson, J. N., Price, P. W. 1977. Plant plasticity, phenology and herbivore dispersion: wild parsnip and the parsnip webworm. *Ecology* 58: 1112–19

236. Thullen, R. J., Keeley, P. E. 1975. Yellow nutsedge sprouting and resprouting potential. *Weed Sci.* 23:333–37

237. Timmer, M. J. G., Van der Valk, G. G. M. 1973. Effects of planting density on the number and weight of tulip daughter bulbs. *Sci. Hort.* 1:193–200

238. Tomlinson, P. B. 1970. Monocotyledons—towards an understanding of their morphology and anatomy. *Adv. Bot. Res.* 3:207–92

239. Tomlinson, P. B. 1974. Vegetative morphology and meristem dependence— the foundation of productivity in seagrasses. *Aquaculture* 4:107–30

240. Tomlinson, P. B., Gill, A. M. 1973. Growth habits of tropical trees: some guiding principles. In *Tropical Forest Ecosystems in Africa & South America: a Comparative Review,* ed. B. J. Meggers, E. S. Ayensu, W. D. Duckworth, pp. 129–43. Washington, DC: Smithsonian Inst. Press

241. Tomlinson, P. B., Soderholm, P. 1975. *Corypha elata* in flower in South Florida. *Principes* 19:83–99

242. Tomlinson, P. B., Zimmermann, M. H. 1978. *Tropical Trees as Living Systems.* Cambridge: Cambridge Univ. Press. 675 pp.

243. Tripathi, R. S., Harper, J. L. 1973. The comparative biology of *Argropyron repens* and *A. caninum.* I. The growth of mixed populations established from tillers and seeds. *J. Ecol.* 61:353–68

244. Troll, W., Hohn, K. 1973. *Allgemeine Botanik.* Stuttgart: Ferdinand Enke. 994 pp.

245. Turrell, F. M. 1961. Growth of the photosynthetic area of citrus. *Bot. Gaz.* 122:285–98

246. Turrell, F. M., Garber, M. J., Jones, W. W., Cooper, W. C., Young, R. H. 1969. Growth equations and curves for citrus trees. *Hilgardia* 39:429–45

247. Uranov, A. A. 1967. *Ontogenesis and Age Composition of Populations of Flowering Plants.* Moscow: Nauka (In Russian). 155 pp.

248. Uranov, A. A. 1968. *Problems in the Morphogenesis of Flowering Plants and their Population Structure.* Moscow: Nauka (In Russian). 231 pp.

249. Valentine, H. T., Hilton, S. J. 1977. Sampling oak foliage by the randomized-branch method. *Can. J. For. Res.* 7:295–98

250. Van Andel, J. 1975. A study on the population dynamics of the perennial plant species *Chamaenerion angustifolium* (L.) Scop. *Oecologia* 19:329–37

251. Van der Pijl, L. 1952. Absciss-joints in the stems and leaves of tropical plants. *Proc. Kon. Ned. Akad. Wetensch.* 50:574–86

252. Van der Pijl, L. 1953. The shedding of leaves and branches of some tropical trees. *Indones. J. Nat. Sci.* 109:11-25

253. Van Valen, L. 1978. Arborescent animals and other colonoids. *Nature* 276:318

254. Vyvyan, M. C., Evans, H. 1932. The leaf relations of fruit trees. I. A morphological analysis of the distribution of leaf surface on two nine-year-old apple trees (Laxton Superb). *J. Pomol. Hort. Sci.* 10:228–70

255. Waisel, Y., Liphschitz, N., Kuller, Z. 1972. Patterns of water movement in trees and shrubs. *Ecology* 53:520–23

256. Wangerman, E. 1965. Longevity and ageing in plants and plant organs. *Encycl. Plant Physiol.* 15(2):1026–57

257. Ward, R. T. 1961. Some aspects of the regeneration habits of the American beech. *Ecology* 42:828–32

258. Waring, R. H., Gholz, H. L., Grier, C. C., Plummer, M. L. 1977. Evaluating stem conducting tissue as an estimator of leaf area in four woody angiosperms. *Can. J. Bot.* 55:1474–77

259. Wells, G. J. 1974. The germination and growth of *Poa annua* sown monthly in the field. *Proc. 12th Br. Weed Control Conf.,* pp. 525–32

260. White, F. 1976. The underground forests of Africa: a preliminary review. *Gard. Bull. Singapore* 29:57–71
261. White, J. 1977. Generalization of self-thinning of plant populations. *Nature* 268:373
262. White, J. 1979. Demographic factors in populations of plants. In *Demography and Dynamics of Plant Populations,* ed. O. T. Solbrig, pp.21–48. Oxford: Blackwells
263. Wilken, D. H. 1977. Local differentiation for phenotypic plasticity in the annual *Collomia linearis* (Polemoniaceae). *Syst. Bot.* 2:99–108
263a. Williams, G. C. 1975. *Sex and Evolution.* Princeton: Princeton University Press. 200 pp.
264. Williams, R. D. 1970. Tillering in grasses cut for conservation, with special reference to perennial ryegrass. *Herb. Abstr.* 40:383–88
265. Williams R. F. 1974. *The Shoot Apex and Leaf Growth. A Study in Quantitative Biology.* Cambridge: Cambridge Univ. Press. 256 pp.
266. Williamson, P. 1976. Above-ground primary production of chalk grassland allowing for leaf death. *J. Ecol.* 64:1059–75
267. Wilson, E. O. 1975. *Sociobiology: the New Synthesis.* Cambridge, Mass.: Harvard Univ. Press. 697 pp.
268. Woodman, J. N. 1971. Variation of net photosynthesis within the crown of a large forest-grown conifer. *Photosynthetica* 5:50–54
269. Worsdell, W. C. 1915–1916. *The Principles of Plant Teratology.* London: The Ray Society. 2 vols: 270, 296 pp.
270. Yuwaka, J., Yamauchi, S., Nagai, S., Tokuhisa, E. 1977. Leaf longevity and the defoliating process in saplings of *Actinodaphne longifolia. Jpn. J. Ecol.* 27:171–75
271. Zimmermann, M. H., Brown, C. L. 1971. *Trees—Structure and Function.* NY: Springer. 336 pp.
272. Zimmermann, M. H., Tomlinson, P. B. 1972. The vascular system of monocotyledonous stems. *Bot. Gaz.* 133:141–55
273. Zimmermann, M. H., Tomlinson, P. B. 1974. Vascular patterns in palm stems: variations of the Rhapis principle. *J. Arnold Arbor. Harv. Univ.* 55:402–24

Ann. Rev. Ecol. Syst. 1979. 10:147–72

FEEDING ECOLOGY
OF STREAM INVERTEBRATES

♦4158

Kenneth W. Cummins

Department of Fisheries and Wildlife, Oregon State University,
Corvallis, Oregon 97331

Michael J. Klug

Kellogg Biological Station, Michigan State University,
Hickory Corners, Michigan 49060

INTRODUCTION

A general conceptual model of small stream ecosystem structure and function has emerged in the last decade (e.g. 34, 56, 77), an important cornerstone of which is the geomorphic view (e.g. 18, 40) of the stream as a subsystem of its watershed (e.g. 46). The intimate relationship between the stream and its riparian zone forms the basis for a significant (often the dominant) portion of the annual energy input. Light, and secondarily nutrients and temperature, in large measure determine changes in the balance between heterotrophy, dependent on allochthonous (terrestrial) organic matter, and autotrophy, based on autochthonous (in-stream) primary production by periphytic algae and macrophytes (e.g. 71, 113). Changes in the proportional balance between terrestrially linked heterotrophy and channel based autotrophy constitute a dominant control of broad scale differences in community structure. In-stream gross primary production approaches a balance with, or exceeds, community respiration with increasing stream size [order (89)], at least through mid-sized rivers (e.g. 36), or in other light-rich watershed settings as at higher latitudes and altitudes and in arid regions (77). Within broad temperature-defined boundaries such trophic relationships (heterotrophy/autotrophy) may be well correlated with stream invertebrate community structure (e.g. 35, 36). Because the relative domi-

147

0066-4162/79/1120-0147$01.00

nance of invertebrate groups shifts with differences in available sources of energy, morpho-behavioral adaptations of food acquisition match well the general resource conditions in a given stream order (size).

Stream invertebrate functional feeding groups have been described according to adaptations for food acquisition rather than food eaten (e.g. 5–7, 33, 36, 66, 75, 108, 116). For exampie, the functional group designated scrapers (= grazers) depend primarily on autochthonous production—i.e. live plant tissue, usually epilithic algae. However, by scraping they also often ingest fine particulate organic matter (FPOM, < 1mm, Table 1) and animals (e.g. 31, 33, 35). The following discussion is organized around the concept of functional feeding groups. Because other reviews cover portions of our general topic (1, 5, 33, 66, 74, 75, 109, 116), we have restricted our literature coverage to the last decade and have concentrated on various aspects of invertebrate-microbial interactions.

FOOD RESOURCES OF STREAM INVERTEBRATES

Food resource categories (Table 1) intergrade and the distinctions are not always clear. The separations correspond generally with the morpho-behavioral adaptations of the invertebrate functional feeding groups discussed below and are approximately separable on microbial and/or biochemical bases. Distinctions between food sources are based on particle size (detrital categories), presence of chlorophyll (periphyton dominated by algae), and the high protein content of typical animal prey.

Detritus

Because of the importance of detritus (particulate organic matter, >0.5 μm) in stream ecosystems, major emphasis has been directed at compartmentalization of this category (Table 1). Although the distinction between the microbial biomass and nonliving portions of detritus is theoretically appealing it is operationally difficult (e.g. 6, 34, 114). For example, the efficient separation of microbes from surface-colonized FPOM and surface and matrix colonized CPOM (coarse particulate organic matter, >1mm, Table 1) by techniques as rigorous as sonification has proven to be extremely difficult (M. J. Klug and L. Louden, unpublished data). In stream environments, nonliving organic matter probably does not occur without associated microorganisms.

Two broad categories of detritus can be defined, CPOM and FPOM [Table 1; (21, 22, 85)]. CPOM has been further divided into wood and nonwoody material (leaves, needles, flowers, bud scales). A portion of the fines (FPOM) has been designated ultrafine particulate organic matter (UPOM, < 50 μm > 0.5 μm). By convention, organic matter smaller than

$0.5\,\mu$m is considered dissolved (DOM) (e.g.22). Woody debris is further divided into coarse ($>$10 cm diameter) and fine ($<$10 cm, approx. $>$1 mm) material (5, 6). The coarse fraction includes branchwood, boles, and root wads, with processing times in decades to centuries. Fine woody debris (twigs, bark, etc) is processed more rapidly—times probably measured in years.

A complex array of biological and abiological interactions involved in detrital processing fractionates detritus into a number of pools (Figures 1–3). CPOM entering stream systems is surface- and matrix-colonized by fungi (primarily aquatic hyphomycetes) and bacteria. Fungi are the initial colonizers of CPOM owing to their invasive enzymatic characteristics; their activity, that of bacteria, and animal shredding result in the continuous release of FPOM, UPOM, and DOM (91, 92). It is estimated (92) that less than 25% of CPOM is mineralized as CPOM; the remainder is transformed into other organic pools of detritus (92).

Bacteria dominate the colonization of FPOM and UPOM; however, the spectrum of sources of FPOM and UPOM (Figures 1–3) leads to a heterogenous colonization of the particles within these detrital compartments. The simple relationship between particle size (surface area) and bacterial activity (49–52) is not always observed in stream systems (21, 22, 61, 111). In general the longer the detritus is recycled within the system the more refractory it becomes to microbial and animal metabolism (91, 94).

Increases in percent lignin and humified (complexed) nitrogen during decomposition of CPOM (94) and FPOM (M. J. Klug, unpublished) increase refractility in stream systems. This is most pronounced in FPOM where the increase in percent lignin (up to 40%) and refractile nitrogen (up to 50%) of the total organic content correlates with decreasing particle sizes (M. J. Klug, unpublished). Because FPOM represents intermediates in the progression from CPOM to DOM in streams (21, 22), smaller particles most likely have the longest residence times and are subject to condensation reactions and selective decomposition of labile compounds, leaving the more refractile materials, such as lignin. The ATP content (microbial biomass) of ground and microbiologically conditioned hickory leaf litter, feces of *Tipula abdominalis* fed on conditioned hickory leaves, and natural stream detritus of the same particle size was 23.6, 2.5, and 1.7 n moles ATP per gram ash free dry weight (gAFDW^{-1}), respectively (111). Interestingly, the C/N ratios of the ground litter and natural detritus, which produced different growth responses, were nearly the same (13.9 and 17.8, respectively), probably reflecting the accumulation of refractile nitrogen in the natural detritus. The ATP and nitrogen differences in the food types support the contention that substrate refractility is directly related to in-stream residence time. Wood CPOM and ultrafines are the most refractory compo-

Table 1 General characteristics of different food resources of stream invertebrates [Modified from (21, 22, 34, 75)]

Food category	Particle diameter size range	Description	Microbiology	Caloric content Kcal · gAFDW^{-1}	Percent ash	Carbon to nitrogen ratio (C/N)
Detritus (particulate organic matter POM)	>0.5 μm	Dead organic matter (primarily vascular plant tissue) plus associated microbiota	Fungi (especially aquatic hyphomycetes, e.g. *Tetracladium, Allotospora Flagellospora* bacteria (e.g. *Cytophaga, Sporocytophaga*); of lesser importance: protozoans, rotifers, nematodes and microarthropods (e.g. harpactacoid copepods).	4.5–4.8[a]	10–50[a]	20–1340:1[b,c]
CPOM (coarse particulate organic matter)	>1 mm					
Leaves, needles, etc.	>1 mm (deciduous leaves generally >16 mm)	Leaf and needle litter and other non-woody plant parts (non-woody stems, flowers, fruits, budscales)	Fungi (esp. aquatic hyphomycetes), bacteria; lesser import. protozoans, rotifers, nematodes and microarthropods. Material surface and matrix colonized.	4.8[a]	10–30[d]	20–80:1[b,c]
Woody material	>1 mm	All wood and woody parts	Same as leaf and needle litter but lower densities. Material primarily colonized in surficial layers until late in decomposition	4.5[a]	40–50[a]	220–1340:1[b]
Coarse wood	>10 cm	Boles, branch wood, root wads				1340:1[b]
Fine wood	<10 cm >1 mm	Twigs, bark, nuts, fine branches and roots				220–235:1[b]
FPOM (fine particulate organic matter)	<1 mm >50 μm (as a general category with UPOM included, <1 mm >0.5 μm)	Leaf, needle and other soft plant fragments, fragments of woody material, feces of aquatic invertebrates (derived from riparian zone), amorphous material derived from flocculation of DOM, free microbial cells, sloughed periphytic algae, mineral particles with adsorbed organic films	Primarily bacteria colonizing surfaces protozoans on larger particles > 250 μm)	4.5[e]		18:1[g] (>250 mm) (<75 mm) 500 μm 9.5–37.8:1[g] 250 μm 7.4–27.5:1 100 μm 9.3–34.0:1 50 μm 8.4–21.2:1 0.5 μm 9.4–18.2:1

Table 1 *(Continued)*

Food category	Particle diameter size range	Description	Microbiology	Caloric content Kcal · gAFDW⁻¹	Percent ash	Carbon to nitrogen ratio (C/N)
UPOM (ultrafine particulate organic matter)	<50 m > 0.5 μm	Fragments of leaf, needle and woody FPOM, amorphous DOM flocculated material and clay particles with adsorbed organic films	Bacteria sparsely colonizing surfaces or associated with amorphous flocculated organics			
Periphyton	<500 μm > 10 μm	Attached, living algal cells (diatoms, green, bluegreen, red and golden brown algae and flagellated forms); FPOM usually associated with periphytic algal communities	Bacteria colonizing the surface of gelatinous matrix covering of some algal colonies	4.6 (diatoms 5.3)[a]	10–40[a]	5.4:1[f] 3.7–10.1:1[h]
Macrophytes	>1 cm (Microalgal colonies, e.g. *Cladophora*, may be in the range of <1 cm > 1 mm)	Mosses and flowering plants (e.g. watercress [*Nasturtium*]) in small ground-water-rich streams; species of *Potamogeton* and others in well lighted streams and midsized rivers. Macroalgae (e.g. *Cladophora*).	Bacteria of periphyton coating on surfaces of submerged portions of plants	4.5 (moss)[a] 4.3–4.6 (*Potamogeton*)[a]	12 (moss)[a]	24.4:1[i] (13.2–69.1:1)
Animals	>50 μm	Macroinvertebrates (early life stages may be very small, defined in terms of full grown size approximately > 1 mm).	Bacteria of the digestive tract and some surface attached forms	5.2–6.4[a]	10–30[a]	<17

[a] Cummins and Wuycheck 1971.
[b] Triska et al 1975, Anderson and Sedell 1978.
[c] Suberkropp and Klug 1976.
[d] K. W. Cummins, unpublished data for basswood, hickory, ash, aspen and oak leaves, stream conditioned.
[e] K. W. Cummins, R. C. Petersen, J. C. Wuycheck unpublished data for stream collected detritus in the range <1 mm > 250 μm.
[f] McCullough 1979b.
[g] Ward and Cummins 1979.
[h] McMahon et al 1974.
[i] Reference h, mean 11 species.

nents of the array of detritus in streams (except ultrafines generated from labile DOM, e.g. by flocculation or clay particle adsorption).

Periphyton

The periphyton (aufwuchs) assemblages include both attached algae and associated detritus, including microfloral and microfaunal elements (e.g. 71, 113). Algal tissues, particularly diatoms, are the critical nutritional component of this resource [Table 1; (7, 27, 70)], though often they do not constitute the major fraction of the organic layer on rock (and wood) surfaces (35, 68). Depending upon their species and associated biochemical diversity, the periphyton assemblages with their low C/N ratios (Table 1) often provide a "balanced" diet for scrapers (72, 83).

Macrophytes

Macrophytes (mosses, flowering plants, and macroalgae) seem to be little used as a food source in a living state in running waters (e.g. 47, 64, 112). [The cell-piercing hydroptilid caddisflies, however, do consume macrophytes (75).] The limited use of macrophytes by aquatic herbivores is attributed to high C/N ratios (Table 1), large quantities of cellulose and lignin, and decreased digestibility of their proteins (23, 72). Thus macrophytes constitute a food resource largely in a moribund state as part of the detritus (POM) (63, 77). They make quantitatively significant contributions to the detrital pool of mid-sized and other autotrophic running waters (36, 77) and may be qualitatively important throughout drainage systems (7).

All rooted (e.g. *Elodea,* some species of *Potamogeton,* etc) and attached macrophytes provide sites of adherence for microfloral and microfaunal assemblages and macroinvertebrates. Mosses (e.g. *Fontinalis*), which are common components of small streams, entrain detrital particles; FPOM consumers (collectors) are usually abundant in moss coverings on rock and wood surfaces. Similarly, clumps of such plants as watercress (*Nasturtium*) in depositional, groundwater-rich zones of streams are trapping areas for POM.

The work of Smirnov (88) with a phryganeid caddisfly is one of the few studies of a stream invertebrate species that consumes living vascular plant tissue. As he showed, biochemical differences (amino acid composition) between plant species can be of critical significance to consumers.

Animals

Because of both high caloric and protein content (Table 1), animal prey have been considered the highest quality food resource in stream environments (7). It is obtained by predators at considerable energy expense or over longer than annual life cycles. Anderson (2, 3, 7) has shown the importance of intake of high-protein-content food by a detritivore during the last phase of rapid growth. Many species of nonpredatory invertebrate functional groups may typically be large enough during the rapid growth phase near the end of the feeding portion of the life cycle to ingest incidentally some live animal food (31, 33).

INVERTEBRATE FUNCTIONAL FEEDING GROUPS

As indicated above, functional group designations stress feeding mechanisms rather than food eaten because a given feeding mode, such as filtering, can result in the intake of all food categories—living or dead plant or animal material (33). On this basis, most stream macroinvertebrates are omnivores (20). However, efficiencies of intake and digestion (assimilation) may vary significantly. For example, shredders exposed to periphyton or scrapers fed conditioned leaf litter in each case represent a poor fit between the morpho-behavioral characteristics of food acquisition and food supply. Significant differences in digestive capability also have been shown (e.g. 12, 13, 17, 28) and are undoubtedly important in species niche separation.

Shredders

The shredder:CPOM:fungal-bacterial system is depicted in Figure 1. Shredders usually prefer CPOM well-colonized by microorganisms (3, 55, 57, 58, 60, 62, 65) and have been shown to select for the most microbially colonized (conditioned) material (4, 12, 13, 15). Thus, those leaves or needles designated as "fast" ($\sim$1–2% loss$\cdot$day^{-1}) with regard to processing rate (80) will be selected over "slow" ($\sim$0.5% loss$\cdot$day^{-1}) species if compared after *equal* conditioning times at the same temperature. Preference for each is at conditioning times when microbial biomass is maximum; fungi dominate the biomass during these periods (e.g. 14, 90, 91, 97). The preference is reflected in high rates of consumption (CI) and relative growth [(RGR) expressed as percentage of body weight per day] and in the efficiency of converting food to growth (ECI) (39, 101). This indicates that microbial biomass (especially fungi) is nutritionally important to the macroinvertebrates, either through the direct assimilation of digested microbial biomass or through utilization of substrate partially transformed by the associated microorganisms (16, 26).

Invertebrate assimilation efficiencies are considerably higher (50–92%) for ingested microorganisms than for detritus (6–35%) (summarized in 20). Although the nutritive value (e.g. low C/N, fiber) of microorganisms reflected in invertebrate assimilation efficiency is higher, percentage microbial biomass of ingested detritus is very low (0.03–10%) [(10, 26, 57); M. J. Klug, K. W. Cummins, unpublished]. Assuming a 100% assimilation of

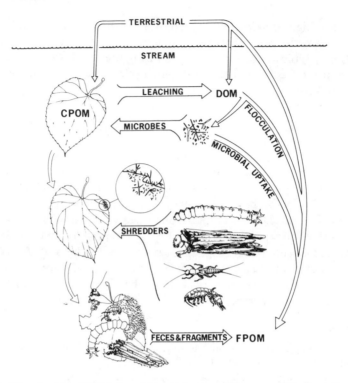

Figure 1 The shredder:CPOM:fungal-bacterial system. The example is for a first-order woodland stream in Michigan. The representative shredders (cranefly and limnephilid caddisfly larvae, nemourid stonefly nymph, and amphipod) are shown invading and skeletonizing leaf litter (represented by a basswood leaf) after colonization by microbes [represented by aquatic hyphomycete fungi (spores) and bacterial cells]. See text for details.

microbial biomass associated with ingested hickory leaves, which yielded the highest CI, RGR, and ECI values, only 8.3% of the observed growth in *T. abdominalis* could be attributed to microbial biomass (M. J. Klug, K. W. Cummins, unpublished). Although the animal may derive little total C and N from it, the microbial biomass may well provide more complete forms of nitrogen, critical growth factors, or other nutritional requirements (e.g. 10, 12, 26). The poor digestibility and incompleteness of senescent terrestrial plant proteins (45), CPOM during early stages of conditioning [(94); M. J. Klug, K. W. Cummins, unpublished], and living and decomposing macrophytes (23) suggest this.

Microbes (particularly fungi) transform CPOM to FPOM-UPOM, DOM, and CO_2 during conditioning through their own metabolism (e.g. 92). Intermediates of polymer metabolism (e.g. cellulose) can more readily be metabolized by shredders, which lack the enzymatic (cellulase) capability to digest the polymers (15). Preference for "predigested" plant and woody tissue by many terrestrial invertebrates is common (e.g. 23).

Partially hydrolyzed (acid treated) leaf tissue and filter paper were consumed at the same rate as well-colonized leaf tissue by *Gammarus pseudolimnaeus* (16), but no growth rates were reported. *T. abdominalis* was found to have nearly the same growth rate on acid-treated (1N HCL) sterile hickory leaves as on well-conditioned leaves (M. J. Klug, K. W. Cummins, unpublished). Differences in the hydrolytic enzymatic capabilities of various aquatic hyphomycetes (92) could explain the preference by *G. pseudolimnaeus* (17) for CPOM colonized by different fungi. Thus, shredders feeding on nutritionally incomplete CPOM (high C/N, fiber) and lacking polymer hydrolyzing digestive enzymes appear to derive their nutritional requirements from ingested microorganisms and partially hydrolyzed detritus polymers. Gut tract symbionts (see below) may also be of significance in shredder nutrition.

Collectors

The collector : FPOM-UPOM : bacterial system (Figure 2) emphasizes primarily surface-colonized small detrital particles (<1mm, Table 1) as a food resource harvested in a size-dependent fashion. Collectors exhibit a wide range of morpho-behavioral adaptations for acquiring fine particle detritus. The distinction between filtering and gathering collectors is intended to

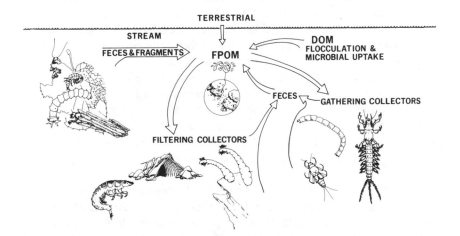

Figure 2 The collector : FPOM-UPOM : bacterial system. The example is for a first-order woodland stream in Michigan. The representative filtering (net-spinning caddisfly and blackfly larvae) and gathering (caenid mayfly nymph, midge larva, and burrowing mayfly nymph) collectors utilize FPOM derived from a number of sources [including shredder feces (Figure 1)] and colonized by bacteria. See text for details.

partition those feeding on FPOM-UPOM in transport (suspension) from those feeding on deposited, sediment-related detritus. The separation is imperfect because some lotic invertebrates living in the sediments maintain a current through their more or less discrete burrows (11) or tubes (midges); this may draw in particles from transport and/or the sedimentary organics deposited around the intake point. In addition, the material deposited on the bottom may continue to move by saltation (bed load). Nevertheless, there are sufficient differences between transport and sediment particulates (85) and between the biological strategies evolved for their harvest by macroinvertebrates to make a conceptual separation.

Although some studies (e.g. 24, 25, 41) have purported to show selective feeding by FPOM-UPOM-collectors on the basis of food quality (e.g. most nitrogen-containing organic matter), some of the assumptions need further testing (see discussion below). The partitioning of the FPOM-UPOM food resource according to particle size is well documented for filtering collectors [filterers, suspension feeders, transport collectors (29, 108, 109, 118)] and has been proposed for gathering collectors [gatherers, deposit feeders, sediment collectors (6, 111)].

There is conflicting evidence concerning the relative importance of the microorganisms vs the detrital particle substrate. Based on a number of assumptions, Baker & Bradnam (10) concluded that at normal field densities the bacterial component was insufficient to support blackfly growth, although Fredeen (48) grew blackflies on high densities of bacteria alone. If bacterial densities on FPOM-UPOM are too low to be compensated by increased feeding rate (as shredders compensate), biochemical alteration of the detrital substrate by the microbes to produce some components utilizable by the animals may be of critical importance.

The importance of microbial biomass (as indicated by ATP levels) to growth rates of gathering collectors was demonstrated for the midges *Paratendipes albimanus* (111) and *Stictochironomus annulicrus* (61). As discussed above, natural stream detritus varies in quality (refractility). Fragmented, well-conditioned CPOM and macroinvertebrate feces (52, 87) probably represent the highest quality components of native stream detritus.

Maintenance of growth on a generally low-quality food could result either from selective ingestion of the highest quality particles available or through increased feeding rate, which might compensate for variability in food quality as in other invertebrates (54, 86). Differential digestion, such as that shown by Brinkhurst & Chua (24) for tubificids on two types of bacteria (*Pseudomonas* and *Aeromonas*), indicates the importance of selective assimilation.

Collectors instigate changes in particle size, either increasing or decreasing it through their feeding activities. For example, *Hexagenia limbata* nymphs gather fine particles and aggregate them before ingestion (119). Although such alterations are small compared to the shredder conversion of CPOM to FPOM-UPOM, there may be a significant effect on resource-partitioning between various collectors feeding in a size-specific fashion. Anderson & Sedell (6) referred to collectors that increase particle size from food to feces as "compactors."

Collector gut loading (filling or passage) times are generally short—e.g. 20–30 minutes for blackflies (63) and less than one hour for a mayfly (69), for a net-spinning caddisfly (70), for gastropods (e.g. 53), and for oligochaetes (8, 25). Gut passage time varied from 4 to 12 hours in *Hexagenia limbata* (Ephemeroptera) nymphs (119). Since gut loading times may be longer in shredders and scrapers (e.g. 33, 35), the collector strategy might be to pass a comparatively low quality food through the gut rapidly. In this case, food quality is defined in terms of the growth that can be produced per unit of ingestion (Fig. 4; 7, 111), while quantity is defined in terms of food density per unit of the animal's environment.

Filtering collectors capture and ingest essentially the entire size range of FPOM-UPOM (0.2 μm–800 μm+) from transport (108, 109). Wallace and co-workers (102–107) have demonstrated partitioning of the fine-ultrafine detritus food resource by particle size according to the mesh dimensions of net-spinning trichopteran capture nets. Various species of blackflies have also been shown to filter discrete, although somewhat broader, particle size ranges (e.g. 29, 76, 118). Different age classes of a given species also differ with regard to the size range they most efficiently capture (118).

The particle size range of the sediment FPOM-UPOM that gathering collectors can ingest is determined primarily by the morphology of the mouth parts (e.g. 61, 110, 111).

Scrapers

The scraper:periphyton system (Figure 3) involves macroinvertebrates with morpho-behavioral adaptations for grazing upon (i.e. shearing off) food that adheres to surfaces—periphyton in particular (Table 1). Morphological adaptations include the scoop-shaped mandibles with a cutting edge of *Glossosoma* (Trichoptera), the inner grinding surfaces of many mayfly mandibles (e.g. *Stenonema*), and the rasping radular structure of gastropods (e.g. *Ferrissia*). Scrapers typically have morpho-behavioral adaptations for maintaining their position on exposed surfaces in rapidly flowing, turbulent water, such as dorso-ventral flattening or the ventral "sucker" formed by the gills of the mayfly, *Ephemerella dodsi* (75).

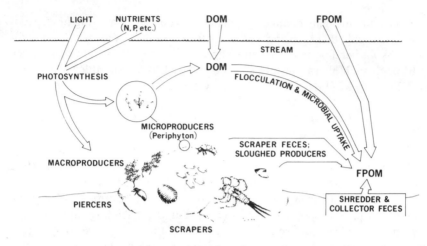

Figure 3   Scraper: periphyton and piercer: macrophyte systems. The example is for a third-order woodland stream in Michigan. The representative scrapers (water penny beetle and glossosomatid caddis larvae, heptageneid mayfly nymph, and limpet) graze periphyton (e.g. diatoms). The representative piercer (hydroptilid caddisfly larva) sucks fluids from macrophytes (e.g. macroalgae or the moss shown). See text for details.

The scraper *Dicosmoecus* (Trichoptera can orient to patches of attached algae (D. S. Hart, personal communication). This is primarily the result of an inverse relationship between locomotion and feeding, in which time spent feeding increases with food density (35). The recognition of food patches (quantity, i.e. density, of food in the environment) is not necessarily qualitative. For example, the ratio of algae to detritus ingested by *Glossosoma* is essentially a function of the relative densities of each within patches of periphyton (35), and larval growth (i.e. maximum size attained) is positively correlated with an algal diet (7).

Piercers

The piercer: macrophyte system (Figure 3) is essentially confined to species of microcaddisflies (Trichoptera: Hydroptilidae) in lotic ecosystems. These larvae are adapted through their mouth-part morphology and small size to climb among the strands of macrophytic algae, pierce individual cells, and imbibe the cell fluids (75, 115).

Predators

The predator category includes all macroinvertebrates that are adapted specifically for the capture of live prey. Representatives of other functional groups (shredders, collectors, scrapers) may ingest live animals in the course of their feeding (31, 33) and, as noted above, such high protein (Table 1) intake may be critical at certain times in their life cycle (2, 7).

Prey abundance was shown to reduce the life cycle from two years to one in *Sialis* (Megaloptera) (7, 9), indicating general food limitation for predators. The study by Johnson (59) on a pond damselfly (Zygoptera) can serve as a general model for stalking predators e.g. nonburrowing odonates and most stoneflies and megalopterans. When prey were abundant, locomotion was reduced. As prey density decreased, movement ("searching") increased up to a very low prey level; then locomotion ceased, presumably an adaptation that conserves energy.

RESOURCE PARTITIONING AND TEMPERATURE-FOOD INTERACTION

The interactions between temperature and quantity and quality of stream macroinvertebrate food resources are complex (7). Temperature can strongly influence quantity (e.g. density of periphytic algae in a stream system) and quality (e.g. microbial populations on detrital particles). These indirect controls of macroinvertebrate growth (and survivorship) are difficult to separate from the direct temperature effects on macroinvertebrate metabolism. Temperature-food and temperature-macroinvertebrate interactions undoubtedly interplay in influencing stream community structure of functional feeding groups.

Community Analysis of Functional Groups

An estimate of the numbers of lotic species (and higher taxa) of aquatic and semi-aquatic insect orders belonging to the various functional feeding groups is summarized in (75). Nonpredaceous stoneflies (formerly the Fili-

Table 2 Comparison of ratios of Nearctic Trichoptera genera categorized by functional feeding group according to biome and stream type. Data from (116)

| Stream type | Ratio of shredder/collector genera | | Ratio of shredder/scraper genera | |
	Eastern deciduous forest biome	Western montane forest biome	Eastern deciduous forest biome	Western montane forest biome
Cool, headwater streams	2:1 (21:12)	3:1 (25:9)	3:2 (21:15)	1:1 (25:26)
Warm, midsized rivers	1:1 (9:8)	1:1 (6:6)	2:3 (9:13)	1:2 (6:13)

palpia), caddisflies—especially the family Limnephilidae (116), and crane-flies (Diptera) comprise most of the CPOM-shredders. Mayflies are largely gathering-collectors and scrapers; mosquitoes and blackflies are filtering-collectors; the midges are gathering-collectors. The Trichoptera are well represented in all functional groups.

An analysis of Nearctic trichopteran genera (116) showed shifts in relative dominance of shredder, collector, and scraper functional groups according to biome and stream type (Table 2). The reduced importance of shredders relative to collectors and especially to scrapers with increasing stream size is a feature of both the eastern deciduous and western montane biomes. However, the greater significance of scrapers in the headwaters of western stream systems agrees with the higher primary-production:community-repiration (P:R) ratios observed in western systems (e.g. 77).

A community analysis by functional groups for a first-order, head-water, woodland stream in Michigan is summarized in Table 3 (39). The standing crop of detritus was relatively constant over the year (standard error = 9% of mean for 15 sample periods), but inputs, largely from FPOM-UPOM, are reflected in the high runoff-related spring estimate. However, although

Table 3 Summary of macroinvertebrate data by season and functional feeding group for a headwater (first-order) tributary of Augusta Creek, Michigan. Summarized and recalculated from (39); macroinvertebrates retained by 75 μm mesh size (all values means or ranges for the periods)

Season	Autumn	Winter	Spring	Summer	Annual mean (or range)
Period	Sept 15– Nov 25	Nov 25– Mar 25	Mar 25– May 25	May 25– Sept 15	—
Temperature (°C)					
Range	6–14	0–6	4–14	14–20	0–20
Degree days	741	210	704	1,696	3,324
Detritus standing crop gAFDW $\cdot$ m^{-2}	366	323	689	408	426
Gross primary production/ community respiration (P/R)[a]	0.5	0.3	0.4	0.8	0.5
Number $\cdot$ m^{-2} × 10^3	22.2	26.6	14.7	34.3	26.1
gDW $\cdot$ m^{-2}	1.9	4.9	3.3	4.5	4.0
Percent of biomass					
Shredders	12	22	40	8	21
Collectors	47	66	20	64	49
Scrapers	12	8	9	13	11
Predators	26	8	2	7	11
Nonfeeding	2	2	30	9	14
Biomass ratios					
Shredders:Collectors	0.25	0.38	1.67	0.11	0.60
Shredders:Scrapers	1.00	2.76	3.33	0.54	1.90

[a]Unpublished data from D. L. King, K. W. Cummins (Kellogg Biological Station).

qualitative differences were probably quite significant between seasons [(110, 111); Table 1, C/N of fines], the total amounts available for detrital feeders appear similar, except for a probable reduction in the winter.

Total numbers and biomass of macroinvertebrates were also constant over the annual cycle (Standard error = 17 and 22% of the mean, respectively, for 15 sample periods). Increased relative importance of non-feeding stages in the spring represents the culmination of fall-winter growth prior to insect pupation and/or emergence (especially collectors). Shredder importance increased from autumn through spring when maximum weights per individual were attained. Relative dominance of scrapers was greatest in summer, correlated with the highest seasonal P : R estimate and warmest temperatures.

The average annual community ratio of shredders to collectors (0.6) is much lower, and that of shredders to scrapers (1.9) slightly higher, than predicted (2.0 and 1.5, respectively) by the generic trichopteran model of Wiggins & Mackay [(116); Table 2]. The discrepancy between trichopteran numeric analysis and biomass ratios indicates the importance of collectors other than caddisflies (especially mayflies and midges) and the smaller average size of the dominant grazer (*Glossosoma*) compared to the dominant shredders (*Pycnopsyche, Tipula*).

Obligate vs. Facultative Representatives of Functional Groups

Restriction to a specific functional group (obligate relationship) implies that only one mode of food acquisition will produce at least the minimal level of growth required for survival and reproduction. Such obligate specialists can be restricted morphologically (e.g. the predator mouth parts are inadequate for filtering FPOM-UPOM) or through behavior patterns that preclude the use of a morphology that might be adequate for harvesting an alternate food resource. An obligate relationship between morpho-behavioral adaptation and a specific food resource presumably maximizes the efficiency of converting food to growth (ECI, 101). In contrast, facultative members of a functional group, or generalists, sacrifice the efficiency gained through a tight linkage between adaptation and resource for the advantages that accrue from an ability to harvest a wider variety of foods. The theoretical differences between obligate (specialist) and facultative (generalist) forms is shown in Figure 4. By tearing out bites of CPOM, shredders ingest more microbial biomass and biochemically altered substrate than would be obtained by scraping alone. For example ECI values for *Tipula* (shredder) and *Stenonema* (scraper) on conditioned leaf litter are 0.17 and 0.05, respectively (38).

Differences in the specificity of resource dependence lead to the prediction that generalists would perform better (i.e. grow and reproduce more)

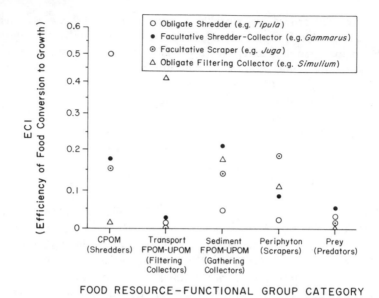

FOOD RESOURCE—FUNCTIONAL GROUP CATEGORY

Figure 4   Predicted differences between obligate (specialist) and facultative (generalist) members of different functional feeding groups. The range of ECI [(101); Table 2] values for a given species would depend upon the specific nature of the food in a particular resource category and temperature.

under disturbed or altered stream conditions in which a particular food resource is reduced or eliminated. Obligate shredders, on the other hand, would be restricted by the loss of CPOM-leaf litter input with the removal of riparian vegetation.

Obligate relationships between feeding adaptation and food resource may involve specific enzymatic capabilities. Also, as discussed below, the presence of a resident gut microbiota could be important in utilization of high-cellulose-content food.

Spatial-Temporal Separation

Seasonal, local (i.e. within a reach), and stream-order (i.e. between streams by size) differences in inputs, production, and storage of food resources provide a spatially and temporally variable system from which macroinvertebrates derive their nutrition. Temporal isolation of potential competitors—i.e. of species belonging to the same functional feeding group—seems to be the general rule among stream macroinvertebrates (e.g. 100). When such species occupy the same general space in the stream, their growth periods are largely nonoverlapping. An example is a group of three (probably obligate) scraper species of caddisflies—*Glossosoma nigrior* (two generations per year), *Goera calcarata,* and *Neophylax oligius* or *concinnus* (one generation per year). The growth of *Goera* occurs in two phases: (*a*) in late summer and early fall, prior to the major autumnal growth of the winter generation of *Glossosoma* and to the late fall-winter growth of *Neophlax;* and (*b*) in late spring, prior to the summer growth of the second generation of *Glossosoma. Neophylax* has a spring-summer diapause; *Glossosoma* grows much more slowly during the winter growth of *Neophylax* [(35, 100); K. W. Cummins, unpublished data].

The timing of life cycles relative to the abundance of a particular food supply is common particularly in shredders. Their fall through early spring growth period overlaps the maximum availability of deciduous leaf litter (e.g. 32, 55, 65).

General correlations between macroinvertebrate distributions and accumulations of detritus have been reported (42–44, 67, 68, 81). Temporal and functional separations within this relationship were not treated. Spatial separation has been demonstrated for some congeners—e.g. for the trichopteran shredder genus *Pycnopsyche* (32, 55, 65), in which one species feeds in erosional, the other in depositional, zones.

Therefore, food resource partitioning by stream macroinvertebrates belonging to the same functional group is largely temporal and secondarily spatial. When both spatial and temporal overlaps of members belonging to the same functional group occur, resource partitioning is probably accomplished by selective feeding—e.g. by particle size in filtering collector species such as blackflies and net-spinning caddisflies. A less subtle example among CPOM-shredders may be observed in the wood gougers *Lara* and *Heteroplectron* (life cycles more than 1 yr) and the leaf-needle feeding *Lepidostoma* (univoltine) (5, 6).

Temperature

The quantity of food resource may not be a sufficient predictor of macroinvertebrate growth because of variations in temperature regime (95, 100) and food quality (111). Temperature exerts a direct influence on the metabolism of macroinvertebrates, though as shown for the shredder *Tipula* (Table 4), some compensation is possible. Larvae fed the same quality food at two temperatures (5° and 10°C) grew at the same rate. This temperature difference represents about 25% of the annual range of the animals' native stream. As Vannote (99) showed for the mayfly *Ephemerella dorothea,* metabolic rate per unit weight was nearly constant during nymphal growth since temperature (directly related to rate) and size (indirectly related to rate) offset each other.

Temperature also regulates the assimilation, respiration, and population growth of the microorganisms (93) that exert primary control over both the

Table 4 Relative growth rate (RGR:% body wt · day^{-1} and coefficient as %) of 15 *Tipula abdominalis* larvae followed individually at two temperatures and fed hickory leaves previously conditioned at 20°C; old leaves replaced with new every two days

Interval	5°C		10°C	
	RGR	ATP of leaves nm · gDW^{-1}	RGR	ATP of leaves nm · gDW^{-1}
First 8 days	0.06 (43)	46.7[a]	0.06 (73)	46.7[a]
Next 7 days	0.02 (82)	—	0.02 (68)	—
Total 15 days	0.04 (48)	68.5	0.03 (74)	61.6

[a] Initial, day zero

nutritional quality of detrital food and (along with light and nutrients) the photosynthetic rates of periphytic primary producers. The dual effect of temperature on macroinvertebrates and their food supply makes it difficult to separate temperature and food quality as regulators of animal growth. Differences in food quality were compensated by increased feeding rate in *Tipula* (Table 5). Ward & Cummins (111) demonstrated that growth of the gathering collector *Paratendipes albimanus* (Chironomidae) was more strongly influenced by food quality than by temperature.

If food quality is defined broadly in terms of macroinvertebrate growth per unit of food intake, then estimates of ECI [RGR/CI (101)] are measures of food quality within a food category (Table 1). (They are also indicators of the obligate or facultative relationship between adaptation and different food categories as discussed above.) If RGR and CI are calculated per temperature time (degree days), then ECI values should reflect differences in food quality (Table 5).

MICROBIAL-ANIMAL ASSOCIATIONS

Symbioses between invertebrates and microorganisms are a common occurrence [extensive review in (26)]. Although most symbioses reported are in terrestrial invertebrates, analogous associations might occur in many freshwater insects evolved from terrestrial ancestors. Symbionts occur extracellularly (i.e. in the gut lumen) or intracellularly in specific animal cells (mycetocytes). Roaches and termites, which have been studied extensively (26), control their microorganisms and have evolved mechanisms for transmission of the symbionts from adults to progeny. Removal of the symbiont from roaches through antibiotic treatment but with no change in diet results in loss in fecundity and smaller size of progeny through loss of critical growth factors (26). Symbiotic associations most likely evolved owing to

Table 5 Consumption rate (CI) and relative growth rate (RGR) of 15 *Tipula abdominalis* larvae followed individually for 68 days at 10°C as a function of two leaf types: high food quality basswood (*Tilia americana*) and medium quality hickory (*Carya glabra*). From (7) based on G. M. Ward, R. Speaker, R. W. Ovink, K. W. Cummins, unpublished data, Kellogg Biological Station

Leaf type	Decay coefficients[a]	Respiration of leaf substrate $(mg \cdot l^{-1} \, gDW^{-1}d^{-1})$	Leaf % N	Percent body weight $\cdot$ day^{-1} ($\pm$ standard deviation) CI	RGR	ECI (RGR/CI)
Basswood	0.018	0.030	3.8	29.7 (18.3)	3.8 (1.8)	0.13
Hickory	0.009	0.018	1.8	50.0 (13.5)	3.2 (1.5)	0.06

[a] From (80); field rates determined autumn through spring (temperature range 0–15°C).

nutriphysiological benefits related to diet. Symbionts, especially endosymbionts, are characteristically found in animals having incomplete diets (26).

Although some preference for nutritionally complete diets exists among aquatic invertebrates (e.g. 27) the animals ingest what their morpho-behavioral feeding mechanisms allow. The spectrum of substrates ingested is highly variable; a collector, for instance, may ingest algae in one mouthful and detritus in another. Because the completeness of these dietary items undoubtedly varies markedly, symbionts could constitute a critical component in the nutrition of stream macroinvertebrates.

Table 6 summarizes investigations from our laboratory on the occurrence and localization of associated (symbiotic) bacteria in the gut tracts of stream invertebrates representative of various functional feeding categories [(5); A. Meitz, M. J. Klug, unpublished]. Of major interest are the bacteria attached to the hindgut wall of the animals in different functional groups and having differing gut morphologies. Both the gut wall of the fermentation chamber of *T. abdominalis* and the straight tube hindgut of *Hydropsyche bronta* were colonized by a morphologically similar microbiota. These associations have been observed in all individuals sampled from a given species population and from a range of streams from various regions.

The attached microbiota of *T. abdominalis* is sloughed during molting but rapidly recolonizes after ecdysis. Attachment of microbiota to the gut wall provides a mechanism for increased retention in the tract and is analogous to the associations observed in higher animals (84). Attachment also indicates a more permanent association rather than chance contamination from ingested food.

In animals having an attached microbiota the bacteria associated with the hindgut were denser than those of the midgut. This can be interpreted as a function of sloughing of bacteria from the walls, multiplication in the

Table 6 The association of bacteria with the gut tracts of stream invertebrates in various feeding categories. From (73) and M. J. Klug, unpublished data

Animal and functional group	Gut morphology[a]	Attached hindgut microbiota	Numbers of bacteria[b]	
			Hindgut lumen	Midgut lumen
Obligate shredders				
Tipula abdominalis	F.C., C	+	C	A
Pycnopsyche guttifer	S, En. H.G.	+	C	A
Hydatophylax argus	S, En. H.G.	+	C	A
Platycentropus radiata	S, En. H.G.	+	C	A
Lepidostoma costalis	S	−	B	A
Facultative shredders				
Pteronarcys picteti	S	−	A	A
Gammarus pseudolimnaeus	Non. Dif.	−	A	A
Wood gougers				
Heteroplectron californicum	S	−	A	A
Lara avara	S	−	A	A
Brillia sp.	S	+ (?)	B	A
Collectors				
Filtering Collectors				
Hydropsyche bronta	S, Pro	+	C	A
Prosimulium sp.	Con	−	A	C
Gathering Collectors				
Taeniopteryx parvula	S	−	A	A
Leptophlebia nebulosa	S	+	C	A
Stictochironomus annulicrus	S	+	C	C
Paratendipes albimanus	S	+	C	C
Brillia flavifrons	S	−	A	A
Scrapers				
Glossosoma nigrior	S	−	A	A
Psephenus herricki	S	−	A	A
Goniobasis sp.	Non. Dif.	−	A	A
Predators				
Paragnetina sp.	S, Pro, C	−	A	A
Nigronia serricornis	Con, Pro, C	+	C	A
Corydalus sp.	Con, Pro, C	+	C	A

[a] Gut morphology: S = Simple-straight tube; F.C. = Ferm. chamber; En. H.G. = Enlarged Hindgut; Con = Convoluted; Pro = Proventriculus; Non. Dif. = Nondifferentiated gut; C = Gastric cecae.
[b] A = 10^5–10^7 bacteria · ml^{-1}; B = 10^8 bacteria · ml^{-1}; C = 10^9–10^{10} bacteria · ml^{-1}.

hindgut lumen, or digestion of bacteria in the midgut. However, bacterial population densities are considerably higher (2–5 orders of magnitude) than densities in ingested food, suggesting the occurrence of multiplication.

In general, attached microbiota are most widespread in the obligate shredders, collectors, and predators that engulf their prey. Scrapers and facultative shredders showed neither attached bacteria nor change in density between midgut and hindgut (Table 6). The associations are therefore found more often in animals ingesting the most refractile diets. However, exceptions occur and it is difficult to speculate further on these relationships until more is known about the nutritional requirements of the stream invertebrates and the nature of the associated bacteria. The only stream detritus feeders examined that have been shown to have a digestive cellulase independent of their symbionts are *Gammarus* (78) [reflecting its marine phylogenetic origin (79)] and certain gastropods (28).

The role of hindgut-associated bacteria is complicated by a lack of detailed information on the role of the hindgut in digestion and absorption (other than that of water and ions) (30, 96, 117). However, low molecular weight compounds have been shown to be absorbed through the hindgut and rectal tissue of a number of terrestrial invertebrates (e.g. 11, 19, 30, 82). Detailed study is required before the nutrition of such stream invertebrates can be adequately understood. In addition, no attention has been directed as yet to the presence and possible role of endosymbionts.

CONCLUSIONS

Classification of functional feeding groups among stream invertebrates provides a useful means for describing the morpho-behavior capacity of stream invertebrates to consume available food resources (e.g. CPOM, FPOM, periphyton). It does not define what portion of an ingested food resource is assimilated; this may depend on short- or long-term physiological adaptations.

Invertebrates that are facultative in their morpho-behavioral feeding have a wider niche breadth (i.e. they can ingest a wider array of substrates) and inhabit a greater diversity of stream habitats. An example of physiological adaptation was described for freshwater gastropods by Calow (27, 28). Species with cellulase can grow in a wide range of habitats because they are able to utilize a greater array of detritus and algal types.

A variety of feeding strategies are used by stream macroinvertebrates to compensate for the changing dietary sufficiency of ingested substrates. Shredders, and possibly some collectors, feed preferentially on particulate organic detritus colonized by microorganisms, utilizing the associated microorganisms and partially hydrolyzed (microbially digested) substrate.

Collectors, scrapers, and facultative shredders increase the consumption of low quality food to compensate for its decreased nutritional benefit.

Animals that eat materials inadequate to support their growth may harbor microbial symbionts that compensate for the dietary deficiency. They are thus analogous to ruminant animals, which take in food of high C:N ratio (>17) and absorb through their gut wall material with a lower C:N ratio (<17) produced by intestinal microbiota that break down polymers and supply essential amino aids (83).

ACKNOWLEDGMENTS

We gratefully acknowledge the help by the following people in various phases of the work reported and in chapter preparation: G. L. Spengler, R. C. Petersen, K. Dacey, G. M. Ward, R. W. Ovink, R. L. Mattingly, R. Speaker, A. Meitz, S. Kotarski, and L. Louden.

The original research reported was supported by contracts 79-EY-10004.000 and EY 76-S-02-2002.A001 from the Division of Ecological Research, Office of Health and Environment, U.S. Department of Energy, and by grants (GB-36069X, BMS-74-20716-A01) from the Ecosystem Analyses Program, National Science Foundation.

Published as Michigan Agricultural Experiment Station Journal Article No. 8938 and Kellogg Biological Station Publication No. 377. Technical Paper No. 5171, Oregon Agricultural Experiment Station.

Literature Cited

1. Anderson, J. M., Macfadyen, A., eds. 1976. *The Role of Terrestrial and Aquatic Organisms in Decomposition Processes.* Oxford: Blackwell 474 pp.
2. Anderson, N. H. 1976. Carnivory by an aquatic detritivore, *Clistoronia magnifica* (Trichoptera:Limnephilidae). *Ecology* 57:1081–85
3. Anderson, N. H. 1978. Continuous rearing of the limnephilid caddisfly *Clistoronia magnifica* (Banks). *Proc. Int. Symp. Trichoptera* 2:317–29
4. Anderson, N. H., Grafius, E. 1975. Utilization and processing of allochthonous material by stream Trichoptera. *Verh. Int. Ver. Limnol.* 19:3083–88
5. Anderson, N. H., Sedell, J. R., Roberts, L. M., Triska, F. J. 1978. The role of aquatic invertebrates in processing of wood debris in coniferous forest streams. *Am. Midl. Natl.* 100:64–82
6. Anderson, N. H., Sedell, J. R. 1979. Detritus processing by macroinvertebrates in stream ecosystems. *Ann. Rev. Entomol.* 24:351–77

7. Anderson, N. H., Cummins, K. W. 1979. The influences of diet on the life histories of aquatic insects. *J. Fish. Res. Bd. Can.* 36:335–42
8. Appleby, A. G., Brinkhurst, R. O. 1971. Defecation rate of three tubificid oligochaetes found in the sediment of Toronto Harbour Ontario. *J. Fish. Res. Bd. Can.* 27:1971–82
9. Azam, K. W., Anderson, N. H. 1969. Life history and production studies of *Sialis californica* Banks and *Sialis rotunda* Banks in western Oregon. *Ann. Entomol. Soc. Am.* 62:549–58
10. Baker, J. H., Bradnam, L. A. 1976. The role of bacteria in the nutrition of aquatic detritivores. *Oecologia* 24:95–104
11. Balshin, M., Phillips, J. E. 1971. Active absorption of amino acids in the rectum of the desert locust *Schistocerca gregaria. Nature New Biol.* 233:53–54
12. Bärlocher, F., Kendrick, B. 1973. Fungi and food preferences of *Gammarus pseudolimnaeus. Arch. Hydrobiol.* 72:501–16

13. Bärlocher, F., Kendrick, B. 1973. Fungi in the diet of *Gammarus pseudolimnaeus* (Amphipoda). *Oikos* 24:295–300
14. Bärlocher, F., Kendrick, B. 1974. Dynamics of the fungal population on leaves in a stream. *J. Ecol.* 62:761–91
15. Bärlocher, F., Kendrick, B. 1975. Assimilation efficiency of *Gammarus pseudolimnaeus* (Amphipoda) feeding on fungal mycelium or autumn-shed leaves. *Oikos* 26:55–59
16. Bärlocher, F., Kendrick, B. 1975. Leaf conditioning by microorganisms. *Oecologia* 20:359–62
17. Bärlocher, F., Kendrick, B. 1976. Hyphomycetes as intermediaries of energy flow in streams. In *Recent Advances in Aquatic Mycology*, ed. E. B. G. Jones, pp. 435–46. London: Paul Elek
18. Beaumont, P. 1975. Hydrology. In *River Ecology*, ed. B. A. Whitton, pp. 1–38. Oxford: Blackwell. 725 pp.
19. Berridge, M. J. 1970. A structural analysis of intestinal absorption. *Symp. R. Entomol. Soc. London* 5:135–51
20. Berrie, A. D. 1976. Detritus, microorganisms and animals in fresh water. See Ref. 1, pp. 323–38
21. Boling, R. H., Goodman, E. D., Zimmer, J. O., Cummins, K. W., Petersen, R. C., Van Sickle, J. A., Reice, S. R. 1975. Toward a model of detritus processing in a woodland stream. *Ecology* 56:141–51
22. Boling, R. H., Petersen. R. C., Cummins, K. W. 1975. Ecosystem modeling for small woodland streams. In *Systems Analysis and Simulation in Ecology*, ed. B. C. Patten, 3:183–204. NY: Academic. 601 pp.
23. Boyd, C. E. 1970. Amino acid, protein, and caloric content of vascular aquatic macrophytes. *Ecology* 51:902–6
24. Brinkhurst, R. O., Chua, K. E. 1969. Preliminary investigations of the exploitation of some potential nutritional resources by three sympatric tubificid oligochaetes. *J. Fish. Res. Bd. Can.* 26:2659–68
25. Brinkhurst, R. O., Chua, K. E., Kaushik, N. K. 1972. Interspecific interactions and selective feeding by tubificid oligochaetes. *Limnol. Oceanogr.* 17: 122–33
26. Buchner, P. 1965. *Endosymbiosis of Animals with Plant Microorganisms.* NY: Interscience. 909 pp.
27. Calow, P. 1975. The feeding strategies of two freshwater gastropods *Ancylus fluviatilis* Mull. and *Planorbis contortus* Linn. (Pulmonata), in terms of inges-

tion rates and absorption efficiencies. *Oecologia* 20:33–49
28. Calow, P., Calow, L. J. 1975. Cellulase activity and niche separation in freshwater gastropods. *Nature* 255:478–80
29. Chance, M. M. 1970. The functional morphology of the mouth parts of blackfly larvae (Diptera:Simuliidae). *Quaest. Entomol.* 6:245–84
30. Cochran, D. G. 1975. Excretion in insects. In *Insect Biochemistry and Function*, ed. D. J. Candy, B. A. Kilby, pp. 177–281. NY: Halsted Press
31. Coffman, W. P., Cummins, K. W., Wuycheck, J. C. 1971. Energy flow in a woodland stream ecosystem: I. Tissue Support trophic structure of the autumnal community. *Arch. Hydrobiol.* 68: 232–76
32. Cummins, K. W. 1964. Factors limiting the microdistribution of the caddisflies *Pycnopsyche lepida* (Hagen) and *Pycnopsyche guttifer* (Walker) in a Michigan stream (Trichoptera: Limnephilidae). *Ecol. Monogr.* 34:271–95
33. Cummins, K. W. 1973. Trophic relations of aquatic insects. *Ann. Rev. Entomol.* 18:183–206
34. Cummins, K. W. 1974. Structure and function of stream ecosystems. *Bio-Science* 24:631–41
35. Cummins, K. W. 1975. Macroinvertebrates. See Ref. 18, pp. 170–98
36. Cummins, K. W. 1975. The ecology of running waters; theory and practice. In *Proc. Sandusky River Basin Symp. Int. Joint. Comm. Int. Ref. Gp. Great Lakes Pollution from Land Use Activities (1976–553–346)*, ed. D. B. Baker, W. B. Jackson, B. L. Prater, pp. 227–93. Washington DC:GPO. 475 pp.
37. Cummins, K. W., Wuycheck, J. C. 1971. Caloric equivalents for investigations in ecological energetics. *Mitt. Int. Ver. Limnol.* 18:1–158
38. Cummins, K. W., Petersen, R. C., Howard, F. O., Wuycheck, J. C., Holt, V. I. 1973. The utilization of leaf litter by stream detritivores. *Ecology* 54: 336–45
39. Cummins, K. W., Petersen, R. C., Spengler, G. L., Ward, G. M., King, R. H., King, D. L. 1979. Microbial-animal processing in a first order woodland stream. *Oikos.* In press
40. Curry, R. R. 1976. Watershed form and process—the elegant balance. *Co-Evol. Q.* 1976:14–21
41. Davies, I. J. 1975. Selective feeding in some arctic Chironomidae. *Verh. Int. Ver. Limnol.* 19:3149–54

42. Egglishaw, H. J. 1964. The distributional relationship between the bottom fauna and plant detritus in streams. *J. Anim. Ecol.* 33:463–76

43. Egglishaw, J. H. 1968. The quantitative relationship between the bottom and plant detritus in streams of different calcium concentrations. *J. Appl. Ecol.* 5:731–40

44. Egglishaw, H. J. 1969. The distribution of benthic invertebrates on substrata in fast-flowing streams. *J. Anim. Ecol.* 38:19–33

45. Feeney, P. 1970. Seasonal changes in oak leaf tannins and nutrients as a cause of spring feeding by winter moth caterpillars. *Ecology* 51:565–81

46. Fisher, S. G., Likens, G. E. 1973. Energy flow in Bear Brook, New Hampshire: an integrative approach to stream ecosystem metabolism. *Ecol. Monogr.* 43:421–39

47. Fisher, S. G., Carpenter, S. R. 1977. Ecosystem and macrophyte primary production of the Fort River, Massachusetts. *Hydrobiologia* 47:175–87

48. Fredeen, F. J. H. 1964. Bacteria as a food for blackfly larvae (Diptera: Simuliidae) in laboratory cultures and in natural streams. *Can. J. Zool.* 42:527–48

49. Hargrave, B. T. 1970. The effect of a deposit feeding amphipod on the metabolism of benthic microflora. *Limnol. Oceanogr.* 15:21–30

50. Hargrave, B. T. 1970. The utilization of benthic microflora by *Hyalella azteca. J. Anim. Ecol.* 39:427–37

51. Hargrave, B. T. 1972. Prediction of egestion by the deposit-feeding amphipoda *Hyalella azteca. Oikos* 23:116–24

52. Hargrave, B. T. 1976. The central role of invertebrate feces in sediment decomposition. See Ref. 1, pp. 301–21

53. Heywood, J., Edwards, R. W. 1962. Some aspects of the ecology of *Potamopyrgus jenkinsi* Smith. *J. Anim. Ecol.* 31:239–50

54. House, H. L. 1965. Effects of low levels of the nutrient content of a food and of nutrient imbalance on the feeding and nutrition of a phytophagous larva, *Celerio euphorbiae. Can. Entomol.* 97:62–68

55. Howard, F. O. 1974. *Natural history and ecology of* Pycnopsyche lepida, P. scabripennis *(Trichoptera:Limnephilidae) in a woodland stream.* PhD thesis. Michigan State Univ., East Lansing, Mich. 115 pp.

56. Hynes, H. B. N. 1975. The stream and its valley. *Verh. Int. Ver. Limnol.* 19:1–15

57. Iversen, T. M. 1973. Decomposition of autumn-shed beech leaves in a springbrook and its significance for the fauna. *Arch. Hydrobiol.* 72:305–12

58. Iversen, T. M. 1975. Disappearance of autumn-shed beech leaves placed in bags in small streams. *Verh. Int. Ver. Limnol.* 19:1687–92

59. Johnson, D. M. 1973. Predation by damselfly naiads on cladoceran populations: fluctuating intensity. *Ecology* 54:251–68

60. Kaushik, N. K., Hynes, H. B. N. 1971. The fate of dead leaves that fall into streams. *Arch. Hydrobiol.* 68:465–515

61. King, R. H. 1978. *Natural history and ecology of* Stictochironomus annulicrus *(Townes)* *(Diptera:Chironomidae), Augusta Creek, Michigan.* PhD thesis. Michigan State Univ., East Lansing, Mich. 156 pp.

62. Kostalos, M., Seymour, L. R. 1976. Role of microbial enriched detritus in the nutrition of *Gammarus minus. Oikos* 512–16

63. Ladle, M., Bass, J. A. B., Jenkins, W. R. 1972. Studies on production and food consumption by the larval Simuliidae (Diptera) of a chalk stream. *Hydrobiologia* 39:429–48

64. Mann, K. H. 1975. Patterns of energy flow. See Ref. 18, pp. 248–63

65. Mackay, R. J., Kalff, J. 1973. Ecology of two related species of caddisfly larvae in the organic substrates of a woodland stream. *Ecology* 54:499–511

66. Mackay, R. J., Wiggins, G. B. 1979. Ecological diversity in Trichoptera. *Ann. Rev. Entomol.* 24:185–208

67. Madsen, B. L. 1968. The distribution of nymphs of *Brachyptera risi* Mort and *Nemoura flexuosa* Aub. (Plecoptera) in relation to oxygen. *Oikos* 19:304–10

68. Madsen, B. L. 1972. Detritus on stones in streams. *Mem. Inst. Ital. Idrobiol.* 29:385–403

69. McCullough, D. A., Minshall, G. W., Cushing, C. E. 1979. Bioenergetics of lotic filter-feeding insects *Simulium* spp. (Diptera) and *Hydropsyche occidentalis* (Trichotera) and their function in controlling organic transport in streams. *Ecology.* In press

70. McCullough, D. A., Minshall, G. W., Cushing, C. E. 1979. Bioenergetics of a stream "Collector": organism *Tricorythodes minutus* (Insecta: Ephemeroptera). *Limnol. Oceanogr.* 24:45–58

71. McIntire, C. D. 1975. Periphyton assemblages in laboratory streams. See Ref. 18, pp. 403–30

72. McMahon, R. F., Hunter, R. D., Russell-Hunter, W. D. 1974. Variation in aufwuchs at six freshwater habitats in terms of carbon biomass and of carbon:-nitrogen ratio. *Hydrobiologia* 45:392–404

73. Meitz, A. K. 1975. *Alimentary tract microbiota of aquatic invertebrates.* MS thesis. Michigan State Univ., East Lansing, Mich. 64 pp.

74. Melchionni-Santolini, U., Hopton, J. W., eds. 1972. Detritus and its role in aquatic ecosystems. *Mem. Inst. Ital. Idrobiol.* 29:Suppl. 1–540

75. Merritt, R. W., Cummins, K. W., eds. 1978. *An Introduction to the Aquatic Insects of North America.* Dubuque, Iowa: Kendall/Hunt. 441 pp.

76. Merritt, R. W., Mortland, M. M., Gersabeck, E. F., Ross, D. H. 1978. X-ray diffraction analysis of particles ingested by filter-feeding animals. *Entomol. Exp. App.* 24:27–34

77. Minshall, G. W. 1978. Autotrophy in stream ecosystems. *BioScience* 28:767–71

78. Monk, D. C. 1976. The distribution of cellulase in freshwater invertebrates of different feeding habits. *Freshwater Biol.* 6:471–75

79. Nielsen, B. 1966. Carbohydrases of some wrack invertebrates. *Natura. Jutl.* 12:191–94

80. Petersen, R. C., Cummins, K. W. 1974. Leaf processing in a woodland stream. *Freshwater Biol.* 4:343–68

81. Rabeni, C. F., Minshall, G. W. 1977. Factors affecting microdistribution of stream benthic insects. *Oikos* 29:33–43

82. Ramsey, J. H. 1958. Excretion by the malpighian tubules of the stick insect *Dixippus morosus* (Orthoptera, Phasmidae): amino acids, sugars and urea. *J. Exp. Biol.* 35:871–91

83. Russell-Hunter, W. D. 1970. *Aquatic Productivity.* NY: Macmillan. 306 pp.

84. Savage, D. C., Blumershire, R. V. H. 1974. Surface-surface associations in microbial communities populating epithelial habitats in the murine gastrointestinal ecosystem: scanning electron microscopy. *Infect. Immun.* 10:240–50

85. Sedell, J. R., Naiman, R. J., Cummins, K. W., Minshall, G. W., Vannote, R. L. 1978. Transport of particulate organic material in streams as a function of physical processes. *Verh. Int. Ver. Limnol.* 20:1366–75

86. Schindler, J. E. 1971. Food quality and zooplankton nutrition. *J. Anim. Ecol.* 40:589–95

87. Short, R. A., Maslin, P. E. 1977. Processing of leaf litter by a stream detritivore: effect on nutient availability to collectors. *Ecology* 58:935–38

88. Smirnov, N. N. 1962. On the nutrition of caddisworms *Phryganea grandis* L. *Hydrobiologia* 19:252–61

89. Strahler, A. N. 1957. Quantitative analysis of watershed geomorphology. *Trans. Am. Geophys. Union* 38:913–20

90. Suberkropp, K., Klug, M. J. 1974. Decomposition of deciduous leaf litter in a woodland stream: I. A scanning electron microscopic study. *Microb. Ecol.* 1:96–103

91. Suberkropp, K., Klug, M. J. 1976. Fungi and bacteria associated with leaves during processing in a woodland stream. *Ecology* 57:707–19

92. Suberkropp, K., Klug, M. J. 1979. The degradation of leaf litter by aquatic hyphomycetes. In *Fungal Ecology,* ed. D. T. Wicklow, G. C. Carroll. NY: Marcel Dekker. In press

93. Suberkropp, K., Klug, M. J., Cummins. K. W. 1975. Community processing of leaf litter in woodland streams. *Verh. Int. Ver. Limnol.* 19:1653–58

94. Suberkropp, K., Godshalk, G. L., Klug, M. J. 1976. Changes in the chemical composition of leaves during processing in a woodland stream. *Ecology* 57:720–27

95. Sweeney, B. W., Vannote, R. L. 1978. Size variation and the distribution of hemimetabolous aquatic insects: two thermal equilibrium hypotheses. *Science* 200:444–46

96. Treherne, J. E. 1967. Gut absorption. *Ann. Rev. Entomol.* 12:43–58

97. Triska, F. J. 1970. *Seasonal distribution of aquatic hyphomycetes in relation to the disappearance of leaf litter from a woodland stream.* PhD thesis. Univ. Pittsburgh, Pennsylvania. 189 pp.

98. Triska, F. J., Sedell, J. R., Buckley, B. 1975. The processing of conifer and hardwood leaves in two coniferous forest streams: II. Biochemical and nutrient changes. *Verh. Int. Ver. Limnol.* 19:1628–40

99. Vannote, R. L. 1978. A geometric model describing a quasi-equilibrium of energy flow in populations of stream insects. *Proc. Natl. Acad. Sci.* 75:381–84

100. Vannote, R. L., Sweeney, B. W. 1979. Geographic analysis of thermal equilibria: A conceptual model for evaluating the effect of natural and modified ther-

mal regimes on aquatic insect communities. *Am. Nat.* In press
101. Waldbauer, G. P. 1968. The consumption and utilization of food by insects. *Adv. Insect Physiol.* 5:229–82
102. Wallace, J. B., Sherberger, F. F. 1974. The larval retreat and feeding net of *Macronemum carolina* Banks (Trichoptera:Hydropsychidae). *Hydrobiologia* 45:177–84
103. Wallace, J. B., Sherberger, F. F. 1975. The larval retreat and feeding net of *Macronemum transversum* Hagen (Trichoptera:Hydropsychidae). *Anim. Behav.* 23:594–96
104. Wallace, J. B. 1975. The larval retreat and food of *Arctopsyche;* with phylogenetic notes on feeding adaptations in Hydropsychidae larvae (Trichoptera). *Ann. Entomol. Soc. Am.* 68:167–73
105. Wallace, J. B. 1975. Food partitioning in net spinning Trichoptera larvae: *Hydropsyche venularis, Cheumatopsyche etrona* and *Macronemum zebratum* (Hydropsychidae). *Ann. Entomol. Soc. Am.* 68:463–72
106. Wallace, J. B., Malas, D. 1976. The fine structure of capture nets of larval *Philopotamidae* (Trichoptera), with special emphasis on *Dolophilodes distinctus. Can. J. Zool.* 54:1788–802
107. Wallace, J. B., Malas, D. 1976. The significance of the elongate, rectangular mesh found in capture nets of fine particle filter feeding Trichoptera larvae. *Arch. Hydrobiol.* 77:205–12
108. Wallace, J. B., Webster, J. R., Woodall, W. R. 1977. The role of filter feeders in flowing waters. *Arch. Hydrobiol.* 79:506–32
109. Wallace, J. B., Merritt, R. W. 1980. Filter feeding ecology of aquatic insects. *Ann. Rev. Entomol.* 25. In press
110. Ward, G. M., Cummins, K. W. 1978. Life history and growth pattern of *Paratendipes albimanus* in a Michigan headwater stream. *Ann. Entomol. Soc. Am.* 71:292–84
111. Ward, G. M., Cummins, K. W. 1979. Effects of food quality on growth of a stream detritivore, *Paratendipes albimanus* (Meigen) (Diptera: Chironomidae). *Ecology.* In press
112. Westlake, D. F. 1975. Macrophytes. See Ref. 18, pp. 106–28
113. Wetzel, R. G. 1975. Primary production. See Ref. 18, pp. 230–47
114. Wetzel, R. G., Rich, P. H. 1973. Carbon in freshwater systems. *Symp. Biol. Brookhaven Natl. Labs.* 24:1–20
115. Wiggins, G. B. 1977. *Larvae of the North American Caddisfly Genera.* Toronto: Univ. Toronto Press. 401 pp.
116. Wiggins, G. B., MacKay, R. J. 1979. Some relationships between systematics and trophic ecology in nearctic aquatic insects, with special reference to Trichoptera. *Ecology.* In press
117. Wigglesworth, V. B. 1974. *The Principles of Insect Physiology.* London: Chapman and Hall. 827 pp.
118. Wotton, R. S. 1977. The size of particles ingested by moorland stream blackfly larvae. *Oikos* 24:332–35
119. Zimmerman, M. C., Wissing, T. E., Rutter, R. P. 1975. Bioenergetics of the burrowing mayfly, *Hexagenia limbata,* in a pond ecosystem. *Verh. Int. Ver. Limnol.* 19:3039–49

Ann. Rev. Ecol. Syst. 1979. 10:173–200

RELATIONSHIPS BETWEEN ❖4159 LIFE HISTORY CHARACTERISTICS AND ELECTROPHORETICALLY DETECTABLE GENETIC VARIATION IN PLANTS

J. L. Hamrick

Department of Systematics and Ecology, University of Kansas,
Lawrence, Kansas 66045

Y. B. Linhart and J. B. Mitton

Department of Environmental, Population and Organismic Biology,
University of Colorado, Boulder, Colorado 80309

INTRODUCTION

Numerous studies of morphological and physiological traits (reviewed in 3, 57, 95, 132, 135) as well as the more recent studies of enzyme loci (reviewed in 21, 52, 63) have generally shown the existence of significant amounts of genetic variation both within and between plant populations. These reviews (21, 52, 63) have also demonstrated considerable variation among species in intrapopulation allozyme variation. For instance, Hamrick (63) found that populations of woody plants contained higher levels of genetic variation than did populations of herbaceous species; Gottlieb (52) and Brown (21) demonstrated that predominantly outbreeding species maintain higher levels of intrapopulation genetic variation than predominantly inbreeding species. Nevo (111) found that habitat generalists and animal species with cosmopolitan or tropical distributions were typically more variable than species with specialized habitat preferences or temperate distributions.

While these results are interesting and informative, they consider the influence of only a few evolutionarily important characteristics on the main-

173

0066-4162/79/1120-0173$01.00

tenance of genetic variation. It would be desirable, therefore, to consider the effect of a large number of life-history characteristics on the genetic structure of plant populations. In this review we examine the relationships between twelve life-history and ecological variables and the levels of genetic variation maintained within populations of 113 taxa of plants. The variables studied are: 1. taxonomic status, 2. geographic range, 3. generation length, 4. mode of reproduction, 5. mating system, 6. pollination mechanism, 7. fecundity, 8. seed dispersal mechanism, 9. chromosome number, 10. successional stage, 11. habitat type, and 12. cultivation status. The results suggest that even though many of these life history parameters are significantly correlated several of them have significant individual effects on genetic variation.

PROCEDURES

Genetic Data

We have attempted to survey all of the higher-plant allozyme literature published prior to June 1978 for which genetic interpretations of the data could be made. Papers that reported no data on population, allele, or genotype frequences were omitted. Suitable population-genetic data were obtained from 113 taxa representing 110 species of higher plants. Where the data allowed, three measures of intrapopulation genetic variation were calculated for each species: the percent polymorphic loci per population (P), the mean number of alleles per locus (A), and a polymorphic index (PI). The PI is equivalent to the heterozygote frequency under Hardy-Weinberg proportions (64). The P, A, and PI values, together with the number of populations studied and the number of loci analyzed for each species, are given in Table 5. Weighted (by the number of loci analyzed) mean P, A, and PI values were calculated for species represented by more than one study. In addition, the mean heterozygosity per individual (H) is given if it was reported.

Population geneticists have long been aware (68, 71, 94) that electrophoretic techniques were capable of resolving only a portion of the genetic variation within populations. Studies using amino acid sequencing (17), heat denaturation (15, 128), and variation in gel pore size (78) have demonstrated that a single electrophoretically determined mobility class may contain more than one allele. However, the underestimation of genetic variation by electrophoresis is not a major problem in comparative studies. We found no evidence to indicate that there is a bias in the detection of variability that is associated with any life-history or ecological variable. However, the levels of genetic variation reported here may be minimum estimates of genetic variation within these plant species.

Life History Variables

Original papers (and publications cited therein) were consulted to obtain information on the life-history characteristics of each species. For categories such as taxonomic status, geographic distribution, generation length, pollination, and seed dispersal, regional floras were consulted (e.g. 38, 109). The major reference for chromosome numbers was the compendium by Bolkhovskikh et al (16), while Fryxell's (42) review and the original papers therein provided most of the information on mating systems. Fecundity estimates were based on plant descriptions from monographs and floras coupled with our experience on what constituted the average size, life span, and frequency of flowering of the species. For a few species, such as *Colobanthus quitensis, Shorea leprosula,* and *Xerospermum intermedium,* much of the basic information is not available.

The classification scheme for each life history or ecological variable and the classifications of each species are given in Table 5.

Statistical Analyses

For each category of the twelve life-history variables, the percent polymorphic loci (P), the effective number of alleles per locus (A), and the polymorphic index (PI) were summarized with a weighted (by the number of loci) mean and its standard error. Categories within each variable were compared for heterogeneity of PI values by single-classification analysis of variance.

The data were also analyzed with multivariate statistics to determine which traits were correlated and to test whether any combinations of variables influenced the levels of genetic variation observed. Multivariate analyses were performed on a subset of the total data set. Species forming ring chromosomes (*Oenothera* spp. and *Gaura triangulata*) and species with fewer than three loci were excluded from these analyses, leaving 100 species. The data were first analyzed with a principal component analysis to describe the major patterns of covariation. Associations noted in the principal component analysis were further explored with stepwise multiple regression. Explanations of these techniques and examples of their application to biological problems may be found in (5, 26, 54).

RESULTS AND DISCUSSION

Mean Levels of Allozyme Variation in Plants

The mean values of P, A, PI, and H obtained by pooling all 113 taxa were 36.8%, 1.69, 0.141, and 0.156, respectively. However, considerable variation existed among the species studied. A relatively high percentage had low or moderate *P, A,* or *PI* values (Figure 1) while a significant minority had extremely high values.

In comparison with animals, plants tend to have levels of genetic variation roughly equivalent to those of the invertebrates [$P = 46.9\%$; $PI = .135$, (126); $P = 39.7\%$, $H = 0.112$, (111)] but considerably higher than those of vertebrate species [$P = 24.7\%$; $PI = 0.061$, (126); $P = 17.3\%$, $H = 0.036$, (111)]. The present estimates of genetic variation are somewhat lower than those reported by Hamrick (63)—e.g. mean $PI = 0.141$ vs 0.160. This decrease appears to be due to the use of weighted means in this study and indicates that a negative relationship may exist between the number of

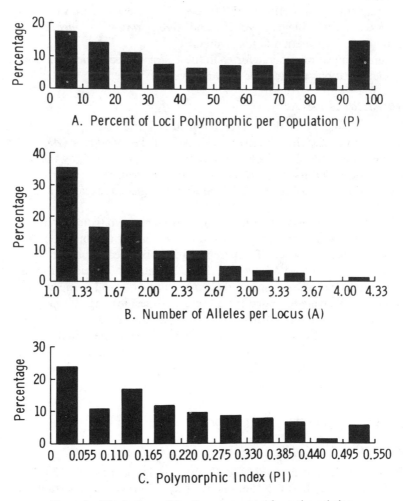

Figure 1   Distributions of the three measures of genetic variation.

enzyme loci surveyed and the magnitude of P, A, and PI. This relationship is undoubtedly caused by the inclusion of studies that report polymorphic loci rather than survey a random sample of enzyme loci. Rather than exclude these studies from our review, we used weighted means to compensate for this problem. A second bias may occur if studies were included that preferentially used highly variable enzyme systems (77). In our study the variable substrate enzymes were the most commonly used functional class (49%) but had lower PI values (0.128) than the regulatory enzymes ($PI = 0.161$) that were the least used functional class (18%). Thus, it appears that the preferential use of variable substrate enzyme systems in many plant genetic studies does not result in an overestimate of genetic variation, as has been suggested by some authors (52).

Few studies of the population genetics of lower plants exist. We know two studies of algal populations (27, 43), three of fungi (45, 131, 137), and one of a non-seed-producing vascular plant (89). No general conclusions can be drawn, but most of the algal and fungal species studied (27, 43, 45, 137) as well as the club moss, *Lycopodium lucidulum* (89), had less variation than was found in the higher plants. An exception is *Neurospora intermedia* (131), with much allozyme variation.

Genetic Variability and Life-History Features

TAXONOMIC STATUS   Although the gymnosperms are more variable than either the monocots or the dicots (Table 1), the existence of inherent differences between gymnosperms and angiosperms is doubtful. The differences in allozyme variation are probably due to the fact that gymnosperms are long-lived trees, are primarily outcrossed, and have high fecundities— all factors associated with high variability (see below). The differences in variability between monocots and dicots are less easily explained. Both groups contain species that represent a wide variety of longevities, mating systems, and fecundities. However, the monocots reviewed are all grasses while the dicots are represented by a wider array of species. The higher variability of the grasses may not be representative of the monocots as a whole.

GEOGRAPHIC RANGE AND DISTRIBUTION   Mean levels of genetic variation are lowest in the endemic category, increase up to the regional category (Table 1), but then decrease in the widespread category. The decrease is probably due to the presence in this category of weedy, predominantly self-pollinated species such as *Avena fatua, Elymus* spp., *Hordeum* spp., *Lolium* spp., *Oenothera biennis,* and *Tragopogon* spp., whose variability is relatively low (see below). In fact, of the 35 widely distributed species, 29 are also categorized as weedy or early successional. Owing to the highly

Table 1 Levels of variability between categories of the 12 life history and ecological variables

Variable	Number of species	Mean number of loci	Polymorphic loci (P) (%) $\bar{x}^a$	S.E.	Number alleles/locus (A) $\bar{x}$	S.E.	Polymorphic index $(PI)^b$ $\bar{x}$	S.E.
1. Taxonomic Status							**	
Gymnospermae	11	9.2	67.01	7.99	2.12	0.20	0.270	0.041
Dicotyledoneae	74	11.4	31.28	3.31	1.46	0.06	0.113	0.014
Monocotyledoneae	28	11.6	39.70	6.02	2.11	0.19	0.165	0.026
2. Geographic Range							*	
Endemic	17	15.1	23.52	5.06	1.43	0.11	0.086	0.019
Narrow	22	11.4	36.73	6.01	1.60	0.14	0.158	0.030
Regional	39	8.3	55.96	5.13	1.85	0.10	0.185	0.025
Widespread	35	12.5	30.36	5.03	1.58	0.15	0.120	0.021
3. Generation Length							***	
Annual	42	11.2	39.47	4.32	1.72	0.11	0.132	0.017
Biennial	13	17.2	15.78	5.12	1.26	0.09	0.060	0.020
Short-lived Perennial	31	12.0	28.09	5.06	1.46	0.09	0.123	0.023
Long-lived Perennial	27	7.6	65.77	5.08	2.07	0.13	0.267	0.027
4. Mode of Reproduction							NS	
Asexual	1	8.0	50.00	0.00	1.91	0.00	0.139	0.000
Sexual	95	11.7	35.64	3.03	1.63	0.07	0.135	0.012
Both	17	8.9	41.71	8.12	1.67	0.14	0.185	0.034
5. Mating System							**	
Primarily Selfed	33	14.2	17.92	3.21	1.27	0.06	0.058	0.014
Mixed	42	8.6	14.16	4.89	1.76	0.10	0.181	0.022
Primarily Outcrossed	36	11.3	51.07	4.95	1.85	0.12	0.185	0.022
6. Pollination Mechanism							***	
Selfed	33	14.2	18.99	3.51	1.31	0.07	0.058	0.016
Animal	55	9.5	38.83	3.94	1.55	0.07	0.130	0.015
Wind	23	10.7	57.45	6.29	2.27	0.17	0.264	0.028
7. Fecundity							***	
$< 10^2$	21	12.0	40.06	6.45	1.72	0.16	0.127	0.026
$10^2 - 10^3$	27	12.4	26.35	3.46	1.44	0.07	0.096	0.013
$10^3 - 10^4$	22	11.9	36.98	6.37	1.64	0.10	0.199	0.034
$> 10^4$	40	9.8	67.99	5.99	2.27	0.17	0.286	0.033
8. Seed Dispersal Mechanism							NS	
Large	27	11.4	37.42	3.73	1.76	0.16	0.156	0.023
Animal-Attached	16	11.1	28.79	5.55	1.55	0.08	0.092	0.020
Small	26	12.4	32.98	5.10	1.51	0.09	0.118	0.018
Winged or Plumose	21	12.2	44.91	7.27	1.86	0.13	0.188	0.029
Animal Ingested	20	7.0	32.98	8.25	1.43	0.10	0.132	0.036
9. Chromosome Number							**	
10 – 20	50	13.1	35.52	3.61	1.55	0.08	0.111	0.014
22 – 30	44	10.0	37.41	5.35	1.73	0.10	0.175	0.023
> 30	16	8.9	41.65	7.20	2.10	0.14	0.224	0.030
10. Stage of Succession							**	
Weedy and Early	54	12.5	29.67	3.82	1.60	0.08	0.116	0.015
Middle	49	9.7	37.90	4.43	1.56	0.08	0.137	0.019
Late	10	12.0	62.76	5.28	2.14	0.19	0.271	0.038

Table 1 *(Continued)*

Variable	Number of species	Mean number of loci	Polymorphic loci (P) (%) $\bar{x}^a$	Polymorphic loci (P) (%) S.E.	Number alleles/locus (A) $\bar{x}$	Number alleles/locus (A) S.E.	Polymorphic index $(PI)^b$ $\bar{x}$	Polymorphic index $(PI)^b$ S.E.
11. Habitat Type							NS	
Xeric	4	8.8	15.39	8.20	1.11	0.09	0.048	0.040
Sub Mesic	19	10.5	43.68	4.86	1.66	0.08	0.140	0.020
Mesic	82	11.4	36.01	3.61	1.65	0.07	0.146	0.016
Hydric	8	13.0	27.71	10.33	1.59	0.22	0.145	0.050
12. Cultivation Status							NS	
Cultivated	21	6.8	38.99	7.15	1.61	0.12	0.172	0.032
Non Cultivated	89	12.6	36.10	3.12	1.63	0.07	0.136	0.013
Both	3	2.7	50.00	—	1.75	0.21	0.209	0.035

[a] Weighted means and standard deviations are given for each measure of variability.

[b] Differences in PI between categories are tested by ANOVA. Significance is indicated by * ($P < 0.05$), ** ($P > 0.01$), and *** ($P < 0.001$). NS = not significant.

disturbed nature of their habitats these weedy and early successional species might be expected to encounter relatively little site-to-site variation regardless of their wide geographic ranges. Thus a limited number of highly plastic "general purpose genotypes" may suffice to adapt these species to their spatially homogeneous but temporally heterogeneous environments (8).

The association of higher levels of variation with wider geographic distribution is supported by specific comparisons of related species—e.g. the genera *Baptisia, Lupinus, Lycopersicon,* and *Stephanomeria;* but other comparisons between closely related species are not consistent with this conclusion—e.g. the genera *Avena, Clarkia, Elymus, Hordeum, Hymenopappus, Phlox,* and *Pinus.* Although this inconsistency may reflect differences in the numbers of individual plants and range of populations sampled by the different groups of workers, it also points out the dangers of making all-encompassing generalizations. Nevertheless, the overall results indicate that nonweedy species with broad distributions have high levels of variability.

Widely distributed nonwoody species might be expected to experience a greater variety of ecological conditions both within and between their populations. Coupled with their generally large population sizes, more continuous distributions, and a greater potential for gene flow this could lead to the maintenance of higher levels of genetic variation. Most endemic species, on the other hand, have smaller populations, exhibit less gene flow, and experience less environmental heterogeneity—conditions that would lead to the maintenance of less genetic variation.

Of the 113 taxa represented in this study, only 26 can be considered tropical or subtropical. These include (see Table 5) species 4, 15–23, 40, 60–68, 78, 81, 96, 99, 109, 110, 112, and 113. With the exception of *Shorea* and *Xerospermum,* all are cultivated or related to cultivated species. Of particular interest are the tropical trees and shrubs (species 15–23, 40, 81, 96, 99, and 110), whose mean values of *P, A,* and *PI* are 67.0%, 1.83, and 0.279, respectively. Thus, although these tropical trees have reproductive biologies and population structures different from those of temperate woody plants, they appear to have rather similar levels of genetic variability. Nevo (111) compared levels of variability in invertebrates and vertebrates and found that in both classes tropical species were more variable than temperate ones. The data for trees are not consistent with his results.

GENERATION LENGTH   Our results agree with those of Hamrick (63) in demonstrating that long-lived woody perennials contain the highest levels of allozyme variation (Table 1). Shorter-lived herbaceous perennials and annuals have intermediate levels of variation, and biennials have the least. The question therefore arises whether the differences observed between trees and herbaceous plants are due to the life history traits of trees or whether something inherent in woody plants leads to the maintenance of high levels of genetic variation. Certainly many of the tree species included in this study combine life history characteristics associated (see below) with high levels of genetic variation: long generation times, an outcrossing mating system, wind pollination, high fecundity, and winged seed dispersal. Conversely, many herbaceous perennials have higher turnover rates, asexual (or a mixture of sexual and asexual) reproduction, a range of pollination mechanisms and mating systems, and somewhat lower fecundities—traits associated with lower amounts of genetic variation. In addition, herbaceous perennials with outcrossing rates, fecundities, and pollination mechanisms similar to those of trees (e.g. *Cucurbita, Lolium perenne, Mimulus, Opuntia,* sexual *Panicum maximum*) are as genetically variable as the tree species. Thus the differences in genetic variation between trees and many of the herbaceous species may be explained by differences in their life history characteristics. A wider spectrum of angiosperm trees and shrubs must be studied before definite conclusions can be reached.

Differences in chromosome numbers might also provide an explanation for differences observed between herbaceous and woody perennials. As discussed below, high chromosome numbers are associated with high variability. Stebbins (134) has shown that woody plants tend to have higher base numbers than herbaceous species. However, this consideration does not explain our results; the correlation between generation length and chromosome number is 0.17 and is not significant. Of the 25 species of woody perennials for which we have chromosome data, six have chromosome numbers of less than 20, 19 have numbers in the 20s, and there are no species with 30 or more. Herbaceous perennials, on the other hand, had relatively higher chromosome numbers but lower levels of genetic variation.

MODE OF REPRODUCTION   Only certain populations of *Panicum maximum* reproduce solely by asexual means. These populations have a *PI* value of 0.139 while the sexual populations have *PI* values of 0.520 (144). We nevertheless find that species with a mixture of asexual and sexual reproduction have more variation than sexually reproducing species (Table 1). The greater variation in the former may be explained in part by the observation that many are obligately outcrossing (e.g. *Agrostis, Liatris, Lolium perenne, Opuntia*) and none is predominantly self-pollinated. This association of vegetative propagation and predominant outcrossing has been noted previously for many plant species (58).

MATING SYSTEM   There is a positive association between the amount of outcrossing and genetic variation (Table 1). This agrees with previous reviews (21, 52) and with studies that compare related species with different breeding systems. Thus predominately selfed species such as *Clarkia franciscana* (47), *Gaura triangulata* (53, 85), and *Lycopersicon parviflorum* (114, 116) are less variable than their predominately outcrossed congeners. Rick et al (115) have also demonstrated that highly outcrossed populations of *L. pimpinellifolium* are more polymorphic than inbred populations. Similarly, obligately outcrossing species of *Phlox* are more variable than self-pollinated species (86, 88). In contrast, *Avena fatua* and *A. barbata* are both highly self-fertilized species with similar habitat requirements, but *A. fatua* has more genetic variation ($PI = 0.138$ vs 0.043). Also, comparisons between the self-fertilized species of *Oenothera* and the outcrossed species, *Oe. hookeri,* show the latter contains less variation (92). Faced by such mixed results Jain (73) came to no definite conclusion regarding the relative levels of variability in inbreeders and outbreeders. On the average, however, the present data tend to support the concept that high outcrossing rates, extensive gene flow, and large neighborhood sizes lead to the maintenance of higher levels of genetic variation.

POLLINATION MECHANISM   Differences in levels of variability among the three categories of pollination are dramatic. Wind pollination is associated with high levels of variability. This result is not unduly biased by the presence of wind-pollinated, outcrossing conifers. Of the 19 wind-pollinated species, the 11 conifers have a mean *PI* of 0.300, while the 8 herbaceous species have a mean *PI* of 0.305. Since the majority of animal-pollinated species reviewed occur in temperate habitats they are primarily pollinated by efficient social insects that provide rather limited pollen dispersal (91). In contrast, wind-pollinated temperate species produce vast quantities of pollen and as a result distribute their pollen longer distances, have larger effective neighborhood sizes, less inbreeding, less fixation of genes by genetic drift and selection, and more intrapopulation variation.

FECUNDITY   The results (Table 1) show quite convincingly that species with high fecundities are highly variable. A possible explanation is that highly fecund species are capable of producing a large variety of recombinant progeny capable of reaching and surviving in a wide variety of microhabitats. This could lead to the maintenance of high levels of genetic variation even if the establishment of new individuals is largely due to chance events. Thus species that produce more variable progeny through a combination of outcrossing, wind pollination, and high fecundity should have high intrapopulation variation.

A second explanation has been offered by Stebbins (133) and Williams (147), who argue that selection in highly fecund populations is far more intense than is appreciated. Although there is some controversy over the number of polymorphic loci that can be maintained by natural selection (81, 94), some models accommodate a vast number of selectively maintained polymorphisms (41, 106, 139, 148). If much of natural selection is balancing, and if species with high fecundities experience high selective mortality, it follows that species with high fecundity should exhibit higher levels of genetic variation. The data presented here are compatible with this prediction.

For an independent test of the relationship between fecundity or reproductive potential and genetic variation we have consulted a review (113) of allozyme variation in animals. Estimates of the maximum number of offspring that a female could bear (given a long life and no mortality of offspring) and average heterozygosities were obtained for 101 animal species (Table 2). There is a highly significant, positive correlation between average heterozygosity and the natural log of the reproductive potential among these animal species ($r = .66$, $P < .001$). The correlation in 100 species of plants is lower but also significant ($r = .35$, $P < .01$).

Table 2 Mean heterozygosities and fecundities of several groups of animals. The correlation between mean heterozygosity and log mean fecundity for these 101 species is 0.66 ($P < 0.001$). Genetic data were obtained from (113).

Group	Number species	Mean heterozygosity	Natural log of mean fecundity
Birds	3	0.031	4.09
Rodents	22	0.095	4.07
Marine Mammals	2	0.015	2.99
Drosophila	41	0.169	7.49
Fishes	18	0.135	6.75
Amphibians	7	0.182	8.82
Lizards	8	0.087	5.01
Total	101		

SEED DISPERSAL   Levels of genetic variation are not strongly associated with seed dispersal distance. This is unexpected since we had predicted that seed dispersal would have much the same effect as pollination: Species with restricted dispersal and therefore limited gene flow would have less genetic variation than those with greater dispersal and extensive gene flow. Our results may be consistent with this prediction since species that disperse their seeds by means of a wing or a plumose pappus seem to have more variation. This result is heavily influenced by the tree species, whose mean *PI* value is 0.301, whereas the mean *PI* value of herbaceous plants in this category is 0.107. It is possible that in trees the combination of greater height and wind dispersal interacts to produce levels of gene flow that can affect the maintenance of genetic variation.

CHROMOSOME NUMBER   Several investigators have discussed the relationship between ploidy levels and genetic variation (58, 134). However, since the exact ploidy status of many of the plant species dealt with here is unknown, we have used the diploid chromosome number in our analyses. We expected that species with high chromosome numbers would have increased recombination and that their progeny would be more variable genotypically. As has been argued above, species that produce a large variety of recombinant progeny might be expected to maintain more genetic variation. The results bear out this expectation. It is also noteworthy that the relationship between chromosome number and variability is not confounded by any other life-history variable (Tables 3 and 4).

STAGE OF SUCCESSION   Open, early-successional habitats are primarily influenced by physical factors and therefore tend to be relatively homoge-

Table 3 Correlations among genetic, life-history, and ecological variables. Sample size is 100, except for occasional missing data points. The critical value ($P = 0.05$) for these correlations is 0.20. See Table 5 for a description of each variable.

	P	A	PI	TAX	GEO	GEN	REP	MAT	POL	FEC	SED	CHM	SUC	HAB
A	.83	—	—	—	—	—	—	—	—	—	—	—	—	—
PI	.88	.80	—	—	—	—	—	—	—	—	—	—	—	—
TAX	-.12	.00	-.07	—	—	—	—	—	—	—	—	—	—	—
GEO	.19	.24	.20	.23	—	—	—	—	—	—	—	—	—	—
GEN	.21	.11	.28	-.35	.21	—	—	—	—	—	—	—	—	—
REP	-.18	.00	-.16	-.04	-.10	-.43	—	—	—	—	—	—	—	—
MAT	.41	.34	.29	-.48	-.03	.28	-.20	—	—	—	—	—	—	—
POL	.39	-.44	.42	-.28	-.18	.33	-.20	.74	—	—	—	—	—	—
FEC	.34	.30	.36	-.40	.25	.88	-.34	.40	.45	—	—	—	—	—
SED	.13	.00	.09	-.49	-.04	.59	-.29	.16	.04	.50	—	—	—	—
CHM	.03	.13	.18	.15	.14	.18	-.11	-.02	-.11	.16	-.10	—	—	—
SUC	.25	.09	.17	-.49	-.28	.37	-.09	.39	.14	.40	.36	-.08	—	—
HAB	.01	.14	.21	.06	.22	-.28	-.26	-.08	.15	.21	-.03	.08	-.12	—
CUL	-.06	.06	-.10	-.02	-.29	-.15	.41	-.01	-.07	-.08	-.19	.06	.07	-.18

neous. As succession proceeds the biotic environment becomes increasingly important and the habitats become more complex and heterogeneous. As noted above, species inhabiting such complex environments may be more genetically variable. Although present results (Table 1) appear to support this expectation, nine of the ten late-successional species are trees. Thus the higher levels of variation observed in the late-successional category may be due to the life history characteristics of trees rather than to the successional stage. On the other hand, early-successional species are the least variable, and the overall trends are toward increased variability in species of more complex communities. This result agrees with data for *Drosophila* (46) and for the gall-forming aphid, *Geoica utricularia* (149). However, the direction of causality is not clear. While it can be argued (46) that species with high levels of variability tend to do well in variable environments, it is equally plausible that heterogeneous environments select for genetic heterogeneity (103).

HABITAT TYPE   Although we had expected that the directional selection pressures of the extreme environments might lead to a reduction in genetic variation, the four habitat types did not differ significantly (Table 1). However, the extreme xeric or hydric habitats were represented by only a few species.

STATUS OF CULTIVATION   Cultivated species had slightly more variation than the noncultivated species (Table 1). Cultivated trees had a *PI* value of 0.215 while the cereals and vegetables had a *PI* value of 0.118; thus, the higher variation in the cultivated species is predominantly due to the large proportion (52%) of trees in this category.

The variability in four pairs of closely related cultivated and noncultivated taxa may be compared. *Hordeum spontaneum* and the noncultivated *Phlox drummondii* have considerably more variation than the cultivated taxa (*H. vulgare* and cultivated *P. drummondii*). In contrast, comparisons of *Lycopersicon esculentum* and *L. e.* var *cerasiformae* and *Zea mays* and *Z. mexicana* indicate that the cultivated species maintain more variation. Thus no definite conclusions involving the levels of variation maintained in populations of cultivated vs noncultivated species can be made at this time.

Analysis of Multivariate Data

Multivariate techniques were employed to identify groups of variables varying in concert and to determine the covariation of genetic variation with the ecological and life-history variables. Approximately half of the correlations among the ecological, life history, and genetic variables were statistically significant (Table 3). The highest correlations are among the three genetic variables, indicating that they contain similar information. Of the 12 ecological and life history variables, pollination mechanism, mating system, and fecundity have the highest correlations with the genetic variables. There are strong positive correlations between fecundity and generation length, and between generation length and seed dispersal mechanisms. Since predominantly self-pollinated species were assigned to category 1 of both the mating-system and pollination-mechanism variables, these variables are also highly correlated.

Fifteen variables from 100 species were subjected to a principal component analysis, and three axes, describing 58% of the variation in the data, were extracted from the correlation matrix (Table 4, Figure 2). The first principal axis has high loadings from the three genetic variables, generation length, mating system, pollination mechanism, fecundity, seed dispersal, and successional status (Table 4). In Figure 2, species located near the positive end of the first principal axis have high levels of genetic variation, are gymnosperms, have long generation times, are primarily outcrossed, are wind pollinated, are highly fecund, have wind-dispersed or animal-ingested seeds, and are species of late successional stages. Species near the other end of this axis have complementary attributes.

The second principal component has high loadings from the genetic variables, taxonomic status, seed dispersal mechanism, and successional stage (Table 4). Species at the positive end of this axis have high genetic variability, are monocots with large seeds or seeds carried by animals, and are species of early successional stages. Species at the opposite end of this continuum have complementary character states.

Variables that load highest on the third principal component are geographic range, mode of reproduction, habitat type, and cultivation status (Table 4). Species at the positive end of this axis have narrow geographic ranges, have sexual reproduction, are found in xeric and submesic environments, and are both wild and under cultivation. This axis is not as well defined as the first two axes and contributes little to our understanding of genetic variation in these species.

Since patterns of variation in these axes may complement one another, it is useful to portray them simultaneously. Figure 2 shows the position of each species in a three dimensional space defined by the three orthogonal axes. Low values for the first two are in the lower right-hand corner of the figure, while high values for these axes are in the upper left-hand corner. Numbers placed on tall wires have high values for the third axis. The figure may be interpreted as follows: There is a cluster of species in the background of the figure (numbers 38, *Eucalyptus obliqua;* 39, *E. pauciflora;* 88, *Picea abies;* 90, *Pinus longaeva;* 95, *P. sylvestris;* and 97, *Pseudotsuga menziesii*).

This group has high values on the first and third axes and intermediate values on the second axis. Species in this portion of the figure have high levels of genetic variation, have long generation times, are outcrossed, and have high fecundity. Another group that may be used as an example are species 27 (*Clarkia franciscana*), 63 (*Lycopersicon cheesmanii*), and 67 (*L. parviflorum*), which are in the lower right-hand corner of the figure (low

Table 4 Variable loadings from a correlation matrix onto the first three principal components in a 15-character set of genetic, ecological, and life history variables for 100 species of plants

| | Principal components | | | |
Data set	I	II	III	Communalities
A. Whole data set				
P	.71	.45	.32	.81
A	.62	.60	.31	.85
PI	.70	.53	.16	.80
TAX	−.45	.61	−.32	.68
GEO	.23	.44	−.49	.49
GEN	.72	−.35	−.39	.80
REP	−.40	.10	.57	.50
MAT	.67	−.12	.31	.56
POL	.68	.15	.13	.50
FEC	.80	−.25	−.22	.76
SED	.47	−.58	−.17	.59
CHM	.14	.27	−.21	.14
SUC	.46	−.51	.35	.60
HAB	.26	.20	.48	.33
CUL	−.18	−.02	.58	.37
Eigenvalues	4.45	2.36	1.97	
% Variance	29.7	15.8	13.1	
B. Ecological and life history variables				
TAX	.61			
GEO	.12			
GEN	.85			
REP	−.48			
MAT	.65			
POL	.60			
FEC	.87			
SED	.64			
CHM	.08			
SUC	.55			
HAB	.24			
CUL	−.22			
Eigenvalue	3.69			
% Variance	30.8			

values for axes 1, 2; high values for axis 3). These three species tend to be short-lived, have low fecundity, small or animal-dispersed seeds, and low levels of genetic variation.

The communalities of the variables in the principal component analysis indicate what proportion of the variance is explained by the three orthogonal principal axes (Table 4). Most of the variation of the genetic variables, generation length, and fecundity are accounted for, but little of the variation in chromosome number is included in these axes. In one respect, the high communalities for the genetic variables are misleading because the high correlations among these variables indicate that the variance of any genetic variable is explained by the other genetic variables, not by the ecological and life history variables. A better perspective on the covariation of the genetic with the ecological and life history variables may be gained by performing a principal component analysis with only ecological and life history variables and testing the correlation of the genetic variables with the first

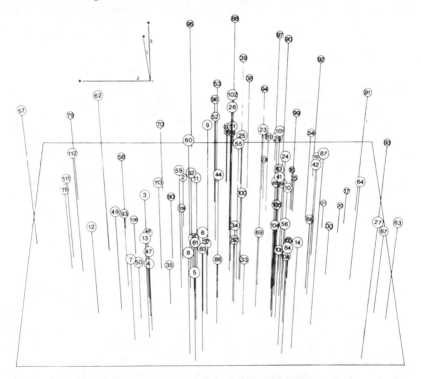

Figure 2 Three dimensional representation of a principal component analysis of genetic, life-history, and ecological variables of 100 species of plants. The labeled axes correspond to the principal axes presented in Table 3A, and the numbers correspond to numbers assigned to the species listed in Table 5. See text for further discussion.

principal axis (Table 4). This axis is similar to the first axis in the previous analysis, since it carries heavy loadings from taxonomic status, generation length, mating system, pollination mechanism, fecundity, seed dispersal, and successional status. This axis is significantly correlated with the variable PI ($r = .40$; $P < .001$).

Although the analyses presented so far indicate that genetic variation is associated with a suite of covarying ecological and life-history variables, it is difficult to say which of these variables are most closely associated with the polymorphic index. To resolve this difficulty a stepwise multiple regression analysis was employed in which PI was the dependent variable and the 12 ecological and life-history variables were independent variables. In this analysis, only two variables were significant: Pollination mechanism is entered first, followed by fecundity. Species that are wind-pollinated and have high fecundity have the most genetic variation, while selfed species with low fecundity have the lowest levels of genetic variation. The multiple regression equation accounts for 21% of the variation in PI across the 100 species in this analysis.

CONCLUSIONS

Our results indicate that seed plants as a group maintain relatively high levels of genetic variation. A number of characteristics of plants should assist in the maintenance of this variation. First, since most plants are unable to move from their site of germination, plants of all kinds might be expected to respond to their environments as coarse-grained. Plants respond to temporal or spatial environmental heterogeneity in at least two ways: They may adjust phenotypically through individual developmental plasticity (8, 18, 19, 67), or they may evolve locally adapted ecotypes (1, 4, 75). The presence of such ecotypes is a common feature of a wide variety of plant species and could lead to the maintenance of large amounts of genetic variation within plant populations (4). Species in which microhabitat variation in allozyme loci is documented include *Avena barbata* (2, 64, 65), *Liatris cylindracea* (121–123), *Helianthus annuus* (141), *Lolium multiflorum* (J. B. Mitton, unpublished data), *Bromus mollis* (23), *Silene maritima* (10), and *P. ponderosa* (108). Studies of localized variation in quantitative traits are more common and include species with a wide variety of life history traits [e.g. *Quercus* spp. (14), *Anthoxanthum odoratum* (130), *Poa annua* (82), and *Veronica peregrina* (97)].

Second, plant population numbers are often large and relatively stable, especially for perennial species in the later stages of succession. Thus, perennial plant populations are less likely to be affected by genetic drift and would be more likely to maintain higher levels of genetic variation than

Table 5 Data on 16 variables for 113 species[a]

Species	Number of populations	Number of loci	Measures of genetic variation				Life history and ecological characteristics												References
			P	A	Pl	H	1	2	3	4	5	6	7	8	9	10	11	12	
1. *Abies lasiocarpa*	3	1	100.0	3.00	.399	—	G	3	4	S	3	W	4	4	24	3	3	N	56
2. *Agrostis stolonifera*	6	12	—	—	.181	—	M	4	3	B	3	W	4	3	28	1	3	N	151
3. *Arabidopsis thaliana*	17	2.5	51.0	2.67	.318	—	D	4	1	S	1	S	1	3	10	1	3	N	61, 72
	(1)	(2)	(2)	(1)	(1)														
4. *Arachis hypogaea*	26	27	22.0	—	—	—	D	4	1	S	1	S	2	1	40	1	3	C	28
5. *Avena barbata*	44	17	26.1	1.30	.043	—	M	3	1	S	1	S	1	2	28	1	2	N	29, 65, 99, 101, 107, 129
	(7)	(7)	(4)	(3)	(7)														
6. *A. canariensis*	14	23	39.0	1.69	—	—	M	1	1	S	1	S	1	2	14	1	2	N	34
7. *A. fatua*	42	11	31.9	—	.138	.055	M	4	1	S	1	S	1	2	42	1	2	N	76, 99, 101
	(3)	(3)	(2)		(2)	(1)													
8. *A. hirtula*	23	28	—	—	.012	—	M	3	1	S	1	S	1	2	14	1	2	N	129
9. *Baptisia leucophaea*	9	5	87.8	1.72	.287	.407	D	3	3	S	2	A	3	1	18	2	3	N	124
10. *B. nuttaliana*	1	5	20.0	1.20	.089	.275	D	2	3	S	2	A	3	1	18	2	3	N	124
11. *B. sphaerocarpa*	3	5	60.0	1.60	.229	.280	D	3	3	S	2	A	3	1	18	1	4	N	124
12. *Bromus mollis*	10	6	92.0	1.92	.204	—	M	4	1	S	1	S	1	2	28	1	3	N	23
13. *B. rubens*	6	3	—	—	.202	—	M	3	1	S	1	S	1	2	28	1	2	N	150
14. *Chenopodium fremontii*	40	6	18.8	1.03	.019	—	D	3	1	S	1	S	3	3	18	1	1	N	35
15. *Citrus aurantifolia*	—	4	50.0	1.50	.250	—	D	3	4	B	2	A	4	5	22V	2	3	C	143
16. *C. aurantium*	—	4	75.0	1.75	.154	—	D	3	4	B	2	A	4	5	18	2	3	C	143
17. *C. grandis*	—	4	25.0	1.25	.049	—	D	3	4	B	3	A	4	5	18V	2	3	C	143
18. *C. jambhiri*	—	4	100.0	2.00	.500	—	D	3	4	B	2	A	4	5	—	2	3	C	143
19. *C. limon*	—	4	100.0	2.00	.499	—	D	3	4	B	2	A	4	5	18V	2	3	C	143
20. *C. medica*	—	4	25.0	1.25	.070	—	D	3	4	B	2	A	4	5	18	2	3	C	143
21. *C. paradisi*	—	4	25.0	1.25	.125	—	D	3	4	B	2	A	4	5	27V	2	3	C	143
22. *C. reticulata*	—	4	100.0	2.75	.361	—	D	3	4	B	2	A	4	5	18V	2	3	C	143
23. *C. sinensis*	—	4	75.0	1.75	.310	—	D	3	4	B	2	A	4	5	27V	2	3	C	143
24. *Clarkia amoena*	1	8	62.0	1.37	.071	—	D	2	1	S	2	A	2	3	14	2	2	N	47

Table 5 *(Continued)*

Species	Number of populations	Number of loci	Measure of genetic variation				Life history and ecological characteristics												References
			P	A	Pl	H	1	2	3	4	5	6	7	8	9	10	11	12	
25. C. biloba	3	8	61.0	2.09	.203	.150	D	2	1	S	2	A	2	3	16	2	2	N	49
26. C. dudleyana	1	8	75.0	2.38	.250	.160	D	1	1	S	2	A	2	3	18	2	2	N	49
27. C. franciscana	1	8	0.0	1.00	.000	.000	D	1	1	S	1	S	2	3	14	2	1	N	47
28. C. lingulata	2	8	56.0	2.06	.175	.080	D	1	1	S	2	A	2	3	18	2	2	N	49
29. C. rubicunda	4	8	59.0	1.69	.177	.110	D	1	1	S	2	A	2	3	14	2	2	N	47
30. Colobanthus quitensis	2	22	0.0	1.00	.000	.000	D	2	3	B	—	—	4	5	—	2	4	N	83
31. Cucurbita foetidissima	9	2	—	2.00	.256	.256	D	3	3	S	3	A	4	5	40	1	2	NC	96
32. Danthonia sericea	10	7	28.6	1.57	—	—	M	3	3	S	3	W	3	1	36	2	2	N	59
33. Elymus canadensis	63	0.5	15.8	1.09	.026	.019	M	4	3	S	1	S	4	2	28	2	3	N	31, 120
	(1)	(2)	(2)	(1)	(1)	(1)													
34. E. hystrix	—	13	7.7	—	—	—	M	2	3	S	1	S	4	2	28	2	3	N	31
35. E. riparius	—	13	0.0	—	—	—	M	4	3	S	1	S	4	2	28	1	4	N	31
36. E. virginicus	—	13	31.0	—	—	—	M	4	3	S	1	S	4	2	28	2	3	N	31
37. E. wiegandii	—	13	7.7	—	—	—	M	2	3	S	1	S	4	2	28	1	3	N	31
38. Eucalyptus obliqua	4	3	100.0	2.42	.351	—	D	3	4	S	2	A	4	4	22	3	3	N	24
39. E. pauciflora	3	7	100.0	3.43	.278	.244	D	2	4	S	2	A	4	4	22	3	3	N	112
40. Ficus carica	4	2	100.0	3.12	.530	.468	D	3	4	A	3	A	4	5	26	2	2	C	145
41. Gaura brachycarpa	—	12	25.0	1.25	—	.060	D	2	1	S	2	A	2	3	14	2	3	N	85
42. G. demareei	2	18	22.0	1.39	.050	.050	D	1	1	S	3	A	2	3	14	2	3	N	53
43. G. longiflora	3	18	25.0	1.40	.074	.074	D	3	1	S	3	A	2	3	14	2	3	N	53
44. G. suffulta	—	12	33.0	1.42	—	.030	D	2	1	S	3	A	2	1	14	1	3	N	85
45. G. triangulata	5	12	8.0	1.08	—	.080	D	2	1	S	1	S	2	3	14	3	3	N	85
46. Helianthus annuus	5	2	50.0	1.50	.162	.176	D	4	1	S	3	A	3	1	34	1	3	NC	141
47. Hordeum distichum	4	5	—	—	.135	.043	M	4	1	S	1	S	2	2	14	1	3	N	60
48. H. jubatum	3	6	61.1	1.67	.192	.015	M	4	3	S	1	S	3	2	28	1	3	N	7
49. H. spontaneum	11	4	66.8	2.33	.282	.010	M	3	1	S	1	S	2	2	14	1	3	N	25
50. H. vulgare	30	4	57.5	1.86	.148	—	M	4	1	S	1	S	2	2	14	1	3	C	79, 80

#	Species																		
51.	*Hymenopappus artemisifolius*	12	7	71.0	2.05	.222	.208	D	2	2	S	3	A	2	4	34	2	3	N	6
52.	*H. scabiosaeus*	14	7	71.0	2.12	.218	.201	D	3	2	S	3	A	2	4	34	2	3	N	6
53.	*Larix decidua*	1	3	66.7	2.67	.347	—	G	4	4	S	3	W	4	4	24	3	3	N	105
54.	*Liatris cylindracea*	1	27	55.5	1.63	.158	.057	D	3	3	B	3	A	4	4	20	3	3	N	121–123
55.	*Limnanthes alba*	6	13	52.2	1.73	.159	.159	D	1	1	S	3	A	2	1	10	2	4	N	36, 74
		(2)		(2)	(2)	(1)														
56.	*L. floccosa*	6	13	17.6	1.19	.130	.130	D	1	1	S	1	S	1	1	10	2	3	N	36, 74
		(2)		(2)	(2)	(1)														
57.	*Lolium multiflorum*	17	15	94.7	3.55	.331	—	M	4	1	S	3	W	1	1	14	1	3	N	33, Mitton unpublished data
58.	*L. perenne*	9	4	75.0	2.25	.280	.267	M	4	3	B	3	W	3	1	14	1	3	N	69
59.	*Lupinus nanus*	1	2	50.0	1.50	.248	.241	D	2	1	S	2	A	1	1	48	2	3	N	125
60.	*L. subcarnosus*	8	8	88.0	1.84	.142	.097	D	2	1	S	2	A	1	1	36V	2	2	N	6
61.	*L. succulentus*	35	4	22.0	1.27	.080	—	D	2	1	S	2	A	1	1	48	1	3	N	66
62.	*L. texensis*	10	8	87.5	3.12	.414	.356	D	2	1	S	2	A	1	1	36	2	3	N	6
63.	*Lycopersicon cheesmanii*	54	14	0.0	1.00	.000	.000	D	1	3	S	1	S	3	5	24	2	2	N	114
64.	*L. chmielewskii*	8	14	13.4	1.14	.053	—	D	1	3	S	2	A	3	5	24	2	3	C	116
65.	*L. esculentum*	7	15	2.8	1.18	.037	—	D	4	3	S	2	A	2	5	24V	1	3	C	114
66.	*L. esculentum var. cerasiformae*	7	15	3.3	1.16	.018	—	D	3	3	S	2	A	3	5	24	1	3	N	114
67.	*L. parviflorum*	8	14	0.0	1.00	.000	—	D	1	3	S	1	S	3	5	24	2	3	N	116
68.	*L. pimpinellifolium*	43	13	36.3	1.54	.128	.074	D	2	3	S	2	A	3	5	24	1	3	N	115
69.	*Lythrum tribracteatum*	7	18	28.0	1.39	—	—	D	2	3	S	1	S	2	3	10	1	4	N	9
70.	*Mimulus guttatus*	11	10	94.1	2.82	.336	—	D	3	3	B	3	A	4	3	28V	4	4	N	102
71.	*Oenothera argillicola*	10	20	20.0	1.25	.075	.080	D	1	2	S	3	A	3	3	24	2	3	N	92
72.	*Oe. biennis*	67	20	13.3	1.31	.083	.152	D	4	2	S	1	S	3	3	14V	1	3	N	84, 90, 93
		(3)		(3)	(3)	(3)	(3)													
73.	*Oe. hookeri*	14	20	0.0	1.00	.000	.000	D	4	2	S	3	A	3	3	14	1	3	N	92
74.	*Oe. parviflora*	29	20	40.0	1.52	.136	.148	D	4	2	S	1	S	3	3	14	1	3	N	92
75.	*Oe. strigosa*	29	20	25.0	1.30	.046	.028	D	4	2	S	1	S	3	3	14	1	3	N	92
76.	*Opuntia basilaris*	1	1	100.0	2.00	.479	.481	D	3	3	B	3	A	4	5	22	2	1	N	136
77.	*Origanum vulgare*	2	1	100.0	2.50	.410	.454	D	4	3	S	3	A	2	3	32	1	2	C	37
78.	*Oryza sativa*	—	3	—	2.33	.383	—	M	3	1	S	2	W	1	1	24	1	4	C	110

Table 5 *(Continued)*

Species	Number of populations	Number of loci	P	A	PI	H	1	2	3	4	5	6	7	8	9	10	11	12	References
79. *Panicum maximum* (sexual)	3	8	100.0	2.69	.377	—	M	4	3	S	2	W	4	3	36V	1	3	N	144
80. *P. maximum* (asexual)	3	8	50.0	1.92	.139	—	M	4	3	A	—		4	3	36V	1	3	N	144
81. *Persea americana*	—	10	80.0	1.90	.195	—	D	3	4	S	3	A	4	5	24	2	3	C	142
82. *Phaseolus coccineus*	15	5	19.3	1.18	.049	—	D	3	2	S	2	A	3	1	22	1	3	C	146
83. *P. vulgaris*	24	5	18.6	1.20	.064	—	D	4	1	S	2	A	2	1	22	1	3	C	146
84. *Phlox cuspidata*	26	18	7.9	1.08	.026	.010	D	2	1	S	1	S	1	1	14	2	3	N	86, 88
	(2)	(2)	(2)		(2)	(2)													
85. *P. drummondii*	42	18	16.5	1.19	.058	.043	D	2	1	S	3	A	1	1	14	2	3	N	86, 88
	(2)	(2)	(2)		(2)	(2)													
86. *P. drummondii* (cultivated)	16	19	11.0	1.11	.036	.013	D	4	1	S	2	A	1	1	14	1	3	C	87
87. *P. roemariana*	15	20	16.3	1.13	.055	.046	D	1	1	S	3	A	1	1	14	2	1	N	88
88. *Picea abies*	10	4	98.7	3.18	.418	.429	G	4	4	S	3	W	4	4	24	3	3	N	12, 13, 98, 140
	(4)	(4)	(4)	(4)	(4)	(1)													
89. *P. englemannii*	3	2	50.0	1.50	.432	—	G	2	4	S	3	W	4	4	24	3	3	N	55
90. *Pinus longaeva*	5	14	78.6	2.35	.364	.327	G	2	4	S	3	W	4	4	24	2	2	N	70
91. *P. pungens*	3	15	40.0	1.33	1.44	—	G	2	4	S	3	W	4	4	24	2	2	N	39
92. *P. ponderosa*	7	22	68.4	2.00	.226	—	G	3	4	S	3	W	4	4	24	3	3	N	Hamrick, unpublished data
93. *P. resinosa*	5	9	0.0	1.00	.000	.000	G	3	4	S	3	W	4	4	24	2	3	N	40
94. *P. rigida*	4	15	96.6	1.96	.170	—	G	3	4	S	3	W	4	4	24	1	3	N	62
95. *P. sylvestris*	17	4	100.0	4.08	.359	.318	G	4	4	S	3	W	4	4	24	2	3	N	118, 119
	(2)	(2)	(1)	(2)	(1)	(1)													
96. *Poncirus trifoliata*	—	4	100.0	2.00	.500	—	D	3	4	S	2	A	4	5	24	2	3	NC	143
	(2)																			
97. *Pseudotsuga menziesii*	8	12	69.3	2.47	.436	—	G	3	4	S	3	W	4	4	24	3	3	N	104, Hamrick unpublished data Morris, unpublished data
	(3)	(3)	(3)	(3)	(2)														

No.	Taxon			P	A	PI	H	1	2	3	4	5	6	7	8	9	10	11	12	Ref
98.	Secale cereale	2	5	—	—	.216	.276	M	4	1	S	2	W	1	2	14	1	3	C	60
99.	Shorea leprosula	3	36	50.0	—	—	—	D	4	4	S	3	A	4	—	14	3	3	N	44
100.	Silene maritima	4	21	29.0	1.49	.140	.153	D	4	3	S	2	A	3	3	24	1	3	N	10
101.	Stephanomeria exigua ssp. carotifera	11	14	57.0	2.10	.156	.092	D	1	1	S	3	A	1	3	16	2	3	N	50
102.	S. exigua ssp. coronaria	1	14	71.4	2.36	.277	—	D	2	1	S	3	A	1	3	16	2	2	N	48, 51
103.	Tragopogon dubius	6	21	8.9	1.09	.028	.001	D	4	2	S	1	S	2	4	12	1	3	N	117
104.	T. mirus	8	21	4.2	—	—	.430	D	1	2	S	1	S	2	4	24	1	3	N	117
105.	T. miscellus	6	21	3.2	—	—	.330	D	1	2	S	1	S	2	4	24	1	3	N	117
106.	T. porrifolius	3	21	6.4	1.07	.014	.001	D	4	2	S	1	S	2	4	12	1	3	N	117
107.	T. pratensis	3	21	0.0	1.00	.000	.000	D	4	2	S	1	S	2	4	12	1	3	N	117
108.	Tripsacum dactyloides	—	20	15.0	—	—	—	M	4	3	B	2	W	4	1	54V	1	3	N	127
109.	T. floridinum	—	20	20.0	—	—	—	M	1	3	B	2	W	4	1	36	1	4	N	127
110.	Xerospermum intermedium	3	6	33.3	1.33	.093	—	D	3	4	S	2	A	—	—	—	3	3	N	44
111.	Zea mays	33	11	67.3	2.65	.301	.420	M	4	1	S	2	W	3	1	20V	1	3	C	20, 22, 127, 138
		(3)		(5)	(5)	(5)	(5)	(1)												
112.	Z. mexicana (annual)	—	20	72.0	2.09	.230	—	M	3	1	S	2	W	2	1	20	1	3	N	127
113.	Z. mexicana (perennial)	—	20	45.0	—	—	—	M	3	3	S	2	W	4	1	20	1	3	N	127

a Data on 16 variables were obtained for the 113 taxa listed above. Four variables describe genetic variation: P = percentage of loci polymorphic per population; A = number of alleles per locus; PI = polymorphic index; and H = observed heterozygosity. Twelve variables describe life-history or ecological characteristics of each species. Categories for each of these 12 variables are indicated either by letters or numbers. Whenever a category is indicated by a letter the number in brackets indicates the number used in the analyses: 1. Taxonomic Status: G = Gymnospermae [1], D = Dicotyledonae [2], M = Monocotyledonae [3]; 2. Geographic Range and Distribution: 1 = endemic, 2 = narrow, 3 = regional, 4 = widespread; 3. Generation Length: 1 = annual, 2 = biennial, 3 = short-lived perennial, 4 = long-lived perennial; 4. Mode of Reproduction: A = asexual, [1], S = sexual [3], B = both [2]; 5. Mating System: 1 = primarily selfed, 2 = mixed, 3 = primarily outcrossed; 6. Pollination Mechanism: S = selfed [1], A = animal [2], W = wind [3]; 7. Fecundity: 1 = $<10^2$, 2 = $1 \times 10^2 - 1 \times 10^3$, 3 = $1 \times 10^3 - 1 \times 10^4$, 4 = $> 1 \times 10^4$; 8. Seed Dispersal: 1 = large, 2 = animal attached, 3 = small, 4 = winged or plumose, 5 = animal ingested; 9. Chromosome Number: Actual number; V = indicates that the species has several chromosome numbers reported for it, in which case the mean or commonly reported number is used. 1 = 10–20 chromosomes, 2 = 22–30, and 3 = > 30 chromosomes; 10. Stage of Succession: 1 = weedy and early, 2 = middle, 3 = late; 11. Habitat Type: 1 = xeric, 2 = submesic, 3 = mesic, 4 = hydric; 12. Cultivation Status: C = cultivated [1], N = noncultivated [2], NC = both [3].

populations of organisms that commonly experience large fluctuations in population sizes. The differences noted above between early- and late-successional species and between endemic and widespread species lend support to this argument.

Also, plant populations, especially those of long-lived perennials, often consist of representatives from many generations. If different alleles or genotypes were favored during the establishment phase of the various generations, plants surviving to maturity will maintain a genetic record of these past evolutionary events. Their continued survival would retard the decay of genetic variation and could lead to its maintenance. The seed carryover abilities of many annual and short-lived perennials would produce a similar genetic record. We know of no detailed studies that document the maintenance of genetic variation by this means. However, Beckman (11) demonstrated genetic differences between old ponderosa pine trees and progeny established in their shade.

Finally, various forms of balancing selection can lead to the maintenance of genetic variation. In species of inbreeding annuals it is common to find an excess of heterozygous individuals (3, 21, 74, 100). Although the mechanisms that produce these excesses are often poorly understood (21), a form of balancing selection can occur when there is selection for certain alleles or genotypes at one stage of the life cycle and selection for other alleles or genotypes at some subsequent stage. The pioneering work of Clegg & Allard (30) on *Avena barbata,* of Clegg et al (32) on *Hordeum vulgare,* and of Schall & Levin (123) on *Liatris cylindracea* all present evidence that fitness values may fluctuate widely between stages of the life cycle.

Perhaps of equal interest are the underlying causes of the differences in genetic variation observed among species or groups of species. We have attempted to explain these differences by examining the effect of 12 life history and habitat variables on genetic variation. However, unknown variables such as past historical events must play a significant role in the genetic structure of plant populations. For example, *Pinus longaeva, P. resinosa,* and *P. rigida* have essentially identical life history traits but vary somewhat in their geographic range and habitat preferences. However, the geographically and environmentally most restricted species, *P. longaeva,* has the highest *PI* value (0.364) while the more widespread species, *P. rigida* and *P. resinosa,* have less variation, *PI* = 0.170 and 0.000, respectively.

The problem of intercorrelation arises in the interpretation of associations between individual life history parameters and genetic variation. The use of multivariate statistics partially resolves this problem. In addition, the multivariate analysis allows two additional questions to be raised: Do species with certain combinations of life history traits have different amounts of genetic variation? Are certain characteristics more important than others? Our results indicate that species with large ranges, high fecundities, an outcrossing mode of reproduction, wind pollination, a long generation time, and from habitats representing later stages of succession have more genetic variation than do species with other combinations of characteristics. The arguments discussed above concerning plant species as a whole can be applied with even more justification to these species. Their longevity and high fecundity tend to ensure large, stable populations resistant to chance fluctuations in gene and genotype frequencies. Longevity also ensures that many generations are represented in the populations. Since each generation may be faced with somewhat different selection regimes, a wide variety of genotypes may be maintained. The high rates of fecundity, outcrossing, and wind pollination ensure large neighborhood sizes and the production of a variety of genotypes. Such an array of genotypes may allow the species to utilize a wider variety of microhabitats. Finally, natural selection could act to maintain this variation through the evolution of locally adapted ecotypes or through various types of balancing selection.

ACKNOWLEDGMENTS

We wish to acknowledge the fine technical assistance of Diane Bowman, Kareen Sturgeon, and Karen Hamrick throughout the production of this paper. We also wish to thank Drs. H. G. Baker, A. H. D. Brown, M. T. Clegg, and J. L. Harper for their comments on earlier drafts of the paper. Portions of this research were supported by NSF Grant DEB 76-01295 to JLH and NSF Grant BMS 75-14050 to YBL and JBM.

Literature Cited

1. Allard, R. W. 1975. The mating system and microevolution. *Genetics* 79: 115–26
2. Allard, R. W., Babbel, G. R., Clegg, M. T., Kahler, A. L. 1972. Evidence for coadaptation in *Avena barbata*. *Proc. Natl. Acad. Sci. USA* 69:3043–48
3. Allard, R. W., Jain, S. K., Workman, P. L. 1968. The genetics of inbreeding populations. *Adv. Genet.* 14:55–131
4. Antonovics, J. 1971. The effects of a heterogeneous environment on the genetics of natural populations. *Am. Sci.* 59:593–99
5. Atcheley, W. R., Bryant, E. H. 1975. *Multivariate Statistical Methods, Among-Groups Covariation.* Stroudsburg, Pa: Halsted Press. 464 pp.
6. Babbel, G. R., Selander, R. K. 1974. Genetic variability in edaphically restricted and widespread plant species. *Evolution* 28:619–30
7. Babbel, G. R., Wain, R. P. 1977. Genetic structure of *Hordeum jubatum.* I. Outcrossing rates and heterozygosity levels. *Can. J. Genet. Cytol.* 19:143–52
8. Baker, H. G. 1965. Characteristics and modes of origin of weeds. In *The Genetics of Colonizing Species,* ed. H. G. Baker, G. I. Stebbins, pp. 147–72. NY: Academic. 588 pp.
9. Baker, I., Baker, H. G. 1976. Variation in an introduced *Lythrum* species in California vernal pools. In *Vernal Pools: Their Ecology and Conservation,* ed. S. Jain, pp. 63–69. Davis. Calif: Inst. Ecol.
10. Baker, J., Maynard Smith, J., Strobeck, C. 1975. Genetic polymorphism in the bladder campion, *Silene maritima*. *Biochem. Genet.* 13:393–410
11. Beckman, J. S. 1977. *Adaptive peroxidase allozyme differentiation between colonizing and established populations of* Pinus ponderosa Laws on the Shanahan

Mesa, Boulder Colo. MA thesis. Univ. Colorado, Boulder.

12. Bergmann, V. F. 1973. Genetische Untersuchungen bei *Picea abies* mit Hilfe der Isoenzyme-Identifizierung. *Silvae Genet.* 22:63–66

13. Bergmann, V. F. 1975. Identification of forest seed origin on the basis of isoenzyme gene frequencies. *Allg. Forst. Jagdztg.* 146:181–85

14. Benson, L., Phillips, E. A., Wilder, P. A. et al. 1967. Evolutionary sorting of characters in a hybrid swarm. I. Direction of slope. *Am. J. Bot.* 54:1017–26

15. Bernstein, S., Thockmorton, L., Hubby, J. L. 1973. Still more genetic variability in natural populations. *Proc. Natl. Acad. Sci. USA* 70:3928–31

16. Bolkhovskikh, Z., Grif, V., Matvejeva, O., Zakharyeva, O. 1969. *Chromosome Numbers of Flowering Plants.* Leningrad: Acad Sci. USSR

17. Boyer, S. H., Noyes, A. N., Timmons, C. F., Young, R. A. 1972. Primate hemoglobins: polymorphisms and evolutionary patterns. *J. Human Evol.* 1: 515–43

18. Bradshaw, A. D. 1965. Evolutionary significance of phenotypic plasticity in plants. *Adv. Genet.* 13:115–55

19. Bradshaw, A. D. 1972. Some of the evolutionary consequences of being a plant. *Evol. Biol.* 5:25–46

20. Brown, A. H. D. 1971. Isozyme variation under selection in *Zea mays. Nature* 232:570

21. Brown, A. H. D. 1979. Enzyme polymorphism in plant populations. *Theor. Pop. Biol.* In press

22. Brown, A. H. D., Allard, R. W. 1971. Effect of reciprocal recurrent selection for yield on isozyme polymorphisms in maize (*Zea mays* L.). *Crop Sci.* 11: 888–93

23. Brown, A. H. D., Marshall, D. R., Albrecht, L. 1974. The maintenance of alcohol dehydrogenase polymorphism in *Bromus mollis* L. *Aust. J. Biol. Sci.* 27:545–59

24. Brown, A. H. D., Matheson, A. C., Eldridge, K. G. 1975. Estimation of the mating system of *Eucalyptus obliqua* L' Herit by using allozyme polymorphisms. *Aust. J. Bot.* 23:931–49

25. Brown, A. H. D., Nevo, E., Zohary, D. 1977. Association of alleles at esterase loci in wild barley *Hordeum spontaneum* L. *Nature* 268:430–31

26. Bryant, E. H., Atcheley, W. R. 1975. *Multivariate Statistical Methods, Within-Groups Covariation.* Stroudsburg, Pa: Halsted Press. 458 pp.

27. Cheney, D. P., Babbel, G. R. 1975. Isoenzyme variation in Florida populations of the red alga *Eucheuma. J. Phycol.* 11: Suppl. pp. 17–18

28. Cherry, J. P., Ory, R. L. 1973. Electrophoretic characterization of six selected enzymes of peanut cultivars. *Phytochemistry* 12:283–89

29. Clegg, M. T., Allard, R. W. 1972. Patterns of gentic differentiation in the slender wild oat species *Avena barbata. Proc. Natl. Acad. Sci. USA* 69:1820–24

30. Clegg, M. T., Allard, R. W. 1973. Viability versus fecundity selection in the slender wild oat, *Avena barbata* L. *Science* 181:667–68

31. Clegg, M. T., Horch, C. R., Church, G. L. 1976. Extreme genetic similarity among northeastern species of wild rye. *Genetics* 83: Suppl., p. 15

32. Clegg, M. T., Kahler, A. L., Allard, R. W. 1978. Estimation of life cycle components of selection in an experimental plant population. *Genetics* 89:765–92

33. Coleman, P. 1977. *Spatial and temporal variation in population structure of Lolium multiflorum Law. (Italian ryegrass).* PhD thesis. Univ. California, Davis.

34. Craig, K. L., Murray, B. E., Rajhathy, T. 1974. *Avena canariensis:* morphological and electrophoretic polymorphism and relationship to the *A. magna-A. murphyi* complex and *A. sterilis. Can. J. Genet. Cytol.* 16:677–89

35. Crawford, D. J., Wilson, H. D. 1977. Allozyme variation in *Chenopodium fremontii. Syst. Bot.* 2:180–90

36. De Arroyo, M. T. K. 1975. Electrophoretic studies of genetic variation in natural populations of allogamous *Limnanthes alba* and autogamous *Limnanthes floccosa* (Limnanthaceae). *Heredity* 35:153–64

37. Elena-Rossello, J. A., Kheyr-Pour, A., Valdeyron, G. 1976. La structure genetique et le regime de la fecondation chez *Origanum vulgare* L.; repartition d'un marqueus enzymatique dans deux populations naturelles. *C. R. Acad. Sci. Paris* 283:1587–89

38. Fernald, M. L. 1950. *Gray's Manual of Botany.* NY: American Book Company. 1632 pp. 8th ed.

39. Feret, P. P. 1974. Genetic differences among three small stands of *Pinus pungens. Theor. Appl. Genet.* 44:173–77

40. Fowler, D. P., Morris, R. W. 1977. Genetic diversity in red pine: evidence for low heterozygosity. *Can. J. For. Res.* 7:341–47

41. Franklin, I., Lewontin, R. C. 1970. Is the gene the unit of selection? *Genetics* 65:707–34

42. Fryxell, P. A. 1957. Mode of reproduction of higher plants. *Bot. Rev.* 23:135–233

43. Gallagher, J. C. 1977. Population genetics of *Skeletonema costatum* (Grev.) Cleve over yearly cycles of abundance in Narragansett Bay. *J. Phycol.* 13(s): Suppl., p. 23

44. Gan, Y. Y., Robertson, F. W., Ashton, P. S. 1977. Genetic variation in wild populations of rain-forest trees. *Nature* 269:323–24

45. Garber, E. D., Rippon, J. W. 1968. Proteins and enzymes as taxonomic tools. *Adv. Appl. Microbiol.* 10:137–54

46. Giesel, J. T. 1976. Reproductive strategies as adaptations to life in temporally heterogeneous environments. *Ann. Rev. Ecol. Syst.* 7:37–80

47. Gottlieb, L. D. 1973. Enzyme differentiation and phylogeny in *Clarkia franciscana, C. rubicunda* and *C. amoena. Evolution* 27:205–14

48. Gottlieb, L. D. 1973. Genetic differentiation, sympatric speciation, and the origin of a diploid species of *Stephanomeria. Am. J. Bot.* 60:545–53

49. Gottlieb, L. D. 1974. Genetic confirmation of the origin of *Clarkia ligulata. Evolution* 28:244–50

50. Gottlieb, L. D. 1975. Allelic diversity in the outcrossing annual plant *Stephanomeria exigua* spp. *carotifera* (Compositae). *Evolution* 29:213–25

51. Gottlieb, L. D. 1977. Genotypic similarity of large and small individuals in a natural population of the annual plant *Stephanomeria exigua* spp. *coronaria* (Compositae). *J. Ecol.* 65:127–34

52. Gottlieb, L. D. 1977. Electrophoretic evidence and plant systematics. *Ann. Mo. Bot. Gard.* 65:164–80

53. Gottlieb, L. D., Pilz, G. 1976. Genetic similarity between *Gaura longiflora* and its obligately outcrossing derivative *G. demoreei. Syst. Bot.* 1:181–87

54. Gould, S. J., Johnston, R. F. 1972. Geographic variation. *Ann. Rev. Ecol. Syst.* 3:457–87

55. Grant, M. C., Mitton, J. B. 1976. Genetic variation associated with morphological variation in engelmann spruce at tree line. *Genetics* 83: Suppl., p. 28

56. Grant, M. C., Mitton, J. B. 1977. Genetic differentiation among growth forms of engelmann spruce and subalpine fir at tree line. *Arct. Alp. Res.* 9:259–63

57. Grant, V. 1958. The regulation of recombination in plants. *Cold Spring Harbor Symp. Quant. Biol.* 23:337–63

58. Grant, V. 1975. *Genetics of Flowering Plants.* NY: Columbia Univ. Press. 514 pp.

59. Gray, J. R., Fairbrothers, D. E., Quinn, J. A. 1973. Biochemical and anatomical population variation in the *Danthonia sericea* complex. *Bot. Gaz.* 134:166–73

60. Griffin, W. B. 1976. Genetic variability in some New Zealand grown cultivars of barley and rye. *Mauri Ora* 4:93–100

61. Grover, N. S., Byrne, O. R. 1975. Genetic control of acid phosphate isozymes in *Arabidopsis thaliana* (L.) Heynh. *Biochem. Genetics* 13:527–31

62. Guries, R. P., Ledig, F. T. 1977. Analysis of population structure from allozyme frequencies. *Proc. So. For. Tree Improv. Conf.* 14:246–53

63. Hamrick, J. L. 1979. Genetic variation and longevity. In *Topics in Plant Population Biology*, ed. O. Solbrig, S. Jain, G. Johnson, P. Raven. NY: Columbia Univ. Press. In press

64. Hamrick, J. L., Allard, R. W. 1972. Microgeographical variation in allozyme frequencies in *Avena barbata. Proc. Natl. Acad. Sci. USA* 65:2100–4

65. Hamrick, J. L., Holden, L. R. 1979. The influence of microhabitat heterogeneity on gene frequency distribution and gametic phase disequilibrium in *Avena barbata. Evolution.* In press

66. Harding, J. Mankinen, C. B. 1972. Genetics of *Lupinus.* IV. Colonization and genetic variability in *Lupinus succulentus. Theor. Appl. Genet.* 42:267–71

67. Harper, J. L. 1977. *Population Biology of Plants.* NY: Academic. 892 pp.

68. Harris, H. 1966. Enzyme polymorphisms in man. *Proc. R. Soc. London Ser. B.* 1964:298–310

69. Hayward, M. D., McAdam, N. J. 1977. Isozyme polymorphism as a measure of distinctiveness and stability in cultivars of *Lolium perenne. Z. Pflanzenzücht.* 79:59–68

70. Hiebert, R. D. 1977. *Ecology and demographic genetics of bristlecone pine* (Pinus longaeva) *of the Great Basin.* PhD thesis. Univ. Kansas, Lawrence. 82 pp.

71. Hubby, J. L., Lewontin, R. C. 1966. A molecular approach to the study of genic heterozygosity in natural populations. I. The number of alleles at different loci in *Drosophila pseudoobscura. Genetics* 54:577–94

72. Jacobs, M., Schwind, F. 1975. Biochemical genetics of *Arabidopsis* acid

phosphatases: polymorphism, tissue expression, and genetics of *Ap1, Ap2,* and *Ap3* loci. In *Isozymes IV.,* ed. C. L. Markert, pp. 349–69. NY: Academic. 965 pp.

73. Jain, S. K. 1976. Evolution of inbreeding in plants. *Ann. Rev. Ecol. Syst.* 7:468–95

74. Jain, S. K. 1976. Evolutionary studies in the meadowfoam genus *Limanthes:* An overview. In *Vernal Pools: Their Ecology and Conservation,* ed. S. K. Jain, pp. 50–57. Davis, Calif: Inst. Ecol. Publ.

75. Jain, S. K., Bradshaw, A. D. 1966. Evolutionary divergence among adjacent plant populations. I. The evidence and its theoretical analysis. *Heredity* 21: 407–41

76. Jain, S. K., Rai, K. N. 1974. Population biology of *Avena.* IV. Polymorphism in small populations of *Avena fatua. Theor. Appl. Genet.* 44:7–11

77. Johnson, G. B. 1974. Enzyme polymorphism and metabolism. *Science* 184: 28–37

78. Johnson, G. B. 1976. Hidden alleles at the α-glycerophosphate dehydrogenase locus in *Colias* butterflies. *Genetics* 83:149–67

79. Kahler, A. L. 1979. Worldwide patterns of variation of esterase allozymes in barley (*Hordeum vulgare* L. and *H. spontaneum*). *Proc. Natl. Acad. Sci. USA.* In press

80. Kahler, A. L., Allard, R. W. 1970. Genetics of isozyme variants in barley. I. Esterases. *Crop Sci.* 10:444–48

81. Kimura, M., Ohta, T. 1971. *Theoretical Aspects of Population Genetics.* Princeton, NJ: Princeton Univ. Press. 219 pp.

82. Law, R., Bradshaw, A. D., Putwain, P. D. 1977. Life-history variation in *Poa annua. Evolution:* 31:233–46

83. Lee, D. W., Postle, R. L. 1975. Isozyme variation in *Colobanthus quitensis* (Kunth) Bartl.: methods and preliminary analysis. *Br. Antarct. Surv. Bull.* 41:133–37

84. Levin, D. A. 1975. Genic heterozygosity and protein polymorphism among local populations of *Oenothera biennis. Genetics* 79:477–91

85. Levin, D. A. 1975. Genetic correlates of translocation heterozygosity in plants. *Bioscience* 25:724–28

86. Levin, D. A. 1975. Interspecific hybridization, heterozygosity and gene exchange in *Phlox. Evolution* 29:37–51

87. Levin, D. A. 1976. Consequences of long-term artificial selection, inbreeding and isolation in *Phlox.* II. The organization of allozymic variability. *Evolution* 30:463–72

88. Levin, D. A. 1978. Genetic variation in annual *Phlox:* Self-compatible versus self-incompatible species. *Evolution* 32:245–63

89. Levin, D. A., Crepet, W. L. 1973. Genetic variation in *Lycopodium lucidulum:* A phylogenetic relic. *Evolution* 27:622–32

90. Levin, D. A., Hawland, G. P., Steiner, E. 1972. Protein polymorphism and genic heterozygosity in a population of the permanent translocation heterozygote, *Oenothera biennis. Proc. Natl. Acad. Sci. USA* 69:1475–77

91. Levin, D. A., Kerster, H. W. 1974. Gene flow in seed plants. *Evol. Biol.* 7:139–220

92. Levy, M., Levin, D. A. 1975. Genic heterozygosity and variation in permanent translocation heterozygotes of the *Oenothera biennis* complex. *Genetics* 79:493–512

93. Levy, M., Winterheimer, P. L. 1977. Allozyme linkage disequilibria among chromosome complexes in the permanent translocation heterozygote *Oenothera biennis. Evolution* 31:465–76

94. Lewontin, R. C., Hubby, J. L. 1966. A molecular approach to the study of genic heterozygosity in natural populations. II. Amount of variation and degree of heterozygosity in natural populations of *Drosophila pseudoobscura. Genetics* 54:594–609

95. Libby, W. J., Stettler, R. F., Seitz, F. W. 1969. Forest genetics and forest tree breeding. *Ann. Rev. Genet.* 3:469–94

96. Lilley, S. R., Wall, J. R. 1972. Determination of geographic variation in allelic frequencies for leucine aminopeptidase and alcohol dehydrogenase isozymes in *Cucurbita foetidissima. Va. J. Sci.* 23: 112

97. Linhart, Y. B. 1974. Intra-population differentiation in annual plants. I. *Veronica peregrina* L. raised under noncompetitive conditions. *Evolution* 28: 232–43

98. Lundkvist, K., Rudin, D. 1977. Genetic variation in eleven populations of *Picea abies* as determined by isozyme analysis. *Hereditas* 85:67–74

99. Marshall, D. R., Allard, R. W. 1970. Isozyme polymorphisms in natural populations of *Avena fatua* and *A. barbata. Heredity* 25:373–82

100. Marshall, D. R., Allard, R. W. 1970. Maintenance of isozyme polymorphisms in natural populations of *Avena barbata. Genetics* 66:393–99

101. Marshall, D. R., Jain, S. K. 1969. Genetic polymorphism in natural populations of *Avena fatua* and *A. barbata*. *Nature* 221:276–78

102. McClure, S. 1973. *Allozyme variability in natural populations of the yellow monkeyflower*, Mimulus guttatus, *located in the North Yuba River Drainge*. PhD thesis. Univ. California, Berkeley. 246 pp.

103. McDonald, J. F., Ayala, F. J. 1974. Genetic response to environmental heterogeneity. *Nature* 250:572–74

104. Mejnartowicz, L. 1976. Genetic investigations on Douglas-fir [*Pseudotsuga menziesii* (Mirb.) Franco] populations. *Arbor. Kornickie* 21:126–87

105. Mejnartowicz, L., Bergmann, F. 1975. Genetic studies on European larch (*Larix decidua* Mill.) employing isoenzyme polymorphisms. *Genet. Pol.* 16: 29–36

106. Milkman, R. D. 1967. Heterosis as a major cause of heterozygosity in nature. *Genetics* 55:493–95

107. Miller, R. D. 1977. *Genetic variability in the slender wild oat* Avena barbata *in California*. PhD thesis. Univ. California, Davis.

108. Mitton, J. B., Linhart, Y. B., Hamrick, J. L., Beckman, J. 1977. Population differentiation and mating system in ponderosa pine of the Colorado Front Range. *Theor. Appl. Genet.* 51:5–14

109. Munz, P. A., Keck, D. D. 1965. *A California Flora*. Berkeley: Univ. California Press. 1681 pp.

110. Nakagahra, M. 1977. Genic analysis for esterase isoenzymes in rice cultivars. *Jpn. J. Breed.* 27:141–48

111. Nevo, E. 1978. Genetic variation in natural populations: patterns and theory. *Theor. Pop. Biol.* 13:121–77

112. Phillips, M. A., Brown, A. H. D. 1977. Mating system and hybridity in *Eucalyptus pauciflora*. *Aust. J. Biol. Sci.* 30:337–44

113. Powell, J. 1975. Protein variation in natural populations of animals. *Evol. Biol.* 8:79–119

114. Rick, C. M., Fobes, J. F. 1975. Allozyme variation in the cultivated tomato and closely related species. *Bull. Torr. Bot. Club* 102:376–84

115. Rick, C. M., Fobes, J. F., Holle, M. 1977. Genetic variation in *Lycopersicon pimpinellifolium*: evidence of evolutionary change in mating systems. *Plant Syst. Evol.* 127:139–70

116. Rick, C. M., Kesichi, E., Fobes, J. F., Holle, M. 1976. Genetic and biosystematic studies on two new sibling species of *Lycopersicon* from interandean Peru. *Theor. Appl. Genet.* 47:55–68

117. Roose, M. L., Gottlieb, L. D. 1976. Genetic and biochemical consequences of polyploidy in *Tragopogon*. *Evolution* 30:818–30

118. Rudin, D. 1974. Gene and genotype frequencies in Swedish scots pine populations studied by the isozyme technique. *Hereditas* 78:325

119. Rudin, D., Eriksson, G., Ekberg, I., Rasmuson, M. 1974. Studies of allele frequencies and inbreeding in scots pine populations by the aid of the isozyme technique. *Silvae Genet.* 23:10–13

120. Sanders, T. B., Hamrick, J. L. 1979. Allozyme variation in *Elymus canadensis* from the tall grass prairie region. I. Geographic variation. *Am. Midl. Nat.* In press

121. Schaal, B. A. 1974. Isolation by distance in *Liatris cylindracea*. *Nature* 252:703

122. Schaal, B. A. 1975. Population structure and local differentiation in *Liatris cylindracea*. *Am. Nat.* 109:511–28

123. Schaal, B. A., Levin, D. A. 1976. The demographic genetics of *Liatris cylindracea* Michv. (Compositae). *Am. Nat.* 110:191–206

124. Scogin, R. 1969. Isoenzyme polymorphism in natural populations of the genus *Baptisia* (Leguminosae). *Phytochemistry* 8:1733–37

125. Scogin, R. 1973. Leucine aminopeptidase polymorphism in the genus *Lupinus* (Leguminosae). *Bot. Gaz.* 134: 73–76

126. Selander, R. K. 1976. Genic variation in natural populations. In *Molecular Evolution*, ed. F. J. Ayala. Sunderland, Mass: Sinauer. 277 pp.

127. Senadhira, D. 1976. *Genetic variation in corn and its relatives*. PhD thesis. Univ. California, Davis.

128. Singh, R., Hubby, J., Lewontin, R. C. 1974. Molecular heterosis for heat sensitive enzyme alleles. *Proc. Natl. Acad. Sci. USA* 71:1808–10

129. Singh, R. S., Jain, S. K. 1971. Population biology of *Avena*. II. Isoenzyme polymorphisms in populations of the Mediterranean region and central California. *Theor. Appl. Genet.* 41:79–84

130. Snaydon, R. W., Davies, M. S. 1972. Rapid population differentiation in a mosaic environment. II. Morphological variation in *Anthoxanthum odoratum*. *Evolution* 26:390–405

131. Spieth, P. T. 1975. Population genetics of allozyme variation in *Neurospora intermedia*. *Genetics* 80:785–805

132. Stebbins, G. L. 1950. *Variation and Evolution in Plants.* NY: Columbia Univ. Press. 643 pp.

133. Stebbins, G. L. 1958. Longevity, habitat and release of genetic variability in higher plants. *Cold Spring Harbor Symp. Quant. Biol.* 23:365–78

134. Stebbins, G. L. 1971. *Chromosomal Evolution in Higher Plants.* London: Arnold. 216 pp.

135. Stern, K., Roche, L. 1974. *Genetics of Forest Ecosystems.* NY: Springer. 330 pp.

136. Sternberg, L., Ting, J. P., Hanscom, Z. 1977. Polymorphism of microbody malate dehydrogenase in *Opuntia basilaria. Plant Physiol.* 59:329–30

137. Strobel, R., Wohrmann, K. 1975. Populationsgenetische Untersuchungen an *Saccharomyces cerevisiae.* II. Der Einfluss von esterase Loci auf genfrequenzveränderne Faktoren. *Genetica* 45: 509–18

138. Stuber, C. S., Goodman, M. M., Johnson, F. M. 1977. Genetic control and racial variation of β-glucosidase isozymes in maize (*Zea mays* L.) *Biochem. Genet.* 15:383–94

139. Taylor, C. E. 1976. Genetic variation in heterogeneous environments. *Genetics* 83:887–94

140. Tigerstedt, P. M. A. 1973. Studies on isozyme variation in marginal and central populations of *Picea abies. Hereditas* 75:47–60

141. Torres, A. M., Diedenhofen, U., Johnstone, I. M. 1977. The early allele of alcohol dehydrogenase in sunflower populations. *J. Hered.* 68:11–16

142. Torres, A. M., Diedenhofen, U., Bergh, B. O., Knight, R. J. 1978. Enzyme polymorphisms as genetic markers in the avocado. *Am. J. Bot.* 65:134–39

143. Torres, A. M., Soost, R. K., Diedenhofen, U. 1978. Leaf isozymes as genetic markers in *Citrus. Am. J. Bot.* 65:869–81

144. Usberti, J. A., Jain, S. K. 1978. Variation in *Panicum maximum;* a comparison of sexual and asexual populations. *Bot. Gaz.* 139:112–16

145. Valizadeh, M. 1977. Esterase and acid phosphatase polymorphism in the fig tree (*Ficus carica* L.). *Biochem. Genet.* 15:1037–48

146. Wall, J. R., Wall, S. W. 1975. Isozyme polymorphisms in the study of evolution in *Phaseolus vulgaris - P. coccineus* complex of Mexico. See Ref. 72, pp. 287–305

147. Williams, G. C. 1975. *Sex and Evolution.* Princeton, NJ: Princeton Univ. Press. 200 pp.

148. Wills, C. 1978. Rank-order selection is capable of maintaining all genetic polymorphisms. *Genetics* 89:403–17

149. Wool, D., Koach, J. 1976. Morphological variation in the gall forming aphid, *Geocia utricularia* (Homoptera) in relation to environmental variation. In *Population Genetics and Ecology,* ed. S. Karlin, E. Nevo, pp. 239–61. NY: Academic. 822 pp.

150. Wu, K. K., Jain, S. K. 1978. Genetic and plastic responses in geographic differentiation of *Bromus rubens* populations. *Canad. J. Bot.* 56:873–79

151. Wu, L. 1976. Esterase isoenzymes in populations of *Agrostis stonifera. L. Bot. Bull. Acad. Sinica* 17:175–84

Ann. Rev. Ecol. Syst. 1979. 10:201–27
Copyright © 1979 by Annual Reviews Inc. All rights reserved

GRANIVORY IN DESERT ECOSYSTEMS

♦4160

James H. Brown

Department of Ecology and Evolutionary Biology, University of Arizona, Tucson, Arizona 85721

O. J. Reichman

Department of Biology, Museum of Northern Arizona, Flagstaff, Arizona 86001

Diane W. Davidson

Department of Biological Sciences, Purdue University, West Lafayette, Indiana 47907

INTRODUCTION

Some of the most challenging questions of contemporary biology concern the diversity and stability of ecological communities. In most natural habitats many coexisting species of plants, animals, and microbes interact as competitors, mutualists, predators, and prey. These biotic processes interact with the physical environment to determine the structure and function of ecological systems. The numerous theoretical and empirical studies of competition and predation have left much to learn about how these processes contribute to the structure and stability of natural communities. Some of the questions may be answered by empirical investigations of systems complex enough to contain diverse species that interact in different ways yet simple enough to permit detailed comparative analyses and experimental manipulations.

The seeds and seed-eaters of desert habitats provide a system of two trophic levels that offers many advantages for community studies. Some investigators have suggested that the struggle to exist in the harsh physical

201

0066-4162/79/1120-0201$01.00

environment so dominates the ecology of desert organisms that interactions with other species are insignificant. Certainly many desert animals and plants possess spectacular morphological, physiological, and behavioral specializations for surviving in an environment of little water and extreme temperature fluctuations (6, 53, 84), but such adaptations hardly diminish the necessity of competing for food, water, and other limited resources or of avoiding predators. Indeed, seeds are diverse and abundant in deserts, and representatives of three major taxa (mammals, birds, and insects) have specialized and radiated to fill seed-eating niches. Within the last decade numerous comparative and experimental studies have clarified the roles of competition and predation in the structure and function of this resource-consumer system. The present review attempts to integrate and summarize what we have recently learned about the patterns and processes of inter-specific interactions in this intensively studied system.

We are concerned (*a*) with the importance of seeds as food resources for granivores, and (*b*) with the role of seed-eaters as competitors for these resources and as predators on plant populations. Most of the published work is based on research conducted in the southwestern United States. We draw on this work to synthesize and summarize what is known about granivory in North American deserts. We then discuss some of the interesting but still poorly understood similarities and differences in the organization of granivore communities among the major deserts of the world.

THE RESOURCES

Seeds play a special role in the ecology of deserts. We include as deserts habitats that purists would classify as semideserts and arid grasslands degraded by livestock grazing. In these xeric habitats the dominant plants, in terms of individual size and community biomass, are perennial shrubs and succulents, which occur as widely spaced individuals. Of equal importance, however, are the annuals that spend most of their lives as seeds hidden in the soil. During the brief periods when sufficient moisture is available these seeds germinate; the resulting vegetative plants rapidly grow, flower, and set seed (4, 6, 93, 94, 101–103). Seeds are so abundant in the soil that in particularly good seasons annuals may literally carpet the ground under and between the sparse perennials. Because of their rapid, facultative response to availability of moisture, annuals are particularly important in coupling primary production to the infrequent, unpredictable precipitation that limits the productivity of most arid ecosystems (6, 34, 78, 94, 103). Life history strategies of annuals dictate that they allocate a large proportion of their energy and materials to seeds. For animal consumers seeds provide nutritious, particulate food resources. They are always available in desert soils and can be collected when abundant and stored for later use.

For granivores seeds represent diverse food resources that vary in morphology, chemical composition, abundance, and spatial distribution. Seeds of annuals typically dominate seed crops and soil seed banks in terms of numbers, biomass, and species diversity. It is not uncommon for 20–30 species of annuals to account for more than 95% of total individual seeds and more than 80% of seed biomass; the remainder is made up largely by seeds of 2–6 species of codominant perennials (60). Seeds of these plants range in size from at least 0.01–150 mg. Most annuals produce seeds weighing <2 mg, whereas seeds of perennials are larger, typically >1 mg (60, 71). Seeds vary greatly in shape, hardness, surface texture, and in the ways they are packaged within seed coats, husks, and fruits. Although seeds in general represent concentrated sources of energy and nutrients, they differ in chemical composition. Patterns of temporal abundance and spatial distribution are complex. Seed production is closely dependent on the quantity and timing of precipitation, which is scanty and unpredictable in desert habitats. Consequently there is tremendous year-to-year variation in seed crops; in the more arid deserts there may be no seed production at all for one or more successive years (5, 6, 93, 95, 103). Once they fall from the parent plant, seeds are dispersed by wind, water, and animals. Wind is particularly important, and locally high densities of seeds accumulate in the wind shadows created by canopies of shrubs and irregularities in the soil surface. Most seeds remain near the surface of the soil and may be redispersed by strong winds or flowing water (60, 73, 75).

Quantities of seeds produced in good years and remaining in the desert soils long afterward are surprisingly high. Maximum values of seed production or soil seed banks range from 180×10^6 to 3700×10^6 seeds ha^{-1} (between 18×10^3 and 370×10^3 seeds m^{-2}) or 80–1480 kg ha^{-1} for various desert habitats (30, 50, 95). Minimum densities of seeds in the soil probably are rarely below 10×10^6 seeds ha^{-1} (1000 seeds m^{-2}) in most habitats, even years after the last seed crop (30, 60, 75, 95).

THE CONSUMERS

The primary consumers of desert seeds are rodents, ants, and birds. Important characteristics of these taxa, which influence and differentiate their roles as granivores, are compared in Table 1 and discussed in more detail below.

Rodents

In terms of consumer biomass and quantity of seeds harvested, rodents are rivaled only by ants as important granivores in North American deserts. The most specialized granivorous rodents belong to the family Heteromyidae, which includes kangaroo rats (*Dipodomys*), kangaroo mice (*Micro-*

Table 1 Some characteristics of the three major classes of granivores in North American deserts

	Rodents	Ants	Birds
Dominant genera of specialized granivores	*Dipodomys* *Perognathus* *Microdipodops*	*Pogonomyrmex* *Pheidole* *Veromessor*	*Lophortyx* *Zenaidura*
Others	*Peromyscus* *Reithrodontomys*	*Novomesser* (*Aphaenogaster*) *Solenopsis*	several genera of finches
Individual body size	7–120 g	0.5–15 mg	10–200 g
Social organization	solitary	large eusocial colonies	flocks (sometimes solitary)
Foraging strategy	central place, multiple load	central place, single load	dispersed, single load
Seed storage	great	great	none
Mobility of individuals	limited	very limited	great
Thermoregulation	endothermic	ectothermic	endothermic
Activity	nocturnal, active throughout the year (*Dipodomys*) or only in warm months (*Perognathus*)	daily and seasonal activity highly dependent on environmental temperature and humidity	diurnal, active throughout the year, but many species are migratory
Maximum life expectancy	few years	established colonies may survive for many years	several years

dipodops), and pocket mice (*Perognathus*). These mammals are largely or exclusively granivorous. Their adaptations for this diet include external, fur-lined cheek pouches, which are used for collecting and transporting seeds, and highly efficient kidneys, which enable them to excrete nitrogenous wastes and maintain osmotic balance on a diet that contains little free water (26, 84). Kangaroo rats and kangaroo mice use their elongated tails and hind limbs for bipedal, saltatorial locomotion (3, 25); they have specialized ears capable of detecting vertebrate predators (99, 100). Pocket mice have more generalized, mouselike morphology and quadrupedal, scansorial locomotion (2, 26). In addition to the heteromyids, several rodents of the family Cricetidae feed heavily on seeds. Desert species of the genera *Reithrodontomys, Peromyscus,* and *Onychomys* tend to be omnivorous in diet and relatively unspecialized in morphology and physiology. Many of them are largely insectivorous in the warm months, but their diets shift to seeds when preferred foods are not available.

Rodents are multi-load, central-place foragers: They use their cheek pouches to harvest many seeds in a single foraging bout, and their activities are confined to a small area centered around a permanent burrow system (26). Most rodents appear to spend the majority of their foraging time searching for seeds distributed on or just below the surface of the soil, but some collect many entire fruits and seedheads from plants before the seeds disperse (45, 58). Rodents are compulsive hoarders; when seeds are available, they collect large quantities and store them either in larders within their burrow system or in small, dispersed, shallowly buried caches outside (57, 77, 88, 97). Kangaroo rats forage largely in the open area between the scattered perennial plants (10, 15, 46, 64, 79, 114). In these microhabitats many seeds occur in clumps, and kangaroo rats use their hopping gait to move rapidly and efficiently between widely spaced aggregations of seeds (3, 63, 73, 75). Pocket mice do most of their foraging under shrub canopies (10, 15, 46, 64, 79, 113) where seed average densities are higher and more uniform than in open microhabitats (60, 75). Compared to that of kangaroo rats, the foraging pattern of pocket mice is fine-grained. In fact some species act almost like filter feeders; they sift rapidly through large quantities of soil and litter, collecting seeds as they encounter them (J. H. Brown, J. S. Findley, O. J. Reichman, unpublished data).

All granivorous desert rodents are nocturnal. They spend the day in burrows and emerge to forage at night (26, 43, 48). Kangaroo rats and some of the cricetids are active throughout the year, but most of the small rodents, including pocket mice and kangaroo mice, hibernate or utilize facultative torpor to remain inactive and minimize energy expenditure during periods when food is scarce or environmental temperatures are low (11, 29, 43, 74, 96). Granivorous desert rodents are preyed upon by owls, mammalian carnivores, and snakes. There is substantial evidence that predation has been an important selective pressure on desert rodent populations, which have responded by evolving specialized ears capable of predator detection (99, 100), restricted periods of nocturnal foraging (43, 48, 80), and particular patterns of microhabitat utilization (10, 15, 46, 64, 79, 114).

Ants

Harvester ants are the major nonmigratory granivores that rival rodents in total biomass and seed consumption in deserts and semi-deserts (12, 49, 58, 59, 104, 110). The most specialized seed-eaters in arid regions of North America belong to the genera *Pogonomyrmex, Pheidole,* and *Veromessor,* all members of the subfamily Myrmicininae, which contains most harvester ants (but see 47). Within the myrmicines granivory has arisen polyphyletically and is associated predominantly with arid habitats, where dependable sources of insect food are lacking. The suggested evolution of harvester ants from primarily carnivorous ancestors whose strong mandibles preadapted them for crushing seeds probably required only modest changes in mandible structure (106) but perhaps more complex modifications in the processes of digestion. In at least one North American species, *Veromessor pergandei,* the larvae appear to function as the primary agents of digestion, converting crushed seed material provided by workers into nutrient solutions that are returned to workers via trophallaxis (104). The generality of this mechanism has not been investigated. To some extent desert-adapted species may have undergone significant evolution with respect to other physiological characteristics, such as their rates of oxygen consumption and their abilities to utilize metabolically produced water (27). Probably the most familiar characteristic of harvester ants world-wide is their tendency to maintain within their soil nests extensive granaries upon which they draw during periods when the standing crop of seed resources has been depleted. Few if any of even the most highly evolved granivorous ants are strict seed specialists, however, and most species avidly forage for insect material when it is available (20, 25, 104, 109–112).

In contrast to rodents and with infrequent exceptions, worker ants typically are single-load, central-place foragers. Some North American harvester ants occasionally use the "beard" or psamnophore projecting from the ventral head region to accommodate a second seed on some foraging excursions (P. Kareiva, personal communication); such behavior has not been widely reported. In other granivorous ants, such as *Pheidologeton ocellifer* in India, especially large seeds may be transported to the nest cooperatively by several workers (116). However, the vast majority of harvester ant workers return their booty individually to a central nesting and caching site from distances ranging from a few to more than 40 m (104).

High-density seed patches represent preferred resources for ants as well as rodents, though for somewhat different reasons. Ants can take advantage of aggregated resources only to the extent that their recruitment systems enable them to draw nestmates to the resource patch and reduce the mean search time for locating food. Desert harvester ants differ markedly in their recruitment abilities and, hence, in their average success at exploiting aggregated resources (21, 35–37, 39). Grossly classified, species are group or individual foragers according to their tendencies to restrict the directionality of foraging for seeds. Group foragers travel to and from the foraging grounds in well-defined columns; consequently, over a limited time interval, searching and seed aquisition take place in a restricted portion of the area surrounding the nest (21, 22, 36, 37, 106). Hölldobler (35, 36, 37) has proposed that in some group-foraging *Pogonomyrmex* species, foraging columns arise gradually from recruitment trails; this hypothesis seems to be supported by the successful elicitation of group foraging at artificial seed baits (22, 24, 36, 37, 39, 92). In colonies of individual foragers, workers search for and collect seeds independently and may rely primarily on visual cues for orientation, though some species occasionally use short-lived chemicals for recruitment over very short distances (36). When nonnative seeds were supplied simultaneously in aggregated and dispersed distributions to colonies of group and individually foraging species, Davidson (22) found that the individual forager (*Novomessor cockerelli*) gathered seeds primarily from the dispersed distribution. The group forager (*Pogonomyrmex barbatus*) was recruited to one or a few clumps of seeds and rapidly depleted these, while taking far fewer seeds from other aggregates of seeds and from the dispersed distribution. Based on the results of these experiments and on the observation that the seasonal activity schedules of group foragers are confined to periods of peak seed production more than are those of individual foragers, group foraging has been interpreted as an adaptation for exploiting high densities of seed resources, and individual foraging as a behavioral specialization for feeding on dispersed or low-density resources [(22); for a conflicting opinion, see (8)]. Recently Hölldobler et al (38) demonstrated that *N. cockerelli* is capable of recruiting small numbers of foragers with short-lived chemicals when the cooperation of several foragers is required to transport a large prey item to the nest. However, the recruitment pheromone of this species lacks any persistent chemical component that would permit the amassing of large numbers of foragers.

It is not surprising that the foraging behaviors of these sedentary ants, like those of other central-place foragers including rodents, should be subject to intense selection. However, in eusocial ants, where inclusive fitness may form the dominant component of a worker's total genetic fitness (62), evolution may act effectively to produce traits that increase the success of the colony as a unit. Among the desert harvester ants, group and individual foraging behaviors appear to represent evolutionarily derived strategies for enhancing colony fitness in environments differing in resource density or dispersion (22). At least some of the experimental evidence demonstrating that foraging behavior of harvester ants supports predictions of optimal foraging models also suggests that selection may operate effectively at the colony level (36, 92).

In contrast to the rigid nocturnality of rodent foraging, ant activity schedules are plastic within the bounds dictated by ectothermy and other physiological constraints (20, 27, 42, 87, 95, 105, 108–110, 113). During colder seasons, activity is predominantly diurnal, but a number of species regularly forage crepuscularly and nocturnally during the hot desert summers. Individually foraging species of *Pogonomyrmex* seem to be entirely restricted to diurnality, possibly because of their reliance on visual orientation (21), but group-foraging *Pogonomyrmex, Pheidole,* and *Solenopsis* species and *Veromessor pergandei* shift from diurnal to nocturnal and crepuscular activity schedules during the hottest months, thereby avoiding climatic stresses or taking advantage of especially abundant resources (22, 95, 106, 111).

Desert ants experience predation or parasitism from a variety of sources including other ants, beetles, spiders, solpugids, lizards, blind snakes, rodents, and birds (28, 33, 106, 115; O. J. Reichman, unpublished data). Probably the most significant and specialized ant predators are diurnal horned lizards of the genus *Phrynosoma* (W. G. Whitford and M. Bryant, unpublished data). Frequently confined by ectothermy to foraging during warmer periods of the day when lizards are also active, virtually all the granivorous North American species sustain losses to *Phrynosoma,* with larger lizards such as *P. cornutum* selectively eating larger ants and smaller lizards such as *P. modestum* specializing on small ants (D. W. Davidson, unpublished data). Unlike the avian predators of rodents, ant predators do not exert differential selection pressures in microhabitats differing in shrub cover. This may be why there is little evidence to suggest that granivorous ant species are specialized to forage in particular microhabitats defined by proximity to shrub cover (22). In addition, we may speculate that predator-avoidance activities may be less apt to constrain foraging decisions in eusocial organisms for which single incidents of predation do not coincide with the death of the evolutionary unit.

Ant life histories probably differ markedly from those of rodents. Once colonies of desert harvester ants are established, they may have a long life expectancy. Colonies of *Novomessor cockerelli* and some species of honey ants (genus *Myrmecocystus*) in the same habitat may persist over 20 years (R. M. Chew, personal communication); for species whose colonies or colony entrances are more mobile, longevities are less readily determined. Ectothermy, dormancy (113), and seed storage permit desert ants to survive periods of climatic stress or resource depression with minimal energetic expenditure, and eusociality may allow additional options for maintaining homeostasis in the face of extrinsic environmental fluctuations. Wilson (115) has suggested that individual colonies may sustain dramatic reductions in populations of workers and brood while retaining the capacity to respond quickly when conditions become more favorable, but this hypothesis has never been confirmed empirically for desert seed-eating ants. In contrast, rodent recolonization following periods of resource scarcity may occur more gradually through reproduction and immigration from refugia of survivors.

Birds

Although seed-eating birds can be temporarily and locally abundant in North American deserts, few species are specialized granivores. Most habitats support one species of quail and another of dove that feed primarily on seeds and are resident throughout the year. Large mixed-species flocks of finches migrate into deserts and consume large quantities of seeds when they are available. Some birds that are primarily insectivorous (e.g. thrashers and wrens) take significant quantitites of seeds when their preferred food is scarce, especially in winter.

Unlike rodents and ants, most birds are not central-place foragers and do not store seeds. Except when they are nesting, birds are highly mobile. Flocks of finches and doves in particular search over large distances and concentrate their foraging in areas where seeds are dense (18, 69). Both the local and the large-scale abundance of finches wintering in desert and grassland habitats in the southwestern United States appear to be closely and positively correlated with the size of seed crops [(66, 69); J. Dunning, personal communication].

Other Granivores

Organisms other than rodents, ants, and birds consume seeds in desert habitats, but unfortunately the ecology of most of them is poorly known. In addition to ants, several kinds of insects, including bruchid and curculionid beetles, lepidopterans, and hemipterans, feed on seeds during some stages of their life histories. Such seed predators and parasites may be qualitatively important in desert communities because of their specialization on dominant perennial plants. Probably the best-studied and most numerous of these seed predators are the bruchids, whose larvae develop inside the large and nutritious seeds of leguminous shrubs. In the Western Hemisphere, some of the bruchid genera known to use the fruits and seeds of mesquites (genus *Prosopis*) are specialists on this genus, but a number also exploit related host species belonging to the genera *Cercidium* and *Acacia*. Kingsolver et al (44) have argued that associations of bruchid species with particular hosts depend largely on the coevolution of behavioral and morphological adaptations of predators and hosts rather than on complex chemical coevolution. While these insects are not strictly host-specific, they can consume major portions of the seed crops of small sets of host species (44). They seem unlikely, however, to have a quantitative impact on seed resources comparable to that of the three major granivore taxa.

INTERSPECIFIC COMPETITION AND COMMUNITY ORGANIZATION

Seeds as Limiting Resources

Several lines of evidence indicate that availability of seeds limits populations of desert granivores. First, the population density, biomass, and species diversity of the primary classes of desert seed-eaters are closely and linearly correlated with mean annual precipitation in environmental gradients where other habitat characteristics are held constant [Figure 1; (9, 10, 21)]. In arid ecosystems precipitation provides an accurate estimate of primary productivity (34, 78), and increasing annual rainfall is correlated with increasing size, frequency, and predictability of seed crops. This pattern holds so long as the habitat remains a desert and annual plants are abundant. Granivore abundance and diversity decrease dramatically once there is sufficient precipitation to produce a perennial grassland, presumably because seed availability also decreases as plants allocate proportionately more carbon and nutrients to vegetative structures. Abramsky (1) found that a species of granivorous rodent (*Dipodomys ordi*) that did not normally occur in short-grass prairie habitat colonized and maintained permanent populations on experimental plots where he regularly supplied supplemental seeds.

Additional evidence that seed resources are limiting comes from temporal fluctuations of local granivore populations in response to variations in seed availability. Reproduction of some heteromyid rodents appears to be triggered by the presence in the diet of the leafy parts of recently germinated annual plants (4, 76). Availability of green annuals should provide a reliable cue that a seed crop will be produced within a few weeks. Several studies have shown that populations of granivorous rodents increase rapidly immediately following large seed crops and decline in periods between seed crops (5, 30, 61, 77, 107). In the most arid habitats, where seed crops are highly unpredictable and may be separated by several years, rodent populations may fluctuate through several orders of magnitude. Ants also fluctuate with availability of seeds. Because ant colonies are long-lived and successful founding of new colonies occurs infrequently, the most pronounced varia-

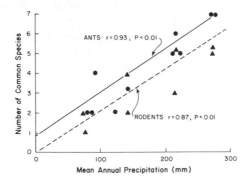

Figure 1   Relationship between the number of common species of seed-eating rodents (triangles) and harvester ants (circles) occurring in sandy flatland habitats and mean annual precipitation for sites in southern California, Arizona, and New Mexico. Correlation coefficients and statistical significance of the plotted linear regressions are indicated. Data replotted from (9, 10, 12, 21).

tion is in the size and activity levels of established colonies (22, 95, 106, 109, 111). Their eusocial system may allow colonies either to increase or reduce the number of workers as well as to regulate the amount of foraging in response to the erratic, pulsile occurrence of seed crops. Granivorous birds show perhaps the most dramatic response to temporal variation in seed availability, though these have been better studied in arid grasslands than in strict deserts. Flocks of migratory finches, in particular, use their mobility to detect and exploit local areas where seed densities are high [(18, 69); H. R. Pulliam, personal communication]. Using Christmas Count censuses, J. Dunning (personal communication) has shown recently that winter densities of virtually all species of migratory finches in the deserts and grasslands of southeastern Arizona are highly correlated with the amount of precipitation, and hence with the quantity of seeds produced, during the previous summer.

Because for sessile, ground-nesting ants the availability of food is contingent upon access to space, data on use of space provide additional evidence that harvester ant populations are limited by resource productivity. One pattern emerges from comparisons of the distributions of species that differ in foraging behavior. Colonies of species characterized by limited worker mobility and small foraging territories become increasingly abundant as seed productivity increases along altitudinal (8) and longitudinal (21) gradients of precipitation. This suggests that colonies with limited foraging areas are unable to acquire sufficient resources for maintenance and reproduction when seed densities are very low. In addition, intraspecific colony spacing is more uniform than random in several ant species (8, 35, 36). Whether such intraspecific territoriality is maintained by killing of colonizing queens [(8); D. W. Davidson, unpublished data], by border skirmishes among workers of established colonies (35–38), or by chemical markers, this behavior probably reflects competition for seed resources. Similar intraspecific spacing might be expected in desert rodents that forage in restricted areas centered around their burrows. The kangaroo rat, *Dipodomys spectabilis,* devotes much time and energy to constructing and maintaining a large, mounded burrow system that may last for many years and several generations of individuals. These mounds are spaced more regularly than at random, presumably reflecting intraspecific territoriality and competition for food (85).

Finally, evidence that seed resources limit granivore populations comes from actual estimates of seed consumption in relation to seed production. Calculations of seed consumption from estimated population densities and metabolic requirements suggest that granivores may consume most of the seeds produced in desert habitats. Either rodents or ants may consume the majority of seeds produced by particular plant species (17, 90, 95, 110), and rodents alone may use over 75% of all seeds produced at certain Mojave and Chihuahuan desert sites (17, 30, 60). In some years finches may consume virtually the entire seed crop in arid grasslands (66).

Competition and Coexistence among Rodents

If seed resources limit granivore populations, then seed-eaters should compete for food and coexistence of species should depend on differential utilization of resources. By influencing the abilities of species to coexist in the same habitats, interspecific competition should play a major role in determining community organization both within the major classes of granivores and in the taxonomically diverse granivore community as a whole.

Strong but circumstantial evidence for the importance of competition among desert rodent species comes from geographic comparisons of rodent communities. As the number of coexisting rodent species increases with increasing seed production along geographic gradients of precipitation, species with particular characteristics are added in a regular order (9, 10). The most striking patterns involve body size (Figure 2) and mode of locomotion. Brown (9) showed that body sizes of coexisting species are more different than would be expected on the basis of random co-occurrence of those species whose geographic ranges were included in his study sites. Furthermore, different species of similar body size and locomotor special-

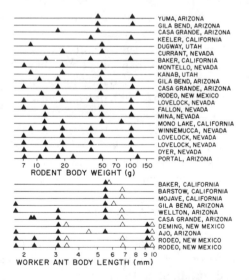

Figure 2 Distribution on a logarithmic scale of body sizes for the common species of coexisting seed-eating rodents (above) and harvester ants (below) for numerous sandy soil habitats in the southwestern United States. Ant species classified as group foragers are indicated by shaded triangles, individual foragers by unshaded triangles. Data replotted from (9, 10, 21).

izations tend to coexist to produce structurally and functionally convergent rodent communities in similar habitats in geographically isolated deserts (10, 52). On the other hand, immediately adjacent but structurally dissimilar habitats, such as areas of sandy and rocky soil, sometimes support rodent communities that differ conspicuously in the number and characteristics of the component species (10, 19, 31, 50, 81, 83). These patterns suggest that coexistence of granivorous desert rodents depends largely on differential utilization of food resources and habitats.

Attempts to measure habitat and seed utilization suggest that at least two mechanisms promote coexistence and provide a functional basis for the morphological differences among species. First, species forage in different habitats. On a large scale, macrohabitat selection may restrict species to particular habitats and prevent coexistence. Frequently pairs of species of extremely similar body size and morphology occur side by side, but with little overlap, in adjacent habitats that can be distinguished by differences in soil texture and/or vegetation (9, 10, 40, 85). On a smaller scale, coexisting species with completely overlapping home ranges often forage in different microhabitats. The most common pattern is for quadrupedal pocket mice to concentrate their activity among dense vegetation and under the canopy of shrubs, whereas bipedal kangaroo rats forage primarily in patches of bare, open ground (10, 15, 46, 64, 79, 114). Rosenzweig (79) has shown by means of an elegant field experiment that the relative abundances of a kangaroo rat, *Dipodomys merriami,* and a pocket mouse, *Perognathus penicillatus,* changed in predictable ways when he manipulated the vegetation structure of the habitat. Such differences in microhabitat utilization may reflect, in part, specializations for avoiding predation, but they also reduce overlap in seed utilization. They may be sufficient to reduce competition and permit coexistence of kangaroo rats and pocket mice in simple, two species communities (46, 64, 79). However, it is unlikely that microhabitat selection alone is sufficient to account for the coexistence of five or six kinds of granivorous rodents in productive habitats, because some of these species appear to forage in similar microhabitats (10, 15).

The conspicuously regular displacement in body size among coexisting species suggests that seed resources are utilized differentially by species of different size and that this is an important mechanism promoting coexistence. Unfortunately it is not yet clear which properties of seeds (if any) are allocated on the basis of size. Tests of the simplest hypothesis—that rodents differentially harvest sizes of seeds positively correlated with their body size (10, 15, 82)—have produced conflicting results (10, 15, 41, 45, 50, 56, 70, 82, 89, 91), and it appears that this idea is too simplistic. However, the idea that the energetic constraints of body size are reflected in differences in foraging behavior and seed selection remains attractive. Recently several investigators have suggested that rodent species of different body size specialize on packets or clumps of seeds that differ in their net energy return (41, 63, 73, 75). They suggest that several small seeds that have not yet been dispersed from a fruit or that have collected in a microsite because of dispersal by wind or water may provide an energetic reward equivalent to that of a single large seed; rodents of large body size and bipedal locomotion may be able to specialize on such rewarding packets because they travel more rapidly and forage over larger areas than their smaller competitors. Although some recent laboratory studies support this hypothesis (41, 63, 75), it remains to be tested rigorously in the field.

Coexisting desert rodent species often differ conspicuously in at least two other important respects. First, species of small body size (e.g. representatives of the genera *Perognathus, Microdipodops, Peromysus,* and *Reithrodontomys*) can enter torpor and remain inactive in their burrows for periods of days, weeks, or months when food is scarce and/or temperatures are low (11, 29, 43, 74). Such bouts of short-term inactivity, hibernation, or estivation reduce energy requirements of the individuals involved and remove them temporarily from competition for seed resources. Often coexisting rodents also differ in adaptations for avoiding predation. The most striking contrast is between the kangaroo rats and kangaroo mice, on the one hand, and the pocket and cricetid mice, on the other. The former have highly specialized morphologies (including internal ears adapted for detecting the low frequency sounds made by their predators); they forage in exposed open microhabitats. The latter are relatively unspecialized; they concentrate their activity in areas that provide dense vegetative cover (2, 3, 26, 99, 100). Since these adaptations for utilizing torpor and avoiding predators affect the timing and location of foraging, they undoubtedly influence competitive interactions; but we have little precise understanding of their effects on community organization.

Perhaps the strongest and most direct evidence for the role of interspecific competition in desert rodent communities comes from comparisons of habitat selection, seed utilization, and population densities when different numbers and combinations of species occur in similar habitats (51). Brown (9, 10) sampled the rodents inhabiting sand dunes in the eastern and western Great Basin; those areas have similar habitats, but there are fewer kinds of rodents in the eastern part because of its historical biogeographic isolation. Brown (9) found that average population densities per species were higher and the range of seed sizes utilized by the kangaroo rat, *Dipodomys ordi,* was greater in the eastern than in the western Great Basin. He suggested that such density compensation and niche expansion were responses to reduced interspecific competition. E. Larsen (unpublished data) used both his own data and a reanalysis of Brown's data to show that in structurally similar macrohabitats particular heteromyid species foraged in a wider range of microhabitats when one or more competing species were absent. Similar results have been obtained from experimental manipulations of species abundances in enclosures (63) and from comparisons of microhabitat utilization by the kangaroo rat, *D. merriami,* in the same habitat between seasons when coexisting pocket mice were either active or in hibernation [(63); O. J. Reichman, unpublished data]. Perhaps the most perplexing experimental results are those of Schroder & Rosenzweig (86), who observed no increase in the population densities of either *D. merriami* or *D. ordi* when its congener was removed in reciprocal experiments on large unfenced plots. Although these authors concluded that the two species exhibit inflexible, nonoverlapping patterns of habitat selection and do not compete, their results should be viewed with caution because of problems with the design and execution of their experiment.

Competition and Community Organization in Ants

Geographic patterns in the diversity and organization of harvester ant communities are strikingly like those already outlined for rodents. Characteristics of the structural and functional organization of ant communities are preserved as both biomass and species diversity increase over a gradient of enhanced precipitation and seed productivity in the semi-deserts of the southwestern United States (21). Species for which similar feeding niches result in strongly overlapping resource requirements appear unlikely to coexist locally within relatively uniform habitat types (21, 112). Two attributes of granivorous ant species that appear to be closely tied to patterns of resource utilization and coexistence are worker body size [Figure 2; (16, 21, 32)] and colony foraging behavior (group or individual strategies, as detailed above). Species characterized by similar worker body sizes and colony foraging strategies tend not to coexist locally in homogeneous habitat types even at sites that lie within the distributional ranges of both species (21, 110). The suggestion that some such species pairs may be intense competitors and behave as ecological replacements for one another is supported by observations of interspecific territorial defense in marginal areas where their populations come into contact (35, 112).

The differences in worker body sizes and species-specific colony foraging behaviors that appear to facilitate coexistence have been related directly to resource allocation by seed size and density (16, 21–23, 32). Specific worker sizes probably impose mechanistic upper limits and energetic lower limits to the sizes of seeds that can be exploited. In addition, among species employing similar foraging strategies, worker body sizes are positively correlated with foraging distance and probably, as a result, with access to preferred resources (8, 22, 32). Ants select seeds partly on the basis of particle sizes both from experimental distributions of nutritionally identical seeds in different size classes (21, 23, 24, 92) and from naturally occurring distributions of native seeds (21, 23). However, considerable variability exists in the latter relationship; additional factors, such as nutritional content of the seed, accessibility of the embryo within the seed or the seed within the fruit, and foraging choices of ants in response to spatial distributions of seeds, may also be important.

Above, we argued that group and individual foraging behaviors are adaptations enabling exploitation of high- and low-density seed resources, respectively; we suggested that ant species with similar body sizes might coexist if they differed in foraging strategy. The potential for resource partitioning by "density specialization" was also invoked in our consideration of the coarse-grained and fine-grained foraging behaviors employed by rodents. In order for interspecific differences in density specialization to qualify as mechanisms of resource allocation, such differences must grant each competitor exclusive access to some subset of resources. To the extent that seed taxa are characterized by species-specific dispersion patterns [(73); H. R. Pulliam, personal communication] high- and low-density seed specialists may exploit different resource populations. If evolutionary responses to predation and competition interact to effect microhabitat separation and the associated differences in rodent foraging behaviors, rodent species feeding under shrub canopies and those feeding in open habitats may share only the subset of seed resources that cross habitat boundaries. This subset may be particularly insignificant for buried seeds that are less likely to be moved by physical forces.

Finally, Davidson (22) has suggested a mechanism by which density-specialization might promote coexistence even in the absence of microhabitat separation of competitors or taxonspecific patterns of seed dispersion. This mechanism depends on spatial heterogeneity in seed density (documented above) and/or temporal alternation of seed abundance and rarity, a distinguishing feature of these deserts. Using a cost-benefit approach, Davidson suggested that group and individually foraging ants with similar worker body sizes could coexist if adaptations such as larger colony sizes, more effective recruitment strategies, and better facilities for seed storage granted group foragers access to most or all of seed resources above some threshold density while preventing them from foraging profitably at lower resource levels. Comparisons of resource depletion by group and individually foraging ant species (22, 108) support the hypothesis that group foragers can reduce high density resources more rapidly than do individual foragers. Assuming lower foraging costs for individual foragers that seldom or never use chemical recruitment, these species might forage profitably at resource densities below the threshold level that permits group foragers to forage economically. The proportionate subdivision of the ant community among group and individually foraging ants should reflect the mean availability of resources at different seed densities, and some evidence supports this prediction (22). Further data are needed to define more accurately the cost-benefit curves that are the basis of this model.

Resource subdivision of temporal displacement of foraging activities has also been suggested as a mechanism that promotes coexistence among food-limited, desert seed-eating ants. Seed resources in deserts are renewed only once or twice annually, and though seasonal differences in activity schedules may enhance coexistence, diurnal differences would not. Differences in activity schedules on a diurnal timescale probably reflect physiological or morphological adaptations that facilitate foraging in specific seasons to which particular species have adapted (7). *Pogonomyrmex desertorum* and *P. californicus,* congeners characterized by similar worker body sizes and colony foraging strategies, may coexist by virtue of species-specific differences in seasonality of foraging. While activity in *P. desertorum* peaks during midsummer, maximum activity of *P. californicus* colonies occurs in late summer and fall in southern New Mexico (87, 109, 111). Some evidence indicates that consistent and distinctive seasonal activity patterns characterize group and individual foragers and that these patterns are adaptive responses related to density specialization (22).

Several investigations of colony dispersion patterns have provided information on interspecific interactions. In the Chihuahuan Desert, the species composition of assemblages of harvester ants may change across habitat boundaries marked by differences in microtopography (16). Within sampling areas chosen for habitat homogeneity, interspecific overlaps in foraging areas were significantly nonrandom for only two of many pairwise comparisons (21). The spatial separation of nesting sites and foraging territories for these congeners apparently results from interference interactions rather than interspecific differences in microhabitat utilization. One marked positive association of colonies of two coexisting harvester ants has been attributed to indirect pathways of competitive interaction in a six species assemblage (25). In this case as well, the primary microhabitat variable to which ant species appear to be responding is the distribution of preestablished colonies, rather than any other measurable habitat variable.

Paralleling increases in precipitation, productivity, and ant species diversity in the arid southwestern United States is a trend toward enhanced morphological, behavioral, and ecological specialization by harvester ants. This is precisely the pattern predicted by MacArthur (51) for assemblages of resource-limited consumers. Dominating the most depauperate ant faunas in the least productive of these deserts is *Veromessor pergandei,* a supreme generalist characterized by an intermediate mean worker body size and marked continuous worker size polymorphism within the colony. The remarkable polymorphism displayed by *V. pergandei* enables colonies of this species to utilize a diversity of seed sizes efficiently (23, 104) and, perhaps also, to forage efficiently at a range of distances from the central nest site. Also a generalist with respect to foraging behavior, *V. pergandei* utilizes a form of group foraging that apparently involves neither the use of scouts nor the chemical recruitment that characterizes other group foragers (104). However, populous columns containing as many as 17,000 workers (106) deplete high-density seed resources, while demonstrating an apparently unique capacity to rotate the foraging column at a rate that is inversely proportional to resource abundance and, consequently, to take advantage of seeds scattered at low densities as well (8, 104, 105). Throughout the distribution of *V. pergandei* the expression of within-colony worker size polymorphism reflects remarkably specific character shifts in response to the presence or absence of competing species (23). At the eastern edge of its distribution in the moderately productive Sonoran Desert of south-central Arizona where *V. pergandei* may coexist with up to four potentially competing species with larger or smaller but approximately monomorphic worker body sizes, the size polymorphism is markedly less well-developed, and workers of intermediate body size dominate individual colonies. In habitats of still greater resource productivity and ant species diversity *V. pergandei* is replaced by ant species whose colonies are specialized to produce workers of uniform body sizes and to forage on narrower ranges of seed densities (21, 22).

Competition among Birds

Desert seed-eaters other than rodents and ants have been so little studied from a community perspective that we know virtually nothing about competition and resource utilization in these taxa. Patterns of interspecific interactions and community organization described for granivorous birds in other habitats (65–68) might also, with some modification, be looked for in desert birds. This will be a profitable area for future research.

Competition between Classes

If seeds are limiting to the extent that interspecific competition has played a major role in determining patterns of coexistence and resource utilization within the major classes of seed-eaters, then we might also expect significant competition between the classes. A necessary condition for competition is overlap in requirements for limited resources. Comparisons of native seeds collected by naturally foraging animals and experiments with domestic seeds indicate that rodents and ants (and probably also birds) overlap broadly in the sizes and species of seeds they harvest and the microhabitats where they forage (12, 14, 55). As documented earlier, quantitative estimates of seed consumption suggest that particular species of granivores may harvest most of the seeds produced by certain plant species. Rodents have been estimated to consume more than 75% of the total seeds available in some desert habitats, and finches may take virtually all of the seeds produced in arid grasslands in relatively poor years. These indirect observations suggest that interclass competition may be severe.

Direct evidence for the importance of competition between rodents and ants comes from our own field experiments in which each class was excluded from experimental plots in the Sonoran Desert (12, 13). When either rodents or ants were removed from 0.1 ha plots of desert habitat we observed reciprocal increases in the remaining taxon. Density of ant colonies increased 71% on plots where rodents had been excluded relative to control plots. Rodents increased 20% in population density and 29% in biomass in the absence of ants. Measurements of quantities of seeds in the soil and densities of annual plants provide corroborative evidence that these reciprocal increases reflect exploitative competition for seeds. Compared to control plots where all classes of granivores were present, densities of seeds and plants were significantly higher on plots where either rodents or ants had been removed, and they were 2–5 times higher on plots where both classes of granivores had been excluded [(13); R. S. Inouye, G. S. Byers, and J. H. Brown, unpublished data).

Rodents and ants differ in many ways that affect population dynamics and seed utilization. It is difficult to make more than speculative comments about the mechanisms that permit stable coexistence of the two classes in natural desert habitats. As a class, ants appear to harvest seed size classes in proportion to their occurrence in the environment; they take smaller seeds on the average than rodents, which seem to forage selectively for large seeds [(70, 72); D. W. Davidson, unpublished data; R. S. Inouye, G. S. Byers, and J. H. Brown, unpublished data]. However, the sizes and species taken by the two classes often overlap and the diets of particular ant and rodent species may be even more similar. Established ant colonies may be less affected by the long, unpredictable periods between seed crops than rodent populations because ants can regulate colony size and foraging activity to resource availability. In addition, because of the small body size of individual workers, ants may be able to forage profitably on lower densities of seeds than rodents (13). Rodents, on the other hand, are more efficient than ants at collecting dense seeds; they can dig for buried seeds, which may remain in locally dense clumps long after most seeds have been removed from the soil surface (14, 73, 75).

While it is likely that both rodents and ants also compete with seed-eating birds and other taxa of desert granivores, the extent and consequences of these interactions remain to be investigated. From the few studies available (18, 66, 69), we suspect that birds that do not store food are much more sensitive than either rodents or ants to temporarily and locally low densities of available seeds, but they use their mobility and flocking behavior to search over large areas and exploit seeds where they are locally abundant.

Comparisons of Communities among Continents

Differences and similarities in the kinds of granivores that inhabit the major deserts of the world suggest that unique historical events, environmental factors, and interspecific competition both within and between closely related taxa have interacted to produce observed patterns of community organization. Extensive deserts occur on all continents except Europe and Antarctica. Birds and ants are highly vagile, and several taxa of both classes have been able to colonize all of the major deserts; but historical events have greatly influenced the evolution of desert rodent faunas on different continents. Australia and South America have long histories of isolation from the other land masses, and both continents have depauperate desert rodent faunas. Two genera (*Notomys* and *Pseudomys*) of specialized granivorous rodents, and one genus (*Rattus*) of generalized, omnivorous rodents (all in the family Muridae) occur in Australia, but the densities of populations and the number of coexisting species in most desert habitats are conspicuously lower than in North America [(57); S. R. Morton, personal communication]. Although South America has many generalized rodents, some of which are ecologically similar to North American cricetids, it has no specialized granivores comparable to the heteromyids. However, the granivorous mammal niche may have been filled until quite recently by small, bipedal marsupials of the extinct family Argyrolagidae (54, 55). On each of the other continents with extensive deserts, specialized seed-eating rodents have evolved independently from unspecialized distantly related ancestral stocks: in North America the genera *Dipodomys, Microdipdops,* and *Perognathus* in the family Heteromyidae; in Asia the genera *Jaculus, Dipus, Salpingotus, Cardiocranus,* and others in the family Dipodidae; and in Africa and the Middle East the genera *Gerbillus, Tatera,* and *Taterillus* in the family Cricetidae (98). The genera *Dipodomys, Jaculus, Dipus, Gerbillus,* and *Notomys* provide some of the most spectacular examples of similar morphology, physiology, and behavior produced by convergent evolution to fill similar ecological niches on different continents.

Although more work is needed in Asian and African deserts, it appears that their granivore communities are convergently similar to those in North America, particularly when habitats of similar climate, vegetation, and soil type are compared. On all three continents, seed-eating rodents, ants, and birds are diverse and abundant; from the limited data available, coexisting species appear to show comparable patterns of body size and foraging behavior. South American and Australian granivore communities are distinctive. In South America, despite the absence of specialized seed-eating rodents, granivorous desert ants and birds are not conspicuously more diverse or abundant than in North America (55). A very different situation obtains in Australia, where the rodent fauna is also depauperate but there are about 3 times more species of harvester ants and 6 times more species of granivorous birds than in North American deserts. The failure of rodents to diversify in Australia might be explained in part by the low frequency of rains. The coefficient of variation in annual precipitation tends to be much greater for Australian than for North American deserts (58). If the interval between seed crops becomes too long, rodent populations may have difficulty surviving. The high diversity of granivorous birds and ants in Australia suggests that these classes have been able to radiate by exploiting seed resources equivalent to those utilized by rodents in North American, Asian, and African deserts.

THE IMPACT OF GRANIVORES ON DESERT ECOSYSTEMS

Aside from interacting among themselves, seed-eaters have two important direct effects on the structure and function of desert ecosystems: As predators they influence the distribution and abundance of plant populations, and as prey they provide food for carnivore populations. We have already cited evidence that granivores may consume large fractions both of the total seed crop produced in an area and of the seeds produced by particular plant species. The exclusion experiments designed to study competition between rodents and ants (12, 13) were used by R. S. Inouye, G. S. Byers, and J. H. Brown (unpublished data) to assess the impact of these taxa on the annual plant community. Compared to control plots where both classes of granivores were present, densities of seeds in the soil and of vegetative annual plants increased significantly on plots where rodents or ants were excluded. Rodents and ants preyed differentially on particular plant species and consequently had different effects on the annual plant community (Table 2). Rodents apparently preyed selectively on large-seeded species (e.g. *Erodium* and *Lotus* spp.); these increased to dominate vegetative plant biomass on plots where rodents were excluded. Ants seem to have preyed most intensely on the most abundant species (*Filago califormia*), and this small-seeded composite increased to dominate the community numerically and reduce species diversity on plots where ants had been removed. The plants also competed intra- and interspecifically for limited water resources, and granivore predation apparently interacted with this competition to

Table 2 Effects of experimental removal of rodents or ants on densities of particular annual plant species and on characteristics of the annual plant community (R. S. Inouye, G. S. Byers, and J. H. Brown, unpublished data).

	Seed weight (mg)	Density[a] (number m^{-2})	Effect of removing Rodents	Effect of removing Ants	Probable explanation
Density of:					
All plants	variable	209.6	increase P < 0.05	increase P < 0.01	seed predation by rodents and ants
Erodium cicutarium[b]	1.6	1.75	increase P < 0.01	not significant	selective predation by rodents on large seeds
Erodium texanum[b]	1.6	0.59	increase P < 0.05	not significant	selective predation by rodents on large seeds
Lotus humistratus	1.5	11.4	increase P < 0.05	not significant	selective predation by rodents on large seeds
Euphorbia polycarpa	0.18	2.8	decrease P < 0.02	not significant	selective predation by rodents on competing *Erodium* spp.
Filago californica	0.04	142.1	not significant	increase P < 0.05	selective predation by ants on most abundant species
Species diversity ($H' = -\Sigma p_i \ln p_i$)			not significant	decrease P < 0.05	selective predation by ants on most abundant species
Biomass (g dry weight m^{-2})			increase P < 0.01	not significant	selective predation by rodents on large seeded species which grow rapidly and dominate biomass[b]

[a] Average density on control plots where both rodents and ants were present.
[b] Biomass of the two *Erodium* spp. constituted 41% of total biomass on control plots and a much greater proportion on plots where rodents were removed.

affect annual community structure. Thus small-seeded *Euphorbia polycarpa* was present in high densities on plots where rodents were present, but it was replaced by large-seeded (and presumably competitively superior) *Erodium* spp. on plots where rodents had been removed. These effects on annual plants were apparent after excluding granivores for only four years, and there is no evidence that the plant community had yet equilibrated to the experimental manipulations. Longer experiments would be required to assess the possible impact of granivores on the long-lived perennial shrubs and succulents.

Because seed-eaters are among the most abundant and diverse sources of food for desert carnivores, predation on granivores must be important in the functional organization of desert ecosystems. Unfortunately, little is known about the effects of granivores as prey on the population dynamics and community structure of carnivores or, conversely, about the influence of predation on granivore abundance and community organization. As indicated earlier, ants are consumed by many kinds of insectivores, including horned lizards of the genus *Phrynosoma,* which are virtually obligate ant eaters. Similarly, many vertebrate predators take granivorous rodents. Among both rodents and ants, coexisting species differ both in the extent to which they are exposed to predation and in their adaptations for avoiding predators. Unlike the relatively generalized pocket mice, kangaroo rats and kangaroo mice forage in open microhabitats and they possess greatly inflated tympanic bullae (the bony chamber surrounding the middle and inner ear), which appear to enhance auditory acuity for low-frequency sounds made by their vertebrate predators (99, 100). The excruciatingly painful venomous stings of some desert harvester ants may serve primarily as a defense against vertebrate predators. These observations suggest that predation may interact with competition to influence the structure of granivore communities and that seed-eaters are important food resources for both specialized and generalized carnivores.

CONCLUSIONS

The seeds and seed-eaters of desert habitats provide an attractive system for investigating the ecological interactions that determine the functional organization of natural communities. In our attempt to summarize what is known about this system, we have drawn the following conclusions:

1. Seeds, particularly those of annuals, provide particulate, nutritious, relatively reliable food resources for desert animals. The size and spacing of seed crops depend on the quantity and timing of unpredictable desert rains.
2. Numerous species of rodents, ants, and birds (and a few kinds of other animals) are specialized to feed on seeds. They possess adaptations for harvesting seeds when they are available and for surviving the periods of food scarcity between crops. Population densities and species diversity of these granivores are determined in large part by the availability of seed resources.
3. Competition for limited food resources has had a major influence on the organization of granivore communities. Differences in body size, habitat utilization, and foraging behavior enable closely related species to subdivide resources and coexist to form stable communities. Distantly related taxa also compete for seeds, but the mechanisms of resource utilization that permit them to coexist remain to be investigated.
4. Comparisons of desert granivores on different continents provide examples of evolutionary convergence of both particular species and entire communities toward similar ecological roles.
5. Granivores as predators have important effects on population densities and community structure of annual plants. As prey, seed-eaters may in turn exert important influences on both the population dynamics and the interspecific interactions of carnivores.

ACKNOWLEDGMENTS

We are grateful to numerous colleagues, students, and assistants for generously sharing their data and ideas and for helping with our research. Our own studies of desert granivores have been supported by the National Science Foundation and the Desert Biome, US/1BP. A grant from the Desert Biome provided partial support for the preparation of this review.

Literature Cited

1. Abramsky, Z. 1978. Small mammal community ecology. Changes in species diversity in response to manipulated productivity. *Oecologia* 34:113–23
2. Bartholomew, G. A., Cary, G. R. 1954. Locomotion in pocket mice. *J. Mammal.* 35:386–92
3. Bartholomew, G. A., Caswell, H. H. 1951. Locomotion in kangaroo rats and its adaptive significance. *J. Mammal.* 32:155–69
4. Beatley, J. C. 1967. Survival of winter annuals in the northern Mojave Desert. *Ecology* 48:745–50
5. Beatley, J. C. 1969. Dependence of desert rodents on winter annuals and precipitation. *Ecology* 50:721–24
6. Beatley, J. C. 1974. Phenological events and their environmental triggers in Mojave Desert ecosystems. *Ecology* 55:856–63
7. Bernstein, R. A. 1974. Seasonal food abundance and foraging activity in some desert ants. *Am. Nat.* 108:490–98
8. Bernstein, R. A. 1975. Foraging strategies of ants in response to variable food density. *Ecology* 56:213–19
9. Brown, J. H. 1973. Species diversity of seed-eating desert rodents in sand dune habitats. *Ecology* 54:775–87
10. Brown, J. H. 1975. Geographical ecology of desert rodents. In *Ecology and Evolution of Communities*, ed. M. L. Cody, J. M. Diamond, pp. 315–41. Cambridge, Mass: Belknap
11. Brown, J. H., Bartholomew, G. A. 1969. Periodicity and energetics of torpor in the kangaroo mouse, *Microdipodops pallidus. Ecology* 50:705–9
12. Brown, J. H., Davidson, D. W. 1977. Competition between seed-eating rodents and ants in desert ecosystems. *Science* 196:880–82
13. Brown, J. H., Davidson, D. W., Reichman, O. J. 1979. An experimental study of competition between seed-eating desert rodents and ants. *Am. Zool.* In press

14. Brown, J. H., Grover, J. J., Davidson, D. W., Lieberman, G. A. 1975. A preliminary study of seed predation in desert and montane habitats. *Ecology* 56:987–92
15. Brown, J. H., Lieberman, G. A. 1973. Resource utilization and coexistence of seed-eating rodents in sand dune habitats. *Ecology* 54:788–97
16. Chew, R. M. 1977. Some ecological characteristics of the ants of a desert-shrub community in southeastern Arizona. *Am. Midl. Nat.* 98:33–49
17. Chew, R. M., Chew, A. E. 1970. Energy relationships of the mammals of a desert shrub (*Larrea tridentata*) community. *Ecol. Monogr.* 40:1–21
18. Cody, M. L. 1971. Finch flocks in the Mohave Desert. *Theor. Pop. Biol.* 2:142–58
19. Congdon, J. 1974. Effect of habitat quality on distributions of three sympatric species of desert rodents. *J. Mammal.* 55:659–62
20. Creighton, W. S. 1953. New data on the habits of the ants of the genus *Veromessor. Am. Mus. Novit.* 1612:1–18
21. Davidson, D. W. 1977. Species diversity and community organization in desert seed-eating ants. *Ecology* 58:711–24
22. Davidson, D. W. 1977. Foraging ecology and community organization in desert seed-eating ants. *Ecology* 58:725–37
23. Davidson, D. W. 1978. Size variability in the worker caste of a social insect (*Veromessor pergandei* Mayr) as a function of the competitive environment. *Am. Nat.* 112:523–32
24. Davidson, D. W. 1978. Experimental tests of the optimal diet in two social insects. *Behav. Ecol. Sociobiol.* 4:35–41
25. Davidson, D. W. 1979. Some consequences of diffuse competition in a desert ant community. *Am. Nat.* In press
26. Eisenberg, J. F. 1963. The behavior of heteromyid rodents. *Univ. Calif. Publ. Zool* 69:1–114

27. Ettershank, G., Whitford, W. G. 1973. Oxygen consumption of two species of Pogonomyrmex harvester ants (Hymenoptera:Formicidae). Comp. Biochem. Physiol. 46A:605–11

28. Fall, H. C. 1937. The North American species of Nemadus Thom., with descriptions of new species (Coleoptera, Silphidae). J. NY Entomol. Soc. 45: 335–40

29. French, A. R. 1976. Selection of high temperatures for hibernation by the pocket mouse, Perognathus longimembris: ecological advantages and energetic consequences. Ecology 57:185–91

30. French, N. R., Maza, B. G., Hill, H. O., Aschwanden, A. P., Kaaz, H. W. 1974. A population study of irradiated desert rodents. Ecol. Monogr. 44:45–72

31. Hafner, M. S. 1977. Density and diversity in Mojave Desert rodent and shrub communities. J. Anim. Ecol. 46:925–38

32. Hansen, S. R. 1978. Resource utilization and coexistence of three species of Pogonomyrmex ants in an Upper Sonoran grassland community Oecologia 35:109–18

33. Hatch, M. H. 1933. Studies on the Leptodiridae (Catopidae) with descriptions of new species. J. NY Entomol. Soc. 41:187–239

34. Hillel, D., Tadmor, N. 1962. Water regime and vegetation in the central Negev highlands of Israel. Ecology 43:33–41

35. Hölldobler, B. 1974. Home range orientation and territoriality in harvesting ants. Proc. Natl. Acad. Sci. USA 71(8): 3274–77

36. Hölldobler, B. 1976. Recruitment behavior, home range orientation and territoriality in harvester ants, Pogonomyrmex. Behav. Ecol. Sociobiol. 1:3–44

37. Hölldobler, B. 1978. Territoriality in ants. Presented at Meet. Am. Philos. Soc., Philadelphia

38. Hölldobler, B., Stanton, R. C., Markl, H. 1978. Recruitment and food retrieving behavior in Novomessor (Formicidae, Hymenoptera). I. Chemical signals. Behav. Ecol. Sociobiol. 4: 163–81

39. Hölldobler, B., Wilson, E. O. 1970. Recruitment trails in the harvester ant, Pogonomyrmex badius. Psyche 77: 385–99

40. Hoover, K. D., Whitford, W. G., Flavill, P. 1977. Factors influencing the distributions of two species of Perognathus. Ecology 58:877–84

41. Hutto, R. L. 1978. A mechanism for resource allocation among sympatric heteromyid rodent species. Oecologia 33:115–26

42. Kay, C. A., Whitford, W. G. 1975. Influences of temperature and humidity on oxygen consumption of five Chihuahuan desert ants. Comp. Biochem. Physiol. 52A:281–86

43. Kenagy, G. J. 1973. Daily and seasonal patterns of activity and energetics in a heteromyid rodent community. Ecology 54:1201–19

44. Kingsolver, J. M., Johnson, C. D., Swier, S. R., Teran, A. 1977. Prosopis fruits as a resource for invertebrates. In Mesquite: Its Biology in Two Desert Scrub Ecosystems, ed. B. B. Simpson, pp. 108–22. Stroudsburg, Pa: Dowden, Hutchinson & Ross

45. Lemen, C. A. 1978. Seed size selection in heteromyids. Oecologia 35:1201–19

46. Lemen, C. A., Rosenzweig, M. L. 1978. Microhabitat selection in two species of heteromyid rodents. Oecologia 33: 13–19

47. Levieux, J., Diomande, T. 1978. The nutrition of granivorous ants II. Cycle of activity and diet of Brachyponera senaarensis (Mayr). Insect Soc. 25: 187–96

48. Lockard, R. B., Owings, D. H. 1974. Seasonal variation in moonlight avoidance by bannertail kangaroo rats. J. Mammal. 55:189–93

49. Ludwig, J. A., Whitford, W. G. 1979. Short-term water and energy flow in arid ecosystems. In Ecosystem Dynamics, ed. I. Noy-Meir. London: Cambridge Univ. Press. In press

50. M'Closkey, R. T. 1978. Niche separation and assembly in four species of Sonoran Desert Rodents. Am. Nat. 112:683–94

51. MacArthur, R. H. 1972. Geographical Ecology. NY: Harper & Row. 269 pp.

52. MacMahon, J. A. 1976. Species and guild similarity of North American desert mammal faunas: a functional analysis of communities. In Evolution of Desert Biota, ed. D. W. Goodall, 133–48. Austin, Tex: Univ. Texas Press. 250 pp.

53. MacMillen, R. E., Lee, A. K. 1967. Australian desert mice: independence of exogenous water. Science 158:383–85

54. Mares, M. A. 1975. South American mammal zoogeography: evidence from convergent evolution in desert rodents. Proc. Natl. Acad. Sci. USA 72:1702–6

55. Mares, M. A., Rosenzweig, M. L. 1978. Granivory in North and South American deserts: rodents, birds, and ants. Ecology 59:235–41

56. Mares, M. A., Williams, D. F. 1977. Experimental support for food particle size resource allocation in heteromyid rodents. *Ecology* 58:1186–90
57. Monson, G. 1943. Food habits of the banner-tailed kangaroo rat in Arizona. *J. Wild. Manag.* 7:98–102
58. Morton, S. R. 1979. Diversity of desert-dwelling mammals: a comparison of Australia and North America. *J. Mammal.* In press
59. Mott, J. J., McKeon, G. M. 1977. A note on the selection of seed types by harvester ants in northern Australia. *Aust. J. Ecol.* 2:231–35
60. Nelson, J. F., Chew, R. M. 1977. Factors affecting seed reserves in the soil of a Mojave Desert ecosystem, Rock Valley, Nye County, Nevada. *Am. Midl. Nat.* 97:300–20
61. O'Farrell, T. P., Olson, R. J., Gilbert, R. O., Hedlund, J. D. 1975. A population of Great Basin pocket mice, *Perognathus parvus,* in the shrub-steppe of south-central Washington. *Ecol. Monogr.* 45:1–28
62. Oster, G., Eshel, I., Cohen, D. 1977. Worker-queen conflict and the evolution of social insects. *Theor. Pop. Biol.* 12:49–85
63. Price, M. V. 1978. Seed dispersion preferences of coexisting desert rodent species. *J. Mammal.* 59:624–26
64. Price, M. V. 1978. The role of microhabitat in structuring desert rodent communities. *Ecology* 59:910–21
65. Pulliam, H. R. 1975. Coexistence of sparrows: a test of community theory. *Science* 189:474–76
66. Pulliam, H. R., Brand, M. R. 1975. The production and utilization of seeds in plains grassland of southeastern Arizona. *Ecology* 56:1158–66
67. Pulliam, H. R., Enders, F. 1971. The feeding ecology of five sympatric finch species. *Ecology* 52:557–66
68. Pulliam, H. R., Mills, G. S. 1977. Use of space by wintering sparrows. *Ecology* 58:1393–99
69. Raitt, R. J., Pimm, S. L. 1976. Dynamics of bird communities in the Chihuahuan Desert, New Mexico. *Condor* 78:427–42
70. Reichman, O. J. 1975. Relation of desert rodent diets to available resources. *J. Mammal.* 56:731–51
71. Reichman, O. J. 1976. Relationships between dimensions, weights, volumes and calories of some Sonoran Desert seeds. *Southwest. Nat.* 20:573–75
72. Reichman, O. J. 1977. Optimization of

diets through food preferences by heteromyid rodents. *Ecology* 58:454–57
73. Reichman, O. J. 1979. Factors influencing foraging patterns in desert rodents. In *Mechanisms of Optimal Foraging,* ed. A. Kamil, T. Sargent. NY: Garland. In press
74. Reichman, O. J., Brown, J. H. 1979. The use of torpor by *Perognathus amplus* in relation to resource distribution. *J. Mammal.* In press
75. Reichman, O. J., Oberstein, D. 1977. Selection of seed distribution types by *Dipodomys merriami* and *Perognathus amplus. Ecology* 58:636–43
76. Reichman, O. J., Van DeGraff, K. 1975. Influence of green vegetation on desert rodent reproduction. *J. Mammal.* 56:503–6
77. Reynolds, H. G. 1958. The ecology of the Merriam kangaroo rat (*Dipodomys merriami* Mearns) on the grazing lands of southern Arizona. *Ecol. Monogr.* 28:111–27
78. Rosenzweig, M. L. 1968. Net primary productivity of terrestrial communities: prediction from climatological data. *Am. Nat.* 102:67–74
79. Rosenzweig, M. L. 1973. Habitat selection experiments with a pair of coexisting heteromyid rodent species. *Ecology* 54:111–17
80. Rosenzweig, M. L. 1974. On the optimal aboveground activity of bannertail kangaroo rats. *J. Mammal.* 55:193–99
81. Rosenzweig, M. L., Smigel, B., Kraft, A. 1975. Patterns of food, space, and diversity. In *Rodents in Desert Environment,* ed. I. Prakash, P. K. Ghosh, pp. 241–68. The Hague: Junk
82. Rosenzweig, M. L., Sterner, P. 1970. Population ecology of desert rodent communities: body size and seed husking as bases for heteromyid existence. *Ecology* 51:217–24
83. Rosenzweig, M. L., Winakur, J. 1969. Population ecology of desert rodent communities: habitats and environmental complexity. *Ecology* 50:558–72
84. Schmidt-Nielsen, K. 1964. *Desert Animals: Physiological Problems of Heat and Water.* Oxford: Clarendon Press. 277 pp.
85. Schroder, G. D., Geluso, K. H. 1975. Spatial distribution of *Dipodomys spectabilis* mounds. *J. Mammal.* 56:363–68
86. Schroder, G. D., Rosenzweig, M. L. 1975. Perturbation analysis of competition and overlap in habitat utilization between *Dipodomys ordii* and *Dipodomys merriami. Oecologia* 19:9–28

87. Schumacher, A., Whitford, W. G. 1976. Spatial and temporal variation in Chihuahuan desert ant faunas. *Southwest. Nat.* 21:1–8

88. Shaw, W. T. 1934. The ability of the giant kangaroo rat as a harvester and storer of seeds. *J. Mammal.* 15:275–86

89. Smigel, B. W., Rosenzweig, M. L. 1974. Seed selection in *Dipodomys merriami* and *Perognathus penicillatus. Ecology* 55:329–39

90. Soholt, L. F. 1973. Consumption of primary production by a population of kangaroo rats (Dipodomys *merriami*) in the Mojave desert. *Ecol. Monogr.* 43:357–76

91. Stamp, N. E., Ohmart, R. D. 1978. Resource utilization by desert rodents in the lower Sonoran Desert. *Ecology* 59:700–7

92. Taylor, F. 1977. Foraging behavior of ants: experiments with two species of Myrmecine ants. *Behav. Ecol. Sociobiol.* 2:147–68

93. Tevis, L. 1958. Germination and growth of ephemerals induced by sprinkling a sandy desert. *Ecology* 39:681–88

94. Tevis, L. 1958. A population of desert ephemerals germinated by less than one inch of rain. *Ecology* 39:688–95

95. Tevis, L. 1958. Interrelations between the harvester ant *Veromessor pergandei* (Mayr) and some desert ephemerals. *Ecology* 39:695–704

96. Tucker, V. A. 1966. Diurnal torpor and its relation to food consumption and weight changes in the California pocket mouse *Perognathus Californieus. Ecology* 47:245–52

97. Vorhies, C. T., Taylor, W. P. 1922. Life history of the kangaroo rat, *Dipodomys spectabilis spectabilis* Merriam. *US Dep. Agric. Bull.* 1091:1–40

98. Walker, E. P. 1968. *Mammals of the World,* Vols. 1, 2. Baltimore: Johns Hopkins Press. 1500 pp.

99. Webster, D. B. 1962. A function of the enlarged middle-ear cavities of the kangaroo rat, *Dipodomys. Physiol. Zool.* 35:248–55

100. Webster, D. B., Webster, M. 1975. Auditory systems of heteromyidae: Functional morphology and evolution of the middle ear. *J. Morphol.* 146:343–76

101. Went, F. W. 1948. Ecology of desert plants. I. Observations on germination in the Joshua Tree National Monument, California. *Ecology* 29:242–53

102. Went, F. W. 1949. Ecology of desert plants. II. The effect of rain and temperature on germination and growth. *Ecology* 30:1–13

103. Went, F. W., Westergaard, M. 1949. Ecology of desert plants. III. Development of plants in the Death Valley National Monument, California. *Ecology* 30:26–38

104. Went, F. W., Wheeler, J., Wheeler, G. C. 1972. Feeding and digestion in some ants (*Veromessor* and *Manica*). *BioScience* 22:82–88

105. Wheeler, J., Rissing, S. W. 1975. Natural history of *Veromessor pergandei.* II. Behavior. *Pan-Pac. Entomol.* 51: 303–14

106. Wheeler, W. M. 1910. *The Ants.* NY: Columbia Univ. Press

107. Whitford, W. G. 1976. Temporal fluctuations in density and diversity of desert rodent populations. *J. Mammal.* 57:351–69

108. Whitford, W. G. 1976. Foraging behavior of Chihuahuan Desert harvester ants. *Am. Midl. Nat.* 95:455–58

109. Whitford, W. G. 1978. Structure and seasonal activity of Chihuahuan Desert ant communities. *Ins. Soc.* 25:79–88

110. Whitford, W. G. 1978. Foraging in seed-harvester ants *Pogonomyrmex* spp. *Ecology* 59:185–89

111. Whitford, W. G., Ettershank, G. 1975. Factors affecting foraging activity in Chihuahuan Desert harvester ants. *Environ. Entomol.* 4:689–96

112. Whitford, W. G., Johnson, P., Ramirez, J. 1976. Comparative ecology of the harvester ants *Pogonomyrmex barbatus* (F. Smith) and *Pogonomyrmex rugosus* (Emery). *Ins. Soc.* 23:117–32

113. Willard, J. R., Crowell, H. H. 1965. Biological activities of the harvester ant, *Pogonomyrmex owyheei,* in central Oregon. *J. Econ. Entomol.* 58:484–89

114. Wondolleck, J. T. 1978. Forage-area separation and overlap in heteromyid rodents. *J. Mammal.* 59:510–18

115. Wilson, E. O. 1971. *The Insect Societies.* Cambridge, Mass: Belknap. 548 pp.

116. Wroughton, R. C. 1892. Our ants. *J. Bomb. Nat. Hist. Soc.* Pts. I & II. 77 pp.

Ann. Rev. Ecol. Syst. 1979. 10:229–45
Copyright © 1979 by Annual Reviews Inc. All rights reserved

THE POPULATION BIOLOGY
OF AUSTRALIAN *DROSOPHILA*

♦4161

P. A. Parsons and I. R. Bock

Australian *Drosophila* Research Unit, Department of Genetics and Human
Variation, La Trobe University, Bundoora, Victoria 3083, Australia

INTRODUCTION

The species of the Australian *Drosophila* fauna fall into two groups. The
smaller group consists of the eight universally known "cosmopolitan" species plus two others of widespread occurrence. All except possibly one or
two of these species have almost certainly been introduced into Australia
and dispersed within the continent by human activities. The larger group
consists of species whose distributions are limited to parts of Australia or
that occur in Australia and neighboring regions only. These species are
native to the continent or are survivors of ancient faunal invasions.

The Australian fauna extends (Figure 1) from 11°S (Cape York Peninsula) to 43°S (southern Tasmania)—i.e. across 32° of latitude, over a wide
range of climates (tropical to temperate) and natural habitats. Several results of general significance for population biology of both the cosmopolitan
and endemic faunas are discussed below after brief consideration of the
components of the fauna.

THE COMPONENTS OF THE FAUNA

The cosmopolitan species are *melanogaster, simulans, ananassae, immigrans, funebris, hydei, repleta,* and *busckii;* they are "cosmopolitan" because
they have been recorded from all of the six commonly recognized zoogeographic faunal realms (Palaearctic, Nearctic, Neotropical, Ethiopian, Oriental, and Australian), though each species is not necessarily widespread
within each realm (41). These species have been transported about the
world in association with human movements and are generally only found

229

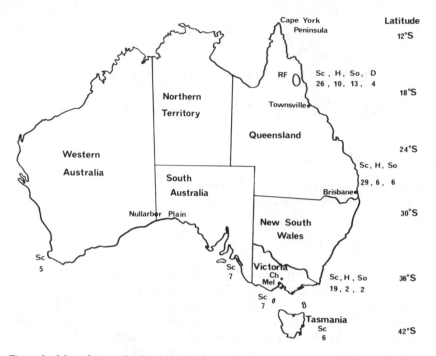

Cape York
Peninsula

Latitude
12°S

RF   Sc , H, So, D
      26 , 10, 13 , 4

18°S

Northern

Townsville

Territory

Queensland

24°S

Western

Sc, H, So

Australia

South

29, 6 , 6

Australia

Brisbane

30°S

Nullarbor Plain

New South

Wales

Sc
5

Sc
7

Victoria
Ch
Mel

Sc,H, So
19, 2 , 2

36°S

Sc
7

Tasmania
Sc
6

42°S

Figure 1   Map of Australia showing localities mentioned in the text. Ch = Chateau Tahbilk, Mel = Melbourne, RF = Rain Forests of the humid tropics of North Queensland. Where we have carried out detailed survey work, known native species numbers are given by subgenus, Sc = *Scaptodrosophila,* H = *Hirtodrosophila,* So = *Sophophora,* D = *Drosophila.* (Cosmopolitan and introduced widespread species are omitted.)

near human habitations. However, their occurrence in parts of the world remote from human habitations may provide clues to their places of evolution [thus *melanogaster* and *simulans* probably originated in Africa (46)]. *D. ananassae* is widespread and abundant in Southeast Asia and New Guinea and, although cosmopolitan, may well be self-introduced into Australia (12). *D. immigrans* might also be self-introduced although it appears to be extremely rare in Southeast Asia and New Guinea. Within Australia, the cosmopolitan species are restricted to human habitats with the exception of *ananassae,* which occurs in rain forests of north Queensland, and *immigrans,* which is found as a rare inhabitant of rain forests from north Queensland to southern New South Wales (8). The other widespread species that have evidently been introduced into Australia are *buzzatii* and *aldrichi.* These inhabit stands of the prickly pear cactus (*Opuntia* spp.) only and were presumably introduced with the latter (4).

Most Australian *Drosophila* species are native to the continent or are limited to Australia and neighboring areas. The taxonomic composition of this fauna indicates that it has been derived from successive invasions from the north. All of the major subgenera of the Australian fauna are well represented in Southeast Asia and New Guinea (5). Only four (excluding cosmopolitan and introduced) species of the typical subgenus occur in Australia; all are limited to north Queensland (8). More species of the subgenus *Sophophora* occur in Australia, mostly of the typically Oriental *melanogaster* species group; the *Sophophora* species extend south into New South Wales and (one or two species) Victoria (8). A few species of the subgenus *Hirtodrosophila* also occur from northeastern Australia southward (in diminishing numbers) to Victoria; some of the *Hirtodrosophila* species show clear affinities to their Oriental relatives, but the group in general is poorly known (9, 11, 38). Over half of the Australian species are members of the subgenus *Scaptodrosophila*. This subgenus occurs in both southeastern and southwestern Australia, and at least one clearly defined species group [the *inornata* group (38, 39)] appears to be native to southern Australia, two species occurring in Queensland at high altitudes only. Many *Scaptodrosophila* species occur in Southeast Asia and New Guinea, and several *Scaptodrosophila* species were presumably among the earliest members of the genus *Drosophila* to reach Australia. The Australian *Drosophila* fauna has been regarded as ecologically unusual because few species are attracted to the fruit baits conventionally used with such success in other parts of the world; but since few fleshy fruits exist in the Australian flora the lack of fruit-baitable species is not surprising. Some species are attracted to mushroom baits, others may be collected directly on forest fungi, and several have been collected in flowers. Most have only been collected by sweeping the foliage of rain forests or sclerophyll forests (8–11, 15, 38, 39). Most Australian *Drosophila* species stringently avoid extremes of heat/desiccation stress. Species in dry sclerophyll habitats are found only in the immediate vicinity of water on all but the least stressful days (32).

The contrast between the introduced and native faunas is illustrated by the results of extensive collections in the Melbourne region (latitude 38°S). These show that the urban/orchard fauna differs almost entirely from that of neighboring temperate rain forests (Table 1). The urban/orchard species can be collected on conventional fermented-fruit baits; with one exception (*fumida*), the natives cannot. Of the eight urban/orchard species, five are cosmopolitans while three are native *Scaptodrosophila*. Two of the latter species, *enigma* and *lativittata*, belong to the *lativittata* complex within the *coracina* species group, all members of which are characterized by being attracted to fruit baits (10). These two species occur in rain forests and neighboring regions from southern New South Wales to southeastern Queensland and presumably have spread to the Melbourne region with the introduction of orchards (36). The third *Scaptodrosophila* species, *D.*

fumida, is one of only two known picture-winged *Scaptodrosophila* and differs in several other respects from more typical species of the subgenus; its distribution is relatively extensive, from southern Queensland to southwestern Western Australia. These three native *Scaptodrosophila* species have apparently spread beyond their "historic" habitats into orchards and may therefore be acquiring the attributes of cosmopolitan species.

The absence of the cosmopolitans in Australia's temperate rain forests is predictable given the absence there of fleshy fruits. The temperate rain forest species of the subgenus *Scaptodrosophila* do not belong to the same species groups as the fruit-baited species; most of the former are members of the *inornata* group [Table 1, (10, 38)]. Few of the southern rain forest *Scaptodrosophila* species are found north of the upland forest regions of the Queensland–New South Wales border (Figure 1), which is the northernmost extension of the floral elements of the southern temperate rain forests (39).

Throughout the remainder of the continent, the urban and rain forest faunas appear to be separate, though studies have been few. In Townsville, north Queensland (latitude 19°S) and in neighboring rain forests, overlaps occur since a few rain forest species normally attracted to fruit baits have invaded neighboring urban regions (6). The reverse process (the invasion of rain forests by the introduced species) has rarely occurred. *D. immigrans*

Table 1 *Drosophila* species collected in the Melbourne region

Subgenus	Urban/Orchard regions (fruit–baited)	Temperate rain forest habitats (swept)
Drosophila	*immigrans*	
	hydei	
Dorsilopha	*busckii*	
Sophophora	*melanogaster*	
	simulans	*dispar*
Scaptodrosophila[a]	*lativittata* ⎱	
	enigma ⎰	
	fumida	*fumida*
		inornata ⎫
		collessi ⎪
		rhabdote ⎬
		fuscithorax ⎪
		obsoleta ⎭
		barkeri ⎫
		louisae ⎪
		minnamurrae ⎬
		notha ⎪
		ehrmanae ⎭
		parsonsi

[a]Species groups in *Scaptodrosophila* are bracketed.

is found as a rare inhabitant of rain forests from north Queensland to New South Wales, but, as explained above, its origins are unclear. Full utilization of rain forest resources by the endemics prevents colonization by cosmopolitans.

Those attracted to fermented fruit baits are the species most likely to emerge from rain forests into urban areas. Other species would not find suitable resources; in addition, since some of the endemic species are highly vulnerable to heat and desiccation, they would probably not survive in urban habitats. Species of the subgenus *Hirtodrosophila* that utilize soft forest fungi as a larval feeding site and the three (at least) species that utilize the undersides of bracket fungi as lek territories (33) are restricted to the forest habitat. In the subgenus *Scaptodrosophila,* except for the widespread species mentioned above, most species are collectable only by methods more or less specific to forest resources.

CLIMATE-RELATED ENVIRONMENTAL EXTREMES

The success of a *Drosophila* population depends on its adaptation to climatic conditions. The annual cycles of temperate zones provide the greatest range of stresses. Population survival depends upon a stage or stages in the life cycle that can survive such seasonal stress. In temperate regions the major density-independent variables to which *Drosophila* species are exposed are (*a*) a combination of heat/desiccation stress, and (*b*) low temperature.

A distinction is necessary between conditions suitable for *resource utilization* (feeding and breeding) and conditions sufficient for mere *survival* (36). Between 6° and 12°C, for example, *D. melanogaster* may survive in a physiological state of quiescence for many months, whereas at temperatures of 12°C and above the species utilizes resources (mates and oviposits) (25). Evidence indicating minimal resource utilization below 12°C for southern temperate zone (Australian) species includes (*a*) the ineffectiveness of fermented fruit and mushroom baiting for cosmopolitan and endemic species below 12°C and the interpretation of population minima in an orchard near Melbourne, (*b*) the lack of success of sweeping endemic *Scaptodrosophila* in Victorian rain forests as 12°C is approached, and (*c*) the cessation of activity and courtship by the endemic picture-winged lek species *D. polypori* below 12°C in the field (33). This limit to resource utilization occurs for members of all four of the major *Drosophila* subgenera (listed in Table 1). It is probable that *Drosophila* originated in the tropics (43); as members of the genus spread from tropical into temperate regions there must have been selection for resource utilization under environmental extremes. This process has obviously occurred many times during the evolution of the genus.

While the lower limit for resource utilization is mainly a function of temperature, the upper limit depends upon temperature and relative humidity. This became clear when foliage was swept for the common southern Australian *D. (Scaptodrosophila) inornata.* Flies were collected in greatest numbers in the range 15°–20°C irrespective of site (32). Small insects have an extremely high ratio of surface area to volume; the amount of water that can be lost by evaporation is large compared with the amount that can be stored. At temperatures above 20°C for RH <80%, flies are found only close to water; at higher relative humidities flies are more widespread.

Such flies are difficult to collect above about 26°C irrespective of humidities (36). How do certain species, especially cosmopolitans, survive the stringent environmental conditions they may experience in temperate regions on a daily basis? Eclosion is impossible at constant temperatures near 30°C and above in the sibling species *D. melanogaster* and *simulans* (44). However, Parsons (37) found that at stressful temperatures in the range 30°–34°C, adults of these species readily survive short periods (>6 hr but <24 hr) at 95% relative humidity and suffer little or no loss of fertility (as compared with 0% RH). If they can find a humid microhabitat these species can survive periods of high temperatures in the wild. In some other species of *Drosophila,* adults must survive several hours daily at high temperatures. The endemic Australian species *D. (Scaptodrosophila) hibisci* utilizes endemic *Hibiscus* flowers as a resource in open forests in the northern half of Australia where temperatures frequently exceed 30°C (15). In this case the base of the corolla tube provides a humid site where adults apparently survive in an otherwise extreme climate. Because their distribution and abundance depend upon daily and seasonal variations in microenvironmental temperature and humidity, some Australian *Drosophila* species exist in regions where on gross climatic criteria they would not be expected. Similar suggestions of habitat selection to avoid extreme stresses have been made for *D. pseudoobscura* (2, 17).

Assuming no temperature fluctuations, *Drosophila* can utilize resources over a range of 14°C (i.e. between 12° and 26°C). Andrewartha & Birch (1) regard a 16°C range as characteristic of aquatic animals. It is reasonable that *Drosophila* should fit into this range since larvae exploit mainly moist microbial degradation products. Thus *Drosophila* are terrestrial insects that utilize "aquatic" resources.

As discussed elsewhere (36), high (and low) temperature sensitive and resistant local strains of *D. funebris, D. melanogaster,* and *D. simulans* are known. Cosmopolitan species may thus be subdivided into local populations with regard to environmental extremes, a result that implies some interpopulation variations in resource utilization. As another example, McKenzie & Parsons (28) studied the genetic basis of resistance to desiccation for strains of *D. melanogaster* and *D. simulans* collected from Brisbane (latitude 27.5°S) and Melbourne (latitude 38°S). Significant additive genetic variation and dominance were found for all populations except the Melbourne *D. simulans* population, where dominance was not significant. When present, dominance is directional for resistance as is usually also true for environmental stress traits (31). When desiccation resistance was studied on an annual basis, only the Melbourne *D. simulans* population showed cyclical changes in mean mortality such that the population is more resistant in summer and becomes less so in cooler weather. Adaptation to desiccation in this population occurs presumably through gene frequency changes; in other populations it is more likely to be a property of the entire genome and to involve interaction components. The genetic basis of desiccation resistance thus differs among populations of *D. simulans*. The unit under consideration is clearly the population rather than the species. A cosmopolitan species may thus be characterized in part by a genome capable of adapting by various genetic means to a variety of environmental stresses.

The ratio of *D. simulans* to *D. melanogaster* is higher in Brisbane than in Melbourne, while temperature variability is lower in Brisbane (29). In general, *D. simulans* appears to outnumber *D. melanogaster* in regions where temperature fluctuations are small. *D. simulans* is more sensitive to environmental extremes than is *D. melanogaster;* thus Melbourne may be regarded as more marginal than Brisbane for *D. simulans*. The summer-winter changes in desiccation resistance in Melbourne *D. simulans* are quite substantial, which argues that rather few genes are involved. The "simpler" genetic basis for resistance to desiccation in Melbourne *D. simulans* may be analogous to lower levels of chromosomal polymorphism at the margins of distributions, as reported for some species of *Drosophila* (14).

Populations from Townsville, north Queensland (latitude 19°S) and from Melbourne (latitude 38°S) were compared for cold-resistance in the two sibling species [in terms of mortality after exposure to -1°C for 48 hr (34)]. The sequence of mortalities was *D. melanogaster* (Townsville) > *D. simulans* (Townville) > *D. simulans* (Melbourne), which were all considerably greater than *D. melanogaster* (Melbourne). In Melbourne, resource utilization effectively ceases during winter months with temperatures mainly below 12°C, so that in early spring few flies are present. There is no such restriction on the Townsville population. The advantage of *D. melanogaster* over *D. simulans* in Melbourne can almost certainly be correlated with the earlier build-up of the former in spring (29), which leads to observed population maxima of the former species in late spring and of the latter in late summer. The genetic basis of cold-resistance was not studied in these populations; however, this stress involves underlying physiological (and genetic) processes different from those underlying high-temperature desiccation stress, since the between-isofemale-strain correlation coefficient for the two stresses is close to zero.

The small interspecific difference for cold-resistance in tropical Townsville is not surprising since there resource utilization would rarely be prevented by temperatures below 12°C. The low temperatures at which significant mortalities occur in Melbourne do not occur in Townsville; natural selection for cold-resistance would be minimal. The similarity of the two *D. simulans* populations suggests that the species lacks the genetic heterogeneity needed to adapt to the more extreme conditions where *D. melanogaster* is frequently dominant. In any case, generalizations at the species level are difficult to make since the population is the unit of study. Population studies from many climatic zones are needed, especially since the physiological and genetic bases of cold-resistance and of resistance to high-temperature-desiccation stresses are different.

COMPARATIVE ECOLOGICAL TOLERANCES AMONG SPECIES

The superiority of *D. melanogaster* over *D. simulans* in tolerance to environmental extremes is shown in Figure 2 (40) where LD_{50} values (in terms of hours required for 50% of the flies to die for two stresses (abscissa: $-1°C$, and ordinate: desiccation at 0% RH at 25°C) for Melbourne-collected strains are plotted. The figure also shows that *D. immigrans* is somewhat more sensitive to desiccation and somewhat more resistant to cold than either of the sibling species. This agrees with our observations that *D. immigrans* but not the sibling species can be collected readily in mid-winter in Melbourne (29); conversely both sibling species but not, apparently, *D. immigrans* occur in heat-stressful urban regions of tropical north Queensland (6). The effects of the stresses under discussion appear to correlate with species abundances in nature. This is reasonable since insects such as cosmopolitan *Drosophila* species are presumably subject to *r*-selection rather than *K*-selection; thus mortalities are expected to be largely density-independent, nondirected, and related to variations in the physical environment (19, 29, 42).

Two fruit-baitable species, *D. (Sophophora) bipectinata* and *D. (Sophophora) birchii*, from the rain forests of the humid tropics of north Queensland and the north of the continent are, by contrast, extremely sensitive to environmental extremes, as is the neotropical species *D. (Sophophora) paulistorum*. On the contrary, the habitat-specific *D. buzzatii* is associated with the cactus genus *Opuntia*. Since its only known breeding sites are rot pockets in the cactus itself (4), these flies are less able than are cosmopolitan species to avoid environmental extremes by means of habitat selection. This

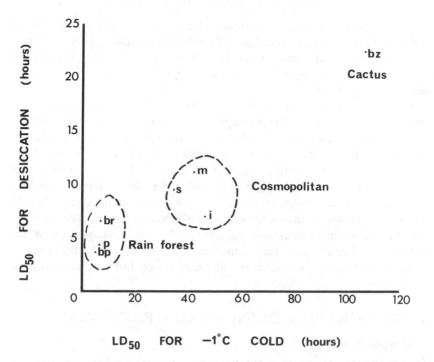

Figure 2   LD₅₀ values expressed as a plot of the numbers of hours at which 50% of flies died of 0% RH desiccation (ordinate) vs –1°C stress (abscissa) for various *Drosophila* species (40)—bp (*bipectinata*), br (*birchii*), bz (*buzzatii*), i (*immigrans*), m (*melanogaster*), p (*paulistorum*), s (*simulans*).

suggests that intense natural selection for survival in environmental extremes would occur because of the specific nature of the resources used. High tolerances to environmental extremes may in any case be adaptive in such species where frequent migrations to new and often distant rot pockets are necessary.

D. lativittata, the widespread *Scaptodrosophila* species of orchards in the Melbourne area, shows a desiccation tolerance similar to that of the cosmopolitan species and is more cold resistant (P. A. Parsons, unpublished data). It may be acquiring the properties of cosmopolitan species within Australia by having a genome capable of adapting to the conditions under which the cosmopolitans flourish. The high cold-resistance is reasonable since winter temperatures in the orchard areas are lower than in Melbourne urban regions. Neither is it surprising that the widespread endemic species of southern temperate rain forests, *D. inornata,* is extremely resistant to cold, surviving several days; its habitats are frequently colder than the neighboring orchard habitats. It is also reasonable that *D. inornata* is less desiccation-resistant than the cosmopolitan species, given the cool moist habitats where it is most abundant (38). It is indeed striking how tolerances to environmental extremes can be directly related to habitats; this fits the assumption that *Drosophila* species are subject mainly to *r*-selection, at least in temperate regions.

Cosmopolitan species do not appear to have unique genetic structures but may have gene pools capable of adapting to greater environmental extremes than, for example, tropical rain forest species. This does not preclude certain species being highly resistant to stresses but is likely to be associated with specialist resource utilization (e.g. *D. buzzatii*). It is of interest to note that another dipteran, the Queensland fruit fly, *Dacus tryoni,* has also apparently extended its geographical range as postulated for certain *Scaptodrosophila* species. In this case there is good evidence for associated physiological adaptation to extreme conditions (23). Our knowledge of the genetics of *Dacus tryoni* (an economic pest) is not good; the study of adaptations in such species may well be aided by fundamental work on *Drosophila.*

BIOGEOGRAPHIC CLINES AND GRADIENTS

Cosmopolitan Species—Alcohol

One of the best examples of evolution in action comes from the study of clines, since they can normally be explained by fitness differences due to environmental gradients leading to the observable phenotypic gradients (18). Indeed microdifferentiation resulting from such geographical discontinuities over remarkably short distances has been observed in some species, especially plants (20). In populations of grasses clines have been studied in response to certain toxic elements—lead, zinc, copper—in the soil near abandoned mines. These clines are maintained by strong selective forces in the face of gene flow through pollen dispersal under random mating. The selective effect of heavy metals in these clines is direct; differential fitnesses result from the environmental stress. An analogous stress upon *Drosophila* is the presence of alcohol in the environment. A combination of laboratory and field experimentation (27) has shown that alcohol is utilized as a resource by *D. melanogaster* but not by *D. simulans.* Both species occur outside the cellar at the Chateau Tahbilk winery 100 km north of Melbourne, while only *D. melanogaster* is inside. During vintage *D. simulans* tends to move away from the cellar and *D. melanogaster* towards it, so that together with differential mortalities, a transient *melanogaster:simulans* gradient occurs, going from 100% *melanogaster* at the cellar to a stable *D. simulans* majority about 30 m from the cellar.

Heterogeneity among isofemale strains was found for both species. However, comparison of populations within the cellar, outside the cellar, and in

Melbourne, demonstrated differentiation in response to alcohol only in *D. melanogaster,* since *D. simulans* is universally alcohol sensitive. This creates a situation of substantial evolutionary significance since, as in plants, the microdifferentiation leading to a cline occurs over a remarkably short distance and is describable in terms of adaptation to an alcohol-associated resource (30). Levels of alcohol tolerance increase with proximity to the cellar during vintage. The cline is transient, since during nonvintage periods when alcohol in the outside environment is greatly reduced, the distribution of tolerance phenotypes outside the cellar is relatively uniform. Thus the cline is *directly* attributable to selection in the alcohol gradient. This cannot be directly related to changes in allelic frequencies at the *Adh* locus as in Spain (13), since if tolerance were involved mainly with this locus, a relationship between tolerance and *Adh* allele frequencies should be apparent; no such relationship was found (26). In conclusion, the level of stress caused by high alcohol concentrations has consequences (*a*) intraspecifically, leading both to the cline across the barrier between the cellar and outside and to the transient cline outside the cellar above during vintage, and (*b*) interspecifically, leading both to the elimination of *D. simulans* in the cellar and to a transient *melanogaster:simulans* gradient outside the cellar during vintage.

Latitudinal clines occur for adult resistance to alcohol: Tolerance increases with latitude in *D. melanogaster* in the northern hemisphere; predictably, it does not in *D. simulans* (16). In the southern hemisphere, preliminary evidence shows that strains of *D. melanogaster* from Melbourne are more tolerant to alcohol than strains from Townsville. This difference may be associated with the relative importance of alcohol as a resource. It suggests that as *D. melanogaster* spread away from equatorial regions, selection favored the exploitation of alcohol as a resource.

The behavior of newly hatched larvae in response to agar containing alcohol gives parallel results: (*a*) Melbourne and Chateau Tahbilk *D. melanogaster* larvae have a strong preference for agar containing alcohol, while sympatric *D. simulans* populations show no such preference; and (*b*) the Townsville *D. melanogaster* population shows a lesser preference than the more southern Melbourne population (35). Interestingly, in the Townsville population some isofemale strains were as extreme as those of the Melbourne population and others showed almost no alcohol preference. In southern Australia there is presumably a greater advantage in alcohol resource exploitation than in the north, and this can be regarded as a process of selection among isofemale strains. Indeed this lower heterogeneity in southern *Drosophila melanogaster* agrees with the general principle of declining biological diversity with increasing latitude, demonstrated here for a response to a chemically defined resource rather than by the more usual species diversity relationships to be discussed below.

Native Species—Diversity Patterns

The study of resource utilization by larvae may have considerable potential since many metabolites can readily be quantified. Unfortunately, such studies of metabolite use and preference have considered only a few species. However, resource utilization may be assessed indirectly by examining the methods used to collect various species (7). This is well illustrated in the humid tropical habitats of north Queensland, which can be subdivided according to rain forest type. In Table 2, species numbers from 6 forest types are given. The first three listed are on basaltic soils: 1a = complex mesophyll vine forest, < 400 m altitude; 1b = complex mesophyll vine forest, 400–800 m altitude; 5a = complex notophyll vine forest, 800–1600 m altitude. The next three are on granitic soils in the same altitudinal sequence: 2a = mesophyll vine forest, < 400 m altitude; 8 = simple notophyll vine forest, 400–800 m altitude; 9 = simple microphyll vine-fern forest, 800–1600 m altitude. Forest types at a given latitude are often adjacent to one another. [The classification is that of Webb (47).]

Four collection methods were used: baiting with fermented fruits, baiting with rotted mushrooms, sweeping from foliage and leaf litter, and aspirating from the undersides of hard bracket fungi. Each takes species that utilize different resources. Comparison of forests at equivalent latitudes (i.e. 1a vs

Table 2 Species numbers from the humid tropics of North Queensland

Habitat type	Locality	Collection method	Scaptodro- sophila	Hirtodro- sophila	Drosophila	Sophophora	Total	Total species in habitat
1a	Mulgrave	Fruit	4	1	5	4	14	
	Forestry	Mushroom	2	1	2	3	8	
	Road	Swept	1	3	1	2	7	19
1b	Lake	Fruit	1	2	3	2	8	
	Eacham	Mushroom	3	1	2	1	7	
		Swept	—	1	—	1	2	
		Bracket fungus	—	3	—	—	3	14
5a	The	Fruit	—	—	2	1	3	
	Crater	Mushroom	3	—	—	3	6	
	National	Swept	—	1	—	—	1	
	Park	Bracket fungus	—	1	—	—	1	10
2a	Mulgrave	Fruit	—	—	2	1	3	
	Forestry	Mushroom	1	—	—	2	3	
	Road	Swept	1	2	—	—	3	8
8	Cardwell	Fruit	1	—	2	1	4	
	Ranges	Mushroom	3	—	—	—	3	
		Swept	—	1	—	—	1	7
9	Mt. Lewis	Fruit	—	—	—	2	2	
		Mushroom	—	—	—	1	1	
		Swept	—	2	—	1	3	4

2a, 1b vs 8, and 5a vs 9) shows that floristically diverse forests of basaltic soils have generally higher species diversities for all subgenera and all collection methods. Altitudinal gradients exist as expected on biogeographic grounds; species diversities occur in the sequence 1a > 1b > 5a for basaltic soil forests, and 2a > 8 > 9 for granitic soil forests.

Most rare (not previously recorded) species are found in the floristically very diverse 1a forests. In general, rare species are not found in the granitic soil forests. Thus *Drosophila* species diversities follow plant species diversities; this is predictable given the dependence of *Drosophila* species upon plants as resources. The need for detailed experimental work on resource utilization is clear.

As shown in Table 2, the *Drosophila* fauna of the humid tropics of north Queensland comprises species of the four major subgenera, *Scaptodrosophila, Hirtodrosophila, Sophophora* and *Drosophila,* the first two subgenera being dominant in number of species. The north Queensland fauna contrasts with that of southern Australia, where the subgenus *Scaptodrosophila* (as collected by sweeping) dominates almost totally (Table 1). Five successful collection methods (Table 3) reveal that the phylogenetically diverse Australian *Drosophila* fauna utilizes a wide range of resources. Only *Hirtodrosophila* exploits the forest fungus resource. The *Hirtodrosophila* species obtained by sweeping are closely related members of a species complex; they are found mainly near the forest floor or close to water. On the other hand, species of *Scaptodrosophila* are collected by all methods listed *except* from forest fungi; however, they are taken on rotted commercial mushrooms. Taxonomic divergence and divergence in resource utilization are correlated between subgenera and between species (groups) within subgenera. Where known, species distributions are more similar within than between species groups (38).

From the considerations so far presented, the existence of species gradients, especially on a north-south basis, is not surprising (see Figure 1). The

Table 3 Number of species in the Australian endemic *Drosophila* fauna classified by resources utilized as indicated by collection method and by subgenus

Collection method[a]	1	2	3	4	5
Scaptodrosophila[b]	6	8		24	1
Hirtodrosophila[b]			8	3	
Sophophora	12			3	
Drosophila	4				

[a] 1. Fermented fruit baiting; 2. Rotted commercial mushroom baiting; 3. Collected on or near fleshy fungi and bracket fungi in the forest; 4. Sweeping foliage, leaf litter, and flowers; 5. Demonstrated as utilizing decaying flowers for a larval resource (*D. hibisci*).

[b] Data on resources utilized are unavailable for many rare species in these subgenera (6).

proportion of *Scaptodrosophila* species increases towards the south from north Queensland to Tasmania; the endemic *Drosophila* fauna of southern and southwestern Australia is also exclusively *Scaptodrosophila,* but is depauperate. The proportion of species in the remaining subgenera shows the reverse trend, decreasing from north to south, and the total number of *Drosophila* species falls dramatically along the same gradient.

Unlike many other zoogeographic surveys, ours finds a southward progressive reduction in the diversity of resources utilized. In the extreme south only sweeping is effective in obtaining flies. In Victoria (except in the extreme east) the forests are inhabited exclusively by temperate species; this absence of subtropical elements is associated with a comparable reduction in floral diversity (38).

CONCLUSIONS AND FUTURE DIRECTIONS

This is a first brief account of the population biology of a *Drosophila* fauna across 32° of latitude. That the native fauna is unique is to be expected because of the dependence of *Drosophila* species on a unique flora. The native fauna is limited to "islands" of vegetation surrounded by physically and biologically unsuitable habitats, most major "islands" being at least 5000 years old (38). During the Pleistocene the climate fluctuated; species presumably migrated up and down the coast to give the current distribution of *Drosophila* in islands of vegetation, including relict southern populations in upland northern areas and relict northern populations in lowland southern areas. The particular species in a given island depend largely upon vegetational type.

Can *Drosophila* species migrate relatively long distances over apparently unsuitable habitats? Cactus-feeding *Drosophila* in the southwestern United States may be able to migrate farther than studies with baits have indicated (22). A similar situation probably occurs in both *D. buzzatii* and the endemic *D. hibisci* in Australia. *D. melanogaster* and *D. simulans* may survive extreme temperatures and remain fertile (37), provided that the humidity is high. In summer in temperate regions, the diurnally active flies probably seek humid microhabitats during the day when environmental stresses are maximal. However, on humid cloudy days, especially just before a storm, *D. inornata* (for example) may be abundant and widely scattered irrespective of the time of day (presumably because the whole habitat has a permissive humidity), and under such circumstances long-distance migrations may be feasible (38). Such events may be rare in temperate regions and difficult to document because of their unpredictability, but if they occur their evolutionary significance could be immense. Circumstantial evidence indicates that contacts cannot be maintained over more than 1600 km (39).

This is the distance across the Nullarbor Plain separating the southwestern from the southeastern Australian fauna. The main elements of both faunas belong to the *inornata* species group; but at least two species are unique to the southwest, suggesting divergence after isolation. One species, however, *D. fuscithorax,* is common on both sides of the Nullarbor and is apparently resistant to environmental extremes; this suggests cross-Nullarbor contacts over greater distances than are feasible for other species (39).

Studies of larval resource utilization in *D. melanogaster* and *D. simulans* indicate that divergences (e.g. in alcohol use) may occur among species commonly regarded as occupying identical niches. The converse, that closely related species select differing habitats, must almost certainly be true. Indeed, larval ability to smell other possible metabolites (e.g. acetic acid, ethyl acetate, and lactic acid) varies somewhat between the two species, although larvae of both are attracted to them (P. A. Parsons, unpublished data). To alcohol and these other metabolites, *D. immigrans* (subgenus *Drosophila*) responds differently from the two sibling species of the subgenus *Sophophora.* This result agrees with findings obtained by studying emergences of domestic *Drosophila* species from various fruits and vegetables in an English market (3). These results suggest a testable relationship between taxonomic and resource-utilization divergences. Apart from evidence discussed above, Stalker (43) has found differing inversion frequencies in *D. melanogaster* emerging from different fruits—a result which may be interpretable in terms of resource utilization heterogeneity.

Australia appears to be an ideal continent for the study of resource heterogeneity in widespread species. Because of a lack of climatic extremes (24) the tropics should be characterized by a fuller spectrum of resources, a lower proportion of resources utilized per species, and more resource overlaps per species than in the temperate zones. Such generalizations, derived from studies on other biotas, can be tested in *Drosophila* at a fine level, especially if New Guinea, where the *Drosophila* fauna is extremely diverse even by comparison with the richest Australian *Drosophila* fauna, is included. The cosmopolitan *D. immigrans* probably occurs in the greatest diversity of habitats; unlike other cosmopolitan species, it frequently occurs, though at low densities, in rain forests throughout the continent. It utilizes fruits and vegetables as resources, whereas *D. melanogaster* and *D. simulans* are basically fruit specialists (3). This great diversity of resources utilized by *D. immigrans* may result in its observed presence in rain forests.

ACKNOWLEDGMENTS

These are given in detail in (5–12, 15, 27–40). The Australian Biological Resources Survey and the Australian Research Grants Committee have provided financial assistance.

Literature Cited

1. Andrewartha, H. G., Birch, L. C. 1954. *The Distribution and Abundance of Animals.* Chicago: Univ. Chicago Press. 782 pp.
2. Arlian, L. G., Eckstrand, I. A. 1975. Water balance in *Drosophila pseudoobscura,* and its ecological implications. *Ann. Entomol. Soc. Am.* 68:827–32
3. Atkinson, W., Shorrocks, B. 1977. Breeding site specificity in the domestic species of *Drosophila. Oecologia* 29: 223–32
4. Barker, J. S. F., Mulley, J. C. 1976. Isozyme variation in natural populations of *Drosophila buzzatii. Evolution* 30: 213–33
5. Bock, I. R. 1976. Drosophilidae of Australia I. *Drosophila* (Insecta: Diptera). *Aust. J. Zool. Suppl. Ser.* 40:1–105
6. Bock, I. R. 1977. Notes on the Drosophilidae (Diptera) of Townsville, Queensland, including four new Australian species records. *J. Aust. Entomol. Soc.* 16:267–72
7. Bock, I. R., Parsons, P. A. 1977. Species diversities in *Drosophila* (Diptera): a dependence upon rain forest type of the Queensland (Australian) humid tropics. *J. Biogeogr.* 4:203–13
8. Bock, I. R., Parsons, P. A. 1978. Australian endemic *Drosophila* IV. Queensland rain forest species collected at fruit baits, with descriptions of two species. *Aust. J. Zool.* 26:91–103
9. Bock, I. R., Parsons, P. A. 1978. Australian endemic *Drosophila* V. Queensland rain forest species associated with fungi, with descriptions of six new species and a redescription of *D. pictipennis* Kertész. *Aust. J. Zool.* 26:331–47
10. Bock, I. R., Parsons, P. A. 1978. The subgenus *Scaptodrosophila* (Diptera: Drosophilidae). *Syst. Entomol.* 3:91–102
11. Bock, I. R., Parsons, P. A. 1979. Australian endemic *Drosophila* VI. Species collected by sweeping in rain forests of Queensland and northern New South Wales with descriptions of four new species. *Aust. J. Zool.* 27:291–301
12. Bock, I. R., Wheeler, M. R. 1972. I. The *Drosophila melanogaster* species group. *Univ. Tex. Publ.* 7213:1–102
13. Briscoe, D. A., Robertson, A., Malpica, J.-M. 1975. Dominance at *Adh* locus in response of adult *Drosophila melanogaster* to environmental alcohol. *Nature* 255:148–9
14. Carson, H. L. 1958. The population genetics of *Drosophila robusta. Adv. Genet.* 9:1–40

15. Cook, R. M., Parsons, P. A., Bock, I. R. 1977. Australian endemic *Drosophila* II. A new *Hibiscus*-breeding species with its description. *Aust. J. Zool.* 25:755–63
16. David, J., Bocquet, C. 1975. Similarities and differences in latitudinal adaptation of two *Drosophila* sibling species. *Nature* 257:588–90
17. Dobzhansky, Th., Epling, C. 1944. Contributions to the genetics, taxonomy, and ecology of *Drosophila pseudoobscura* and its relatives. *Carnegie Inst. Wash. Publ. 544.* Carnegie Inst.: Washington DC
18. Endler, J. 1977. *Geographic Variation, Speciation and Clines.* Princeton NJ: Princeton Univ. Press. 246 pp.
19. Gadgil, M., Solbrig, O. T. 1972. The concept of r- and K-selection; evidence from wild flowers and some theoretical considerations. *Am. Nat.* 102:389–90
20. Jain, S. K., Bradshaw, A. D. 1966. Evolutionary divergence among adjacent plant populations. I. The evidence and its theoretical analysis. *Heredity* 21: 407–41
21. Johnson, G. B. 1976. Genetic polymorphism and enzyme function. In *Molecular Evolution,* ed. F. J. Ayala, pp. 46–59. Sunderland, Mass.: Sinauer
22. Johnston, J. S., Heed, W. B. 1976. Dispersal of desert-adapted *Drosophila:* the saguaro-breeding *D. nigrospiracula. Am. Nat.* 110:629–51
23. Lewontin, R. C., Birch, L. C. 1966. Hybridization as a source of variation for adaptation to new environments. *Evolution* 20:315–36
24. McArthur, R. H. 1972. *Geographical Ecology.* NY: Harper & Row. 269 pp.
25. McKenzie, J. A. 1975. The influence of low temperature on survival and reproduction in populations of *Drosophila melanogaster. Aust. J. Zool.* 23:237–47
26. McKenzie, J. A., McKechnie, S. W. 1978. Ethanol tolerance and the *Adh* polymorphism in a natural population of *Drosophila melanogaster. Nature* 272:75–76
27. McKenzie, J. A., Parsons, P. A. 1972. Alcohol tolerance: an ecological parameter in the relative success of *Drosophila melanogaster* and *Drosophila simulans. Oecologia* 10:373–88
28. McKenzie, J. A., Parsons, P. A. 1974. The genetic architecture of desiccation resistance in populations of *Drosophila melanogaster* and *Drosophila simulans. Aust. J. Biol. Sci.* 27:441–56

29. McKenzie, J. A., Parsons, P. A. 1974. Numerical changes and environmental utilization in natural populations of *Drosophila. Aust. J. Zool.* 22:175–87

30. McKenzie, J. A., Parsons, P. A. 1974. Microdifferentiation in a natural population of *Drosophila melanogaster* to alcohol in the environment. *Genetics* 77:385–94

31. Parsons, P. A. 1973. Genetics of resistance to environmental stresses in *Drosophila* populations. *Ann. Rev. Genet.* 7:239–65

32. Parsons, P. A. 1975. The effect of temperature and humidity on the distribution patterns of *Drosophila inornata* in Victoria, Australia. *Environ. Entomol.* 4:961–64

33. Parsons, P. A. 1977. Lek behavior in *Drosophila (Hirtodrosophila) polypori* Malloch—an Australian rainforest species. *Evolution* 31:223–25

34. Parsons, P. A. 1977. Resistance to cold temperature stress in populations of *Drosophila melanogaster* and *D. simulans. Aust. J. Zool.* 25:693–98

35. Parsons, P. A. 1977. Larval reaction to alcohol as an indicator of resource utilization differences between *Drosophila melanogaster* and *D. simulans. Oecologia* 30:141–46

36. Parsons, P. A. 1977. Cosmopolitan, exotic and endemic *Drosophila:* their comparative evolutionary biology especially in southern Australia. *Symp. Ecol. Soc. Aust.* 10:62–75

37. Parsons, P. A. 1979. Resistance of the sibling species *Drosophila melanogaster* and *D. simulans* to high temperatures in relation to humidity: Evolutionary implications. *Evolution* 33:131–36

38. Parsons, P. A., Bock, I. R. 1977. Australian endemic *Drosophila* I. Tasmania and Victoria, including descriptions of two new species. *Aust. J. Zool.* 25:249–68

39. Parsons, P. A., Bock, I. R. 1978. Australian endemic *Drosophila* III. The *inornata* species group. *Aust. J. Zool.* 26:83–90

40. Parsons, P. A., McDonald, J. 1978. What distinguishes cosmopolitan and endemic *Drosophila* species? *Experientia* 34:1445–46

41. Patterson, J. T., Stone, W. S. 1962. *Evolution in the Genus* Drosophila. NY: MacMillan. 610 pp.

42. Southwood, T. R. E. 1977. Habitat, the templet for ecological strategies? *J. Anim. Ecol.* 46:337–65

43. Stalker, H. D. 1976. Chromosome studies in wild populations of *D. melanogaster. Genetics* 82:323–47

44. Tantawy, A. O., Mallah, G. S. 1961. Studies on natural populations of *Drosophila.* I. Heat resistance and geographical variation in *Drosophila melanogaster* and *D. simulans. Evolution* 15:1–14

45. Deleted in proof

46. Tsacas, L., Lachaise, D. 1974. Quatre nouvelles espèces de la Cote-D'ivoire du genre *Drosophila,* groupe *melanogaster,* et discussion de l'origine du sous-groupe *melanogaster* (Diptera: Drosophilidae). *Ann. Univ. Abidjan, Série E (Ecologie), Tome VII, Fasc.* 1:193–211

47. Webb, L. J. 1968. Environmental relationships of the structural types of Australian rain forest vegetation. *Ecology* 49:296–311

Ann. Rev. Ecol. Syst. 1979. 10:247–67
Copyright © 1979 by Annual Reviews Inc. All rights reserved

LATE CENOZOIC FOSSIL COLEOPTERA: Evolution, Biogeography, and Ecology

♦4162

G. R. Coope

Department of Geological Sciences, University of Birmingham, Birmingham, England

INTRODUCTION

The recent revival of interest in the fossil insects of the Quaternary Period [the last 1.6 million years (68, 69)] has been summarized in the *Annual Review of Entomology* (21). The present essay was originally intended as a second installment to that earlier review, drawing attention to significant developments since that time. However, the important discoveries of late Miocene fossil insects in Alaska and the Canadian Arctic Archipelago (52) have such a direct bearing on the evolution of the Quaternary and hence the present day fauna that I widened my scope to include these faunas. Studies of the Late Cenozoic biota straddle the boundary between classical palaeontology and neontology, providing vital evidence on the course and rates of evolution, changes in biogeography, and the ecological history of the present biota.

While it may be that fossils from the remote past have only marginal relevance to the understanding of the present day biota and its environment, the fossil record from the Late Cenozoic cannot be brushed aside so easily. In many cases these fossils represent the immediate ancestors of our living species, and the ecosystems in which they lived were the precursors of those of the present day. The relatively recent fossil history of insects, long a no-man's-land between biology and classical palaeontology, has been badly neglected. Only in the past two decades has a systematic attempt been made to unravel the Late Cenozoic fossil record of the insects, the most abundant, taxonomically diverse, and ecologically varied component of terrestrial fossil assemblages.

247

For a number of reasons I focus attention here on the Coleoptera. First, their remarkably robust exoskeletons may resist the agents of decomposition for hundreds of thousands or even millions of years, provided that the deposits in which they are entombed remain damp and air is excluded. Almost invariably the original exoskeleton has been preserved, with no mineralization whatsoever. Most frequently the fossils occur as disarticulated skeletal elements that are recovered from organic silts, detritus muds, or peats by wet sieving of the disaggregated sediment and sorting through the residues under a binocular microscope. The specimens must be stored either under alcohol or on dry card mounts; otherwise, fungal attack can destroy them rapidly. Sometimes entire animals may be recovered from peats or, in the case of wood boring species, from their galleries in subfossil wood (35, 71). The most extraordinarily exact preservation we have yet encountered is in fossils embedded in ozocerite (natural paraffin wax) obtained from excavations to recover the more or less intact carcass of a woolly rhinoceros in Starunia, Western Ukraine (R. B. Angus & G. R. Coope, unpublished data). Here the complete beetles were preserved down to the tarsal and antennal joints; when the elytra were raised, the wings could be unfolded and mounted; and parasitic mites, both larvae and adults, were found underneath the wings. Although this was quite exceptional preservation, it is common to find intact abdomens from which the genitalia can be dissected; the frequently transparent integument often reveals detailed structures of the internal sclerites. Preservation is frequently adequate to enable details of the microstructure of the surface of the hairs and scales to be examined with scanning electron microscopy (57). This extraordinary quality of preservation is not restricted to the Coleoptera but includes Hymenoptera, Hemiptera, and Trichoptera. Diptera and Odonata are apparently frail and when their fossils are found, they are usually crushed and difficult to identify. It is curious that Orthoptera are so rare in Late Cenozoic sediments, a peculiarity that is amplified by their frequent occurrence in classical palaeontological contexts (16). [Here, again, however, the Starunia site was exceptional (74).]

The second reason for concentrating on Coleoptera is that their taxonomy, biogeography, and ecology are well known. In all attempts to understand relatively recent fossils, the palaeontologist must rely upon systematists and ecologists working on the present day species with which he must compare his fossils.

A third reason for selecting the Coleoptera is their enormous diversity, both of species and of environmental preferences. These are important factors when making palaeoenvironmental reconstructions.

Finally, fossil Coleoptera are abundant. In sediments where plant macroscopic fossils occur (i.e. seeds, fruits etc), remains of beetles are by far the commonest insect fossils and are at times the most plentiful of all animal fossils.

Thus my emphasis on fossil Coleoptera results from their abundance, recognizability, and the availability of data on modern beetles, and in no way implies anything biologically exceptional about them. The richness of their fossil record permits biological inferences applicable to other orders of insects whose fossil history is yet to be worked out.

EVOLUTION

The earlier review (21) indicated that modern work on fossil Coleoptera had found no evidence of morphological change during the latter part of the Quaternary, nor was there reason to believe that many species became extinct during this period. Since the earlier review, numerous other fossiliferous sites have been studied across the whole of the northern hemisphere by widely scattered investigators and I see no reason to modify the original statement [for case histories see (12, 44, 48)].

A single example of specific stability will demonstrate the precision with which fossil material can now be processed by a competent present day systematist. I refer to *Helophorus aquaticus,* a hydrophilid beetle that is abundant in shallow pools over most of Europe and which has been intensively studied by Angus (2). This species has two geographical races. One, ranging widely over western Europe, has straight outer margins to the parameres of the male genitalia; the second race, found in eastern Europe and patchily in south and central Europe where it is usually confined to high ground, has distinctly bowed-out margins to the parameres. These races are linked by a relatively narrow clinal zone of intermediates. Dead individuals become buried in sediment accumulating on the pond bottoms and are frequently found as Quaternary fossils. Because they have undergone little *post mortem* transport, the fossils are relatively intact, and it is a fairly simple operation to remove the genitalia. It has now become clear that the geographical races have maintained the same degree of difference from one another for at least 120,000 years, or, if you prefer, 120,000 generations; the cline has remained morphologically stable for this period but the geographical location of the different races has changed with time. During temperate episodes the form with the straight parameres lived in Britain; during the periods of cold, more continental climate the race with curved parameres took its place. Even the zone of intermediate forms has been recognized in the Starunia fossil assemblage (2). In this case the geographical races clearly represent morphological reflections of physiological differences shown not only by their differing present ranges but also by the shift in their ranges as conditions changed. This physiological difference presumably is the basis for selection that counteracts hybridization pressures in maintaining the integrity of the cline over such a long period of time, a balance that could theoretically be kept up almost indefinitely. To view geographical variation as imminent speciation may thus be an oversimplification; species that can maintain some degree of physiological/morphological diversity without breaking off genetic relations with neighboring populations must be at an advantage over those that become fragmented into reproductively isolationist groups. Speciation is a risky business.

But how far back in time does this constancy of species extend? To resolve this problem we have to turn to the Lower Pleistocene (i.e. early Quaternary) and upper Tertiary fossil-bearing deposits in Alaska and the Canadian Arctic Archipelago, now being intensively studied by J. V. Matthews of the Canadian Geological Survey. In these deposits the state of preservation is as good as, or even better than, in many Late Quaternary deposits. The discovery of these fossiliferous deposits of this critical age is one of the most exciting recent events in this field of investigation.

The early Pleistocene fossils, probably dating from over a million years ago, are referable to living species (50), and some existing species extend well back into the late Tertiary (47, 53). However, among the early Pleistocene fossils there is a hint of evolutionary change for at least one species. The staphylinid beetle *Tachinus apterus* has today only rudimentary flying wings. In a well preserved fossil from Cape Deceit on the Seward Peninsula, Alaska, the flying wings were twice the length of the modern representatives of this species (54), although still too small for actual flight. Furthermore, a statistical analysis of numerous fossil elytra showed that early Pleistocene elytra were on the whole larger and proportionately longer than those from Pleistocene assemblages. Matthews's claim that this evidence "seems to represent an intraspecific evolutionary trend" may indeed be true; but it would be good to have more than two points on our graph, especially in the light of known wing polymorphism and variation in elytral length in populations of other present day beetle species (39).

I turn now to the older (i.e. Late Tertiary) insect fossils. Although some are conspecific with living species (in spite of the passage of several millions of years), most of the fossils from the Late Miocene deposits of the Canadian arctic are outside the range of variation of living species (52). They are nevertheless close to modern forms—so close that Matthews (49) at first believed that they belonged to living species. Some of these are almost certainly direct ancestors of present day species; others appear to represent extinct species (or perhaps merely species that have not yet been described) that may have evolved cladistically from the same stocks that gave rise to living species. In the interpretation of the interrelationships of insect species the Late Cenozoic fossils must play an increasingly significant role. I give two examples.

In the late Miocene succession at Lava Camp Mine in Western Alaska, insect-bearing deposits are overlain by a lava flow that has been dated radiometrically as 5.7 million years old. From these sediments and among numerous other fossil insects were recovered four species of the staphylinid genus *Micropeplus*. Two of these are identical with living species; the other two, though highly characteristic, did not match any known living species. These fossil insects enabled Matthews (47) to erect a tentative phylogeny for the Micropeplinae based on the nature of the elytral striae and the interstrial puncturation of both the living and fossil representatives of the sub-family. This is surely a more securely founded phylogeny than any based entirely on species that happened to have survived to the present day.

The second phylogenetic study concerned *Helophorus tuberculatus,* a hydrophilid water beetle that seems at present to be taxonomically rather isolated from other species of the genus. The species has well-marked tubercles on the elytral interstices. Matthews (52) records fossil elytra (from both the Lava Camp deposits and from Miocene sediments on Meighan Island in the northwest Canadian arctic) that resemble those of *H. tuberculatus* but have subdued elytral tubercles. The two fossil insects were sufficiently different from one another and from their nearest modern relative to merit specific names of their own. Two plausible evolutionary sequences involving these three species may be suggested. First, they could represent stages of evolution in which tubercle size gradually increased in a single phylogenetic sequence. Second, the three species might have arisen cladistically through several episodes of geographical isolation, the two species with smaller elytral tubercles becoming extinct at some time between the Miocene and the present. Which of these conflicting views of phylogeny is the more acceptable? It is customary in the adjudication of such contests to apply 'Occams razor' [*"Frustra fit per pleura quot potest fieri per pauciora"* (58)]. Of the two hypothetical sequences it is clearly the second that is unnecessarily elaborate in the light of the available information; the simple gradualist view wins on the grounds of economy. This does not diminish the important role that cladistics must play in systematics. It does indicate that to be of real value phylogenetic diagrams must include all the available data, both living and fossil. What would we make of a phylogenetic diagram of, say, the Perissodactyls if it were constructed without taking the fossil record into consideration?

The extremely slow evolutionary rate of Coleoptera throughout the Late Cenozoic is contrary to what we might have expected in the absence of a fossil record. I outline here some of the reasons that, in the past encouraged us to believe that the Coleoptera (and, by implication, many other orders of insects) might have evolved relatively rapidly during this period.

First, in the evolution of mammals during the Late Cenozoic (and their fossil record is far too spectacular to be ignored) we see abundant evidence for rapid change and numerous extinctions (42). If we look at mammalian diversity at present we see marked intra-specific geographical variation connected by clinal zones. The Coleoptera of the present day show even greater diversity and detailed adaptation to environmental conditions. Their tendency to form geographical races seems as well developed as that of mammals. If we can so easily accept that the mammals have evolved and are evolving rapidly, why should we not view Coleoptera similarly? Yet the fossil record of specific constancy shows that intra-specific geographical variation, regardless of how sophisticated the measurements or subtle the mapping, is not by itself adequate evidence of evolution in action today.

Second, the extraordinarily slow rate of evolution shown by Late Cenozoic Coleoptera might also be unexpected in the light of numerous large-scale climatic oscillations that characteristized the period. This environmental instability might have been expected to have caused widespread extinctions of thermophilous species as glacial conditions swept down into the temperate latitudes from the north, and the polar species might have suffered when the climate ameliorated. We now know that these changes of climate took place very rapidly (65). Further, we might have expected episodes of rapid speciation during colonization of the vast tracts of land made newly available by the retreating glaciers. Species driven by circumstance into isolated refuges might have been subjected to intense selection pressures which, coupled with the smallness of the populations, also could have promoted rapid speciation. Yet in the areas most affected by fluctuations in climate we find abundant evidence of specific constancy and no evidence of widespread extinctions. To resolve this apparent paradox it is necessary to know something of the biogeographic history of Late Cenozoic Coleoptera.

BIOGEOGRAPHY

Late Cenozoic fossils provide factual data on where species lived in the relatively recent past. Many species changed their geographical distributions on a far larger scale than could have been anticipated from a study of their modern biogeography. These changes of range follow an orderly pattern that is closely related to the large-scale climatic fluctuations—i.e. the glacial-interglacial cycles that have been shown in recent years to have been far more numerous and to have begun earlier than was formerly suspected [well before the accepted beginning of the Pleistocene period (68, 69)]. Within the major cycles other lesser oscillations occurred (interstadials and stadials), each of which involved climatic changes greater than any that have taken place in historic time. The geographical movements of insect species reflect these minor episodes with the same precision as they do the major ones (25, 27, 29, 30, 60). This intimate relationship between changes in climate and changes in distribution has been well documented in northwest Europe, in particular for the British Isles, for the period since the last truly interglacial climatic phase (120,000 years ago).

Last Interglacial

Fossil assemblages of insects are known from three localities in Britain dating from the warmest phase of the Last Interglacial (which is known by a variety of names—Ipswichian in Britain, Eemian in continental Europe, and Sangomanian in North America). Two of these faunas are remarkably similar: that from Bobbitshole, Ipswich (23), the type section for the interglacial in Britain, and that from foundation excavations in Trafalgar Square, London (G. R. Coope, unpublished data). Both of these fossil insect assemblages contained a high proportion of species that live today in central and southern Europe but that do not now extend as far north as the British Isles. Among these were numerous species of scarab beetles such as *Oniticellus fulvus, Caccobius schreberi, Onthophagus furcatus,* and *Onthophagus opacicollis.* These occurred in great profusion in the Trafalgar Square deposits where they were associated with bones of straight-tusked elephant, hippopotamus, and lion. But no doubt the most interesting record from the biogeographical point of view is the presence at Trafalgar Square of the genus *Drepanocerus,* which is known today from Africa south of the Sahara, from the Indian subcontinent, and from Southeast Asis (10). Its present day range is clearly a fragmented remnant of a once wide distribution. The British fossils suggest that the fragmentation occurred as late as the Last Glacial period, a suggestion that can only be confirmed by finding further fossils from widespread localities.

The third locality that yielded a Last Interglacial insect fauna was Tattershall, Lincolnshire (36). Numerous large tree trunks in the deposit indicated a mature forest, and with the forest came the timber beetles *Xestobium rufovillosum* [the Death Watch beetle, confined to indoor structural timber in Britain today (14)], *Anobium fulvicorne,* and *A. punctatum* (the familiar furniture beetles). Two species, *Isorhipis melasoides* and *Pycnomerus terebrans,* now occur in the European old forest. Their ranges have contracted in historic times because their original habitat has been greatly reduced by human activities. Both species have been recorded as fossils in postglacial deposits in Britain, but they have now apparently become extinct there (15, 59). Thus some of the old forest insects have been able to exploit new habitats provided by human activities. Their ranges have been extended both as a result of commerce and by the provision of protected environments within buildings. Other insects of the primordial forests have found no such analogs of their original habitats and have gradually been driven by humans into isolated remnants of their former ranges, where they have become the entomological equivalents of the wolf and bear.

Last Glaciation

A long climatically complex period, predominantly much colder than the present day, followed the Last Interglacial discussed above and lasted until about 10,000 years ago, the conventional beginning of the present interglacial. This long cold period will be referred to here as the Last Glaciation (it is called the Devensian in Britain, the Weichselian or Würm in continental Europe, and the Wisconsinan in North America). During this period temperatures fluctuated considerably. During stadials the British Isles knew climates of arctic severity; during brief milder spells (interstadials) conditions were at times similar to those of the present day. Associated changes in the specific composition of the beetle faunas of northwest Europe have now been documented.

EARLY LAST GLACIATION   We know little about the insect fauna of northwest Europe from the early part of the Last Glaciation. Deposits from this period are uncommon. Two fossil assemblages from Britain come from an early interstandial, probably equivalent in age to the Brörup interstadial of Denmark, established on palaeobotanical grounds by Andersen (1). The first of these came from Chelford, Chesire, where a large assemblage of fossil insects was obtained from a peaty deposit that represented the accumulate on a forest floor complete with logs of pine, fir, and birch (17). The second fauna was obtained from a small organic lens in a gravel pit at Four Ashes, Staffordshire, and was described by Anne Morgan (57). These faunas are very similar and represent an association of species that can be found today living in the coniferous forests of central Finland—an interpretation that is amply supported by the palaeobotanical evidence (70).

In central Sweden insect faunas have been obtained that apparently date from the period just described. They indicate a very different environment. When these fossils (from four separate sites) were originally described by Lindroth (43) they were thought to be of interglacial age, but recent work (45) suggests that these deposits also date from the time of the Brörup Interstadial. Lindroth's work was pioneering in two respects. First, he demonstrated clearly that the fossils were exactly comparable with living species, in spite of earlier efforts to ascribe some of them to new excinct species (55). Second, he established for the first time that the true tundra insect fauna had occupied relatively low latitudes during the glacial periods.

If this tundra fauna from central Sweden is indeed of the same age as the forest fauna from central England, the tree line must have been well to the south of its present position during this early interstadial of the Last Glaciation.

MIDDLE LAST GLACIATION   The insect faunas from the climatically complex middle phase of the Last Glaciation are relatively well understood. This is the period between 50,000 years ago and about 15,000 years ago when the climate of Western Europe was predominantly cold, interrupted by a single, but brief, temperate interlude at about 43,000 years ago. This interlude was followed by a period of about 15,000 years when the climate was cool and continental in style but nevertheless still able to support a rich flora and fauna. The temperate interlude plus the cool period that followed has been called in Britain the Upton Warren Interstadial Complex. On the adjacent continent this period has been subdivided into three discrete interstadials, the Moershoofd, Hengelo, and Denekamp Interstadials. After this interstadial complex the climate deteriorated still further, ushering in the phase of maximum southward extension of the glaciers which covered most of northern Europe as far south as central England and northern Germany about 20,000 years ago. (In North America the climatic changes during this period followed much the same pattern, with differences in details of timing and intensity; the ice sheets reached their southernmost limited in Ohio by about 18,000 years ago.) The biogeographical repercussions of these environmental upheavals were profound. The fossil record shows large-scale changes of insect distribution directly related to the sequence of climatic changes outlined above.

Pre-Upton Warren Interstadial   Fossil insect faunas are known from three sites in central England that apparently date from just before the Upton Warren Interstadial (24). Their precise timing is difficult because their ages fall near the limit of radiocarbon dating techniques, but one fauna was positively dated at just over 44,000 years ago. These faunas are characterized by presence of a rather restricted suite of forty named species of Coleoptera, many of which have at present boreal or boreo-montane distributions. Such species include *Diacheila polita, Diacheila arctica, Bembidion fellmanni, Pterostichus blandulus, Helophorus obscurellus, Ochthebius kaninensis, Boreaphilus nordenskioeldi, Olophrum boreale,* and *Simplocaria metallica.* Though none of these species now lives in Britain, they all occur in arctic Russia. *Octhebius kaninensis* is of particular interest because it is known at present only from the Kanin Peninsula in arctic Russia; in Britain it is an abundant fossil found only in deposits of this early cold period.

The thermal maximum of the Upton Warren Interstadial   The temperate episode that followed rapidly after the cold period discussed above reached its thermal maximum about 43,000 years ago. Fossil insect assemblages from this episode provide no evidence for survival in Britain of the tundra element. In its place is a rich fauna of over three hundred named species of varied ecological preferences, many of which are now found in central and southern Europe (29). The following selection suggests the ecological diversity of these relatively southern species: *Calosoma reticulatum, Hydrochus flavipennis, Hister funestus, Aphodius bonvouloiri, Opetiopalpus scutellaris, Chaetocnema obesa, Rhynchites pubescens, Otiorrhynchus mandibularis,* and *Pseudocleonus cinereus.* None of these lives today in Britain. The temperate interlude that permitted this complex fauna to become established in Britain and that was presumably responsible for the extermination, certainly in the lowland areas, of the earlier tundra fauna, may have lasted for as little as 1000 years. If this is indeed the case, the fossil record demonstrates how rapidly insect species can respond to climatic changes. Until we have more information on the distances involved in these geographical changes we cannot give useful estimates of the speed of movement involved.

The cool later part of the Upton Warren Interstadial   The climate deterioration after the temperate period was evidently fast, though not as fast as the amelioration at its beginning, so that by 40,000 years ago almost the entire southern element of the beetle fauna of the British Isles had been exterminated and the boreal species (with the exception of *Ochthebius kaninensis*) that were so characteristic of the pre-Interstadial fauna had returned (24). In addition to these northern species, the British Isles were invaded by many species that are today found no nearer than eastern Asia —e.g. *Carabus maeander, Agabus clavicornis, Coelambus mongolicus, Helophorus aspericollis, Helophorus jacutus, Helophorus mongoliensis, Tachinus jacuticus, Tachinus caelatus,* and *Aphodius holdereri.* Some of these species are, in terms of numbers of individuals, among the most abundant components of our fossil assemblages. Their presence in western Europe as recently as the middle of the Last Glaciation has numerous biogeographical implications. I comment upon two.

   Helophorus aspericollis was recently described from Eastern Siberia (2). It is extremely similar to *Helophorus brevipalpis,* a species common and widespread in most of Europe. The two species can only be distinguished with certainty on the basis of detailed differences in the male genitalia, though less reliable exoskeletal characters can also be used to differentiate them. Their present day distributions give the appearance of two species that have originated cladistically in relatively recent time, since the areas now occupied by the European and Siberian forms look like a once continuous range riven by the expansion of the ice sheets. Yet the fossil record tells a quite different story. The Siberian *H. aspericollis* occurred abundantly in Britain during the cold parts of the Last Glaciation (identifications were made by R. B. Angus on numerous examples with the male genitalia intact), and *H. brevipalpis* lived here during the temperate interludes. *H. aspericollis* was also found in deposits associated with carcass of a wooly rhinoceros at Starunia in Western Ukraine, where it was associated with other Siberian species (R. B. Angus & G. R. Coope, unpublished data). The present day ranges of these two species thus reflect the different geographical locations of the environments that each finds acceptable; their modern distributions are not an indication of their recent evolutionary history.

Aphodius holdereri is by far the most abundant large dung beetle in fossil assemblages from the British Isles that date from the middle of the Last Glaciation (22). This is the species which I called '*Aphodius sp A* of Upton Warren' (34), and under this name it has appeared in records from numerous fossil sites. The species long defied identification because today it is restricted to the high plateau of Tibet above 10,000 feet (10). In the absence of the fossil record *Aphodius holdereri* might have been construed as an endemic Tibetan species adapted to high altitude environments. But the fossil evidence suggests that this simplistic view is misleading. The species could just as well have originated in Britain. In many cases such a restricted modern range may be the 'last stand' of a species from which, if the next glaciation comes in time, it may sally out to regain its former territory. As to its altitudinal restriction, it is difficult to reconcile its abundance at sea level in Britain during the Last Glaciation with any obligatory high altitude adaptations. However, high altitude Tibet and sea level glacial Britain could have had a common thermal environment and the restricted range of *A. holdereri* at the present day may be largely a matter of restriction to acceptable thermal conditions. Of course it may be argued that the fossils represent a race adapted to different environmental conditions from those of the modern remnant population and it is possible that the western European populations differed slightly in their habitat requirements from the modern representatives. Nevertheless, in its heyday *Aphodius holdereri* in glacial Britain was accompanied by other species with varied ecological preferences that are today characteristic of the high cold steppes of central Asia. There is no need, therefore, to invoke any major difference in adaptations of modern and fossil representatives of this species.

The period of maximum ice advance   Insect faunas from the British Isles that date from the period when the glaciers reached their maxima are, not unexpectedly, rare. Large areas of western Europe south of the ice sheets were at this time probably reduced to conditions of near polar desert; there is evidence for the survival of meager biota of moss and a few insect species. Two insect faunas from England can be dated with certainty to this period —one from Dimlington in Yorkshire (64a), the other from Barnwell Station near Cambridge (18, 19). Both faunas are impoverished in terms of the variety of species and consist of Coleoptera that are today able to survive in areas of extreme cold. Bearing in mind the logical weakness of negative evidence, an interesting absence of the exclusively eastern species from these two faunas contrasts with their abundance (in terms at least of individual numbers) in the fossil assemblages from the immediately earlier period. It is tempting to take this as evidence for a decreased climatic continentality at the time, perhaps a reflection of the increased precipitation that was necessary to feed the glaciers in their advance into the lowlands.

THE LAST GLACIATION IN EUROPE   On the adjacent continent of Europe much less is known about the changing insect fauna of the Last Glaciation apart from the work of Lindroth on the beetles from the Jamptland interstadial (43). Little is known about the early stages of the Glaciation; deposits of this age from Sweden and north Germany are now being investigated for insect fossils. Arctic beetles are known from the middle phases of this glaciation (3, 20); in Holland at this time one finds the arctic species *Pterostichus vermiculosus, Helophorus spendidus,* and *Tachinus arcticus* [the first species from (20), the last two species from G. R. Coope, unpublished data]. Work now in progress on sites in Belgium and northern Germany that date from the episode of maximum glaciation is providing abundant evidence of the presence in lowland Europe of many species that are today restricted to arctic regions or the high mountains. Of particular biogeographical interest is the presence of several species of the dominantly tundra ground beetle *Pterostichus* (subgenus *Cryobius*) that is today entirely absent from western Europe. Today its species are chiefly concentrated in Alaska and northeast Siberia. This pattern of distribution might well have been thought to indicate 'a center of dispersal' in that region, but again the fossil occurences of a number of species of *Cyrobius* in lowland western Europe in the geologically recent past urges caution in the acceptance of concepts based on analyses of present day ranges alone. The concentration of related species in certain areas may be a measure of their common inheritance of similar environmental requirements rather than of their biogeographic history.

The "Lateglacial"   The term "Lateglacial" is used in western Europe to denote the period between about 14,000 and 10,000 years ago, after which a marked climatic amelioration brought the Last Glaciation to an abrupt end. In Britain the faunas of this period are now well known. They present a pattern of change in insect biogeography associated with an intense but short-lived climatic oscillation, whose warm episode is called the Windermere Interstadial in the British Isles (broadly equivalent to the Bölling and Allerød episodes of continental authors) (33). I will briefly describe the insect faunas under three headings: (a) pre-Interstadial, (b) Windermere Interstadial and (c) post-Interstadial.

Pre-Windermere Interstadial   The pre-interstadial fauna is essentially the one that followed the retreating ice and colonized the newly bared areas. It was dominated by arctic/alpine species, many of which are now absent from the British Isles—e.g. *Bembidion fellmanni, Bembidion hasti, Bemidion lapponicum, Helophorus obscurellus, Helophorus spendidus, Boreaphilus henningianus, Simplocaria metallica,* and *Phytonomus obovatus* (30).

The Windermere Interstadial   The insect fauna of the first part of the Windermere Interstadial contrasts sharply with the one just described. All the arctic/alpine species disappeared from lowland Britain and were replaced almost instantly by an ecologically elaborate suite of species that today live in temperate regions of western Europe. Some of these barely range as far north as the extreme south of the British Isles. Such species include *Asaphidion cyanicorne, Bembidion callosum, Bembidion octomaculatum, Metabletus minutulus, Berosus signaticollis, Bruchidius debilis,* and *Larinus planus* (30, 38, 60). About 12,000 years ago this entirely temperate fauna was suddenly replaced. The lowlands of Britain were reinvaded by more northern species, the southern element of our insect fauna disappeared. Thus the second half of the Windermere Interstadial is characterized by cool-temperate species, particularly in central and northern Britain.

Post-Windermere Interstadial   In its turn the Windermere Interstadial was abruptly terminated by a further climatic deterioration at about 11,000 years ago. The numerous arctic/alpine species characteristic of fully glacial conditions returned (31, 60, 61, 64). This Lateglacial fauna differed from that of the middle phase of the glaciation by its poverty of eastern species; only *Helophorus jacutus* and *Tachinus jacuticus* reappeared at this time, and even they were represented only by isolated individuals in contrast to their profusion in the full glacial faunas of Britain. One new Siberian species occurred in England at this time, namely *Pterostichus magus,* found in Standlake near Oxford (26). From the biogeographic point of view it is interesting to note that almost 40% of the coleopterous fauna of this period is no longer to be found in the British Isles, but almost all of the species concerned are members of the present day Fennoscandian fauna. Little is known of Lateglacial insect faunas of Europe, but sites in Jutland, north Denmark, and Holland are being studied, and resemble those of the British Isles during this period.

Postglacial Insect Biogeography.

At or very close to 10,000 years ago the arctic/alpine assemblages of Coleoptera disappeared suddenly from lowland Britain and were immediately replaced by a suite of thoroughly temperate species, many of which have ranges that today lie to the south of their fossil localities (4, 5, 11, 62). It seems likely that this complete faunal switch took place in a matter of at most a few centuries. Even in the Highlands of Scotland in the middle of Rannoch Moor the most abundant beetle in a fossil assemblage obtained from the first sediments to be laid down after the retreat of the ice was the small hydrophilid *Ochthebius foveolatus,* a central European and Mediterranean species at the present day (G. R. Coope, unpublished). The prompt reappearance at this time of relatively thermophilous species even in the far north of Britain suggests an almost immediate return of warm, fully interglacial climatic conditions. The appearance of a gradual build up towards 'the climatic optimum' (hypsithermality) of the present day interglacial may well be merely a reflection of the tardiness with which those elements of the biota hitherto used as climatic indicators responded to a sudden change.

The postglacial period is unique in Quaternary history because it includes the rise of civilized man, who now dominates almost the entire earth. The biogeographic consequences of this event are incalculable. In Europe the clearance of the primordial forests was as inimical to the survival of many insect species as it was to the survival of many of the more spectacular mammals. In particular, man's activities curtailed the ranges of beetles associated with dead or dying mature timber, and the fossil record from excavations in swamps shows that many species lived in the old primordial forests of Britain that are today extinct in these Islands (15, 40).

Not all the biogeographic effects of human activity are on the debit side, however. The removal of the forest has contributed to the expansion of open grassland species (58a). Furthermore, the habitats provided by human habitations have enabled many species of Coleoptera to extend their ranges. For example *Aglenus brunneus* was abundant in medieval contexts in Britain (41), and the exclusively east Mediterranean longhorn beetle *Hesperophanas fasciculatus* has been found (59) in a Roman excavation in central England. There can be little doubt that the latter species was imported accidentally into Britain inside timber objects (structural timbers would hardly have been imported into a country so well forested as Britain at that time). The beetles must have survived the rigors of our climate in the cosy interior of a Roman dwelling.

The peculiar environments associated with stored products have also led to widespread dissemination of many pest species. In our fossil lists from natural sites such species are almost unknown. They clearly made up a very small component of the fauna, but their day was to come with the arrival of civilization. The familiar scourges of *Oryzaephilus surinamensis* and *Sitophilus granarius* were as much a headache to the Roman storeman as they are to the present day grain warehouseman (13, 32, 63).

North American Faunas During the Last Glaciation

In North America the study of the beetle faunas of the Last Glaciation is in a more embryonic state than in Britain, but it seems that a broadly similar pattern of zoogeographic change is emerging. I give here the briefest outline of some recent work. Probably the oldest deposits that can be dated with certainty to the Last Glaciation occur at Scarborough Bluffs near Toronto. These deposits yielded the rich fauna originally worked on by Scudder (66, 67), who credited most of the fossils with new specific names. These fossils are being studied by Anne Morgan [(56) and communication], who can show that they should be assigned to living species. Two records from this site are of particular biogeographic interest: the tundra ground beetle, *Diacheila polita,* and the hydrophilid beetle, *Helophorus sibiricus,* both limited in their modern North American occurrences to the extreme northwest. These species show that even within the Last Glaciation large-scale changes in insect biogeography have occurred, and present day arctic species occupy ranges well to the south.

A large fossil assemblage from the middle, interstadial, phase of the Last Glaciation has been discovered at Titusville, Pennyslvania, and is being investigated by S. M. Totten [(72) and unpublished data]. Here a fauna typical of the northern boreal forest (the Hudsonian zone) also includes the exclusively tundra species *Boreaphilus nordenskioeldi.* This suggests that the environmental zonation south of the ice front may have been more compressed in glacial than in modern times.

Insect fossils from near Cleveland, Ohio (19) date from almost the maximum period of the expansion of the Wisconsinian glaciers. Here again the suite of species is more indicative of the northern edge of the boreal forest than of the true tundra. These Wisconsinian faunas thus indicate a zonation south of the Laurentide ice sheets rather different from that in Europe, where a vast expanse of tundra fringed the Fennoscandian glaciers.

In Alaska and the Canadian northwest, the insect fossils indicate (46, 51) that for much of the Wisconsinian glacial period wide expanses of steppe-tundra were inhabited by a complex fauna and flora.

In the more arid areas of the southern United States, Ashworth (6) has been investigating a seemingly unprepossessing source of fossil insects: the urine-soaked middens of packrats. Apparently these represent hoards of biological materials and can be shown by radiometric measurements to date back as much as 40,000 years. This material is often well-preserved and gives us a potentially valuable insight into changes in the insect fauna in low latitudes where insect fossils have rarely been investigated.

The insects species that followed the rapidly retreating Laurentide Ice sheet were biogeographically heterogeneous, with various distributional styles (4, 8, 9). This suggests that during this phase of fast transition from glacial to interglacial conditions, peculiar environments arose and faunal assemblages came together that have no precise analogs at present.

ECOLOGICAL IMPLICATIONS

Several important ecological implications arise from this investigation of Late Cenozoic fossil Coleoptera. First, although morphological stability at the species level is now an established fact, a similar degree of physiological stability cannot be demonstrated directly. However, the fact that fossil assemblages even as far back as the Miocene faunas of arctic Canada strongly resemble living communities suggests that the ecological requirements of the fossil species were in the remote past largely the same as those of the living representatives. Had physiological evolution been progressing in a clandestine fashion under the cloak of morphological constancy, we would have expected a gradually increasing distortion of this picture of ecological conformity with increasing age of our faunas. No such distortion has been found. Apart from the unlikely possibility that physiological change took place in all species independently, to the same degree and in the same direction, it seems inescapable that for the most part morphological stability was accompanied by physiological stability.

We now return to the apparent paradox left unresolved earlier (page 252): An extraordinary degree of specific stability coincided with a period in the earth's history characterized by numerous large-scale fluctuations in climate, a situation that leads one to expect rapid speciation and numerous extinctions (28). I believe the explanation lies in the readiness with which insect species shifted their geographical ranges as the climate changed. Three options are open to species when changing environments become intolerable: they may adapt to the new conditions, they may become extinct, or they may move to an area where conditions are still acceptable. Of these three, the last was the most usual response. As the climate changed rapidly, the opportunity to 'evolve out of trouble' seems to have been beyond their genetic agility. Rather they simply tracked the tolerable environment across the continents. These repeated forced marches must have continuously broken down the geographical barriers that separated populations, permitting genetic mixing and keeping the gene pools well stirred. Speciation under these circumstances must have been well nigh impossible. We may thus view specific constancy in a rather different light—as a result of, rather than in spite of, the vicissitudes of the Late Cenozoic climates.

The readiness with which so many insect species changed latitudes as they tracked the shifting climatic zones has further ecological implications. While general features of the thermal environment could be followed in this way, latitude-dependent periodicities could not: A species could alter its geographical distribution to keep its crude thermal environment fairly constant, but the changes of latitude must have affected its biorhythms in one of two ways. First, and I believe the most probable, as it changed latitude the species must have been able to adjust its rhythm to changing periodicities of its environment. Such adjustments of biorhythm must have been accomplished with great rapidity—at times in a matter of a few centuries —to accommodate the sudden biogeographic alterations that can be shown from the fossil record.

Did these adjustments entail changes in the genetic composition of the population, or has the biorhythmic control a built-in adjustability that is itself an adaptation to frequent changes in its past biogeography? Such a notion involves the concept that, because environments change and recur again and again in glacial/interglacial cycles, a species must be adapted to the environments of its immediate past as well to that of the present day. The fact that the Cenozoic cycles of climatic change did not result in numerous extinctions of insect species testifies such long-term adaptability.

On the other hand, a species might not alter its biorhythms at all as it changed latitude, provided that poor adjustment would not be inimical to its survival. Thus we might find species, say in the arctic regions today, that retain rhythms more appropriate to lower latitudes—vestigial biorhythms indicative of the recent history of the species. This is not as improbable as it seems at first sight since the present interglacial period is but a brief temperate interlude in an otherwise dominantly cold period. It has recently been shown that, during the past half-million years, interglacial episodes have been short and the intervening glacials prolonged (68, 69) and that the present global disposition of climate and biota is thus exceptional; a more glacial pattern is the norm to which many of our middle- and high-latitude species have been adapted and to which they will return when the present interglacial comes to an end.

Speculations such as these that arise from investigation of the Late Cenozoic insect fossils need not (like many palaeontological hypotheses) remain in the limbo of conjecture for ever. They are subject to experimental testing based on specifically identical living representatives of the fossil species in question. Furthermore, the evolutionary and biogeographic speculation of the present are likewise subject to arbitration based on the fossil record. In this interdisciplinary field between palaeontology and neontology, progress lies in cooperation; we ignore the evidence of our sister sciences at our peril.

CONCLUSIONS

1. Insect fossils, particularly those of Coleoptera, are abundant in many Late Cenozoic terrestrial and freshwater deposits.
2. Fossils from the Quaternary (last 1.6 million years) are identical with living species.
3. Fossils from the Late Miocene (5.7 million years ago) fall outside the range of variation of living species but are so close that many of them may be ancestral to present day forms.
4. In response to climatic oscillations of the Quaternary, species of Coleoptera have altered their geographical ranges on an enormous scale so that the whereabouts of a species today may give us little evidence of its biogeographical history.
5. Geographical races, connected by clinal zones, can be shown to have remained morphologically stable for many tens of thousands of years (generations), but the geographical distributions of the races themselves have changed radically with the climatic oscillations.
6. The readiness with which insect species track the changing environment from place to place suggests that they were able to keep the conditions in which they lived more or less constant in spite of the glacial/interglacial cycles.
7. This mobility of insects coupled with the frequency of large-scale climatic fluctuations during the Quaternary period must have continuously broken down geographical barriers between populations. The gene pools were thus kept well stirred. Speciation under such circumstances must have been difficult.
8. In isolated situations such as islands or caves, from which escape is all but impossible, environmental change must have been endured on the spot. Under these circumstances rapid evolutionary change is the only alternative to extinction. Rates of evolution in such situations are likely to be atypically rapid.
9. As a corolary to conclusion 7, evolution may be expected to have been more rapid where the environment has been stable for long periods—e.g. at tropical latitudes where arid and humid periods may not have forced insect populations to change their geographical distributions on the same scale as in the middle latitudes.
10. The fact that insect species have frequently changed latitudes in response to the glacial/interglacial cycles means that some of their biorhythms (photoperiodism for instance) must be readily adjustable.

Literature Cited

1. Andersen, S. T. 1961. Vegetation and its environment in Denmark in the early Weicheselian Glacial (Last Glacial). *Dan. Geol. Unders. Afh. Raekke 2.* 75:1–75
2. Angus, R. B. 1973. Pleistocene *Helophorus* (Coleoptera, Hydrophilidae) from Borislav and Starunia in the western Ukraine, with a reinterpretation of M. Kommicki's species, description of a new Siberian species, and comparison with British Weicheslian faunas. *Philos. Trans. R. Soc. Lond. Ser. B.* 265 (No. 869):299–326
3. Angus, R. B. 1975. Fossil Coleoptera from Weichselian Deposits at Voorthuizen, the Netherlands. *Geol. Mijnbouwk.* 54:211–24
4. Ashworth, A. C. 1972. A late-glacial insect fauna from Red Moss, Lancashire, England. *Entomol. Scand.* 3:211–24
5. Ashworth, A. C. 1973. The climatic significance of a late Quaternary insect fauna from Rodbaston Hall, Staffordshire, England. *Entomol. Scand.* 4:191–205
6. Ashworth, A. C. 1973. Fossil beetles from a fossil wood rat midden in Western Texas. *Coleopt. Bull.* 27(3):139–40
7. Ashworth, A. C. 1977. A late Wisconsinian coleopterous assemblage from Southern Ontario and its environmental significance. *Can. J. Earth Sci.* 14:7:1625–34
8. Ashworth, A. C., Brophy, J. A. 1972. Late Quaternary fossil beetle assemblage from the Missouri Coteau, North Dakota. *Geol. Soc. Am. Bull.* 83:2981–88
9. Ashworth, A. C., Clayton, L., Bickley, W. B. 1972. The Mosbeck site: a palaeoenvironmental interpretation of the late Quaternary history of Lake Agassiz based on fossil insect and mollusk remains. *Quat. Res.* 2:177–88
10. Balthasar, V. 1963/64. *Monographie der Scarabaeidae und Aphodiidae der palaearktischen und Orientalischen Region.* Prague: Tschechoslowak. Akad. Wiss. 3 vols., 391, 627, 652 pp.
11. Bishop, W. W., Coope, G. R. 1977. Stratigraphical and faunal evidence for late glacial and early Flandrian environments in South West Scotland. In *Studies in Scottish Late Glacial Environment,* ed. J. M. Gray, J. J. Lowe, pp. 61–88. Oxford: Pergamon
12. Briggs, D. J., Gilbertson, D. D., Goudie, A. S., Osborne, P. J., Osmaston, H. A., Pettitt, M. E., Shotton, F. W.,

Stuart, A. J. 1975. A new interglacial site at Sugworth. *Nature* 257:477–79
13. Buckland, P. C. 1974. Archaeology and environment in York. *J. Archaeol. Sci.* 1:303–16
14. Buckland, P. C. 1975. Synanthropy and the death-watch; a discussion. *Naturalist* April/June 1975:37–41
15. Buckland, P. C., Kenward, H. K. 1973. Thorne moor: a palaeo-ecological study of a Bronze Age site. *Nature* 241:405–6
16. Carpenter, F. M. 1953. The evolution of insects. *Am. Sci.* 41:256–70
17. Coope, G. R. 1959. Late Pleistocene insect fauna from Chelford, Cheshire. *Proc. R. Soc. Lond. Ser. B* 151:70–86
18. Coope, G. R. 1967. Diachila (Col., Carabidae) from the glacial deposits at Barnwell Station, Cambridge. *Entomol. Mon. Mag.* 102:119–20
19. Coope, G. R. 1968. Coleoptera from the arctic bed at Barnwell Station, Cambridge. *Geol. Mag.* 105:482–86
20. Coope, G. R. 1969. Insect remains from mid-Weichselian deposits at Peelo, The Netherlands. *Mededel. Rijks. Geol. Dienst.* 20:79–83
21. Coope, G. R. 1970. Interpretations of Quaternary insect fossils. *Ann. Rev. Entomol.* 15:97–120
22. Coope, G. R. 1973. Tibetan species of Dung Beetle from late Pleistocene deposits in England. *Nature,* 245:335–36
23. Coope, G. R. 1974. Interglacial Coleoptera from Bobbitshole, Ipswich, Suffolk. *Geol. Soc.* 130:333–40
24. Coope, G. R. 1975. Mid-Weichselian climatic changes in Western Europe, reinterpreted from coleopteran assemblages. In *Quaternary Studies,* ed. R. P. Suggate, M. M. Cresswell, pp. 101–8. R. Soc. New Zealand
25. Coope, G. R. 1975. Climatic functions in North West Europe since the last interglacial indicated by fossil assemblages of Coleoptera. Ice Ages: Ancient and Modern, ed. A. E. Wright, & F. Moseley. *Geol. J.* Spec. Iss. No. 6:153–68
26. Coope, G. R. 1976. Assemblages of fossil Coleoptera from terraces of the Upper Thames near Oxford. In *Field Guide to the Oxford Region,* ed. D. Roe, pp. 20–23. Oxford: Quaternary Research Association
27. Coope, G. R. 1977. Fossil coleopteran assemblages as sensitive indicators of climatic changes during the Devensian (last) Cold Stage. *Philos. Trans. R. Soc. Lond. Ser. B.* 280:313–40

28. Coope, G. R. 1978. Constancy of insect species versus inconstancy of Quaternary environments. *Diversity of Insect Faunas Symposium No, 9,* pp. 176–87. London: R. Entomol. Soc. Lond.

29. Coope, G. R., Angus, R. B. 1975. An ecological study of a temperate interlude of the middle of the last glaciation, based on fossil Coleoptera from Isleworth, Middlesex. *J. Anim. Ecol.* 44:365–91

30. Coope, G. R., Brophy, J. A. 1972. Late glacial environmental changes indicated by coleopteran succession from North Wales. *Boreas* 1(2):97–142

31. Coope, G. R., Dickson, J., McKutcheon, J., Mitchell, G. F. 1978. Lateglacial flora & faunas from Drumurcher Co., Monnaghan. *Proc. R. Irish. Acad.* In press

32. Coope, G. R., Osborne, P. J. 1967. Report on the coleopterous fauna of the Roman Well at Barnsley Park, Gloucestershire. *Trans. Bristol Gloucester Archaeol. Soc.* 86:84–87

33. Coope, G. R., Pennington, W. 1977. The Windermere interstadial of the late Devensian. *Philos. Trans. R. Soc. Lond. Ser. B* 280:337–39

34. Coope, G. R., Shotton, F. W., Strachan, I. 1961. A late Pleistocene fauna & flora from Upton Watten, Worcestershire. *Philos. Trans. R. Soc. Lond. Ser. B.* 244:379–421

35. Duffy, E. A. J. 1968. The status of *Cerambyx* L. (Col., Cerambycidae) in Britain. *Entomol. Gaz.* 19(3):164–66

36. Girling, M. A. 1974. Evidence from Lincolnshire of the age and intensity of the Mid-Devensian temperature episode. *Nature* 250(5463):270

37. Hallam, A., ed. 1977. *Patterns of Evolution as Illustrated by the Fossil Record. Developments in Palaeontology and Stratigraphy,* 5:1–591. Amsterdam: Elsevier

38. Joachim, M. J. 1978. *Late-glacial coleopteran assemblages from the west coast of the Isle of Man.* PhD thesis. Univ. of Birmingham.

39. Johnson, C. G. 1969. *Migration and Dispersal of Insects by Fight.* London: Methuen. 704 pp.

40. Kelly, M., Osborne, P. J. 1963. Two faunas and floras from the alluvium at Shustoke, Warwickshire. *Proc. Linn. Soc. Lond.* 176:37–65

41. Kenward, H. K. 1975. The biological and archaeological implications of the beetle, *Aglenus brunneus* (Gyll), in ancient faunas. *J. Archaeol. Sci.* 2:63–69

42. Kurten, B. 1968. *Pleistocene Mammals of Europe.* London: Weiderfeld & Nicolson. 317 pp.

43. Lindroth, C. H. 1948. Interglacial insect remains from Sweden. *Sver. Geol. Under. Ser. C.* 42 (1):3–29

44. Lindroth, C. H., Coope, G. R. 1971. The insects from the interglacial deposits at Leveahiemi. *Sver. Geol. Unders. Ser. C.* 65:44–55

45. Lundqvist, J. 1967. Submorana Sediment I Jamtlands Lan. C 618. *Sver. Geol. Unders.* 61:1–267

46. Matthews, J. V. Jr. 1968. A palaeoenvironmental analysis of three late Pleistocene coleopterous assemblages from Fairbanks, Alaska. *Quaest. Entomol.* 4:202–24

47. Matthews, J. V. Jr. 1970. Two new species of *Micropeplus* from the Pliocene of Western Alaska with remarks on the evolution of Micropeplinae (Coleoptera: Staphylinidae). *Can. J. Zool.* 48:779–88

48. Matthews, J. V. Jr. 1974. Fossil insects from the early Pleistocene olyor suite (Chukochya River: Kolymian Lowland, USSR) *Geol. Surv. Can. Pap.* 74–1A, pp. 207–11

49. Matthews, J. V. Jr. 1974. A preliminary list of insect fossils from the Beaufort Formation, Meighen Island, District of Franklin. *Geol. Surv. Can., Pap.* 74–1A, pp. 203–6

50. Matthews, J. V. Jr. 1974. Quaternary environments at Cape Deceit (Seward Peninsula, Alaska): evolution of a tundra ecosystem. *Geol. Soc. Am. Bull.* 85:1353–84

51. Matthews, J. V. Jr. 1975. Insects and plant macrofossils from two quaternary exposures in the Old Crow-Porcupine Region, Yukon Territory, Canada. *Arct. Alp. Res.* 7:3:249–59

52. Matthews, J. V. Jr. 1976. Evolution of the Subgenus *Cyphelophorus* (Genus *Helophorus:* Hydrophilidae, Coleoptera): Description of two new fossil species and discussion of *Helophorus tuberculatus* Gyll. *Can. J. Zool.* 54:652–73

53. Matthews, J. V. Jr. 1976. Insect fossils from the Beaufort Formation: geological and biological significance. *Geol. Surv. Can. Pap.* 76–1B, pp. 217–27

54. Matthews, J. V. Jr. 1977. Coleoptera fossils: their potential value for dating and correlation of late Cenozoic sediments. *Can. J. Earth. Sci.* 14(10): 2339–47

55. Mjöberg, E. 1916. Über die Insektenreste der sogenannten "Harnogytta" im

nordlichen Schweden. *Sver. Geol. Unders. Ser. C.,* vol. 9

56. Morgan, A. 1972. The fossil occurrence of *Helophorus Arcticus* Brown (Coleoptera, Hydrophilidae) in Pleistocene deposits of the Scarborough Bluffs, Ontario. *Can. J. Zool.* 50:555–58

57. Morgan, A. 1973. Late Pleistocene environmental changes indicated by fossil insect faunas of the English Midlands. *Boreas* 2(4):173–212

58. Occam, W. 1324. *Summa Logicae Paris Prima,* ed. P. Bochner, 1951. New York and Louvian: Franciscan Inst. Publ.

58a. Osborne, P. J. 1969. An insect fauna of Late Bronze Age date from Wilsford, Wiltshire. *J. Anim. Ecol.* 38:555–66

59. Osborne, P. J. 1971. An insect fauna from the Roman Site at Alcester, Warwickshire. *Britannia* 2:156–65

60. Osborne, P. J. 1972. Insect faunas of late Devensian and Flandrian age from Church Stretton, Shropshire. *Philos. Trans. R. Soc. London. Ser. B* 263:327–67

61. Osborne, P. J. 1973. A late-glacial insect fauna from Lea Marston, Warwickshire. *Proc. Conventry Nat. Hist. Sci. Soc.,* vol. 4(7)

62. Osborne, P. J. 1974. An insect assemblage of early Flandarian age from Lea Marston, Warwickshire, and its bearing on the contemporary climate and ecology. *Quat. Res.* 4:471–86

63. Osborne, P. J. 1977. Stored product beetles from a Roman site at Droitwich, England. *J. Stored Prod. Res.* 13:203–4

64. Peake, D. S., Osborne, P. J. 1971. The Wandle Gravels in the vicinity of Croydon. *Proc. Croydon Nat. Hist. Sci. Soc.* 14(7):145–76

64a. Penny, L. F., Coope, G. R., Catt, J. A. Age and insect fauna of the Dimlington Silts, East Yorkshire. *Nature* 224:65–67

65. Ruddiman, W. F., Sancetta, C. D., McIntyre, A. 1977. Glacial/interglacial response rate to subpolar north Atlantic waters to climatic change: the record in oceanic sediments. *Philos. Trans. R. Soc. Lond. Ser. B* 280:119–142

66. Scudder, S. H. 1895. The Coleoptera hitherto found fossil in Canada. *Geol. Surv. Can. Contrib. Can. Palaeontol.* 2:27–56

67. Scudder, S. H. 1900. Canadian fossil insects. *Geol. Surv. Can. Contrib. Can. Palaeontol.* 2:67–92

68. Shackleton, N. J., Opdyke, N. D. 1973. Oxygen isotope and palaeomagnetic stratigraphy of equatorial Pacific core V28.238: Oxygen isotope temperatures and ice volumes on a 10^5 yr and 10^6 yr scale. *Quat. Res.* 3:39–55

69. Shackleton, N. J., Opdyke, N. D. 1976. Oxygen isotope and palaeomagnetic stratigraphy of equatorial Pacific core V28.239 Late Pliocene to Latest Pleistocene. An investigation of late Quaternary palaeoceanography and palaeoclimatology. *Mem. Geol. Soc. Am.* 145:449–64

70. Simpson, I. M., West, R. G. 1959. On the stratigraphy and palaeobotany of a Late Pleistocene organic deposit at Chelford Cheshire. *New Phytol.* 57: 239–50

71. Thomsen, M., Krog, H. 1949. *Cerambyx cerdo* L. (= *heros* Scop.) Fra Subboreal Tid I Danmark. *Vidensk. Medd. Naturhist. Foren. Khobenhavn* 111: 131–48

72. Totten, S. M. 1971. The occurrence of beetle remains in Pleistocene deposits, East Central United States. *Abstr Progr. Ann. Meet. Geol. Soc. Am. Washington DC* 3(7):733–34

73. Ullrich, W. G., Coope, G. R. 1974. Occurrence of the East Palaeartic beetle *Tachinus jacuticus.* Poppius (Col. Staphylinidae) in deposits of the Last Glacial period in England. *J. Entomol.* (B) 42(2):207–12

74. Zeuner, F. 1934. Die Orthopteren aus der diluvialen Nashornschicht von Starunia (Polnische Karpathen). *Pol. Akad. Umiejet. "Starunia"* 3:1–17

Ann. Rev. Ecol. Syst. 1979. 10:269–308
Copyright © 1979 by Annual Reviews Inc. All rights reserved

ADAPTIVE CONVERGENCE AND DIVERGENCE OF SUBTERRANEAN MAMMALS

♦4163

Eviatar Nevo

Institute of Evolution, University of Haifa, Haifa, Israel

INTRODUCTION

Evolutionary theory should ideally specify the necessary and sufficient determinants of the convergent and divergent patterns that have evolved in subterranean mammals (i.e. mammals that have radiated over space and time into the subterranean ecological zone). These include fossorial species that spend most of their lives in sealed burrows and come to the surface only incidentally (Table 1). The physical and biotic uniqueness of subterranean ecology provides an excellent evolutionary theater where adaptive convergent evolution molds populations, species, and communities in similar but geographically distinct environments throughout the world on all levels of organization: genetic, biochemical, physiological, anatomical, behavioral, populational, life-historical, and speciational. The objective of this paper is to compare and contrast the evolutionary patterns and adaptive strategies of completely subterranean mammals and to suggest a theory specifying the selective forces operating in this unique environment.

Which mammals radiated underground? Three of the 19 orders of mammals have completely subterranean representatives: the rodents, the insectivores, and the marsupials. Distribution patterns, geographical and paleontological, are given in the Table and Figures. The taxonomic distribution of subterranean mammals is extensive: 10 of 132 mammalian families (~7.5%), comprising some 35 of 1004 mammalian genera (~3.5%), and 144 of 4060 mammalian species (~3.5%), based on classical taxonomy (2). The new discoveries of widespread chromosomal sibling species in mammals may increase manyfold the number of subterranean

269

species: Herbivore speciation underground was prolific and gave rise to an unusually large number of species, few of which are sympatric. Throughout the world the subterranean environment is divided basically into two trophic modalities, herbivorous (148) and insectivorous, as is suggested by the occurrence of sympatric distributions only between the two in contrast to the allopatric and parapatric patterns within each. The adaptive convergence of unrelated subterranean mammals in size, structure, and function (39, 42) discussed in this paper seems to be linked intimately with the

Table 1 Subterranean mammals: taxonomic, geographic and geologic ranges

Order	Family and common name	Extant Genera	Species[a] (no.)	Geographic distribution	Geologic range and major fossil groups
Marsupialia	*Notoryctidae* (Marsupial moles)	(1) *Notoryctes*	2	Australia	Recent. No fossils known.
Insectivora	*Chrysochloridae* (Golden moles)	(5) *Amblysomus*	5	Africa	Early Miocene to Recent in Africa.
		Chrysochloris	2		
		Chrysospalax	2		*Prochrysochloris* early
		Cryptochloris	1		Miocene Kenya.
		Eremitalpa	1		
	Talpidae (moles)	(7) *Talpa*	2	America (North)	Late Eocene to Recent in
		Mogera	1	Europe	Europe.
		Parascaptor	1	Asia	Late Oligocene to Recent
		Scaptochirus	1		in North America.
		Parascalops	1		Recent in Asia.
		Scapanus	3		7 extinct genera.
		Scalopus	3		
Rodentia	*Geomyidae* (Pocket gophers)	(8) *Cratogeomys*	10	America (North and Central)	Early Miocene to Recent in North America.
		Geomys	7		9 extinct genera.
		Heterogeomys	2		
		Macrogeomys	6		
		Orthogeomys	3		
		Pappogeomys	2		
		Thomomys	9[a]		
		Zygogeomys	1		
	Cricetidae (Voles)	(3) *Myospalax*	5	Asia	Oligocene to Recent in Asia.
		Ellobius	3		
		Prometeomys	1		
	Spalacidae (Mole rats)	(1) *Spalax*	3[a]	Europe (SE) Asia Africa (North)	Upper Pliocene to Recent in Europe. Middle Pleistocene to Recent in Israel. 2 extinct genera.
	Rhizomyidae (Bamboo rats)	(3) *Cannomys*	1	Asia (SE) Africa (East)	Late Oligocene in Europe. Late Miocene to Recent in Asia. Pleistocene to Recent in Africa; 5 extinct genera.
		Rhizomys	3		
		Tachyoryctes	14		
	Octodontidae (Octodonts)	(1) *Spalacopus* (Coruro)	1	America (South)	
	Ctenomyidae (Tuco-tucos)	(1) *Ctenomys*	26[a]	America (South)	Pliocene to Recent in South America. 4 extinct genera.
	Bathyergidae (Mole rats)	(5) *Bathyergus*	2	Africa	Oligocene in Mongolia. Early Miocene and Pleistocene to Recent in Africa.
		Cryptomys	15		
		Georychus	1		
		Heliophobius	3		
		Heterocephalus	1		

[a] Based on classical systematics.

physical and biotic structures of the underground environment. Considerations of the origin, geological history, and environmental dimensions of the subterranean ecological zone are thus essential to an understanding of the adaptive convergent evolution of subterranean mammals.

Evolutionary History and Origins of Subterranean Mammals

Evolutionary convergence and worldwide recurrent radiations of unrelated mammals into the fossorial ecological zone seem to be related causally to the evolution of open country biota in the Cenozoic due to mountain formation, extensive sea recessions, and climatic changes (207). The mid-to-late Cenozoic cycles of increasing aridity caused the development of nonforest biomes from forests, through savannas, to steppes, grasslands, and deserts. New open country environments provided for rapid evolution and diversification of an open country fauna. This included large cursorial herbivores (e.g. horses, camels, and ruminants), small fossorial sedentary herbivores (e.g. voles, pocket gophers, and mole rats), and insectivores (e.g. moles). Mammalian adaptive radiation of cursoriality and fossoriality as adaptations to open country biota emerged as early as the late Paleocene (207). However, impressive open country adaptations in rodents evolved only in the Miocene.

The Plio-Pleistocene evolutionary scenery involved drier climatic conditions followed by a decrease in woodlands and an extension of grasslands, steppes, and deserts. The increasing aridity caused massive extinctions of large ungulates and the survival of small to medium-sized xeric species. New adaptive radiations of small fossorial rodents followed, including cricetids in the Pliocene and spalacids and ctenomyids primarily in the Pleistocene.

Rodent evolution was determined chiefly by the development of evergrowing gnawing incisors. It seems to have involved three evolutionary levels or grades and at least 11 clades (that have passed independently from grade one to grade two) instead of the three suborders that were formerly recognized (167, 179, 217–219). A graphic representation of rodent phylogeny, including subterranean taxa, appears in Romer [(167), Figure 435, p. 303]. Originating from Paleocene rodents, the first grade radiated in the Eocene and involves well-developed gnawing animals with primitive mammalian jaw musculature. Grade two includes animals that have modified the jaw musculature, and grade three includes animals with very high-crowned or ever-growing cheek teeth. Burrowing adaptations originated in derivatives of the first grade during the Eocene-Oligocene (e.g. Cylindrodontidae, Tsaganomyinae, and bathyergoids). Burrowing trends proceeded independently during the Miocene-Pliocene by parallel and convergent evolution during the second (Mylagulidae and Geomyidae) and third (Spalacidae) evolutionary grades, where extreme hypsodonty was involved not only in grazing in grassland communities but also in burrowing modes of living. Extremely high crowns, as burrowing adaptations, characterize such un- related rodents as bathyergoids, spalacids, rhizomyids, and microtine crice- tids, among others.

Burrowing adaptations also evolved independently among the insec- tivores in the Talpidae (which radiated into subterranean habitats as early as Eocene times) and in the Chrysochloridae (which followed the same pattern in the early Miocene). Fossorial marsupials display parallel subter- ranean evolution in the Notoryctidae, which are unknown as fossils—the one genus and two extant species are known only from recent times (2)

The Subterranean Environment

The subterranean ecotope (211) is structurally simple. It is essentially a sealed system (45), microclimatically relatively stable (i.e. permanent and predictable), highly specialized, and presumably low in productivity [see (103, 184) for definitions of terms]. According to current theory (97, 98, 103) the greater buffering and predictability of underground as compared to overground environments, both physically (microclimate) and biotically (food supply, low predation, parapatry), should lead to a greater degree of specialization (i.e. narrow niches) in the former. The relative constancy and periodicity of physical factors result in high predictability underground (27). Predictability of temperature, relative humidity (daily or seasonal), darkness, and air currents are consistently greater in underground than in overground habitats [(39, 87, 108) and their references], primarily because of constancy [i.e. the levels characteristic of the different phases in the cycle are equal to one another; see (27)]. Other factors (such as oxygen content, range 6–21%, carbon dioxide content, range 0.5–4.8%, and at times soil moisture) display not only a greater range but often a faster shift in values than the respective surface factors do (5, 32, 87, 108, 213a). However, even the fluctuating patterns of O_2 and CO_2 may still be predictable underground as a result of equalities in cyclic phases (27)—i.e. hypoxia and hypercapnia increase underground primarily after rains or active digging (4). Likewise, the gaseous environment underground rapidly approaches equilibrium con- ditions (213a).

A Verbal Model of Adaptive Evolution of Subterranean Mammals

THE ECOLOGICAL BACKGROUND STRUCTURES The subterranean ecotope is relatively *simple, stable, specialized,* and *predictable.* It is pre- sumably *poor in productivity* and *carrying capacity,* buffered against massive predation, and *discontinuous* in spatial structure. It is essentially *fine-*

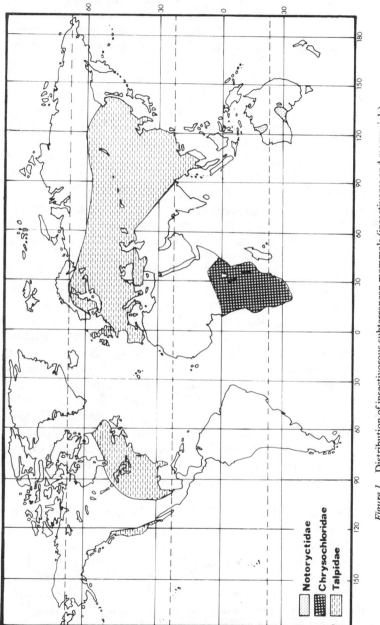

Figure 1   Distribution of insectivorous subterranean mammals (insectivores and marsupials).

grained—a mosaic of unequally distributed sparse resources in both space and time. Consequently, massive convergence ensues in the subterranean ecotope. The major evolutionary determinants of this convergence are *specialization, competition* and *isolation* both within and between species.

THE EVOLUTIONARY DETERMINANTS

Specialization Subterranean mammals adaptively converge on a variety of levels of organization in specialized patterns. Specialization involves (*a*) relatively low genetic variation (heterozygosity) [(127, 129) and this paper], (*b*) stenothermicity and stenohygrobicity, (*c*) cylindrical body, (*d*) anatomical reductions (of tail, limbs, eyes, ears, etc) and hypertrophies (acoustic and tactile sensitivities), (*e*) food generalism whether herbivorous (rodents) or insectivorous (insectivores and marsupials)—i.e. monomorphic populations of food generalists (168) eating a wide range of foods (176, 177), and (*f*) 24 hour activity patterns. Populations and species are narrow habitat specialists.

Competition Competition for similar resources may favor convergence (104). This may be particularly true for the subterranean ecotope, where resource competition (which generally takes the form of both intra- and interspecific aggression) is keen. The results of intraspecific competition are (*a*) solitariness and high territoriality generated by aggressive behavior, (*b*) low-density populations fairly constant and saturated in time near to the carrying capacity of their environments, (*c*) competitive exclusion as an extreme of MacArthur's broken stick model (90), (*d*) keen competition within and between species increasing with genetic relatedness (132), and (*e*) *K*-strategy (152, 184). *K*-selection in subterranean mammals favors great competitive ability and overall individual fitness reinforced by a relatively slow development, relatively large size, high longevity and low reproductive rate, food generalism, effective predator escape, and low recruitment and mortality rates.

The intense interspecific competition results in largely parapatric distributions, where each species is better adapted to, and more efficient in, its preferred microhabitats. The more finely distinct the competing species are genetically and ecologically, the smaller their zone of geographic or habitat overlap and the sharper their abutting ranges. This pattern supports the niche-overlap hypothesis, which predicts that maximal tolerable niche overlap should decrease with increasing intensity of competition (153). Consequently species diversity of subterranean mammals in a given area is low, and the species are distributed in specific microhabitats due to limiting resources. Territoriality within a population and parapatry or allopatry

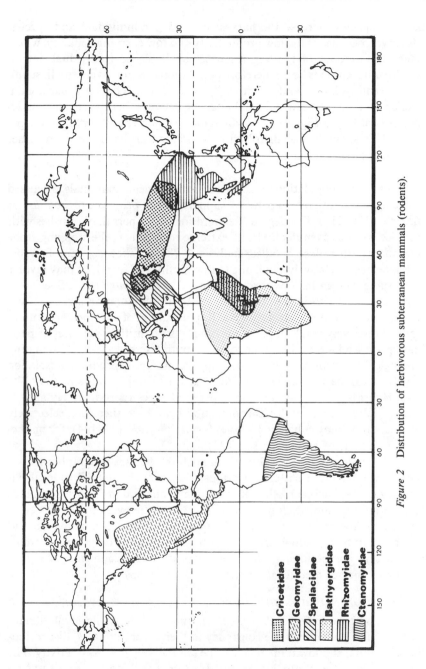

Figure 2 Distribution of herbivorous subterranean mammals (rodents).

between species increase the harvest of food per individual and species. Moreover, because of the low productivity and low carrying capacity, which reflect the amount of limiting resources available per individual, intraspecific competition is keen; therefore populations are relatively small, subdivided, semi-isolated, and territorially structured. The only substantial overlap in subterranean mammals occurs between herbivores and insectivores (Table and Figures), for which food resources are completely separated. This suggests that distribution is dictated primarily by food rather than competition for space.

Isolation   The population structure of subterranean mammals discussed above displays emigration patterns supporting the principle of *isolation by distance* (214, 215): The degree of emigration, or dispersal, diminishes with distance from a given deme. In its extreme form this is the stepping stone migration pattern [see detailed analysis in (84)]. Likewise, gene flow between demes is relatively small due to low vagility. [Exceptions may involve some subterranean insectivores and colonial *Spalacopus* (163)]. Speciation is prolific (124) and is greatly facilitated by the population structure and geographic isolation discussed above. Rapid fixation of chromosomal mutations in relatively small populations often results in the evolution of postmating reproductive isolation, which initiates speciation. Premating ethological isolation is superimposed and reinforces reproductive isolation largely in allopatry incidental to local adaptive differentiation. (68, 125, 134). Hybrids are mostly inferior to parental types reproductively and/or in their viabilities and ecological compatibility and are, therefore, selectively eliminated, thereby enhancing species formation [(93, 133, 144); for mathematical models of speciation see (122)].

The homogeneity of the subterranean habitat dictates geographic speciation. In the resulting pattern, vicarious species, each adapted to local microhabitats, replace each other in space due to continuous or discontinuous (e.g. on mountaintops) climatic variation. Since the evolving species compete for a fine-grained mixture of *similar resources* they converge and generalize in their feeding strategies on broad resource curves with extreme interspecific competition leading to competitive exclusion in space. This reinforces species identification and divergence (133).

Comparison of Cave and Burrow Ecology and Evolution

Burrow and cave ecologies (9) are similar and result in similar evolutionary patterns. They share relative constancy in their microclimates, discontinuity of habitats, and low total biomass due to relatively low productivity. The resultant shared evolutionary adaptive patterns include (*a*) structural reductions and hypertrophied sensory receptors to compensate for the loss of vision, (b) reduction in genetic variation (93), (c) stenothermicity and stenohygrobicity, (d) absence of definitive diurnal activity rhythms, (e) low population density, (f) food generalism leading to high intra- and inter-specific competition, (g) low reproductive rate (mostly K- strategies), (h) subdivided population structure and low vagility, (i) richness in sibling species (92), (j) relatively small geographic range of species, and (k) prevalence of allopatry or parapatry.

MORPHOLOGICAL ADAPTIVE STRATEGIES

General Patterns

Subterranean mammals are highly specialized and adaptively convergent structurally and functionally for burrowing and living underground in relatively permanent sealed burrow systems (39, 42, 108, 148). Structural reductions (of limbs, tail, eyes, external ears) and structural developments (of incisors, forelimbs, pectoral girdle, claws, sense organs, pineal gland), both presumably mediated by directional selection (154, 216), complement each other to optimize burrowing capacities and efficiency in subterranean existence. Similar structural reductions in the subterranean ecotope found in a variety of unrelated amphibians and reptiles emphasize their adaptive nature and origin (12). Conversely, extreme adaptive geographic variation exists in body characters and pelage color. I review these convergent and divergent adaptations only briefly since they have been elaborated in an extensive review (39) and dealt with in numerous publications either general (70) or specific for the different taxa: Notoryctidae (187), Chrysochloridae (155), Talpidae (20, 22, 159–161), Spalacidae (23, 48, 138), Geomyidae (60, 71, 139), and Bathyergidae (37, 44, 72, 82). [For diagnosis, general characters, habits, and habitats of all 10 families see (2).]

Variation

Morphologic geographic variation in a wide variety of habitats from arid tropical lowlands to alpine biota is extreme in subterranean rodents as reflected in the bewildering array of named forms. For example, nearly 300 forms of *Thomomys* (213 subspecies of *T. umbrinus* alone) and 220 of *Geomys* have been described in North America based on body and skull characters (60). Regardless of their taxonomic status and validity, whether species, subspecies, or intraspecific varieties (1), they reflect numerous ecotypic adaptations to local environments involving geographic, ecologic, and edaphic populations and races (e.g. 117). In contrast to such extreme geographic variation, some sibling species—e.g. the four karyotypes of *Spalax ehrenbergi* in Israel—are morphologically indistinguishable on about 50 body and skull characters (E. Nevo & E. Tchernov, unpublished).

Structure and Function

Body form varies from a basically rat-like (*Tachyoryctes*) to a dorso-ventrally flattened and sausage-like shape with flat head and short massive neck (*Spalax*), while tail length varies from long (Ctenomyidae, *Heterocephalus*), to short (*Notoryctes,* Geomyidae). Tails are absent in *Spalax* and Chrysochloridae. Weight and body length range from 35 g and 80 mm (*Heterocephalus glaber,* 81) to 1500 g. and 300 mm [*Bathyergus suillus* (79)] or greater maximum total length [535 mm, *Rhizomys* (2)]. Size varies with species, sex (males mostly larger than females), age, locality, and trophic level [herbivores mostly larger than insectivores (39)]. Snout varies from elongate (insectivores) to wide, flat, and cornified (utilized in pressing soil) (*Notoryctes, Spalax,* Bathyergidae). Limbs and claws are short in teeth diggers (*Spalax,* Bathyergidae), medium-length in teeth-limb diggers (Geomyidae, Ctenomyidae), and extremely large in forelimb diggers (*Notoryctes,* insectivores). Lip modifications prevent soil from entering the mouth during digging. Cheek pouches opening externally and lined with fur are unique to Geomyidae and are primarily used for carrying food (74). Teeth structures reflect phylogenetic origins and dietary specializations: in the insectivorous Notoryctidae 44 teeth, Talpidae and Chrysochloridae 34–40 teeth, and in the herbivorous rodents 12–20 teeth.

The most remarkable morphofossorial adaptations involve the structure (bones and musculature) and function of the incisors, head, neck, forelimbs and pectoral girdle, which comprise the major locomotive and burrowing structures (39, 91, 96, 159). The pelvic girdle, in contrast to epigean species, is the least developed region of the body. Digging is done primarily either by (*a*) robust, short, heavily muscled, and large-clawed forelimbs that maximize the generated force (*Notoryctes,* insectivores, *Myospalax*), (*b*) prominent, rapidly evergrowing, curved incisors [*Ellobius, Spalax,* Rhizomyidae, Bathyergidae excepting *Bathyergus,* which digs with its foreclaws (42)], or (*c*) a combination of both (Geomyidae, Ctenomyidae).

The animal clears excavated soil behind it by using both sets of limbs; it ejects soil outside by means either of forefeet [*Talpa* (54)], hindfeet [Bathyergidae (82)], turning around and pushing by headside and forefoot [*Tachyoryctes* (82), Geomyidae (174)], or by bulldozing with the head [*Spalax* (123)]. The various methods for disposing of soil may reflect phylogenetic differences. A unique burrowing procedure, in which colony members work together in relay, characterizes *Heterocephalus* (82).

Pelage is short; in hot climates (85) hairlessness in rear body parts may adaptively develop. Permanent nakedness characterizes *Heterocephalus glaber,* which lives in hot African deserts (72). Pelage color varies cryptically regionally (with the humidity index) and locally (with substrate color).

Regionally, in altitudinal color clines in *Thomomys* (51) or latitudinal ones in *Spalax* (E. Nevo, unpublished), populations are darker in mesic and lighter in xeric environments in accordance with Gloger's rule. That pelage color varies locally with substrate color (77, 85, 86) suggests differential overground predation of disharmonious types by owls and other predators.

Sense Organs

External ears are either reduced (*Tachyoryctes,* Geomyidae, Ctenomyidae) or absent (*Notoryctes,* Chrysochloridae, *Spalax*). Hearing and vocal communication, however, may be important (21). Eyes display all regressive stages from medium (*Spalacopus*), to small (Geomyidae, Bathyergidae, *Tachyoryctes, Ctenomys, Myospalax, Ellobius*), to minute (*Talpa europea*), to completely covered by skin (*Notoryctes,* Chrysochloridae, *Talpa caeca, Spalax*). Structurally, eyes vary from almost normal [*Talpa* (39)], to partly degenerated [*Tachyoryctes* (23) and Bathyergidae (24, 44, 72)], to completely degenerated [*Spalax* (23), Chrysochloridae, and ultimately *Notoryctes*]. Vision must decrease accordingly. The few critical tests available indicate that photosensitivity exists in *Tachyoryctes* (80) but not in bathyergids (44). The sense of smell and olfactory communication are well-developed (e.g. 135). The tactile sense is well-developed and permits efficient spatial orientation in the burrows (43, 53), particularly during rapid backward movement. It involves special body and tail hairs (Talpidae, Geomyidae) and forefeet vibrissae (Talpidae, Geomyidae, Ctenomyidae). *Heterocephalus* has head, limb, and genital vibrissae plus a sensory tail (72). The pineal gland synchronizes seasonal reproductive activity and affects endocrine and cellular activity. Its function is controlled photoperiodically and enhanced in darkness. Pineal structures studied in *Talpa europea* and *Spalax ehrenbergi* (151) show adaptively convergent, unique, and intensively synthesized proteinaceous substances. These are probably associated with darkness-reinforced antigonadotropic activity provoked by blindness and by the total darkness in which these subterranean mammals live.

ECOLOGICAL AND BEHAVIORAL ADAPTIVE STRATEGIES

Population Structure and Dynamics

HOME RANGE AND TERRITORIALITY   The optimal area hypothesis assumes that the home range of animals, the area they know and patrol (17), is large enough to yield an adequate supply of energy. The home ranges of subterranean mammals are generally also their exclusive and defended territories (15), except for brief periods during the breeding season when multiple occupancies by both sexes occur. This pattern is found in pocket gophers (e.g. 41, 46, 63, 64, 74, 78, 116, 117, 174, 198, 199, 212), mole rats (40, 123, 138, 170), some tuco-tucos (149), and moles (6, 54, 59, 94, 222). Territories, once established and used for one breeding season, remain fixed for life (except for minor boundary changes). Exceptions usually involve subadults living in marginal habitats (54, 74, 117). In general, territories of males or females do not overlap, whereas partial overlap occurs between male-female territories (74, 212).

In accord with the predictions of the optimal area hypothesis, the territory sizes of subterranean mammals vary with age, sex (=body size), habitat, population density (54, 74, 117), and diet. The territories of subadults are considerably smaller than those of adults, and females' territories are smaller than those of males. The size and shape of territory are more constant at high densities and more variable at low densities. Finally, insectivorous ranges are significantly larger than those of herbivorous subterranean mammals. The maximum territory size of *Talpa* was 2400 m² in England (varying between 1000 m² in low resource arable land and 400 m² in rich pasture lands) (54) and 488 m² in Holland (59). Minimal territories of *Talpa* in Germany were 2400 m² (186); ranges for some American moles are 3540 m² (6) and 10,900 m² for male and 2800 m² for female *Scalopus* (67). These territories are several times larger than those of subterranean rodents. Thus average territory size for *Spalax microphthalmus* in Russia was 150 m² (40); for *S. leucodon* in Yugoslavia, 452 m² (range: 194–1000 m²) (170); for *S. ehrenbergi* in Israel, 341 m² (range: 100–769 m²) (123); for several species of pocket gophers in Colorado, 200 m² (64) and in California, 50–200 m² (74, 78); and for the South American tuco-tuco, *Ctenomys talarum*, 50 m² (149). The larger territories of insectivores as compared to herbivores of the same size strongly implicates resources and energetics in the selection for territory size (15, 107).

Spacing patterns may shift from individual-territorial to colonial as environmental conditions associated with food density change (15). Coloniality occurs, though infrequently, in subterranean mammals—e.g. in *Ctenomys peruanus* (148), *C. minutus, Spalacopus cyanus* (163), *Cryptomys hottentotus* (49), and *Heterocephalus glaber* (72). *Heterocephalus* has also evolved an adaptive hierarchic caste structure (81). Some cases (e.g. *Heterocephalus* and *Spalacopus*) are associated with unfavorable climatic and/or resource conditions. Thus territoriality and coloniality can be viewed as responses to spatiotemporal changes of exploitable resources. Both patterns are consistent with a time-and-energy-budget model (15).

POPULATION DENSITY   Territoriality affects population density and spacing through aggressive behavior of individually selected competitors, first causing dispersion in optimal and suboptimal habitats and then, when population densities are high, preventing some individuals from breeding.

Consequently, populations become optimally spaced, stabilized, and adjusted to resource availability (15). In optimal habitats where soil and moisture conditions are uniform and food is abundant and regularly spaced, territories tend to be constant in size and distribution. However, environmental variation in soil and vegetation results in heterogeneous and variable population densities. Thus the number of individuals per acre varies within and between species and populations—e.g. *Ctenomys opimus,* 1/acre; *C. peruanus,* 17/acre (148); *Geomys bursarius,* 6.8/acre (35) and 5.5/acre (4.6–7.1) (212); *Thomomys bottae,* in rich alfalfa fields 50–60/acre, otherwise 10/acre (74), or 30/acre (76); *Geomys bursarius,* 1–15/acre, mostly 4–7/acre (33, 212); *Geomys personatus,* 1–4.6/acre (85); *Geomys breviceps,* 1.35/acre (0.67–6.8) (33); *Thomomys talpoides,* 5.8–10.3/acre (202) but in more favorable environments, 28/acre (range, 9–74) (61, 64); *Spalax microphthalmus,* 6–8/acre (40); *Spalax ehrenbergi,* 11.8/acre (range: 3.2–27.2), correlated ($P < 0.05$) with plant cover (E. Nevo, unpublished); *Cryptomys hottentotus,* 3/acre (49); *Talpa europea,* 8/acre (range: 1–20) (54, 59, 94, 186). In the Caucasus, Folitarek (in 54) reports that a two year campaign against *Talpa* produced 50,000 moles from 5000 hectares, or 3.7/acre. Trapping-out of *Ctenomys talarum* in South America (149), of *Tachyoryctes splendens* in Kenya (80), and of *Thomomys bottae* in California (76) yielded 207, 201, and 75 animals per hectare ($=10,000$ m^2), respectively. Colonial *Spalacopus* in Chile consists of contiguous subdivided populations each comprising some 15 members and occupying an area of 800 m^2 (163). Colonial *Heterocephalus* in Kenya also consists of subdivided populations comprising about 20 members each, in about 100 m^2 and separated by 80 m (72, 82). Evidently population densities are extremely variable, ranging from few individuals up to 85 (149) per acre; they reflect species variation as well as spatiotemporal changes in productivity—including those caused by modern agriculture and competition with other species (64, 65, 74, 80, 117). Thus effective population size in subterranean mammals may vary between tens and thousands.

POPULATION FLUCTUATIONS   In general, populations are stabilized by territoriality and other density-dependent factors that prevent overcrowding and safeguard against environmental damage, plagues, food shortage, or increase of predators. This is true for gophers (212), mole rats [(123), E. Nevo, unpublished] and moles (54, 59). Density fluctuations are regulated primarily by food supply and plant cover and only secondarily by predation, parasites and disease, or by density-independent and random factors (73, 74, 197).

AGE STRUCTURE   The equilibrium populations of subterranean mammals contain a relatively high proportion of breeding adults. At a stable density, the age composition of the population of the mole, *Talpa europea,* in England was 47% juveniles, 40% 1–2 year olds, and 13% 2–3 year olds. This confirms that mortality and/or emigration are high in the first year of life (54). Similarly, the analysis of *Geomys bursarius* in Texas shows an influx of subadults following the dispersal of the young. In *Cryptomys hottentotus* in Rhodesia almost a third of the population consisted of large breeding individuals (49). In *Tachyoryctes splendens* in Kenya from a trapped-out garden sample of 201 animals, 67% were adults and only 33% subadults (80). In *Spalax leucodon* in Yugoslavia one population consisted chiefly of mature individuals from August onwards (170). In *S. ehrenbergi* in Israel, of 386 animals caught from 1967–1978, 19% were juveniles, 40% 1–2 years old, and 41% 2–3 years old (E. Nevo, unpublished). In *Thomomys monticola* in the autumn in California, 33% were juveniles; of the adults, 20–31% were males, 38–44% females (78). In colonial *Spalacopus* in Chile, of a 15-member colony, 4 were juveniles and the remainder adults (163).

SEX STRUCTURE   Adult sex ratios (expressed below as male/female ratios) unbalanced in favor of females have been commonly recorded, particularly in subterranean herbivores—e.g. *Ctenomys peruanus,* 0.45 (N = 71), but 0.5 in sparse populations (148); *Geomys bursarius,* 0.33 (N = 1218) (199); *Thomomys bottae,* 0.25 in highly dense populations, but 0.5 in sparse populations (74); *Spalax ehrenbergi,* 0.39 (N = 1091) (E. Nevo, unpublished); *S. leucodon,* 0.45 (N = 672) (170); *Geomys breviceps,* 0.40 (N = 585) (220); *Geomys bursarius,* 0.43 (N = 1218) (199); *Pappogeomys castanops,* 0.23 (N = 77) and *Tomomys bottae,* 0.32 (N = 122) (213); *Ctenomys opimus,* 0.37 (N = 94) (148); and *Tachyoryctes splendens,* 0.42 (N = 803) (80). Reverse trends are also recorded—e.g. *Cryptomys hottentotus,* 0.56 (N = 87) (49). The sex ratios in moles are apparently in favor of males— e.g. in *Talpa europea,* 0.56 (N = 1617) (54) and *Scalopus aquaticus,* 0.66 (N = 800) (222).

Sex ratio in herbivorous subterranean mammals appears to be density-dependent. In *Ctenomys peruanus,* females are nonterritorial in contrast to males; in *Spalax,* while both sexes are territorial, males are more aggressive than females (132). Thus, the higher territorial aggression in males may cause sexually differential mortality. Several studies indicate a sex ratio closer to 1:1 in subadults than in adults (80, 123, 220), suggesting that the unbalance increases with time. The reverse trends in moles may be related to the relatively greater activity of males—hence their higher trapping proportion (54).

POPULATION SUBDIVISION   Population size, structure, and distribution have substantial correlates to both the phyletic past and trends in speciation

(106, 214, 215). Theory predicts that evolutionary differentiation is faster in a subdivided population with varying sizes and degrees of isolation of local demes than in a homogeneous population of comparable total size, owing to local differential selective pressures and gene flows (84,214). Populations of subterranean herbivores often conform to the "island model" type of distribution (215)—e.g. *Cratogeomys castanops* and *Thomomys bottae* (64), and *Geomys bursarius* and *G. personatus* (34, 85). In *Thomomys bottae* (213), groups were isolated from each other and the main population without exchange of individuals between these groups, a pattern characterizing *Thomomys* populations (144, 188, 201). This pattern, however, seems to be rare in central populations of *Spalax* in Israel (E. Nevo, unpublished) and in subterranean insectivores (222), where continuous population structures may abound.

Life History Patterns and Population Parameters

ECOTOPE STABILITY   The relatively stable and predictable subterranean ecotope presumably leads to convergent K-selection in subterranean mammals. The latter comprise "equilibrium species"—i.e. those at or near the carrying capacity of the environment, selected for their ability to harvest food efficiently through resource competition but without overshoot and resource destruction. Even when perturbed they return quickly to equilibrium. They display the expected K-correlated parameters (152), involving (*a*) equilibrium numbers near carrying capacity; (*b*) fast return time to equilibrium, and (*c*) slow growth rate (99, 184).

Equilibrium populations in subterranean mammals are achieved by an optimized gain-loss balance based on the following: *gain factors*—maximizing breeding age, duration of breeding season, and minimizing offspring number, fraction of breeding population and immigration rate; and *loss factors*—minimizing mortality rate, emigration, and predation. Most overground mammals of similar size have higher indexes of reproductive efforts than do subterranean mammals (80, 114).

GAINS   Offspring number, a prime adaptive feature of mammalian reproduction (114), is very low in both herbivorous and insectivorous subterranean mammals. In general, they start breeding during their second year and have only one breeding season lasting about 4–5 months (range: 2–8). They usually produce one small litter per year averaging 3 young (range: 1–6, N = 19 species of several families). This pattern has been found in *Thomomys, Geomys* (31, 62, 65, 74, 199, 212, 220), *Spalax* (40, 123, 138, 170), *Bathyergus, Tachyoryctes* (79, but see also 80), *Ctenomys* (148), Talpidae (6, 28, 54, 59, 222), *Notoryctes,* and Chrysochloridae (2). Furthermore, many females are presumably excluded from breeding in equilibrium populations. These can boost the birth rate if population numbers fall [e.g. in *Spalax microphthalmus* (40) and in *Geomys breviceps* (46)].

Reproductive rate in subterranean mammals, as in other mammals (114), varies with local climates and habitat productivities. For example, *Geomys bursarius* in Colorado has one litter with 3.4 young per year, and females breed the year after birth (199); in Texas it has a protracted 8 months breeding season with two litters averaging 2.5 young each, and females may breed in the year they are born (212). At different altitudes *Thomomys talpoides* in Colorado has the following litter sizes: 4.5 ± 0.4, 4.4 ± 0.2, 4.8 ± 0.3 in alpine tundra, subalpine meadow, and shrub-grassland communities, respectively (62). In California foothills *Thomomys bottae* breeds once from January to May with a mean of 4.6 young per litter (74), whereas in irrigated fields, apparently due to high productivity, it breeds almost the year around, and has 1.5–2.5 litters per female with a mean of 4.92 young per litter (115). Similarly, in moles mean litter size is lowest in Britain [3.8 ± 0.13 (54)], medium in Germany [4.54 ± 0.14 (186)] and Holland [4.6 ± 0.23 (59)], and highest in Russia [5.08 ± 0.39 in the Urals, and 5.73 ± 0.06 in the Ukraine (54, 59)]. The extreme adaptation in subterranean mammals to minimize offspring and maximize their genetic fitness through a complex hierarchic social structure was found in *Heterocephalus glaber* as an adaptation to the semi-desert environments in East Africa (81). Noteworthy is the fact that both *Tachyoryctes* and *Heterocephalus* stop breeding completely in adverse external conditions (79).

LOSSES

Prenatal and litter mortality High intra-uterine mortality and embryo reabsorption were recorded in *Talpa europea* [6–9% (94), and 3% (54)], *Heterocephalus glaber* (79), *Tachyoryctes splendens* [up to 50% (80)], *Thomomys bottae* (115), and *Geomys bursarius* (199). Litter mortality also contributes to losses [e.g. in moles (54)].

Predation, disease, and extreme weather The population density of small mammals is partly reduced by predators and diseases. Predators include birds (e.g. owls, hawks, crows, herons, storks), snakes, and mammals (e.g. foxes, weasels, badgers, cats, dogs, and humans). Predation has been recorded or suspected in moles (53, 54, 59, 94, 186), mole rats (40, 72, 80, 123, 138, 170), *Notoryctes* (2), and pocket gophers (31, 46, 65, 74, 212). Also, predation may not be only density-proportional but may become density-dependent in gophers (74). The subterranean ecotope is relatively sheltered from predators; the main toll is taken during subadult dispersal. Diseases and parasites may also be implicated in regulating population densities (30, 64, 72, 119, 170, 172, 208). Finally, mortality also is caused by extreme weather conditions (e.g. 54, 59, 64, 78, 123, 213). It is noteworthy that mortality factors such as parasites, diseases, territorial fighting, and food shortage are density-dependent (e.g. 80). The low annual recruitment of subterranean and fossorial mammals contrasts strongly with that of nonfossorial forms (80).

OPTIMAL DISPERSAL THEORY   Optimal dispersal theory assumes the existence and uniqueness of an optimal strategy of division of progeny into dispersers and nondispersers, presumably maximizing fitness in either the short or long run (101, 120). As expected, the optimal proportion of dispersers is inversely related to both risks of movement and the proportion of vacant territories. Except during environmental perturbations (e.g. 73), populations of subterranean mammals are in equilibrium with their habitat's carrying capacity. Consequently, yearly emigration of surplus subadults must complement territoriality and mortality as a density-regulatory mechanism. Equilibrium densities may be established by both pre-saturation and saturation dispersal (101). In subterranean mammals, only a little is known about the former (e.g. 54, 200,213) but quite a lot about the latter (primarily juvenile) dispersal. Dispersal of subadults into new territories usually takes place one to two months after birth, shortly after weaning.

Dispersal distance tends to be limited in subterranean mammals, which reduces gene flow between populations in accordance with physical (soil type, depth, and moisture conditions) and biotic (competitors, predators, resources) factors (6, 54, 59, 117). Furthermore, philopatry or homing [described in both moles (186) and pocket gophers (74)] may also restrict dispersal. Nevertheless, precise quantitative studies indicate various vagility levels in different species, ages, and sexes of Geomyidae (distance traveled ranging from mean 8 m to 257 m) (213). Similarly, introduced *Thomomys bottae* and *T. talpoides* moved in one year mean 60 m and 237 m and maximal 273 m and 784 m, respectively (200). Likewise, a two-year observation in Texas has revealed range expansion and contraction of 300 m and 160 m in *Pappogeomys castanops* and *Thomomys bottae*, respectively (162). Finally, swimming (69) may enhance vagility.

After territorial establishment most subterranean mammals remain sedentary throughout their lifetimes; except for small boundary changes, hardly any adult movement occurs (54, 74). Habitat stability for any animal is conveniently expressed by the ratio of generation time to the time the habitat remains suitable for food harvesting (184). This ratio is small in subterranean mammals, in which generation time is usually one year and burrows may be used throughout a lifetime averaging three years. This is true for moles (54), pocket gophers (64, 78, 86, 212), mole rats (123, 170), and tuco-tucos (149).

Behavioral Structure and Dynamics

BURROW STRUCTURE   Burrows of subterranean mammals consist of a radial or longitudinal superficial network of feeding tunnels (80% of the total volume) (116) connected to a frequently deeper [or upper in flooded regions (123)] central permanent systems of chambers used for nesting, food storage, sanitation (123), and retreat (54, 82, 174). The closed burrows provide (*a*) a relatively constant microclimate, (*b*) protection from predators and competitors, and (*c*) access to food (74, 87, 212). Tunnels are plugged to keep predators out, while new ones may be constantly dug and repaired in search of food (212). Activity depth is lowered during the hot and dry season to an optimal temperature-moisture regime (39, 82, 123). Deserted burrows are rapidly invaded (74, 212). Male gopher burrows are larger than those of females (86, 212), but the opposite is true in bathyergids and *Tachyoryctes* (82). Burrow dimensions vary with the species, individual, sex, age, and local habitat conditions [(46, 85, 116) for gophers; (123, 138, 170) for *Spalax;* (82) for Bathyergidae]. Minimal and maximal burrow dimensions are 30–250 m total length; 10–40 cm depth of feeding tunnels; 400 cm for maximal depth of deep tunnel in *Spalax* (170), 3–10cm diameter, 10–200 mounds per individual burrow, and up to 3 tons soil displaced. Burrows have been mapped for *Talpa* (53, 54, 59), American mole (6), *Spalax* (123, 138, 170), *Tachyoryctes* (82), *Geomys* (174), *Thomomys* (31, 116, 204), *Heterocephalus* (72, 82), *Cryptomys* (49, 82), *Heliophobius* (82), *Bathyergus* (37), *Spalacopus* (163), and *Ctenomys* (39, 148).

FEEDING STRATEGIES   Optimal foraging theory assumes that natural selection maximizes fitness by optimizing net energy gained per unit feeding time (156, 177). The predictions of optimal foraging theory are testable by analyzing four key aspects of feeding strategies: (*a*) the optimal diet, (*b*) the optimal foraging space, (*c*) the optimal foraging period, and (*d*) the optimal foraging group size. These aspects can be analyzed individually or (preferably) in combination. A further refinement is found in the "central place foraging" model (140), which describes an animal's exit from and return to a central place during foraging. This model may be most applicable to the sedentary subterranean mammals, the majority of those discussed here.

Optimal diet   Theory predicts that in a fine-grain situation (*a*) animals should never specialize on a lower-rank food type regardless of its abundance, (*b*) increasing food abundance should lead to greater food specialization, and (*c*) animals should never exhibit partial preferences unless dietary constraints or random variation in food abundances occurs (156). The data available for subterranean mammals cannot be used to test precisely the optimal-diet predictions. However, some qualitative predictions, such as positive correlation between increased selectivity and high food abundance, are supported.

Theory predicts that at low food abundances food generalists are favored over food specialists (176, 177). Most subterranean mammals are generalized feeders on a wide range and variance of food types. This applies to the insectivorous forms, *Notoryctes* (29, 75), *Talpa* (54, 59), *Scalopus* (181, 222), and other American moles (6), as well as to the herbivorous rodents, *Thomomys* (31, 64, 65, 74, 117, 201, 202), *Geomys* (46, 64, 117, 212), *Heterocephalus* (72, 81), *Spalax* (40, 123, 138, 170), *Ctenomys* (148), *Cryptomys* (49), and *Tachyoryctes* (82). Food generalism in subterranean mammals seems to be related primarily to the low net energy harvestable in the subterranean ecotope.

The theoretical prediction that increasing food abundance or nutritional rank leads to greater food specialization is borne out in several cases of subterranean mammals (39, 54, 64, 80, 181). For example, in habitats consisting of 50% grasses, 42% forbs, and 8% shrubs, the summer diet of *T. talpoides* was 6% grasses, 93% forbs, and 1% shrubs (206). Furthermore, in different habitats, 50% of the diet consisted of either *Opuntia* (201) or *Lupinus* alone (202). In cultivated fields, food storage reflects specialization related to local crops—e.g. alfalfa in *Thomomys bottae* stores (116) and potatoes or carrots in stores of *Spalax ehrenbergi* (123). Likewise, in the insectivore *Talpa*, specialization increases with soil productivity. Earthworms are the most important food, particularly in rich humid soils (54, 180).

Subterranean mammals eat a wide variety of plant and animal species even when their preferred foods are abundant. Thus, *Thomomys talpoides* consumed 21 species of forbs, 9 grasses, and one shrub despite the prevalence of five forb species in their diets (206). Similar patterns occur in the insectivore *Talpa* (6, 53, 54, 59, 94, 180). *Talpa* is a food generalist: examined stomachs contain earthworms, insects, myriapods, and molluscs. In the stomachs of a Suffolk, England, population, representatives of 23 invertebrate families and 37 genera were found. In general, over 40 families of insects have been found among the stomach contents of European moles. Prey length varies from a few millimeters (ants) to many centimeters (earthworms). Regional, seasonal, and yearly dietary differences reflect local habitat and climatic variations (53, 54, 64, 123, 170, 202).

Optimal foraging space   Theory predicts that fitness can be maximized by an optimal foraging space—i.e. optimal home range size and structure, patch selection, foraging path, and exclusiveness of feeding area (177). Predictive models of home range size based on energy requirements, food density, and selectivity (107) are partly supported in some lizards, birds, and mammals (177). All show increased home range or territory size with increased body weight (hence with energy requirements). For mammals of similar size, herbivores have smaller home ranges than carnivores (177). The average home range of a male *Scalopus aquaticus* (a subterranean insectivore) is almost 23 times as large as that of a male *Geomys bursarius* and 42 times as large as that of a male *Thomomys bottae* (both subterranean herbivores) (222). The large mole territories are explicable by the moles' high energy demands. A *Talpa* weighing 100 g consumes daily 40–70 g (53, 54).

Optimal foraging space theory was tested in the subterranean mammal *Thomomys bottae* by analyzing (*a*) the energy cost of burrowing as a foraging technique and (*b*) the way this cost varies with burrow structure. Vleck (204) presented a model predicting the cost of burrowing as a function of burrow geometry. The model (based on soil type, burrowing efficiency, and burrow geometry) permits calculation of the energy expenditure during construction of the burrow system. *Thomomys bottae* was found to minimize the energy expended per meter burrowed and thereby to maximize foraging efficiency. Since energy cost for a given length of tunnel increases with cross-sectional area, larger subterranean mammals have lower foraging efficiencies than small ones (204). Hence body size is a critical factor in adaptive radiation underground. The generality of the model depends on confirmation from other subterranean mammals.

Optimal feeding period   Endogenous rhythmicity appears to be synchronized with, and selected by, the environment as an adaptive strategy increasing fitness (47). Subterranean mammals, both herbivores and insectivores, are active day and night, and their activity is linked with feeding periods. This is true for *Geomys* (203), *Talpa, Parascalopus* (54), and *Scalopus* (67); in some species daytime activity is either greater [e.g. *Spalax* (171)], preferable [e.g. *Eremitalpa, Ellobius* (39)], distinct [e.g. *Tachyoryctes,* but little less so in *Heliophobius* (79a)], or absolute [e.g. *Ctenomys* (148) and *Heterocephalus* (72)]. Extensive standardized laboratory tests of 98 *Spalax ehrenbergi* showed polyphasic activity patterns higher during daytime then at night and varying significantly among karyotypes (136b). Activity patterns in subterranean mammals vary with the species, individual, day, season, and habitat. Significantly higher activity was recorded for *Talpa* in poor sandy arable fields as compared with rich fen pastures in England (54). Subterranean mammals, both herbivores and insectivores, forage and store food extensively in anticipation of unfavorable feeding conditions [i.e. (31, 39, 40, 46, 49, 64, 81, 123, 163, 170, 180), but see exceptions in Bathyergidae (82)]. The generality of environmental correlates to optimality of feeding periods in subterranean mammals awaits explanation and testing.

Optimal foraging group Sociality in subterranean mammals (discussed earlier) is at least partly correlated with ecologically marginal environments. The extreme case is that of *Heterocephalus glaber* (81) living in hot, arid, low-resource semi-deserts in east Africa. It has a highly differentiated social structure involving small worker and nonworker classes and a dominant breeding female. Jarvis (81) suggested that the cost of burrowing can be shared by the workers and the chance of locating food enhanced by having more than one animal searching the area.

PHYSIOLOGICAL ADAPTIVE STRATEGIES

Energetics: Metabolism and Temperature Regulation

The energetics of endotherms is influenced by body size, climate, and food habits (110, 112); hence it is adaptive (111). Low basal metabolic rates (BMR) are shared by desert species and species experiencing poor, periodic, and/or low-energy-content foods. On the other hand, high thermal conductances characterize tropical endotherms (110). The unique subterranean ecotope is characterized by poor and periodic food supply and by atmospheres saturated with moisture.

Consequently, most subterranean rodents share a linkage of low BMRs, high thermal conductances, high ranges of thermoneutrality, low body temperatures (35°–37°C), and relatively poor thermoregulation. This syndrome is adaptive since it may reduce heat storage and water exchange (108, 113) and minimize energy expenditure (204). This pattern has been found in *Spalax* (55, 108, 131, 170), *Geomys, Tachyoryctes, Heliophobius* (108), *Ctenomys* (163), *Cannomys* (108, 112), *Thomomys* (50, 204), and *Geomys* (13). These trends are exaggerated at high burrow temperatures; they culminate in *Heterocephalus* (81, 108). Low basal rates may also reduce gas exchange in the hypoxic hypercapnic burrow conditions (32). However, high basal rates have been reported for *Thomomys talpoides* (50). McNab (112) has recently suggested that basal rates higher than expected from the Kleiber relation can occur in species larger than 80 gr that live in cool to cold burrows.

In contrast, subterranean insectivores (e.g. *Scalopus aquaticus*) have high basal rates and low conductances. These may correlate with their carnivory and may relate to the high cost of predation or to the temporal dependability on soil invertebrates as a food resource. These high rates require a body mass smaller than that of subterranean rodents to reduce overheating in close burrow systems (112).

The BMR varies in accord with environmental variation both between and within species. The BMR of three chromosomal species of *Spalax ehrenbergi* (with diploid numbers at 2n = 52, 58, 60) decrease progressively and significantly toward the desert (131). This physiological cline along an increasingly arid gradient suggests that the BMR is adapted both to subterranean microclimates and to the macroclimate aridity index.

The reduction in BMR and the increase in conductance in subterranean mammals are directly related to mean burrow temperature. These adaptations reduce overheating in the closed burrow system where evaporative and convective cooling are greatly reduced (108, 112). *Spalax* stretches over the substrate and spreads saliva on its body (E. Nevo, unpublished). *Tachyoryctes* and *Heliophobius* use forced evaporative cooling for emergency thermoregulation in spite of the saturated burrow atmosphere. *Heterocephalus* facilitates heat loss during heat stress by increasing peripheral circulation over its naked body. *Geomys* does this by increasing the circulation to its naked tail and, possibly, to its naked feet. *Geomys* may lose up to 30% of its heat production through its tail. Consequently, the lethal ambient temperature is proportional to the mean burrow temperature (108).

Body size of subterranean mammals depends on the productivity of the habitat (33, 86, 204), the cost of burrowing (204), and thermoregulation (108, 112). Size decreases in response to geographic variation in heat loading. A positive correlation of weight with latitude (Bergmann's rule) was found in *Spalax ehrenbergi* (E. Nevo and E. Tchernov, unpublished), of which dwarf populations occur in the northern Negev desert in Israel, and in *Geomys bursarius* and *Scalopus aquaticus* in the United States. [Significant correlations for size and latitude were 0.83 and 0.53, respectively (108).] A negative correlation of body-extremity size with latitude (Allen's rule) was found in *Geomys bursarius,* which responds to heat loading by an increase in tail length (109). Finally, the extreme adaptive response to heat loading is found in *Heterocephalus glaber,* which is not only small and naked but also has the lowest body temperature (about 32°C) and the poorest capacity for thermoregulation of any known mammal (108). These extreme adaptations not only reduce overheating but also reduce the daily energy budget in a harsh environment where food resources are limited (81, 113). The sexual difference in size of subterranean mammals, males being mostly larger than nonbreeding females, has likewise been explained on the basis of heat flux (108).

Respiration: Blood-Gas Properties and Function

Subterranean mammals must adapt not only to the unique problems of food and heat energetics (113) but also to the extreme hypoxia and hypercapnia they may encounter in their sealed burrows due to the oxygen consumption and the limited permeability to gases of the soil barrier (4, 32, 87, 108, 213a). The following description of the extreme respiratory adaptations in subterranean mammals is based mostly on the extensive work on *Spalax ehren-*

bergi (3–5), complemented by findings from other subterranean mammals whenever available (7, 10, 25, 32, 95, 157).

The metabolism of *Spalax* was found to be relatively unaffected by hypoxic-hypercapnic conditions. Critical partial pressure of oxygen (Po_2) for *Spalax* is lower than for any other rodent studied so far. In their ability to withstand hypoxic conditions they exceed even known mammals of the highest altitudes. The adaptive mechanisms to cope with hypoxic-hypercapnic conditions involve facilitation of gas transport from inspired air to blood by changes in respiration and heart rate, blood properties that enable efficient gas transport, tissues that remain optimally functional by a high loading-unloading saturation difference, and high muscular myoglobin content that facilitates diffusion.

GAS TRANSPORT   Gas transport from inspired air to blood is facilitated under hypoxic-hypercapnic conditions by the following mechanisms: (*a*) Normoxic respiratory frequency in both *Spalax* and *Talpa* is lower 40% (4, 157) than the expected (185); (*b*) *Spalax,* like pocket gophers, display increased ventilation during hypercapnia; (*c*) response threshold of ventilation to hypoxia in *Spalax* is very low compared to other mammals, and ventilation rises steeply at a very low Po_2; (*d*) heart rate of *Spalax* (but not its weight) is ⅓ of the expected value (185) and is arrhythmic. In hypoxia, the heart rate of *Spalax* rises up to 230% at 41 and 32 torr O_2, almost twice as much as that of the white rat under the same conditions. Arrhythmic and lower-than-expected heart rate characterizes *Talpa* as well (7). Low heart rate is permitted by high gas transport of the blood and may be considered an adaptive convergence of subterranean mammals to hypoxic conditions, its function being to keep the potential of the cardiovascular system during hypoxia or stress when high oxygen demand is critical (4). The heart of the subterranean mammal may be more resistant to respiratory acidosis than that of the rat (16) and may therefore better withstand hypercapnia.

BLOOD PROPERTIES   Blood properties in *Spalax* (3) involve high erythrocyte count and small corpuscular volume; both characteristics facilitate O_2 exchange, as in the high altitude llama. *Spalax* shares with other fossorial mammals high O_2 affinity of blood, which facilitates oxygen loading in hypoxia. However, *Spalax* maintains wide O_2 dissociation curves and wide pH range in blood at the expense of acid-base regulation; pocket gophers (25, 95), on the other hand, maintain a narrow pH range in blood at the expense of O_2 supply. In *Spalax,* 2–3 diphosphoglycerate : hemoglobin ratio adapts blood O_2 dissociation curves to burrow atmospheres (3). A drop in this ratio is associated with increased O_2 affinity, or better loading; increased Bohr effect, or better unloading; and probably increased

Haldane effect, or better CO_2 transport. In *Heterocephalus*, the high O_2 blood affinity and low variable body temperature relate to hemoglobin proper (83).

TISSUE PROPERTIES; MYOGLOBIN   The tissues of subterranean mammals are functionally active at a constant low Po_2 and high Pco_2. Both *Spalax* (3, 4) and pocket gophers (95) have high skeletal-muscle myoglobin content as in high-altitude and marine mammals. The high myoglobin content may facilitate O_2 diffusion from the capillaries to the mitochondria and may represent convergent physiological adaptation to hypoxic conditions both in high altitudes and underground.

In sum, respiratory adaptations are essential to subterranean life. While some solutions are convergent in pocket gophers (32), mole rats (4), moles (10), and other burrowing mammals [e.g. armadillo, prairie dog, and echidna (3)], others are divergent.

GENETIC ADAPTIVE STRATEGIES

Chromosomal Evolution

PATTERNS   Chromosomal evolution through major structural rearrangements may lead either to speciation through reproductive isolation caused by hybrid sterility and extensive homozygous fixation and/or to local adaptive genetic polymorphism through reorganization of new coadapted supergenes (209, 210). Chromosomal sibling species, based on Robertsonian whole-arm rearrangements and/or pericentric inversions and reciprocal translocations, are widespread in unrelated subterranean herbivores.

To date over 30 karyotypes have been described for mole rats *Spalax* ($2n = 38-62$; $FN = 74-120$) from Russia, the Balkans, Asia Minor, Israel, and North Africa (102, 158, 173, 183, 205); three karyotypes for *Myospalax* ($2n = 44-64$; $FN = 80-108$) from Russia (102, 105); six karyotypes for the microtine cricetid *Ellobius* ($2n = 17-54$; $FN = 54-56$) in Iran and Russia (102); over 20 karyotypes, largely reproductively isolated from one another, in pocket gophers, *Thomomys talpoides*, from the southern Rockies (188, 192–194), for *Geomys* ($2n = 38-72$; $FN = 68-102$) in central and southeastern United States (8, 36, 66, 88), for *Pappogeomys* ($2n = 36-46$; $FN = 66-86$) in Mexico and the United States (11); and in the highly polytypic *Ctenomys* about 11 karyotypes have been described, out of the 60 known species in South America ($2n = 22-68$; $FN = 44-122$) (164). In contrast, colonial *Spalacopus cyanus* ($2n = 58$, $FN = 116$) in Central Chile displays chromosome uniformity and low taxonomic diversification (163, 165).

Intraspecific chromosomal polymorphisms within and between populations involving inversions, translocations, or added/deleted blocks of heterochromatin etc as well as chromosome number caused by Robertsonian changes are also prevalent in subterranean rodents. For example, over 40 different karyotypes, largely between populations, with increasing proportions of acrocentrics eastward and southward, have been recorded in the widely ranging *Thomomys bottae* in the western United States (141, 142, 144–146). This extreme polytypic and polymorphic karyotypic diversity lies not in diploid number (2n = 76) but almost totally in chromosome morphology consisting of arm-number changes involving whole-arm additions/deletions of constitutive heterochromatin (146). Similar patterns occur in other members of the "heavy-rostrum" group including *T. umbrinus* and *T. townsendii* (2n = 74–78) (144, 145, 192). In contrast, the "slender rostrum" pocket gophers, including *Thomomys talpoides, T. monticola,* and *T. mazama* have distinct diversity of diploid numbers (2n = 40–60) (188, 190, 192, 194). Accordingly, extensive apparently balanced chromosome polymorphism, based on Robertsonian changes (2n = 48–51), was found in *T. talpoides agrestis* in southern Colorado, which also exhibits a regional trend of increase in 2n southwards (195). Similarly, intrapopulation chromosomal polymorphism involving 2n = 70, 71, 72 was found in *Geomys bursarius* from New Mexico and West Texas (8, 36, 66) and in *G. personatus* (36) being apparently maintained, as in the other cases, by selective advantage. Two species of *Ctenomys* (164) and at least 2n = 54 in *Spalax ehrenbergi* (205) exhibit intrapopulational polymorphism, presumably for pericentric inversions.

In contrast to the extreme chromosomal diversity of subterranean rodents, the Talpidae, representing most subterranean insectivores, exhibit a striking chromosome uniformity. With the exception of *Talpa caeca* (2n = 36), *Mogera insularis* (2n = 32), and *Neurotrichus gibbsii* (2n = 38), all species have a 2n = 34; with the exception of *Parascalops breweri, Talpa romana* FN = 62), and *Neurotrichus gibbsii* (FN = 72), all species have FN = 64 (58, 196, 221). These data support the hypothesis that pericentric inversions and translocations, rather than Robertsonian changes, have been important in talpid karyotypic evolution (58).

ADAPTIVE NATURE   Adaptive chromosomal variation involving chromosome number and morphology has been postulated in subterranean mammals based on clinal patterns and ecological correlates of physical and biotic factors. In *Spalax ehrenbergi* in Israel, diploid number increases clinally southwards with increasing aridity (2n = 52 → 58 → 60) (205) paralleled by an adaptive physiological cline of decreasing basic metabolic rates (131). Similarly, in *S. micropthalmus* diploid numbers of 2n = 60 and 62 are found widely in the Russian steppes (102). In *Thomomys umbrinus* 2n = 78 is found largely from low to middle elevations in Mexico and Arizona, whereas 2n = 76 occurs only in Mexican boreal forests (147). Likewise, *Thomomys bottae* in Arizona displays an altitudinal cline in chromosome polymorphism strongly correlated with soil and vegetation types and moisture gradient (141). The extreme karyotypic diversity of *T. bottae* across its vast ecological regime suggests the local adaptability and viability of heterozygotes not impaired by meiotic load. This pattern is correlated with historical events coupled with the uniqueness of the subterranean ecotope (146). Furthermore, chromosomal variation is maximal in largely isolated insular montane populations but minimal in adjacent plains populations inhabiting similar ecological zones (142). Similar patterns occur in *Spalax* (102, 183). Variation in both chromosome number and morphology may be adaptive by varying the recombination index and modifying linkage groups via rearrangement of the genome (131, 143).

Genic Evolution

PATTERNS   The environmental theory of genetic variation, or niche-width variation hypothesis, which assumes that environmental heterogeneity is a major factor in maintaining and structuring genetic variation in natural populations, seems to be supported by the evidence presented in a recent review (127). Levels of polymorphism (P) and heterozygosity (H) are positively correlated with ecological heterogeneity. Thus habitat generalists are significantly more polymorphic and heterozygous than habitat specialists.

Living in a relatively stable, simple, and specific environment, subterranean mammals are habitat specialists; they are expected to exhibit generally lower levels of P and H than overground species of similar size living apparently under more fluctuating and complex conditions. This prediction has been largely confirmed in fossorial mammals despite contrary assertions [(146, 147, 178); see Table 1 in (127)] and individual exceptions (e.g. 146). Comparative estimates of P and H for subterranean, fossorial, and other small rodents is certainly unsatisfactory at present both in terms of species and loci compared; and, further critical testing, particularly of subterranean insectivores and of small mammals living under fluctuating environments, is crucial for theory. The following comparison is therefore tentative. I categorized 49 mainland species of small mammals into subterranean, fossorial, and others and compared their H values. For most H estimates see Table 1, pp. 139–41 in (127) and Table 1 in (182). I have added H estimates for *Mus musculus brevirostris* (14) and mainland *Rattus fuscipes* (175), revised estimates of *Spalax* after (136), adjusted values of *Dipodomys* and

Sigmodon for absence of esterase loci, (see 178) and added recent estimates for *Thomomys bottae* (146) and *T. umbrinus* (147). Mean H and standard deviations are: (*a*) *Subterranean* (S), 17 species of *Spalax, Thomomys, Geomys, H* = 0.045, s.d. 0.021; (*b*) *Fossorial* (F), 13 species of *Dipodomys, Acomys, H* = 0.032, s.d. 0.031; (*c*) *Subterranean + Fossorial* (S + F), 30 species, H = 0.039, s.d. 0.026; (*d*) *Other small rodents* (O), 19 species of *Mus, Rattus, Peromyscus,* and *Sigmodon, H* = 0.060, s.d. 0.033. t-tests between all three pairs (with 46 d.f.) are: (S-F), t = 1.22, p $\cong$ 0.20; (S-O), t = 1.52, p $\cong$ 0.15; (F-O), t = 2.68, p $\cong$ 0.01; for [(S + F)–O] t = 2.40, p $\cong$ 0.02. H of the combined subterranean and fossorial rodents, is significantly lower than H of other small mammals.

Heterozygosity is low in a variety of unrelated fossorial and cave taxa when compared to their epigean relatives, including amphisbaenians and lizards (56), skinks (89), frogs (127), insects such as *Gryllotalpa,* landsnails species of *Buliminus* and *Sphincterochila,* and habitat specialist species in general (127); and cave species such as fish [Avise & Selander, 1972 in (127)], crickets (26), beetles (93) and spiders [Johnston & Carmody in (93)]. Exceptions with relatively high H occur both in subterranean rodents (e.g. *Thomomys bottae*) and in cave beetles (52).

ADAPTIVE AND STOCHASTIC PATTERNS   Genetic variation may be structured by both deterministic and stochastic processes and the problem of data interpretation suffers from lack of critical experiments that can either unequivocally exclude or recombine hypotheses. Two current non-mutually-exclusive hypotheses explain low variation in fossorial forms: (*a*) the niche-width variation hypothesis [(127) and references therein]; and (*b*) the population origin, size, and structure hypotheses, all predicting positive correlation between effective population size, N_e, and genetic variation (56). Consequently species differences in H cannot be used to test the niche-width variation hypothesis unless effective population size and structure, as well as time (T) since last bottlenecking and/or origin are controlled, because H may also be a function of N_e and T (121).

Indeed since subterranean mammals are sedentary and frequently subdivided into relatively small, isolated, inbred populations, their genetic structure has been largely attributed to stochastic factors such as founder effect (178), random genetic drift (150), and high inbreeding with little gene flow (E. G. Zimmerman, N. A. Gayden, unpublished) or, alternatively, to lack of severe historical bottlenecking and high gene flow affecting population density and structure (146, 147), despite the apparent low vagility of fossorial rodents and regardless of whether or not allozymes are structured by natural selection. Unfortunately, N_e, T, and migration rate are largely unknown. Moreover, N_e is often related to ecological predictability and thus both N_e and H may be causally related not only to stochastic events but also to ecological heterogeneity and uncertainty. Therefore, total exclusion between the environmental and population hypotheses may be neither theoretically justifiable nor practically feasible. Nevertheless, critical testing might indicate their relative contributions.

Genetic population structure in the recently evolving *Spalax ehrenbergi* (128, 136) and *Thomomys talpoides* (129) complexes is determined primarily by ecological correlates and apparently not by N_e or T (127). First, H is relatively low in both unrelated complexes in accord with the niche-width variation hypothesis. Second, the genic similarity in both complexes due to a general uniformity of alleles predominates across populations and sometimes across reproductively and geographically isolated populations. Third, some allelic frequencies and H vary clinally and are predictable by a combination of ecological variables (129) and spatial environmental heterogeneity (136).

The general pattern of low H that characterizes most fossorial rodents seems to be primarily an adaptive strategy for homozygosity in the relatively uniform narrow-niche subterranean environment. Strong selection for homozygosity may be operating particularly on alleles that did not diverge across the range despite reproductive and ecogeographical isolation (126–129). However, since fluctuations do occur in some variables underground (e.g. burrow atmospheres, food items, predation) a certain level of H may be advantageous. Also, macroclimatic changes may select for high H such as found in *S. ehrenbergi,* 2n = 60 (136), *T. talpoides,* 2n = 60 (129), and *T. bottae* (146). This conclusion does not rule out the additional contributions of stochastic processes to genetic population structure of subterranean mammals [(146, 150, 178); E. G. Zimmerman, N.A. Gayden, unpublished].

An extreme contrast to the low H and implied homoselection of fossorial mammals is displayed by *Peromyscus maniculatus* (38). H (mean 0.155) varies along an altitudinal transect from 0.117 to 0.181 and seasonally at a single site from 0.138 in spring to 0.179 in summer. These ranges of H are as broad as those found across the entire range of *S. ehrenbergi, T. talpoides* and in fact most mammals (127, 182), which implies an extreme case of environmental heteroselection.

MODES OF SPECIATION

Patterns

Speciation theory (106) recently incorporated quantifiable chromosomal and genic variables. This permits testing of critical hypotheses and thereby increases the theory's predictive and explanatory power with new genetic, geographical, and time patterns (18, 100, 209, 210). Chromosomal speciation, including allopatric, parapatric, and stasipatric modes, is generally prevalent in mammals, and particularly in the subterranean rodent genera *Spalax, Thomomys, Geomys,* and *Ctenomys.* Remarkably, classical species have recently been shown to involve many cryptic, sibling chromosomal species. Thus the three classical species of *Spalax* involve over 30 karyotypes (173), most of which seem to be distinct species. Similarly, *Thomomys talpoides* may consist of 20 chromosomal species (195) and *Clenomys* of 60, some of which have been confirmed serologically (164, 166).

Rapid explosive speciation in fossorial rodents was strongly correlated with high rate of chromosomal evolution (19), presumably due to frequently subdivided, small, and semi-isolated populations (214, 215), patchy distributions, strong territoriality, and low vagility. These patterns increase inbreeding and interdeme selection, thereby enhancing homozygous chromosome fixation. However, wherever gene flow is high, e.g. in the subterranean rodent *Spalacopus* (163, 165) and in subterranean insectivores (221), chromosomal evolution and speciation seem to be restricted. Rapid chromosomal speciation proceeded with relatively few identifiable genomic changes in both *Spalax* [mean genetic similarity between chromosomal species $I = 0.966$, range 0.931–0.988 (128, 136)] and *Thomomys talpoides* [mean $I = 0.876$, range 0.727–0.996 (129)]. This is probably a general pattern associated with rapid speciation. Lower genetic similarities, as between five *Geomys* species [mean 0.720, range 0.514–0.735 (150)], may reflect postspeciation divergence.

Isolating Mechanisms

Chromosomal rearrangements may initiate speciation by providing postmating sterility barriers (210). However, at least in *Spalax ehrenbergi* and probably generally in subterranean mammals, chromosomal evolution was followed by reinforcement of ethological mechanisms. The nature, origin, and evolution of premating isolating mechanisms has been extensively tested in the four chromosomal species of *Spalax ehrenbergi.*

Mate selection was indicated in early mating experiments involving all four karyotypes of *S. ehrenbergi* (125). Later (134), estrous females of two parapatric karyotypes (2n = 52 and 58) significantly preferred homochromosomal mates. Positive assortative mating in *Spalax* was shown to be mediated through olfaction (135), vocalization [(21, 125), (E. Nevo et al, unpublished] and aggression (132). Finally (68), mate selection was shown to be similar among central and near-hybrid zone populations. This supported the allopatric theory of the origin of premating isolating mechanisms (106) in *Spalax.* Moreover, the last derivative of speciation, 2n = 60, presumably 75,000 years old (136), displays either no or low mate selection, which suggests that (*a*) postmating mechanisms preceded premating ones, and (*b*) incidental accumulation of effective ethological mechanisms requires at least tens of thousands of years. The ethological mechanisms thus contribute to species identification and divergence.

STAGES   Early to late stages of speciation are indicated by the varying extent of gene flow through natural hybridization in both the *Spalax ehrenbergi* and *Thomomys talpoides* evolving complexes. In *S. ehrenbergi* hybrid-zone widths decrease progressively with chromosomal differences between parental types (133). The nature and extent of hybridization reinforces the hypothesis that the four karyotypes are indeed young, closely related sibling species at early stages of evolutionary divergence as is also indicated by electrophoretic (136) and immunologic (130) analyses. Similar patterns have been found in the *Thomomys talpoides* complex in Colorado where interbreeding is either extensive, limited, or nonexistent (193). Limited hybridization was also described between *Thomomys bottae* and *T. umbrinus* (143) as well as between *T. bottae* and *T. townsendii* (189).

RATES   Chromosomal speciation is rapid and datable in the evolving complexes of *Spalax* and *Thomomys*. Electrophoretic (136) and immunologic (130) evidence suggests that the *Spalax ehrenbergi* complex speciated in Israel within the last $250,000 \pm 20,000$ years, the last derivative, 2n = 60, being 75,000 years old. This is in accord with the hybrid-zone (133) and fossil (in 136) evidence. Likewise, the electrophoretic evidence suggests that speciation in *Thomomys talpoides* occurred from late Pleistocene to recent times (129), approximately within the last 250,000 years, and some karyotypes, e.g. 2n = 44, may be less than 10,000 years old. While *Thomomys* fossils appear in early Pleistocene times, the *T. talpoides* complex is known only from late Pleistocene. It became abundant in Wisconsin time (60,000–10,000 years ago) across North America (169).

Parapatric Distribution

Parapatric or allopatric ranges generally characterize the distributions of subterranean mammals (41, 64, 86, 117, 148, 193). Parapatry has been largely confirmed recently for the different karyotypes of *Spalax* (102, 183, 205), *Thomomys* (188, 193), and *Ctenomys* (164). Exceptions of narrow to broad sympatry have also been described (e.g. 162, 190, 213).

Parapatry, either secondary (106) or primary (18, 19), reflects the combined operation of nearly or totally complete reproductive isolation as well as competitive exclusion. It prevails in species having similar ecological requirements where barriers in the contact zone are vegetational or climatic rather than geographical. Notably, broad sympatry can exist between subterranean herbivores and insectivores, indicating that food rather than space competition is largely responsible for parapatric distributions in subterranean mammals. Sympatry may also occur between herbivore families (e.g. Rhizomyidae and Bathyergidae in East Africa, and Cricitidae with both Spalacidae and Rhizomyidae in Asia) as indicated by the distribution maps. This situation awaits critical clarification.

Competitive Exclusion

Competition theory (118) considers parapatric contacts among rodents (56a) and subterranean mammals (148) as due primarily to interspecific competition. Resource competition among subterranean mammals living in structurally simple and indivisible niches is extreme (117, 148) and is mediated by aggression resulting in strong territoriality within species and largely parapatric distribution between species (132). In Colorado gophers, the species with the strictest niche requirements is the superior competitor and is able to displace the other species into less favorable habitats. The order of competitive ability is *G. bursarius* > *C. castanops* > *T. bottae* > *T. talpoides*. The factors involved in competitive interactions are size, territory, aggression, and dispersal (117). Competitive exclusion has been widely documented and tested (e.g. 203a) in gophers, in which soil depth, texture, and moisture are major critical factors in geographic and habitat distributions (33, 34, 41, 74, 85, 86, 108, 117, 144, 162, 169, 188, 191, 201, 212, 213). The sympatric pairs of large and dwarf types in gophers (190, 191) and moles (137) may reflect their habitat separation and competitive interactions. Likewise, significant population differences in habitat selection have been found within and between karyotypes of *S. ehrenbergi* in accord with their respective ecological regimes (136a).

Speciation and Adaptive Radiation

Chromosomal speciation may facilitate adaptive evolution by providing (*a*) cytogenetic reproductive isolation due to meiotic imbalances (209, 210), (*b*) new superior supergenes for ecotypic adaptation (57), (*c*) changes in the recombination index and hence the potential for new variants in populations (131), and/or (*d*) alteration in gene expression (19).

CONCLUSIONS AND PROSPECTS

The ecological theater of open country biotas that opened up progressively in the Cenozoic due to expanding terrestrialism and increasing aridity set the stage for a rapid evolutionary play of recurrent adaptive radiation of unrelated mammals on all continents into the subterranean ecotope. The latter is relatively simple, stable, specialized, low in productivity, predictable, and discontinuous. Its major evolutionary determinants are specialization, competition, and isolation. This ecotope involves essentially two nonoverlapping niches—the herbivorous, colonized by rodents, and the insectivorous, colonized by the marsupials and insectivores. All subterranean mammals share convergent adaptations to their common unique ecotope and divergent adaptations to their separated niches of herbivory and insectivory and to their different phylogenies.

Adaptive convergence involves burrowing adaptations, structural reductions/hypertrophies and respiratory adaptations for underground life, intraspecific territoriality and aggressive competition, 24-hr activity patterns, food generalism, equilibrium populations and K-strategy, low genetic variation and homoselection (untested yet in insectivores), interspecific competitive exclusion, and parapatric distributions between ecologically and genetically similar species. Adaptive divergence involves alternative patterns in herbivores and insectivores, respectively: large vs small body size, low vs high basic metabolic rates, high vs low thermal conductances, small vs large territories, small vs high gene flow, high vs low chromosomal speciation and taxonomic diversity. Major phylogenetic differences relate to tooth structure and dietary specialization.

Numerous problems remain unresolved and invite critical testing. These involve, among others, genetic variation in insectivorous and many untested rodent forms, ethological isolating mechanisms in most species, migration patterns, the relative contributions of deterministic and stochastic processes to genetic population structure and evolution, competitive interactions in overlap zones between rodent families and between rodents and insectivores, etc. Thus, the subterranean ecotope remains a fruitful field for investigating cardinal unresolved problems of evolutionary biology.

ACKNOWLEDGMENTS

My deep gratitude is extended for critical comments on parts of the manuscript to R. Arieli, A. Beiles, E. Golenberg, R. C. Lewontin, E. Mayr, M. Marmori, B. K. McNab, O. P. Pearson, C. A. Reed, T. W. Schoener, R. Schuster, E. Tchernov, and A. E. Wood. Thanks are also extended to M. Avrahami and G. Heth, long-time collaborators in studying *Spalax,* and to N. Friedman, Y. Fischer, and G. Schönfeld for their constant help in materializing this paper. Special thanks are due to E. Golenberg for many stimulating conversations and arguments as well as for valuable advice for improvement of the manuscript. This study was supported by grants from the United-States-Israel Binational Science Foundation, BSF, Jerusalem, Israel, the Israeli National Academy of Sciences, and the Guggenheim Foundation.

Literature Cited

1. Anderson, S. 1966. Taxonomy of gophers, especially *Thomomys* in Chihuahua, Mexico. *Syst. Zool.* 15:189–98
2. Anderson, S., Jones, J. K. Jr. 1967. *Recent Mammals of the World. A Synopsis of Families.* NY: The Ronald Press. 453 pp.
3. Ar, A., Arieli, R., Shkolnik, A. 1977. Blood-gas properties and function in the fossorial mole rat under normal and hypoxic-hypercapnic atmospheric conditions. *Resp. Physiol.* 30:201–18
4. Arieli, R. 1978. *The mole rat* (Spalax ehrenbergi): *Adaptations to fossorial life.* PhD thesis. Tel-Aviv Univ. (In Hebrew with English summary)
5. Arieli, R., Ar, A., Shkolnik, A. 1977. Metabolic responses of a fossorial rodent (*Spalax ehrenbergi*) to simulated burrow conditions. *Physiol. Zool.* 50: 61–75
6. Arlton, A. V. 1936. An ecological study of the mole. *J. Mammal.* 17:349–71
7. Armsby, A., Quilliam, T. A., Soehnle, H. 1966. Some observations on the ecology of the mole. *Proc. Zool. Soc. London* 149:110–12
8. Baker, R. J., Williams, S. L., Patton, J. C. 1973. Chromosomal variation in the plains pocket gopher. *Geomys bursarius major. J. Mammal.* 54:765–69
9. Barr, T. C. Jr. 1968. Cave ecology and the evolution of troglobites. In *Evolutionary Biology,* ed. Th. Dobzhansky, M. K. Hecht, W. C. Steere, 2:35–102. Amsterdam: North Holland
10. Bartels, H., Schmelzle, R., Ulrich, S. 1969. Comparative studies of the respiratory function of mammalian blood. V. Insectivora: shrew, mole and nonhibernating and hibernating hedgehog. *Resp. Physiol.* 7:278–86
11. Berry, D. L., Baker, R. J. 1972. Chromosomes of pocket gophers of the genus *Poppageomys,* subgenus *Cratogeomys. J. Mammal.* 53:303–9
12. Bezy, R. L., Gorman, G. C., Kim, Y. J., Wright, J. W. 1977. Chromosomal and genetic divergence in the fossorial lizards of the family Anniellidae. *Syst. Zool.* 26:57–71
13. Bradley, W. G., Yousef, M. K. 1975. Thermoregulatory responses in the plains pocket gopher, *Geomys bursarius. Comp. Biochem. Physiol. A* 52:35–38
14. Britton, J., Thaler, L. 1978. Evidence for the presence of two sympatric species of mice (Genus *Mus* L.) in southern France based on biochemical genetics. *Biochem. Genet.* 16:213–25
15. Brown, J. L., Orians, G. H. 1970. Spacing patterns in mobile animals. *Ann. Rev. Ecol. Syst.* 1:239–62
16. Bullard, R. W. 1972. Vertebrates at altitude. In *Physiological Adaptation: Desert and Mountain,* ed. M. K. Yousef, S. M. Horvath, R. W. Bullard. London: Academic
17. Burt, W. H. 1943. Territoriality and home range concepts as applied to mammals. *J. Mammal.* 24:346–52
18. Bush, G. L. 1975. Modes of animal speciation. *Ann. Rev. Ecol. Syst.* 6: 339–64
19. Bush, G. L., Case, S. M., Wilson, A. C., Patton, J. L. 1977. Rapid speciation and chromosomal evolution in mammals. *Proc. Natl. Acad. Sci. USA* 74:3942–46
20. Campbell, B. 1939. The shoulder anatomy of the moles. A study in phylogeny and adaptation. *Am. J. Anat.* 64:1–39
21. Capranica, R. R., Moffat, J., Nevo, E. 1973. Vocal repertoire of a subterranean rodent, *Spalax.* Presented at Acoust. Soc. Am. Ann. Meet., Los Angeles
22. Cei, G. 1946. Morfologia degli organi della vista negli Insettivori. *Arch. Anat. Embriol.* 52:1–42
23. Cei, G. 1946. Ortogenesi parallela e degradazione degli organi della vista negli Spalacidi. *Monit. Zool. Ital., Firenze* 55:69–88
24. Cei, G. 1946. L'occhio di *"Heterocephalus glaber"* Rupp. Note anatomodescrittive e istologiche. *Monit. Zool. Ital., Firenze* 55:89–96
25. Chapman, R. C., Bennett, A. F. 1975. Physiological correlates of burrowing in rodents. *Comp. Biochem. Physiol. A* 51:599–603
26. Cockley, D. E., Gooch, J. L., Weston, D. P. 1977. Genetic diversity in cave-dwelling crickets (*Ceuthophilus gracilipes*). *Evolution* 31:313–18
27. Colwell, R. K. 1974. Predictability, constancy and contingency of periodic phenomena. *Ecology* 55:1148–53
28. Conaway, C. H. 1959. The reproductive cycle of the eastern mole. *J. Mammal.* 40:180–94
29. Corbett, L. K. 1975. Geographical distribution and habitat of the marsupial mole, *Notoryctes typhlops. Aust. Mammal.* 1:375–78
30. Costa, M., Nevo, E. 1969. Nidicolous arthropods associated with different chromosome types of *Spalax ehrenbergi, Nehring. Zool. J. Linn. Soc.* 48:199–215

31. Criddle, S. 1930. The prairie pocket gopher *Thomomys talpoides rufescens*. *J. Mammal.* 11:265–80

32. Darden, T. R. 1972. Respiratory adaptations of a fossorial mammal, the pocket gopher (*Thomomys bottae*). *J. Comp. Physiol.* 78:121–37

33. Davis, W. B. 1938. Relation of size of pocket gophers to soil and altitude. *J. Mammal.* 19:338–42

34. Davis, W. B. 1940. Distribution and variation of pocket gophers (genus *Geomys*) in the southwestern United States. *Texas Agric. Exp. Stn. Bull.* 590:5–38

35. Davis, W. B., Ramsey, R. R., Arendale, J. M. Jr. 1938. Distribution of pocket gophers (*Geomys breviceps*) in relation to soils. *J. Mammal.* 19:412–18

36. Davis, B. L., Williams, S. L., Lopez, G. 1971. Chromosomal studies of *Geomys*. *J. Mammal.* 52:617–20

37. DeGraaff, G. 1964. *A systematic revision of the Bathyergidae (Rodentia) of Southern Africa.* PhD thesis. Univ. Pretoria, S. Africa

38. Dubach, J. M. 1975. *Genetic variation in altitudinally allopatric populations of the deermouse Peromyscus maniculatus.* MS thesis, Univ. Colorado, Denver

39. Dubost, G. 1968. Les mammiferes souterrains. *Rev. Ecol. Biol. Sol.* 5:99–197

40. Dukel'skaya, N. M. 1935. The mole rat *Spalax* and mole rat trapping. In *The Thin Toed Ground Squirrel, the Fat Dormouse, the Mole Rat and the Chipmunk,* ed. S. P. Naumov, N. P. Lavarov, E. P. Spangenberg, N. M. Dukel'skaya, I. M. Zaleskii, M. D. Zverev, pp. 71–79. Moscow: All-Union Coop. Un. Pub. House. (In Russian. English transl. 142 F 395 D Bur. Anim. Pop., Oxford Univ.)

41. Durrant, S. D. 1946. The pocket gophers (genus *Thomomys*) of Utah. *Univ. Kansas Publ. Mus. Nat. Hist.* 1:1–82

42. Ellerman, J. R. 1956. The subterranean mammals of the world. *Trans. R. Soc. S. Afr.* 35:11–20

43. Eloff, G. 1951. Orientation in the mole-rat *Cryptomys. Brit. J. Psychol.* 42:134–45

44. Eloff, G. 1958. The functional and structural degeneration of the eye of the South African rodent moles *Cryptomys bigalkei* and *Bathyergus maritimus. S. Afr. J. Sci.* 54:293–302

45. Elton, C., Miller, R. S. 1954. The ecological survey of animal communities: with a practical system of classifying habitats by structural characters. *J. Ecol.* 42:460–96

46. English, P. F. 1932. Some habits of the pocket gopher *Geomys breviceps breviceps. J. Mammal.* 13:126–32

47. Enright, J. T. 1970. Ecological aspects of endogenous rhythmicity. *Ann. Rev. Ecol. Syst.* 1:221–38

48. Gambarian, P. P. 1953. Adaptive specializations of the forelimb of blind mole rats (*Spalax leucodon nehringi* Saturnin), *Zool. Inst. Akad. Nauk. Armianskoi SSSR* 8:67–125 (In Russian)

49. Genelly, R. E. 1965. Ecology of the common mole rat *Cryptomys hottentotus* in Rhodesia. *J. Mammal.* 46:647–65

50. Gettinger, R. D. 1975. Metabolism and thermoregulation of a fossorial rodent, the northern pocket gopher (*Thomomys talpoides*). *Physiol. Zool.* 48:311–22

51. Getz, L. L. 1957. Color variation in pocket gophers, *Thomomys. J. Mammal.* 38:523–26

52. Giuseffi, S., Kane, C. T., Duggleby, W. F. 1978. Genetic variability in the Kentucky cave beetle *Neaphaenops tellkampfii* (Coleoptera: Carabidae). *Evolution* 32:679–81

53. Godet, R. 1951. Contribution a l'ethologie de la taupe (*Talpa europaea* L.) *Bul. Soc. Zool. Fr.* 76:107–28

54. Godfrey, G., Crowcroft, P. 1960. *The Life of the Mole (Talpa europaea L.)* London: Museum Press. 152 pp.

55. Gorecki, A., Christov, L. 1969. Metabolic rate of the lesser mole rat. *Acta Theriol.* 14:441–48

56. Gorman, G. C., Kim, Y. J., Taylor, C. E. 1977. Genetic variation in irradiated and control populations of *Cnemidophorus tigris* (Sauria, Teiidae) from Mercury, Nevada with a discussion of genetic variability in lizards. *Theor. Appl. Genet.* 49:9–14

56a. Grant, P. R. 1972. Interspecific competition among rodents. *Ann. Rev. Ecol. Syst.* 3:79–106

57. Grant, V. 1971. *Plant Speciation.* NY: Columbia Univ. Press. 435 pp.

58. Gropp, A. 1969. Cytologic mechanisms of karyotype evolution in insectivores. In *Comparative Mammalian Cytogenetics,* ed. K. Benirschke, pp. 247–66. NY: Springer

59. Haeck, J. 1969. Colonization of the mole (*Talpa europaea* L) in the Ijsselmeerpolders. *Nether. J. Zool.* 19:145–248

60. Hall, E. R., Kelson, K. R. 1959. *The Mammals of North America,* Vol. 1. NY: Ronald Press Co. 625 pp.

61. Hansen, R. M. 1965. Pocket gopher

density in an enclosure of native habitat. *J. Mammal.* 46:508–9

62. Hansen, R. M., Bear, G. D. 1965. Comparison of pocket gophers from alpine, subalpine and shrub-grassland habitats. *J. Mammal.* 45:638–40

63. Hansen, R. M., Miller, R. S. 1959. Observations on plural occupancy of pocket gopher burrow system. *J. Mammal.* 40:577–84

64. Hansen, R. M., Vaughan, T. A., Hervey, D. F., Harris, T. V., Hegdal, P. L., Johnson, A. M., Ward, A. L., Reid, E. H., Keith, J. O. 1960. Pocket gophers in Colorado. *Colorado State Univ. Exp. Stn. Bull. 508-S1-26,* 25 pp.

65. Hansen, R. M., Ward, A. L. 1966. Some relations of pocket gophers to rangelands on Grand Mesa, Colorado. *Tech. Bull.* 88:1–21

66. Hart, E. B. 1971. *Karyology and evolution of the plains pocket gopher, Geomys bursarius.* PhD thesis. Univ. Oklahoma, Norman. 111 pp.

67. Harvey, M. J. 1976. Home range, movements, and diel activity of the eastern mole, *Scalopus aquaticus. Am. Midl. Nat.* 95:436–45

68. Heth, G. 1980. *The origin and evolution of ethological isolating mechanisms in the superspecies complex of* Spalax ehrenbergi. PhD thesis. Hebrew Univ., Jerusalem (In Hebrew with English summary)

69. Hickman, G. C. 1977. Swimming behavior in representative species of the three genera of North American Geomyids. *Southwest. Nat.* 21:531–38

70. Hildebrand, M. 1974. *Analysis of Vertebrate Structure.* NY: Wiley

71. Hill, J. E. 1937. Morphology of the pocket gopher, mammalian genus *Thomomys. Univ. Calif. Publ. Zool.* 42:81–172

72. Hill, W. C. O., Porter, A., Bloom, R. T., Seago, J., Southwick, M. D. 1957. Field and laboratory studies on the naked mole rat, *Heterocephalus glaber. Proc. Zool. Soc. London* 128:455–514

73. Howard, W. E. 1961. A pocket gopher population crash. *J. Mammal.* 42:258–60

74. Howard, W. E., Childs, H. E. Jr. 1959. Ecology of pocket gophers with emphasis on *Thomomys bottae mewa. Hilgardia* 29:277–358

75. Howe, D. 1975. Observations on the behaviour of a captive marsupial mole (*Notoryctes typhlops*). *Aust. Mammal.* 1:361–65

76. Howell, A. B. 1922. Surface wanderings

of fossorial mammals. *J. Mammal.* 3:19–22

77. Ingles, L. G. 1950. Pigmental variations in populations of pocket gophers. *Evolution* 4:353–57

78. Ingles, L. G. 1952. The ecology of the mountain pocket gopher, *Thomomys monticola. Ecology* 33:87–95

79. Jarvis, J. U. M. 1969. The breeding season and litter size of African mole-rats. *J. Reprod. Fert. Suppl.* 6:237–48

79a. Jarvis, J. U. M. 1978. Activity patterns in the mole-rats *Tachyoryctes splendens* and *Heliophobius argenteocinereus. Zool. Afr.* 8:101–19

80. Jarvis, J. U. M. 1973. The structure of a population of mole-rats, *Tachyoryctes splendens* (Rodentia: Rhizomyidae). *J. Zool. London* 171:1–14

81. Jarvis, J. U. M. 1978. The energetics of survival in *Heterocephalus glaber* (Ruppell) the naked mole rat (Rodentia: Bathyergidae). *Bull. Carnegie Mus. Nat. Hist.* 6:81–87

82. Jarvis, J. U. M., Sale, J. B. 1971. Burrowing and burrow patterns of East African mole-rats *Tachyoryctes, Heliophobius* and *Heterocephalus. J. Zool. London.* 163:451–79

83. Johansen, K., Lykkeboe, G., Weber, R. E., Maloiy, G. M. O. 1976. Blood respiratory properties in the naked mole rat (*Heterocephalus glaber*) a mammal of low body temperature. *Resp. Physiol.* 28:303–14

84. Karlin, A. 1976. Population subdivision and selection migration interaction. In *Population Genetics and Ecology,* ed. S. Karlin, E. Nevo, pp. 617–57. NY: Academic

85. Kennerly, T. E. Jr. 1954. Local differentiation in the pocket gopher (*Geomys personatus*) in southern Texas. *Texas J. Sci.* 6:297–329

86. Kennerly, T. E. Jr. 1959. Contact between the ranges of two allopatric species of pocket gophers *Evolution* 13:247–63

87. Kennerly, T. E. Jr. 1964. Microenvironmental conditions of the pocket gopher burrow. *Texas J. Sci.* 16:395–441

88. Kim, Y. J. 1972. *Studies of biochemical genetics and karyotypes in pocket gophers (family Geomyidae).* PhD thesis. Univ. Texas, Austin. 112 pp.

89. Kim, Y. J., Gorman, G. C., Huey, R. B. 1978. Genetic variation and differentiation in two species of the fossorial African skink *Typhlosaurus* (Sauria: Scincidae). *Herpetologica* 34:192–94

90. King, C. E. 1964. Relative abundance of

species and MacArthur's model. *Ecology* 45:716–27

91. Krapp, F. 1965. Schädel und Kaumuskulatur von *Spalax leucodon* (Nordmann, 1840). *Z. Wiss. Zool.* 173:1–71

92. Laing, C. D., Carmody, G. R., Peck, S. B. 1976. How common are sibling species in cave inhabiting invertebrates? *Am. Nat.* 110:184–89

93. Laing, C. D., Carmody, G. R., Peck, S. B. 1976. Population genetics and evolutionary biology of the cave beetle *Ptomaphagus hirtus. Evolution* 30:484–98

94. Larkin, R. A. 1948. *The ecology of mole* (Talpa europaea *L.) populations.* PhD thesis. Oxford Univ.

95. Lechner, A. J. 1976. Respiratory adaptations in burrowing pocket gophers from sea level and high altitude. *J. Appl. Physiol.* 41:168–73

96. Lehmann, W. H. 1963. The forelimb architecture of some fossorial rodents. *J. Morphol.* 113:59–76

97. Leigh, E. G. 1975. Population fluctuations, community stability, and environmental variability. In *Ecology and Evolution of Communities,* ed. M. L. Cody, J. M. Diamond, pp. 51–73. Cambridge, Mass: Belknap

98. Levins, R. 1968. *Evolution in Changing Environments. Some Theoretical Explorations.* Princeton, NJ: Princeton Univ. Press. 120 pp.

99. Lewontin, R. C. 1965. Selection for colonizing ability. In *The Genetics of Colonizing Species,* ed. H. G. Baker, G. L. Stebbins, pp. 77–91. NY: Academic. 588 pp.

100. Lewontin, R. C. 1974. *The Genetic Basis of Evolutionary Change.* NY: Columbia Univ. Press. 346 pp.

101. Lidicker, W. Z. Jr. 1975. The role of dispersal in the demography of small mammals. In *Small Mammals: Their Productivity and Population Dynamics,* ed. F. B. Golley, K. Petrusewicz, L. Ryszkowski, pp. 103–28. Cambridge: Cambridge Univ. Press

102. Lyapunova, E. A., Vorontsov, N. N., Martynova, L. Y. 1974. Cytogenetical differentiation of burrowing mammals in the Palaearctic. *Proc. Int. Symp. Species and Zoogeography of Eur. Mammals, Brno 1971,* pp. 203–15

103. MacArthur, R. H. 1972. *Geographical Ecology. Patterns in the Distribution of Species.* NY: Harper & Row. 269 pp.

104. MacArthur, R. H., Wilson, E. O. 1967. *The Theory of Island Biogeography.* Princeton, NJ: Princeton Univ. Press. 203 pp.

105. Martynova, L. Y. 1976. Chromosomal differentiation in three species of zokors (Rodentia, Myospalacinae) *Acad. Sci. USSR* 55:1265–68

106. Mayr, E. 1970. *Populations, Species, and Evolution.* Cambridge, Mass: Harvard Univ. Press. 453 pp.

107. McNab, B. K. 1963. Bioenergetics and the determination of home range size. *Am. Nat.* 97:133–40

108. McNab, B. K. 1966. The metabolism of fossorial rodents: A study of convergence. *Ecology* 47:712–33

109. McNab, B. K. 1971. On the ecological significance of Bergmann's rule. *Ecology* 52:845–54

110. McNab, B. K. 1974. The energetics of endotherms. *Ohio J. Sci.* 74:370–80

111. McNab, B. K. 1978. The evolution of endothermy in the phylogeny of mammals. *Am. Nat.* 112:1–21

112. McNab, B. K. 1979. The influence of body size on the energetics and distribution of fossorial and burrowing mammals. *Ecology.* In press

113. McNab, B. K. 1979. Food habits, energetics and the population biology of small mammals. *Am. Nat.* In press

114. Millar, J. S. 1977. Adaptive features of mammalian reproduction. *Evolution* 31:370–86

115. Miller, M. A. 1946. Reproductive rates and cycles in the pocket gopher. *J. Mammal.* 27:335–58

116. Miller, M. A. 1957. Burrows of the Sacramento Valley pocket gopher in flood-irrigated alfalfa fields. *Hilgardia* 26:431–52

117. Miller, R. S. 1964. Ecology and distribution of pocket gophers (Geomyidae) in Colorado. *Ecology* 45:256–72

118. Miller, R. S. 1967. Patterns and process in competition. In *Adv. Ecol. Res.* 4:1–74

119. Miller, R. S., Ward, R. A. 1960. Ectoparasites of pocket gophers from Colorado. *Am. Midl. Nat.* 64:382–91

120. Motro, U. 1978. *On optimal rates of dispersion and migration in biological populations.* PhD thesis. Tel Aviv Univ. 116 pp. (In Hebrew with English summary)

121. Nei, M. 1975. *Molecular Population Genetics and Evolution.* Amsterdam: North Holland. 288 pp.

122. Nei, M. 1976. Mathematical models of speciation and genetic distance. In *Population Genetics and Ecology,* ed. S. Karlin, E. Nevo, pp. 723–65. NY: Academic

123. Nevo, E. 1961. Observations on Israeli populations of the mole-rat *Spalax e.*

ehrenbergi. Nehring 1898. *Mammalia* 25:127–44

124. Nevo, E. 1969. Mole-rat *Spalax:* A model of prolific speciation. *Proc. NATO. Adv. Study Inst. Istanbul, Turkey, Conf. Vertebrate Evolution, August, 1969*

125. Nevo, E. 1969. Mole-rat *Spalax ehrenbergi:* Mating behavior and its evolutionary significance. *Science* 163: 484–86

126. Nevo, E. 1973. Test of selection and neutrality in natural populations. *Nature* 244:573–75

127. Nevo, E. 1978. Genetic variation in natural populations: patterns and theory. *Theor. Pop. Biol.* 13:121–77

128. Nevo, E., Shaw, C. 1972. Genetic variation in a subterranean mammal, *Spalax ehrenbergi. Biochem. Genet.* 7:235–41

129. Nevo, E., Kim, Y. J., Shaw, C., Thaeler, C. S. Jr. 1974. Genetic variation, selection and speciation in *Thomomys talpoides* pocket gophers. *Evolution* 28: 1–23

130. Nevo, E., Sarich, V. 1974. Immunology and evolution in the mole rat, *Spalax. Isr. J. Zool.* 23:210–11

131. Nevo, E., Shkolnik, A. 1974. Adaptive metabolic variation of chromosome forms in mole rats, *Spalax. Experientia* 30:724–26

132. Nevo, E., Naftali, G., Guttman, R. 1975. Aggression patterns and speciation. *Proc. Natl. Acad. Sci. USA* 72:3250–54

133. Nevo, E., Bar-El, H. 1976. Hybridization and speciation in fossorial mole-rats. *Evolution* 30:831–40

134. Nevo, E., Heth, G. 1976. Assortative mating between chromosome forms of the mole rat, *Spalax ehrenbergi. Experientia* 32:1509–10

135. Nevo, E., Heth, G., Bodmer, M. 1976. Olfactory discrimination as an isolating mechanism in speciating mole rats. *Experientia* 32:1511–12

136. Nevo, E., Cleve, H. 1978. Genetic differentiation during speciation. *Nature* 275:125–26

136a. Nevo, E., Guttman, R., Haber, M., Erez, E. 1979. Habitat selection in evolving mole rats. *Oecologia.* In press

136b. Nevo, E., Guttman, R., Haber, M., Erez, E. 1979. Activity patterns in evolving mole rats. (Submitted)

137. Niethammer, J. 1969. Zur taxonomie europäischer Zwergmaulwürfe (*Talpa "mizura"*). *Bonn. Zool. Beitr.* 4:360–72

138. Ognev, S. I. 1947. *Mammals of U.S.S.R. and Adjacent Countries. Rodents, Vol. 5.*

Transl. from Russian, 1963. Jerusalem: Isr. Prog. Sci. Transl. 662 pp.

139. Orcutt, E. E. 1940. Studies on the muscles of the head, neck and pectoral appendages of *Geomys bursarius. J. Mammal.* 21:37–52

140. Orians, G. H., Pearson, N. E. 1978. On the theory of central place foraging. In *Analyses of Ecological Systems,* ed D. F. Horn. Columbus, Ohio: Ohio State Univ. Press

141. Patton, J. L. 1970. Karyotypic variation following an elevational gradient in the pocket gopher, *Thomomys bottae grahamensis* Goldman. *Chromosoma* 31:41–50

142. Patton, J. L. 1972. Patterns of geographic variation in karyotype in the pocket gopher, *Thomomys bottae* (Eydoux and Gervais). *Evolution* 26: 574–86

143. Patton, J. L. 1973. An analysis of natural hybridization between the pocket gophers, *Thomomys bottae* and *Thomomys umbriuns,* in Arizona. *J. Mammal.* 54:561–84

144. Patton, J. L., Dingman, R. E. 1968. Chromosome studies of pocket gophers, genus *Thomomys.* I. The specific status of *Thomomys umbrinus* (Richardson) in Arizona *J. Mammal.* 49:1–13

145. Patton, J. L., Dingman, R. E. 1970. Chromosome studies of pocket gophers, genus *Thomomys.* II. Variation in *T. bottae* in the American Southwest. *Cytogenetics* 9:139–51

146. Patton, J. L., Yang, S. Y. 1977. Genetic variation in *Thomomys bottae* pocket gophers: Macrogeographic patterns. *Evolution* 31:697–720

147. Patton, J. L., Feder, J. H. 1978. Genetic divergence between populations of the pocket gopher, *Thomomys umbrinus* (Richardson). *Saugetierkunde* 43:12–30

148. Pearson, O. P. 1960. Biology of the subterranean rodents, *Ctenomys,* in Peru. *Mem. Mus. Hist. Nat. Javier Prado* 9:1–56

149. Pearson, O. P., Binsztein, N., Boiry, L., Busch, C., Dipace, M., Gallopin, G., Penchaszadeh, P., Piantanida, M. 1968. Estructura social, distribucion espacial, y composicion por edades de una poblacion de tuco-tuco (*Ctenomys talarum*). *Inv. Zool. Chilenas* 13:47–80

150. Penny, D. E., Zimmerman, E. G. 1976. Genic divergence and local population differentiation by random drift in the pocket gopher genus *Geomys. Evolution* 30:473–83

151. Pevet, P., Kappers, J. A., Nevo, E. 1976. The pineal gland of the mole-rat (*Spalax ehrenbergi*, Nehring). *Cell Tissue Res.* 174:1–24

152. Pianka, E. R. 1970. On r-and K-selection. *Am. Nat.* 104:592–97

153. Pianka, E. R. 1974. Niche overlap and diffuse competition. *Proc. Natl. Acad. Sci. USA* 71:2141–45

154. Prout, T. 1964. Observations on structural reduction in evolution. *Am. Nat.* 98:239–49

155. Puttick, G. M., Jarvis, U. M. 1977. The functional anatomy of the neck and forelimbs of the cape golden mole, *Chrysochloris asiatica* (Lipotyphla: Chrysochloridae). *Zool. Afr.* 12:445–58

156. Pyke, G. H., Pulliam, H. R., Charnov, E. L. 1977. Optimal foraging: A selective review of theory and tests. *Q. Rev. Biol.* 52:137–54

157. Quilliam, T. A., Clarke, J. A., Salsbury, A. J. 1971. The ecological significance of certain new hematological findings in the mole and the hedgehog. *Comp. Biochem. Physiol. A.* 40:98–102

158. Raicu, P., Bratosin, S., Hamar, M. 1968. Study on the karyotype of *Spalax leucodon* Nordm. and *S. microphthalmus* Guld. *Caryologia* 21:127–35

159. Reed, C. A. 1951. Locomotion and appendicular anatomy in three soricoid insectivores. *Am. Midl. Nat.* 45:513–671

160. Reed, C. A. 1954. The origin of a familial character. A study in the evolutionary anatomy of moles. *Anat. Rec.* 118:343

161. Reed, C. A. 1958. Evolution of a functional anatomical system: burrowing mechanisms in the *Talpidae. Anat. Rec.* 132:491–92

162. Reichman, O. J., Baker, R. J. 1972. Distribution and movements of two species of pocket gophers (Geomyidae) in an area of sympatry in the Davis Mountains, Texas. *J. Mammal.* 53:21–33

163. Reig, O. A. 1970. Ecological notes on the fossorial octodont rodent *Spalacopus cyanus* (Molina). *J. Mammal.* 51:592–601

164. Reig, O. A., Kiblisky, P. 1969. Chromosome multiformity in the genus *Ctenomys* (Rodentia, Octodontidae). *Chromosoma* 28:211–44

165. Reig, O. A., Spotorno, A. O., Fernandez, D. R. 1972. A preliminary survey of chromosomes in populations of the Chilean burrowing octodont rodent *Spalacopus cyanus* Molina (Caviomorpha, Octodontidae). *Biol. J. Linn. Soc. London* 4:29–38

166. Roig, V. G., Reig, O. A. 1969. Precipitin test relationships among Argentinian species of the genus *Ctenomys* (Rodentia, Octodontidae). *Comp. Biochem. Physiol.* 30:665–72

167. Romer, A. S. 1966. *Vertebrate Paleontology.* Chicago/London: Univ. Chicago Press. 468 pp. 3rd ed.

168. Roughgarden, J. 1974. Niche width: Biogeographic patterns among *Anolis* lizard population. *Am. Nat.* 108:429–42

169. Russell, R. F. 1968. Evolution and classification of the pocket gophers of the subfamily Geomyinae. *Univ. Kans. Publ. Mus. Nat. Hist.* 16:473–579

170. Savić, I. R. 1973. Ecology of the species *Spalax leucodon* Nordm. in Yugoslavia. *Nat. Sci.* 44:5–70 (In Serbo-Kroatian with English summary)

171. Savić, I. R., Mikes, M. 1967. Zur Kenntnis des 24-Stunden Rhythmus von *Spalax leucodon* Nordmann, 1840. *Z. Säugetierk.* 32:233–38

172. Savić, I. R., Ryba, J. 1977. Contribution to the study of fleas from different chromosomal forms of the mole rat, *Spalax*, in Yugoslavia. *Arh. Biol. Nauka, Beograd* 27:145–53

173. Savić, I. R., Soldatović, B. 1978. Contribution to the knowledge of the genus *Spalax* (*Microspalax*) karyotype from Asia Minor. Presented at *2nd Int. Congr. Theriol. Brno, 1978.*

174. Scheffer, T. H. 1940. Excavation of a runway of the pocket gopher (*Geomys bursarius*). *Trans. Kans. Acad. Sci.* 43:473–78

175. Schmitt, L. H. 1978. Genetic variation in isolated populations of the Australian bush-rat, *Rattus fuscipes. Evolution* 32:1–14

176. Schoener, T. W. 1969. Optimal size and specialization in constant and fluctuating environments: an energy-time approach. *Brookhaven Symp. Biol.* 22:103–14

177. Schoener, T. W. 1971. Theory of feeding strategies. *Ann. Rev. Ecol. Syst.* 2:369–404

178. Selander, R. K., Kaufman, D. W., Baker, R. J., Williams, S. L. 1974. Genic and chromosomal differentiation in pocket gophers of the *Geomys bursarius* group. *Evolution* 28:557–64

179. Simpson, G. G. 1959. The nature and origin of supraspecific taxa. *Cold Spring Harbor Symp. Quant. Biol.* 24:255–71

180. Skoczen, S. 1961. On food storage of the mole, *Talpa europaea* L. 1758. *Acta Theriol.* 5:23–43

181. Slonaker, J. R. 1920. Some morphologi-

cal changes for adaptation in the mole. *J. Morphol.* 34:355–73

182. Smith, M. H., Manlove, M. N., Joule, J. 1977. Spatial and temporal dynamics of the genetic organization of small mammals populations. *Pymatuning Lab. Ecol. Spec. Publ.* 5, pp. 99–113

183. Soldatović, B., Savić, I. R. 1974. The karyotype forms of the genus *Spalax* Guld in Yugoslavia and their areas of distribution. *Proc. Int. Symp. Species and Zoogeography of Eur. Mammals, Brno, 1971,* pp. 125–30

184. Southwood, T. R. E., May, R. M., Hassell, M. P., Conway, G. R. 1974. Ecological strategies and populations parameters. *Am. Nat.* 108:791–804

185. Stahl, W. R. 1967. Scaling of respiratory variables in mammals. *J. Appl. Physiol.* 22:453–60

186. Stein, G. H. W. 1950. Zur Biologie des Maulwurfs, *Talpa europaea* L. *Bonn. Zool. Beitr.* 1:97–116

187. Sweet, G. 1904. Contribution to our knowledge of the anatomy of *Notoryctes typhlops.* Ps. I & II. *Proc. R. Soc. Victoria* 17:76–111

188. Thaeler, C. S. Jr. 1968. Karyotypes of sixteen populations of the *Thomomys talpoides* complex of pocket gophers (Rodentia: Geomyidae). *Chromosoma* 25:172–83

189. Thaeler, C. S. Jr. 1968. An analysis of three hybrid populations of pocket gophers (genus *Thomomys*). *Evolution* 22:543–55

190. Thaeler, C. S. Jr. 1968. An analysis of the distribution of pocket gopher species in northeastern California (genus *Thomomys*). *Univ. Calif. Publ. Zool.* 86:1–46

191. Thaeler, C. S. Jr. 1972. Taxonomic status of the pocket gophers, *Thomomys idahoensis* and *Thomomys pygmaeus* (Rodentia: Geomyidae) *J. Mammal.* 53:417–28

192. Thaeler, C. S. Jr. 1973. The karyotypes of *Thomomys townsendii similis* and comments on the karyotypes of *Thomomys. Southw. Nat.* 17:327–31

193. Thaeler, C. S. Jr. 1974. Four contacts between the ranges of different chromosome forms of the *Thomomys talpoides* complex (Rodentia-Geomyidae). *Syst. Zool.* 23:343–54

194. Thaeler, C. S. Jr. 1974. Karyotypes of the *Thomomys talpoides* complex (Rodentia: Geomyidae) from New Mexico. *J. Mammal.* 55:855–59

195. Thaeler, C. S. Jr. 1976. Chromosome polymorphism in *Thomomys talpoides*

agrestis Merriam (Rodentia: Geomyidae). *Southwest. Nat.* 21:105–16

196. Todorović, M., Soldatović, B., Dunderski, Z. 1972. Characteristics of the karyotype of the populations of the genus *Talpa* from Macedonia and Montenegro. *Arch. Sci. Biol.* 24:131–39 (In Serbo-Kroatian with English summary)

197. Tryon, C. A., Cunningham, H. N. 1968. Characteristics of pocket gophers along an altitudinal transect. *J. Mammal.* 49:699–705

198. Turner, G. T., Hansen, R. M., Reid, V. H., Tietjen, H. P., Ward, A. L. 1973. Pocket gophers and Colorado mountain rangeland. *Colo. State Univ. Exp. Stn. Bull. 554S.* 90 pp.

199. Vaughan, T. A. 1962. Reproduction in the plains pocket gopher in Colorado. *J. Mammal.* 43:1–13

200. Vaughan, T. A. 1963. Movements made by two species of pocket gophers. *Am. Midl. Nat.* 69:367–72

201. Vaughan, T. A. 1967. Two parapatric species of pocket gophers. *Evolution* 21:148–58

202. Vaughan, T. A. 1974. Resource allocation in some sympatric, subalpine rodents. *J. Mammal.* 55:764–95

203. Vaughan, T. A., Hansen, R. M. 1961. Activity rhythm of the plains pocket gopher. *J. Mammal.* 42:541–43

203a. Vaughan, T. A., Hansen, R. M. 1964. Experiments on interspecific competition between two species of pocket gophers. *Am. Midl. Nat.* 72:444–52

204. Vleck, D. J. A. 1978. *The energetics of activity and growth: The energy cost of burrowing by pocket gophers and energy metabolism during growth and hatching of emu, rhea, and ostrich eggs,* chap. 1–2. PhD thesis. Univ. Calif., Los Angeles

205. Wahrman, J., Goitein, R., Nevo, E. 1969. Mole rat *Spalax:* Evolutionary significance of chromosome variation. *Science* 164:82–84

206. Ward, A. L., Keith, J. O. 1963. Feeding habits of pocket gophers on mountain grasslands, Black Mesa, Colorado. *Ecology* 43:744–49

207. Webb, S. D. 1977–1978. A history of savanna vertebrates in the New World. Parts I & II. *Ann. Rev. Ecol. Syst.* 8:355–80; 9:393–426

208. Wertheim, G., Nevo, E. 1971. Helminths of birds and mammals from Israel. III. Helminths from chromosomal forms of mole-rat, *Spalax ehrenbergi. J. Helminthol.* 45:161–69

209. White, M. J. D. 1973. *Animal Cytology*

and Evolution. Cambridge: Cambridge Univ. Press. 961 pp. 3rd ed.

210. White, M. J. D. 1978. *Modes of Speciation.* San Francisco: Freeman 455 pp.

211. Whittaker, R. H., Levin, S. A., Root, R. B. 1973. Niche, habitat, and ecotope. *Am. Nat.* 107:321–38

212. Wilks, B. J. 1963. Some aspects of the ecology and population dynamics of the pocket gopher *Geomys bursarius* in Southern Texas. *Texas J. Sci.* 15:241–83

213. Williams, S. L., Baker, R. J. 1976. Vagility and local movements of pocket gophers (Geomyidae: Rodentia). *Am. Midl. Nat.* 96:303–16

213a. Withers, P. C. 1978. Models of diffusion mediated gas exchange in animal burrows. *Am. Natl.* 112:1101–12

214. Wright, S. 1931. Evolution in Mendelian populations. *Genetics* 16:97–159

215. Wright, S. 1943. Isolation by distance. *Genetics* 28:114–38

216. Wright, S. 1964. Pleiotropy in the evolution of structural reductions and of dominance. *Am. Nat.* 98:65–69

217. Wood, A. E. 1955. A revised classification of the rodents. *J. Mammal.* 36:165–87

218. Wood, A. E. 1959. Are there rodent suborders? *Syst. Zool.* 7:169–73

219. Wood, A. E. 1965. Grades and clades among rodents. *Evolution* 19:115–30

220. Wood, J. E. 1949. Reproductive pattern of the pocket gopher *Geomys breviceps brazensis. J. Mammal.* 30:36–44

221. Yates, T. L., Schmidly, D. J. 1975. Karyotype of the eastern mole (*Scalopus aquaticus*), with comments on the karyology of the family Talpidae. *J. Mammal.* 56:902–5

222. Yates, T. L., Schmidly, D. J. 1977. Systematics of *Scalopus aquaticus* (Linnaeus) in Texas and adjacent states. *Occ. Pap. Mus. Texas Tech. Univ.* 45:1–36

Ann. Rev. Ecol. Syst. 1979. 10:309–26

ORIGINS OF SOME CULTIVATED NEW WORLD PLANTS

♦4164

Charles B. Heiser

Department of Biology, Indiana University, Bloomington, Indiana 47405

I shall not enter into so much detail on the variability of cultivated plants, as in the case of domesticated animals. The subject is involved in much difficulty. Botanists have generally neglected cultivated varieties, as beneath their notice. In several cases the wild prototype is known or doubtfully known, and in other cases it is hardly possible to distinguish between escaped seedlings and truly wild plants, so that there is no safe standard of comparison by which to judge of any supposed amount of change. Not a few botanists believe that several of our anciently cultivated plants have become so profoundly modified that it is not possible now to recognise their aboriginal parent-forms. Equally perplexing are the doubts whether some of them are descended from one species, or from several inextricably commingled by crossing and variation.

Charles Darwin (21)

For none of the major crops can we point with certainty to the exact species (or combination of species) from which it was derived: for some we can make guesses; for a number we can point to closely related weeds. This merely complicates the problem. We then have to determine the origin of the crop, the origin of the weed, and the history of their relationships.

Edgar Anderson (2)

Before he had written the passage cited above, Anderson (1) had called attention to the "scandalous condition" of our knowledge of the origin of cultivated plants and pointed out that "two or three decades of good hard work" would do much to put the house in order. More than two decades have now passed and during that time cultivated plants have received increasing attention from both the botanist and anthropologist. A great deal has been learned about the evolution of some domesticated plants. Wild ancestors have been identified for many of them, but the progenitors of

309

others have remained elusive; and questions have been raised about the ancestry of some whose origins had been considered solved. The general area where domestication took place has been determined for many species, but for some it remains unknown. For example, it is still not agreed whether the sweet potato, *Ipomoea batatas,* was domesticated in Middle America or South America, or perhaps independently in both regions.

The increased research activity on cultivated plants has already resulted in a number of symposia and reviews. Flannery (26) has provided a detailed review of the archaeological record for both the Old and New World, and Smith (84) has treated the American species in an archaeological context. Several papers dealing with American cultivated plants are found in the symposium volumes edited by Hutchinson (53) and Reed (77). The papers in the latter are based on a symposium held in 1973. The most up-to-date review is that of Pickersgill (72). In addition to these works a recent book (79) treats the origin and development of all major crop plants and a number of minor ones as well. The present review discusses papers that have appeared in the last three years or crops that have not been adequately covered in the other works.

MAIZE (ZEA MAYS)

No American plant has received more study than maize; and although today there is some agreement concerning its evolution following domestication, its progenitor remains a matter of controversy. It wasn't always thus. Following its publication in 1939, the tripartite theory of Mangelsdorf & Reeves (64) gained fairly wide acceptance. Their hypothesis held (*a*) that cultivated maize originated from a wild form of pod corn once indigenous to the lowlands of South America; (*b*) that teosinte (*Zea mexicana*), the closest relative of maize, is the product of a natural hydridization of *Zea* and *Tripsacum* which occurred after cultivated maize had been introduced by man into Central America; and (*c*) that new types of corn originating directly or indirectly from this cross and exhibiting admixture with *Tripsacum* or teosinte comprise the majority of modern Central American and North American varieties (65).

Fossil pollen from Mexico and archaeological discoveries from Mexico and the American Southwest in the next two decades, however, pointed to Mexico rather than South America as a more likely place of the origin of maize.

Recently Beadle (8, 9) has revived the old idea that teosinte (*Zea mexicana*) is the progenitor, rather than a derivative, of maize, a view that is finding an increasing number of supporters (28). Mangelsdorf (63) eventually modified the tripartite hypothesis. While still maintaining that the ancestor of maize was a wild maize, now extinct, he now considers teosinte to be a mutant derivative of maize and recognizes domestication of maize in Mexico as well as South America.

Certainly any consideration of the origin of maize must account for the remarkable remains discovered at Tehuacán, Mexico, and dated to around 5000 BC. Mangelsdorf (63) has interpreted these as wild maize, whereas Galinat (28) feels that they represent an early domesticated form. They do not, Galinat admits, form an indisputable link with teosinte.

That introgression of teosinte into maize has been important in the development of the latter plant is generally accepted (93, 94). Whether introgression of *Tripsacum* played a role is far less certain. De Wet et al (23) have produced experimental evidence demonstrating that such introgression is possible, but they conclude that natural introgression is unlikely. The so-called tripsacoid races of South America, they feel, could originally have been teosintoid ones introduced from Middle America or the result of introgression from a race of teosinte that once may have occurred in South America.

Not only the ancestor of maize, but also whether it was domesticated first in one or in several places is still in doubt. While it seems fairly evident that there was at least one origin in Mexico, the possibility that there may have been independent domestications in South America is once again alive. Mangelsdorf (63) has postulated that the six lineages of the races of maize stem from separate domestications, two in Mexico, one in Colombia, and three in Peru. Zevallos et al (100) maintain that by 3000 BC maize cultivation in Ecuador had reached an efficiency not yet known in Mexico at that time. Although their direct evidence—a single grain of maize—is less than a botanist might desire, considerable indirect evidence also substantiates their belief. More recently Pearsall (69) has reported phytoliths of maize in the same area. Of course a primitive maize may still have reached Ecuador very early from Middle America rather than having been domesticated somewhere in South America. If indeed maize descended from a wild maize, it would be much easier to accept the idea of the latter's extinction if the domesticated plant had but a single origin (74).

Another view of the origin of maize may be expected from Randolph, but as of this writing only the first (76) of his projected three papers on the subject has appeared. Probably only future archaeological discoveries can help to determine the ancestor of maize (94). At present teosinte must be regarded as a more likely progenitor than a wild maize, and it seems most likely that the domestication occurred only in Mexico.[1]

[1]The diploid perennial teosinte recently discovered by Guzman (53a) may have implications bearing on the origin of maize. H. G. Wilkes (unpublished) points out that perennial teosinte could not have arisen from maize and that annual teosinte must have originated from hybridization of maize and perennial teosinte.

SQUASHES AND PUMPKINS (CUCURBITA SPP.)

The genus *Cucurbita,* which has provided us with five domesticated species, has also received considerable study and yet wild progenitors are not definitely known for any of the species. Although Whitaker & Bemis (92) state that "*C. moschata* is the indispensable cog through which the [cultivated] species of *Cucurbita* are related," the morphological evidence and the sterility in crosses between the species are such as to suggest that different wild species participated in the origin of each cultigen. The archaeological record and the study of the pollinating insects indicate that domestication of different species occurred in both Middle and South America (52, 91). Feral species or races have been associated with three of the cultigens, *C. texana* with *C. pepo, C. andreana* with *C. maxima,* and an unidentified species with *C. mixta.* Whitaker & Bemis (92) suggest that the feral species may be escapes from cultivation; alternatively, they may be the ancestors of the domesticated forms. *Cucurbita texana* occurs in central Texas, an unlikely place, it must be admitted, for an ancestor of *C. pepo* since the archaeological record indicates that the latter was most likely domesticated in southwestern Mexico. Of course *C. texana* may once have had a much more extensive distribution. A truly wild species, *C. lundelliana* of Guatemala and British Honduras, may be implicated in the origin of the cultivated species, particularly of *C. moschata;* it can be crossed with all five of the domesticated species (92). Thus, while candidates are available for ancestors of the four annual domesticates, none at all has yet been proposed for the perennial *C. ficifolia,* a widely cultivated highland species.

Remains of *C. pepo* from three sites in eastern North America have been dated at 2300 BC. This is considerably earlier than dates known for any of the native domesticates in this area. It suggests that agriculture was introduced from Mexico rather than being an indigenous development in eastern North America as had been previously postulated (20).

BEANS (PHASEOLUS SPP.)

In contrast to the situation in *Cucurbita,* wild, presumably ancestral, forms have been identified for all four species of *Phaseolus* that were domesticated in the Americas. There is, however, some disagreement as to which wild type gave rise to the common bean, *Ph. vulgaris.* In 1952 Burkart (18) described *Ph. aborigineus* from Argentina, which he postulated was the wild ancestor of *Ph. vulgaris.* In the next year Burkart & Brücher (19) made it a subspecies of *Ph. vulgaris* and reported other occurrences of it in South America as well as a small seeded variety from Honduras. They also pointed out that seeds of it were sometimes used for food by man and that hybrids had been obtained between it and the domesticated bean. In 1969 Gentry (31) reported a wild bean whose distribution extended from the state of Sinaloa, Mexico, to Honduras. On the basis of several lines of evidence he claimed that it was the ancestor of the common bean. He also agreed that the domesticated bean had multiple origins in Middle America, a hypothesis put forward earlier by Kaplan (56), based upon archaeological investigations. Gentry suggested that *Ph. aborigineus* might comprise nothing more than escapes from cultivation, "as their seeds suggest."

Berglund-Brücher & Brücher have recently (14) added new data bearing on *Ph. aborigineus* as the progenitor of the common bean. Strong support for a South American origin of the common bean had been provided by an archaeological find in highland Peru dated at 5500 BC or earlier (57), but this, as they concede, does not exclude the possibility of domestication in other regions. They attempt to show that *Ph. aborigineus* is truly wild, not an escape, and that it has not been modified by hybridization with the cultivated bean. They state that it occurs in truly natural, not disturbed, sites, though one locality is given as "right on the access road to Merida [Venezuela]" and another as "in the area surrounding the German brewery outside the town of Cochabamba [Bolivia]." Their bean is a strict annual, the dehiscence mechanism of the pods is "less well developed than in other wild growing Phaseolineae;" the seeds average 5–10 mm long, 3.5–7 mm wide, and 3–4.5 mm thick and have hard seed coats. The hard seed coats, characteristic of the wild species, usually function to delay germination, but they report that in one test 50% of soaked seeds started to germinate in two days. In contrast, Gentry's wild beans of Mexico are mostly perennial; the pods are "nearly all twisting dehiscent, ejecting the seeds violently;" the seeds are hard and average about 5 mm long, 4 mm wide, and 2–2.5 mm thick, with about one in a hundred having much larger seeds. No details are given on germination. The above and the other information suggests that the two wild beans are hardly more than geographical races of one species. Moreover, the Mexican bean with its perennial habit, strongly dehiscent pods, and smaller seeds is the more likely candidate for a truly wild plant. It seems probable that *Ph. aborigineus* is at least in part an ancient escape and has been modified by hybridization with the cultivated forms.

No really important disagreement exists except over the priority of the discovery of the ancestor of the common bean, for elements of the same wild species could have been developed into a domesticated bean in both Middle America and South America. In 1965 Heiser (44) raised the possibility of such independent domestications. At that time the archaeological record showed the common bean to be more ancient in Mexico (4000 BC) than in Peru (500 BC). Recent discoveries that place the common bean at 5000 and

5500 BC in Mexico and Peru, respectively, make a double origin even more likely. The wild bean may originally have been confined to Mexico, and then it may have been carried by man as a wild plant far into South America. Such a process could explain the extensive distribution of the wild beans.

A recent review of the wild and cultivated tepary bean (*Ph. acutifolius*) of the American Southwest and Mexico (66) does not alter previous knowledge about the origin of cultivated forms. Fernández (25) has completed an ecological study of the scarlet-runner bean (*Ph. coccineus*), finding in it a high level of plasticity. Smartt (83) has raised the possibility that the domesticated runner bean, *Ph. coccineus* ssp. *darwinianus,* may represent a distinct species. He (82) has also provided a detailed review of the changes that occurred in the species of *Phaseolus* with domestication.

AVOCADO (PERSEA AMERICANA)

Multiple origins for the avocado have also been postulated (44). The archaeological record reveals it to be one of the oldest food plants of Mexico (8000 BC); it was also well established in Peru in early times (1500 BC). Three major races of the avocado have long been recognized: the Mexican, the Guatemalan, and the West Indian. Many modern varieties are hybrids between the races. While all recent taxonomic treatments agree on scientific names for the Mexican (*P. americana* var. *drymifolia*) and the West Indian (*P. americana* var. *americana*) races, opinion differs about how to treat the Guatemalan race. Kopp (60) used *P. americana* var. *nubigena* to include both cultivated and wild forms of this race. Bergh (11–13) proposed subspecific status under *P. americana* as "variety *guatemalensis,*" though this has yet to be validly published; and he suggested that the wild form, *P. nubigena,* would also be better treated as a variety of *P. americana.* In the most recent treatment Williams (95) uses *P. nubigena* var. *nubigena* for the wild forms and *P. nubigena* var. *guatemalensis* for the cultivated race. In a comparison of leaf terpenes Bergh et al (13) found that cultivars of the West Indian and Mexican races had similar distribution patterns, whereas the Guatemalan race had a different pattern, which was similar to that of *P. nubigena.* They used this, along with other lines of evidence, such as the lack of sterility barriers between the races, to support the concept of a single species. Williams might also have used the leaf terpenes to support the recognition of two species. Bergh et al (13) also suggest that another wild species, *P. floccosa,* recognized by both Kopp and Williams, might also be considered merely a variety of *P. americana.*

The Mexican and Guatemalan races seem to have originated in the regions from which they received their names. It is generally agreed that the West Indian race was not observed in the West Indies when the first

Europeans arrived, so its origin poses more of a problem. Williams (95) states that it was grown by peoples of pre-Conquest Mexico and Guatemala and was developed from selections of the Mexican avocado. Bergh is of the opinion that primitive types of the West Indian race are known from Colombia; the implication is that it was developed from wild forms in that area. Kopp (60) cites specimens of *P. americana* var. *americana* from South America in her monograph, but it is impossible to determine whether these are wild or cultivated. Williams (95) does not mention a wild form in South America. If, as Williams supposes, the West Indian race originated in Mexico, one must postulate an introduction of the avocado into South America by man. If, on the other hand, the West Indian race originated in Colombia (10, 12), the avocado may have been domesticated in both Middle and South America. One question to be answered is: Were there wild avocados in South America in pre-historic times? Patiño (68) records that the Spanish in the 16th Century observed the avocado both as a cultivated fruit and as one of the fruits "de la tierra" in Colombia. While the latter could represent escapes from cultivation, it is possible that they were truly wild. Although it is notoriously difficult to interpret linguistic evidence in attempting to trace origins, it may be significant that there are two native names, *curo* and *okze,* for the avocado in Colombia. The Mexican Indian name, *aguacate,* now widely used in South America, probably stems from the Spanish introduction of the name. *Palta* is the word used throughout Peru, and this name has been attributed to the Jivaros of Ecuador (68).

Isozyme studies now under way (87) may eventually aid in interpreting the origins of the races of the avocado. On the basis of comparisons of peroxidase zymograms of a number of strains, Garcia & Tsunewaki (29) have suggested that the Puebla-Veracruz area was the center of origin of the Mexican race.

CHILI PEPPERS (CAPSICUM SPP.)

Five species of domesticated chili peppers, *C. annuum, C. baccatum, C. chinense, C. frutescens,* and *C. pubescens,* have been recognized in recent years; wild, weed, or spontaneous races, presumably similar to the ancestral forms, have been identified for the first four. A numerical taxonomic analysis of the species by Pickersgill et al (75), excluding the very distinctive *C. pubescens,* has shown that the wild and domesticated forms of *C. baccatum* are distinct from the other three species. Analyses of domesticated forms of *C. annuum* and *C. chinense* produce distinct clusters; domesticated forms of *C. frutescens* produce less satisfactory clusters. The spontaneous forms of these three species, however, tend to cluster more or less together. These results pose a taxonomic dilemma: The three domesticates, which are probably entitled to specific rank, are not sharply delimited from their respective spontaneous forms; but the spontaneous forms intergrade to such an extent that they can hardly be differentiated as separate species. Isozyme studies (24, 62) show that the five species fall into three groups: *C. pubescens* in one; *C. baccatum* in another; and *C. annuum, C. chinense,* and *C. frutescens* in still another, the last three species being indistinguishable. Cytological studies (73) have shown that all of the species have the same basic karyotype, consisting of 11 pairs of metacentric or submetacentric pairs and one pair of acrocentric chromosomes. Karyotypes within *C. annuum,* however, vary. Most cultivated forms have two pairs of acrocentric chromosomes. Because spontaneous forms of this species with two pairs of acrocentrics are largely confined to Mexico, Pickersgill (71) has suggested that country as the original site of domestication—a suggestion supported by the long archaeological record of the species in that region. Although the recent studies have shown that the evolution and relationship of the chili peppers are far more complex than previously thought, they do not falsify the hypothesis that several independent domestications occurred. *Capsicum baccatum, C. chinense,* and *C. pubescens* were domesticated in different parts of South America, *C. annuum* in Mexico, and *C. frutescens* most likely somewhere in southern Middle America.

IRISH POTATO (SOLANUM TUBEROSUM)

The potato, a tetraploid species, arose in the Andes. Various species have been postulated as the diploid progenitor (80); Hawkes (40, 41) has proposed *S. stenotomum* introgressed with *S. sparsipilum,* while Brücher (15) has proposed *S. vernei.* A study by Grun et al (35) revealed that both *S. stenotomum* and *S. vernei* are similar to *S. tuberosum* ssp. *andigena* in their cytoplasmic factors and that either could have given rise to it with minor (if any) cytoplasmic changes. Results of a study of variation in Fraction 1 protein (30) were consistent with an origin of ssp. *andigena* from chromosome doubling of *S. stenotomum* and did not support an involvement of *S. sparsipilum; S. vernei* was not included in the study.

Solanum tuberosum ssp. *andigena* was introduced to Europe in the 16th Century and persisted there until the late blight struck in the middle of the 19th Century. According to one group of investigators (35) catastrophic selection among the survivors of the blight gave rise to *Solanum tuberosum* ssp. *tuberosum.* Experimental evidence shows that ssp. *tuberosum*-like plants can be produced through selection from ssp. *andigena* (32, 78). Grun (34) has proposed as an alternative that at the time of the late blight ssp. *tuberosum* was imported to Europe from islands off the coast of Chile. In their study of cytoplasmic factors Grun et al (35) found that plants of ssp.

tuberosum from the northern hemisphere and coastal Chile shared all or nearly all of the same plasmon sensitivities, whereas they differed from ssp. *andigena* in eight of the nine plasmon sensitivities tested. Thus ssp. *tuberosum,* now cultivated in the northern hemisphere, may be derived directly from Chile rather than through the transformation of ssp. *andigena* (see also 30).

NARANJILLA (SOLANUM QUITOENSE), TOPIRO (SOLANUM TOPIRO), AND PEPINO (SOLANUM MURICATUM)

Several species of *Solanum* were domesticated in the Americas for their fruits rather than their tubers. The cultivation of the *naranjilla* or *lulo* (*S. quitoense*) until recently in this century was apparently confined to Colombia and Ecuador. The presence in Colombia of a spiny form of the species (probably more primitive than the spineless form) and the persistence of the Indian name lulo there suggest that domestication may have originated in Colombia. A comparative study with six related wild species (47) found that on morphological grounds the nearest relative of *S. quitoense* was *S. tequilense. Solanum quitoense* crossed readily with *S. hirtum* when the latter was used as the female parent, and the resulting hybrids showed around 50% fertility. On the other hand, crosses between *S. quitoense* and *S. tequilense* were very difficult to obtain, and the few hybrids secured showed a very low fertility. However, for several reasons it was thought unlikely that either *S. hirtum* or *S. tequilense* was the wild progenitor of the naranjilla.

The same study postulated that the progenitor of the *topiro* or *cocona, S. topiro* var. *topiro,* widely cultivated in the Amazon basin, was *S. topiro* var. *georgicum,* previously recognized as a distinct species. The two share a number of morphological features; the characters differentiating the topiro from the wild variety, such as its larger and more variable fruit, might be the result of human selection. Moreover, the two were found to give fertile hybrids when crossed.

The *pepino* or *cachon, S. muricatum,* is widely cultivated in the Andes. Heiser (42) suggested that the wild species, *S. caripense,* whose fruits are sometimes eaten and which in some combinations gives fertile hybrids with *S. muricatum,* might be the ancestral species. He also pointed out that the morphological similarities of *S. muricatum* and *S. tabanoense* might implicate the latter species in the origin of the pepino. Brücher (16) gave no detailed reasons for his belief that *S. caripense* could not be the wild type of *S. muricatum;* he suggested that *S. tabacoense* (sic) was a more likely candidate. In a follow-up study Heiser (45) pointed out that hybrids of *S.*

tabanoense with *S. muricatum* had reduced fertility; he advanced further reasons for considering *S. caripense* the likely progenitor. Anderson (3) has recently suggested *S. basendopogon* as the possible progenitor. It is morphologically similar to the pepino, but the fruits thus far secured in hybrids between the two species are seedless.

CHENOPODS (CHENOPODIUM SPP.)

Three species of *Chenopodium* are now cultivated in the Americas—*kaniwa* or *canihua* (*Ch. pallidicaule*) and *quinoa* (*Ch. quinoa*) in the Andes, and *huauzontle* (*Ch. nutalliae*) in Mexico. *Chenopodium pallidicaule,* best considered a semi-domesticate, is a diploid and is not closely related to the other two species. It is cultivated today in the Altiplano at high altitudes, growing at 3800–4340 m in the Vilcanota Valley of Peru (27). It probably originated as a crop in the Altiplano. Quinoa and huauzontle, on the other hand, are closely related tetraploids. Simmonds (81) has suggested that they may be conspecific; they could have originated in the areas where they are still grown, or huauzontle could have originated from northern migrations of quinoa. Wilson & Heiser (97), on the other hand, find that although the morphological differences between the two are slight, F_1 hybrids between them show considerable sterility, isozyme differences (96) exist, and thus they should be recognized as distinct species. On the basis of morphology, crossing results, and isozyme analysis, a Mexican tetraploid weed, *Ch. berlandieri,* is very closely related to huauzontle. Wilson & Heiser conclude that huauzontle is no more than a subspecies of *Ch. berlandieri* and that the latter gave rise to the former in Mexico.

The origin of quinoa is not yet as clear cut. It too has a companion weed, *Ch. hircinum,* which occurs with it in the Andes. Morphological, hybridizational, and isozyme analyses indicate that the weed and the domesticate are conspecific. Whether the weed is the progenitor of quinoa or is derived from it requires more study. Quinoa and huauzontle are usually characterized by pale fruits, but a few cultivars have dark fruits, a wild-type characteristic. Since the pale fruit in the two domesticates is controlled by different genes (50), it is unlikely that quinoa is derived from a pale-fruited form of huauzontle. It may have derived from a dark-fruited form of huauzontle that was carried by man to the Andes, where speciation occurred and the pale fruits developed independently. If the last suggestion is correct, then the companion weed of quinoa derives from the crop plant.

A chenopod may also have been cultivated in eastern North America in prehistoric times (98). Asch & Asch (4) found no differences in fruit measurements of archaeological material and modern wild species in the eastern United States, but H. D. Wilson (unpublished) has found certain fruit characters that suggest a domesticated chenopod may have been cultivated in that region. Whether this was *C. nuttalliae* or a native eastern North American species is still open to question.

SWEET POTATO (IPOMOEA BATATAS)

The sweet potato arose somewhere in tropical America (99); nothing more is certain. Austin (5) feels that the few varieties in Mexico and the unpopularity of the sweet potato there make Mexico an unlikely place of origin. He has also suggested (6) that some of the great variability in the domesticate, a hexaploid, stems from its hybridization with diploid species to produce tetraploids with subsequent backcrossing of the latter to the sweet potato. His revision (7) of the "batatas complex" of the genus should aid in future studies of the origin.

CHOCHO (LUPINUS MUTABILIS)

Species of *Lupinus* were domesticated in both the Old and New Worlds. The New World species, *chocho* or *tarui, L. mutabilis,* is apparently an old domesticate of the Andes. Apparently following Kazimierski & Nowacki (59), Smith (85) recently stated that two primitive North American lupins were probably carried by man to South America as semi-domesticated plants where, after a phase of hybridization, they gave rise to *L. mutabilis.* Such an elaborate hypothesis appears unnecessary. Heiser (43) has pointed out several native species in both Bolivia and Peru that might qualify as progenitors of *L. mutabilis,* and Brücher (17) has noted other large-seeded wild species that might have been involved. *Lupinus mutabilis* is known to hybridize with *L. pubescens* in Ecuador (43). The present confused taxonomic situation of the South American wild species of *Lupinus* will probably have to be clarified before we can arrive at a better understanding of the origin of the chocho.

TOTORA (SCIRPUS CALIFORNICUS)

A wild bulrush, ranging from California to Chile and known as *totora* in South America and as *tule* in Mexico, was widely used in the construction of mats, boats, and other items throughout much of its range; it was a minor source of food in the area of Lake Titicaca. The plant is also found on Easter Island, and Heyerdahl (51) uses its presence there as support for his argument that the island was originally settled by ancient Peruvians. He maintains that the plant is identical to that found in Peru, that it was cultivated in Peru in prehistoric times, and that because it reproduces only vegetatively it must have been carried to Easter Island by man. A recent study (48) supports some of Heyerdahl's claims but arrives at a different conclusion. *S. californicus* from Easter Island is virtually identical to that from Peru. Although the works cited by Heyerdahl do not support his claim that the plant was cultivated in Peru in prehistoric times, a recent archaeological investigation (58) has recovered material of the totora from sunken gardens on the north coast of Peru. These gardens were used for growing crop plants. The authors' inference that the totora was actually cultivated does not necessarily follow, for the totora could have invaded the sunken gardens as a weed by natural means. Finally, it has been shown that the totora both in the Americas and Easter Island reproduces by seeds; birds rather than men more likely introduced it onto Easter Island. Many sedges not known to have been used by man have extensive distributions, including both the Americas and Pacific islands; it is generally held that their broad ranges are the result of dispersal by birds.

Today the totora is cultivated in both Peru and Ecuador, but the practice apparently began in this century. The cultivated plants do not differ from those in the wild, so they can not yet be considered domesticated. Prehistoric man may be responsible for the extensive altitudinal distribution of this species, from sea level to 3750 m, but a natural dispersal by birds cannot be ruled out.

BOTTLE GOURD (LAGENARIA SICERARIA)

Although the bottle gourd is generally conceded to be native to Africa, it has had a long history in the Americas. It has been found in archaeological deposits dating to 7000 BC or perhaps earlier. An archaeological report dating the bottle gourd at between 6000 and 10,000 BC in Thailand (33) is in error (49). The earliest remains from Africa come from levels dated at around 3000 BC. Two schools of thought have developed as to how the bottle gourd reached the Americas from Africa, one holding that it was carried by man and the other that fruits floated across the Atlantic. Lathrap (61), the most recent champion of the former view, feels that certain leguminous plants used as fish poison and a linted cotton made the same trip with the bottle gourd. Heiser (49), however, feels that ocean drifting of gourds is the more likely explanation. He points out that the current status of the taxonomy of the leguminous fish poisons makes it difficult to discuss the relationship of the African and American species; furthermore, it is now considered unlikely that man figured in the origin of the New World cottons (70).

Harlan (38) gives the bottle gourd as an example of a species with a very wide distribution that likely was domesticated more than once. While multiple domestications of the bottle gourd may well have occurred, the curious fact remains that truly wild or weedy forms, as opposed to escapes from cultivation, have yet to be documented.

WEEDS AND DOMESTICATES

The relation of a weed to a domesticated plant—whether the weed is the progenitor of the cultivated plant or *vice versa*—is not a simple matter, as Anderson (2) pointed out some time ago. Harlan (36) has expanded Anderson's ideas, and de Wet & Harlan (22) have recently reviewed some of the complex relations of weeds and domesticates. The progenitor, of course, may have changed considerably since giving rise to a domesticated plant; it may have been modified by hybridization with the domesticate or with other species, and the hundreds or thousands of generations through which it has subsequently passed may have allowed new mutants to appear and be selected.

Certain weeds have descended from crop plants. The weedy shatter and chicken canes in the United States come from *Sorghum bicolor* which, although cultivated in the United States, is a native of Africa (38). As indicated by *Sorghum* (39) and by *Solanum quitoense*, this process may take place quite rapidly. The latter was established in Costa Rica less than a quarter of century ago and already a weedy race has arisen there (46). Weedy races of it may exist in its homeland of northern South America; if so, they have yet to be recognized.

In some plants of Old World origin, both wild and weed races are known that belong to the same species as the domesticate, as has been reported for *Sorghum bicolor* and pearl millet, *Pennisetum americanum*, in Africa (38). In many American plants, however, it is often difficult to make a clear distinction between wild and weedy races related to the domesticated plant. An attempt can be made to define those plants growing in undisturbed sites as wild plants, those growing in naturally disturbed sites as weedy wild types, and those growing in man-disturbed sites as true weeds; but site type may not always be a reliable guide. Standley (86) has written of the difficulty of deciding whether trees of cacao, avocado, and sapote found in an apparently undisturbed site of Honduras are truly wild. The vegetation appears primeval, but one finds abundant potsherds upon the exposed slopes. Is it not possible, he asks, that these trees are remnants of plantations of long ago? *Capsicum annuum* var. *glabriusculum*, thought to be the progenitor of the chili pepper, extends from the southern United States to Colombia, and throughout most of that area it is best regarded as a weed. However, in Florida, where it is found growing in apparently natural vegetation, it is nearly always associated with old Indian shell mounds. Perhaps the wild

types of some species had already disappeared before man began to domesticate plants. Would man then have chosen first to cultivate the nearest
weeds? Certainly in some instances the wild species far from the homes of
the people would have had greater appeal. Nevertheless, as supposed by
Vavilov (88), some weeds must have been progenitors of domesticated
plants. Jenkins (54) asked whether the widespread tropical weed, *Lycopersicon esculentum* var. *cerasiforme,* was a reversion from the domesticated
tomato or the ancestor of it; he concluded that the rarity of the reverse
mutation from the three-loculed fruit of the domesticated tomato to the
dominant two-loculed fruit of var. *cerasiforme* argued against considering
it a derivative of the domesticated plant. Similarly, the karyotype changes
in *Capsicum annuum,* already discussed, supports the argument that the
spontaneous type represents a progenitor rather than a descendant of the
cultivated chili pepper. With some domesticates [e.g. the sweet potato (55,
67)] disagreements continue over whether a weed is a progenitor or a
descendant of the domesticate.

CENTERS OF ORIGINS AND DIVERSITY

Vavilov (88) originally proposed Mexico and South America as centers of
origin of cultivated plants in the New World. Later he (89) looked upon
Mexico as the principal center and admitted that two plants, the sunflower
(*Helianthus annuus*) and the Jerusalem artichoke (*H. tuberosus*), had been
domesticated to the north of Mexico. In his final paper on the subject (90)
he recognized two subcenters, Chiloe and Brazil-Paraguay in South America, in addition to the Middle American center and South American center
(Peru, Ecuador, and Bolivia). Harlan (37) has proposed that agriculture
originated in three different parts of the world; in the New World a Mexican
center and South American "non-center" (in which domestication occurred
over an extensive area) interacted. On the other hand, Pickersgill (72) feels
that agriculture may have developed independently in four different parts
of the New World—Middle America, the Andes, the humid lowland tropics, and eastern North America. Pickersgill & Heiser (74) have pointed out
that an impressive number of plants were domesticated in both Middle
America and South America; there may be no justification for regarding one
as a center and the other as "a non-center." Although plants were domesticated over a broader area of South America than of Middle America, they
were domesticated in more than one area of Middle America. It was once
thought that the Mexican center developed earlier than any in South America, but recent discoveries, particularly the finding of two species of domesticated beans in Peru dated at 5500 BC or earlier (57), now make that priority
questionable. One might conclude with Harlan (37) that "crops did not

necessarily originate in centers (in any conventional sense of the term), nor did agriculture necessarily develop in a geographical 'center.' " The idea of centers of origin may have outlived its usefulness, at least in the Americas.

The last statement may also apply to Vavilov's concept that centers of diversity indicate centers of origin, for both the New World and the Old. The criticisms of this concept and other interpretations of centers of diversity have been brought together by Harlan (37) and Pickersgill (72). Pickersgill cautions that in view of the few thorough investigations of cultivated plants inside and outside the range of their relatives it may still be too early to determine how generally applicable Vavilov's criteria are.

ACKNOWLEDGMENTS

I would like to thank Barbara Pickersgill for her constructive comments on earlier drafts of this paper, Bruce Serlin for calling my attention to some of the literature on *Persea,* and Peter Bretting and Lewis Johnson for their helpful comments. The interpretations and conclusions are my own. My work with domesticated plants has been aided by grants from the National Science Foundation.

Literature Cited

1. Anderson, E. 1952. *Plants, Man and Life,* p. 152. Boston: Little, Brown. 245 pp.
2. Anderson, E. 1956. Man as a maker of new plants and new plant communities. In *Man's Role in Changing the Face of the Earth,* ed. W. Thomas, pp. 736–77. Chicago: Univ. Chicago Press. 1193 pp.
3. Anderson, G. J. 1979. Systematics and evolutionary considerations of species of *Solanum* section Basarthrum. In *The Biology and Taxonomy of the Solanaceae,* ed. J. G. Hawkes, K. N. Lester. London-Academic. In press
4. Asch, D. L., Asch, N. B. 1977. Chenopod as cultigen: a reevaluation of some prehistoric collections from eastern North America. *Mid-Cont. J. Archaeol.* 2:3–45
5. Austin, D. F. 1973. The camotes de Santa Clara. *Econ. Bot.* 27:343–47
6. Austin, D. F. 1977. Hybrid polyploids in *Ipomoea* section *batatas. J. Hered.* 68:259–60
7. Austin, D. F. 1978. The *Ipomoea batatas* complex—I. taxonomy. *Bull. Torrey Club* 105:114–29
8. Beadle, G. W. 1972. The mystery of maize. *Field Mus. Nat. Hist. Bull.* 43(10):2–11
9. Beadle, G. W. 1977. The origins of *Zea mays.* See Ref. 77, pp. 615–35
10. Bergh, B. O. 1969. Avocado. In *Outlines of Perennial Crop Breeding in the Tropics,* ed. F. P. Ferwerda, F. Wit, pp. 23–51. Wageningen, The Netherlands: Misc. Pap. 4, Landbouwhogeschool. 511 pp.
11. Bergh, B. O. 1975. Avocados. In *Advances in Fruit Breeding,* ed. J. Janick, J. N. Moore, pp. 541–67. W. Lafayette, Ind: Purdue Univ. Press.
12. Bergh, B. O. 1976. See Ref. 79, pp. 148–51
13. Bergh, B. O., Scora, R. W., Storey, W. B. 1973. A comparison of leaf terpenes in *Persea* subgenus *Persea. Bot. Gaz.* 13:130–34
14. Berglund-Brücher, O., Brücher, H. 1976. The South American wild bean (*Phaseolus aborigineus* Burk.) as ancestor of the common bean. *Econ. Bot.* 30:257–72
15. Brücher, H. 1964. El origen de la papa (*Solanum tuberosum*). *Physis* 24:439–52
16. Brücher, H. 1968. Die genetischen Reserven Südamerikas für die Kulturpflanzenzuchtung. *Theor. Appl. Genet.* 38:9–22
17. Brücher, H. 1970. Beitrag zur Domestikation proteinreicher und alkaloidarmer Lupinen in Südamerika. *Angew. Bot.* 44:7–27

18. Burkart, A. 1952. *Las Leguminosas Argentinas.* Buenos Aires: Acme. 569 pp. 2nd ed.
19. Burkart, A., Brücher, H. 1953. *Phaseolus aborigineus* Burkart, die mutmassliche andine Stammform der Kulturbohne. *Zuchter* 23:65–72
20. Chomko, S. A., Crawford, G. W. 1978. Plant husbandry in prehistoric eastern North America: new evidence for its development. *Am. Antiq.* 43:405–8
21. Darwin, C. 1890. *The Variation of Plants and Animals under Domestication,* Vol. 1. NY: Appleton, 473 pp. 2nd ed. rev.
22. de Wet, J. M. J., Harlan, J. R. 1975. Weeds and domesticates: evolution in the man-made habitat. *Econ. Bot.* 29:99–107
23. de Wet, J. M. J., Harlan, J. R., Stalker, H. T., Randrianasolo, A. V. 1978. The origin of tripsacoid maize (*Zea mays* L.) *Evolution* 32:233–44
24. Eshbaugh, W. H. 1977. The taxonomy of the genus *Capsicum* (Solanaceae). In *"Capsicum 77," C. R. 3ᵐᵉ Congr. Eucarpia Piment, Avignon-Montfavet (France),* ed. E. Pochard, pp. 13–26.
25. Fernàndez, M. P. 1978. *Estudio comparativo de patrones de ciclo de vida de poblaciones de* Phaseolus coccinius (*Leguminosae*). Doctoral thesis. Univ. Nac. Aut. Mexico, Mexico, D. F.
26. Flannery, K. V. 1973. The origins of agriculture. *Ann. Rev. Anthropol.* 2:271–310
27. Gade, D. W. 1975. *Plants, Man and the Land in the Vilcanota Valley of Peru.* The Hague: Junk. 240 pp.
28. Galinat, W. C. 1977. The origin of corn. In *Corn and Corn Improvement,* ed. G. F. Sprague. Madison, Wis: Am. Soc. Agron. 774 pp.
29. Garcia, A., Tsunewaki, K. 1977. Cytogenetical studies in the genus *Persea* (Lauraceae). III. Electrophoretical studies on peroxidase isozymes. *Jpn. J. Genet.* 52:379–86
30. Gateby, A. A., Cocking, E. C. 1978. Fraction 1 protein and the origin of the European potato. *Plant Sci. Lett.* 12:177–81
31. Gentry, H. S. 1969. Origin of the common bean, *Phaseolus vulgaris. Econ. Bot.* 23:55–69
32. Glendinning, D. R. 1975. Neo-Tuberosum: new potato breeding material. 1. The origin, composition and development of the Tuberosum and Neo-Tuberosum gene pools. 2. A comparison of Neo-Tuberosum with unse-

lected Andigena and with Tuberosum. *Potato Res.* 18:256–61
33. Gorman, C. 1969. Hoabinhian: a pebble-tool complex with early plant association in southeast Asia. *Science* 113:671–73
34. Grun, P. 1974. Cytoplasmic sterilities that separate the group Tuberosum cultivated potatoes from its putative tetraploid ancestor. *Evolution* 27:633–43
35. Grun, P., Ochoa, C., Capage, D. 1977. Evolution of cytoplasmic factors in tetraploid cultivated potatoes (Solanaceae). *Am. J. Bot.* 64:412–20
36. Harlan, J. R. 1965. The possible role of weed races in the evolution of cultivated plants. *Euphytica* 14:173–76
37. Harlan, J. R. 1971. Agricultural origins: centers and non-centers. *Science* 174:468–74
38. Harlan, J. R. 1976. Plant and animal distribution in relation to domestication. See Ref. 53, pp. 13–25
39. Harlan, J. R., de Wet, J. M. J. 1974. Sympatric evolution in *Sorghum. Genetics* 78:473–74
40. Hawkes, J. G. 1956. Taxonomic studies on the tuber-bearing Solanums. I. *Solanum tuberosum* and the tetraploid species concept. *Proc. Linn. Soc. London.* 166:97–144
41. Hawkes, J. G. 1972. Evolution of the cultivated potato, *Solanum tuberosum* L. *Symp. Biol. Hung.* 12:183–88
42. Heiser, C. B. 1964. Origin and variability of the pepino: a preliminary report. *Baileya* 12:151–58
43. Heiser, C. B. 1964. Chochos and other lupines in Ecuador. *Bull. Mo. Bot. Gard.* 52(10):8–13
44. Heiser, C. B. 1965. Cultivated plants and cultural diffusion in nuclear America. *Am. Anthropol.* 67:930–49
45. Heiser, C. B. 1969. *Solanum caripense* y el origen de *Solanum muricatum. Politechnica* 1(3):5–11
46. Heiser, C. B. 1971. Notes on some species of *Solanum* (Sect. Leptostemonum). *Baileya* 18:59–65
47. Heiser, C. B. 1972. The relationships of the naranjilla, *Solanum quitoense. Biotropica* 4:77–84
48. Heiser, C. B. 1979. The totora (*Scirpus californicus*) in Ecuador and Peru. *Econ. Bot.* 32. In press
49. Heiser, C. B. 1979. *The Gourd Book.* Norman, OK: Univ. Oklahoma Press.
50. Heiser, C. B., Nelson, D. 1974. On the origin of the cultivated chenopods (*Chenopodium*) *Genetics* 78:503–5
51. Heyerdahl, T., Ferdon, E. N. Jr. 1961. *Archaeology of Easter Island,* Vol. I.

Santa Fe, NM: Sch. Am. Res. Mus. New Mexico.

52. Hurd, P. D., Linsley, E. G., Whitaker, T. W. 1971. Squash and gourd bees (*Peponapis, Xenoglossa*) and the origin of the cultivated *Cucurbita. Evolution* 25:218–34

53. Hutchinson, J., ed. 1977. *The Early History of Agriculture.* Oxford: Oxford Univ. Press. 213 pp. (reprint. from *Philos. Trans. R. Soc. Lond. Ser. B.* 275:1–213)

53a. Iltis, H., Doebley, J. F., Guzmán, M. R., Pazy, B. 1079. *Zea diploperennis* (Gramineae): a new teosinte from Mexico. *Science* 203:186–87

54. Jenkins, J. A. 1948. The origin of the cultivated tomato. *Econ. Bot.* 2:379–92

55. Jones, A. 1967. Should Nishiyama's K123 (*Ipomoea trifida*) be designated as *I. batatas? Econ. Bot.* 21:163–66

56. Kaplan, L. E. 1965. Archaeology and domestication in American *Phaseolus* (Beans). *Econ. Bot.* 19:358–68

57. Kaplan, L. E., Lynch, T. F., Smith, C. E. 1973. Early cultivated beans (*Phaseolus vulgaris*) from an intermontane Peruvian valley. *Science* 179:76–77

58. Kautz, R. R., Keatinge, R. W. 1977. Determining site function: a north Peruvian coastal example. *Am. Antiq.* 42:86–97

59. Kazimierski, J., Nowacki, E. 1961. Indigenous species of lupins regarded as initial forms of the cultivated species, *Lupinus albus* and *Lupinus mutabilis. Flora* 151:202–9

60. Kopp, L. E. 1966. A taxonomic revision of the genus *Persea* in the western hemisphere. *Mem. NY Bot. Gard.* 14:1–117

61. Lathrap, D. W. 1977. Our father the cayman, our mother the gourd: Spinden revisited, or a unitary model for the emergence of agriculture in the New World. See Ref. 77, pp. 713–51

62. McLeod, M. J. 1977. *A systematic and evolutionary study of the genus* Capsicum. Ph.D. thesis. Miami University, Oxford, Ohio. 103 pp.

63. Mangelsdorf, P. C. 1974. *Corn: Its Origin, Evolution and Improvement.* Cambridge, Mass: Harvard Univ. Press. 262 pp.

64. Mangelsdorf, P. C., Reeves, R. G. 1939. The origin of Indian corn and its relatives. *Tex. Agric. Exp. Stn. Bull.* 574: 315 pp.

65. Mangelsdorf, P. C., Reeves, R. G. 1959. The origin of corn. I. Pod corn, the ancestral form. *Bot. Mus. Leafl., Harv. Univ.* 18:329–56

66. Nabhan, G. P., Felger, R. S. 1978. Teparies in southwestern North America —a biogeographical and ethnohistorial study of *Phaseolus acutifolius. Econ. Bot.* 32:2–19

67. Nishiyama, I. 1971. Evolution and domestication of the sweet potato. *Bot. Mag. Tokyo* 84:377–87

68. Patiño, V. M. 1963. *Plantas Cultivadas y Animales Domesticos en America Equinoccial,* Vol. I. Cali, Colombia: Imprenta Departmental. 547 pp.

69. Pearsall, D. M. 1978. Phytolith analysis of archaeological soils: evidence for maize cultivation in Formative Ecuador. *Science* 199:177–78

70. Phillips, L. L. 1976. Cotton. See Ref. 79, pp. 196–200

71. Pickersgill, B. 1971. Relationships between weedy and cultivated forms in some species of chili peppers (genus *Capsicum*). *Evolution* 25:683–91

72. Pickersgill, B. 1977. Taxonomy and the origin and evolution of cultivated plants in the New World. *Nature* 268:591–95

73. Pickersgill, B. 1977. Chromosomes and evolution. See Ref. 24, pp. 27–37

74. Pickersgill, B., Heiser, C. B. 1977. Origins and distribution of plants domesticated in the New World tropics. See Ref. 77, pp. 803–35

75. Pickersgill, B., Heiser, C. B., McNeill, J. 1979. Numerical studies on variation and domestication in some species of *Capsicum.* See Ref. 3. In press

76. Randolph, L. F. 1976. Contributions of wild relatives of maize to the evolutionary history of maize: a synthesis of divergent hypotheses. I. *Econ. Bot.* 30: 321–45

77. Reed, C., ed. 1977. *Origins of Agriculture.* The Hague: Mouton. 1013 pp.

78. Simmonds, N. W. 1966. Studies in the tetraploid potatoes. III. Progress in the experimental re-creation of the Tuberosum group. *J. Linn. Soc. Bot.* 59:279–88

79. Simmonds, N. W., ed. 1976. *Evolution of Crop Plants.* London: Longmans. 339 pp.

80. Simmonds, N. W. 1976. Potatoes. See Ref. 79, pp. 279–83

81. Simmonds, N. W. 1976. Quinoa and relatives, See Ref. 79, pp. 29–30

82. Smartt, J. 1978. The evolution of pulse crops. *Econ. Bot.* 32:185–98

83. Smartt, J. 1972. The possible status of *Phaseolus coccineus* L. ssp. *darwinianus* Hdz. et Miranda C. as a distinct species and cultigen of the genus *Phaseolus. Euphytica* 22:424–26

84. Smith, C. E. 1977. Recent evidence in support of the tropical origin of New

World crops. In *Crop Resources,* ed. D. S. Seigler, pp. 79–95. NY: Academic.

85. Smith, P. M. 1976. Lupin. See Ref. 79, pp. 312–13

86. Standley, P. C. 1931. Flora of Lancetilla Valley, Honduras. *Field Mus. Nat. Hist. Bot. Ser.* 10:418 pp.

87. Torres, A. M., Diedenhofen, U., Bergh, B. O., Knight, R. J. 1978. Enzyme polymorphisms as genetic markers in the avocado. *Am. J. Bot.* 65:134–39

88. Vavilov, N. I. 1926. Studies on the origin of cultivated plants. *Bull. Appl. Bot. Genet. Pl. Breed.* 16:139–248

89. Vavilov, N. I. 1931. Mexico and Central America as the principal centre of origin of cultivated plants in the New World. *Bull. Appl. Bot. Genet. Pl. Breed.* 26:179–99

90. Vavilov, N. I. 1950. The origin, variation and immunity and breeding of cultivated plants. *Chron. Bot.* 13:1–366

91. West, M., Whitaker, T. W. 1979. Prehistoric cultivated cucurbits from the Viru Valley, Peru. *Econ. Bot.* In press

92. Whitaker, T. W., Bemis, W. P. 1965. Evolution in the genus *Cucurbita. Evolution* 18:553–59

93. Wilkes, H. G. 1977. Hybridization of maize and teosinte in Mexico and Guatemala and the improvement of maize. *Econ. Bot.* 31:254–93

94. Wilkes, H. G. 1977. The origin of corn —studies of the last hundred years. See Ref. 84, pp. 211–23

95. Williams, L. O. 1977. The avocados, a synopsis of the genus *Persea,* subg. *Persea. Econ. Bot.* 31:315–20

96. Wilson, H. D. 1976. Genetic control and distribution of leucine aminopeptidase in the cultivated chenopods (*Chenopodium*) and related weed taxa. *Biochem. Gen.* 14:913–19

97. Wilson, H. D., Heiser, C. B. 1978. Origin and evolutionary relationships of huauzontle (*Chenopodium nuttalliae*), a domesticated chenopod of Mexico. *Am. J. Bot.* 66:198–206

98. Yarnell, R. A. 1977. Native plant husbandry north of Mexico. See Ref. 77, pp. 861–75

99. Yen, D. E. 1976. Sweet potato. See Ref. 79, pp. 42–45

100. Zevallos M., C., Galinat, W. C., Lathrap, D. W., Leng. E. R., Marcos, J. G., Klumpp, K. M. 1977. The San Pablo corn kernel and its friends. *Science* 196:385–89

Ann. Rev. Ecol. Syst. 1979. 10:327–49

THE BURGESS SHALE (MIDDLE CAMBRIAN) FAUNA

❖4165

Simon Conway Morris

Department of Earth Sciences, The Open University,
Milton Keynes MK7 6AA, England

INTRODUCTION

The preservation of soft parts in fossils is rare because fossilization usually occurs long after decay has destroyed soft tissues. A notable exception is the soft-bodied fauna from the Middle Cambrian Burgess Shale (about 530 million years old) located near Field in southern British Columbia, where both completely soft-bodied groups (e.g. polychaetes) and the soft parts of creatures with resistant skeletons (e.g. trilobites) are beautifully preserved. In addition, this fauna includes animals with fragile skeletons of thin cuticle that normally do not fossilize. The Burgess Shale fauna is of special importance because it permits a unique glimpse of the period shortly after the upper Precambrian–lowermost Cambrian radiation of the Metazoa (26).

In 1909 Charles Doolittle Walcott (Secretary of the Smithsonian Institution), returning from a field season, stopped to split open a rock that blocked a trail on the western slopes between Wapta Mountain and Mount Field. The rock contained soft-bodied fossils. The following year Walcott and his two sons located the original stratum: the Burgess Shale. Quarrying continued for several seasons (1910–13, 1917), and more than 40,000 specimens were shipped to the Smithsonian Institution (USNM). Subsequent expeditions by Harvard University (MCZ) in 1930 (92, 94), the Geological Survey of Canada (GSC) in 1966 and 1967 (153), and the Royal Ontario Museum (Toronto) in 1975 collected more material. After Walcott's preliminary publications (135–137, 139–146, 148), a much needed reinvestigation was undertaken by the GSC, with H. B. Whittington directing the paleontological work.

327

0066-4162/79/1120-0327$01.00

PALEOENVIRONMENTAL SETTING

The Burgess Shale is a predominantly shale unit within the thick succession of shales and impure limestones that forms the Stephen Formation. Although an integral part of the Stephen Formation, the Burgess Shale is singled out for recognition because of its tectonic isolation and unique fauna (46). The Stephen Formation was deposited in a deep-water basin southwest of an algal reef with a precipitous escarpment that can be traced for at least 16 km along a northwest trend (Figure 3B) (77–79). In addition to the Burgess Shale a number of other prolific faunas have been discovered within the Stephen Formation. All are located at the foot of the reef; basinwards all become impoverished. Only the Burgess Shale contains abundant soft-bodied fossils; the other faunas are dominated by trilobites (78). One of the latter, the famous *Ogygopsis* Shale exposed on Mount Stephen (Figure 3B), has yielded a rich fauna with some exceptionally well-preserved species (14, 30, 54, 75, 76, 89, 93, 97, 107, 108, 112, 131, 133, 134, 136, 137, 139, 141, 144, 146, 147, 152, 154).

Two levels within the Burgess Shale yield abundant soft-bodied fossils (138). The lower Phyllopod bed (2.31 m thick) is exposed in the Walcott quarry. About 20 m higher another excavation (the Raymond quarry) has yielded a sparser and less well-preserved fauna (36).

The Phyllopod bed, at least, was deposited from turbidite flows (86, 87, 153, 154). It is therefore possible to distinguish between a pre-slide environment where the benthonic fauna lived and a deeper post-slide environment to which the fauna was transported, there to be buried and preserved. The distance of transport was probably not more than a few km; the direction was probably parallel to the reef (36, 37). The vertical displacement of trilobite zones from basin to reef indicates that the Phyllopod bed was deposited at a depth of about 160 m (46). The superb preservation and almost complete absence of scavenging or bioturbation suggest that the post-slide environment was inimical to metazoan life; poisonous hydrogen sulphide may have extended above the sediment-water interface.

PRESERVATION

The soft-parts of the Burgess Shale fossils are now composed of very thin films which in part are highly reflective (Figure 1) (153). The fossil film is composed of calcium aluminosilicates with the reflective areas containing additional magnesium (30). Owing to the seeping of sediment during transport, specimens with appendages (e.g. arthropods, polychaetes) have usually been preserved on several levels of microbedding. The split tends to jump from one level to another across a specimen (156). Various factors control the level of splitting and thus the exposure of different parts of the body (37, 156). Transport also resulted in the specimens' adopting different orientations upon burial (153, 154); this facilitates an understanding of their morphology. The reasons for the remarkable preservation are obscure. The fauna was rapidly buried in fine sediment (86, 87) under anaerobic conditions (36, 37, 153), so decay was greatly retarded. However, some additional influence must be invoked to explain the cessation of decay and consequent astonishing preservation. Other examples of arrested decay in fossil (5, 20, 21, 39, 47, 81, 82, 85, 100, 121–124, 127, 128, 150) and modern situations (10, 40, 49, 118) are known, but they throw little light on the Burgess Shale preservation. No other Cambrian locality matches the diversity of the Burgess Shale. However, a few exceptionally well-preserved fossils, some of which are comparable to Burgess Shale genera, have been found in other Cambrian rocks (12, 15, 16, 30, 37, 42, 48, 96–99, 103, 104, 129, 130, 132, 137).

THE FAUNA

The fauna apparently lived on the edge of the open ocean close to the paleo-equator; it seems neither atypical nor aberrant. It consists of about 119 genera (140 species). The approximate composition of the fauna by genera is shown in Figure 3C, a scheme that differs slightly from two earlier compilations (33, 34, 36) in its organization and the values given. Research on the fauna is still unfinished, but the final values should not differ greatly from those given here. A rich flora of uncalcified algae (108, 143, 149) has also been partially restudied (111).

Arthropods

Arthropods account for the largest fraction of the fauna (37%). In addition to typical Cambrian trilobites there is a remarkable assemblage of lightly sclerotized nontrilobitic arthropods. The trilobites are mostly benthonic (about 75% of genera), but pelagic agnostids and eodiscids are numerically abundant (89, 137). With the possible exception of *Elrathina cordillerae* and *Elrathia permulta,* soft parts are known only from *Olenoides serratus* and a single specimen of *Kootenia burgessensis* (91, 137, 141, 157). The other benthonic trilobites lack their appendages, probably because they were originally scarce (89, 141) as living members of the fauna. Although the exoskeletons survived, by chance no living specimens were caught in the turbidite flows. *Olenoides* bore uniramous anterior antennae and posterior cerci. The intervening biramous appendages consisted of an inner jointed and spinose walking leg and an outer filamentous gill. The leg and gill arose from a large gnathobasic coxa (157). In contrast to the dorsal exoskeleton, the biramous appendages of this and other trilobites are notable for showing no distinct tagmosis (9, 21, 73, 122). *Nathorstia transitans* (137) is an exuvium of *O. serratus,* as Raymond (91) originally suggested (H. B. Whittington, in preparation). *Naraoia compacta* (91, 117, 137, 148) had an anterior pair of antennae and trilobitan biramous appendages. The carapace, however, consisted of two subcircular shields. *Naraoia* is thus comparable to an enormous larval trilobite and may be neotenic (158).

The nontrilobitic arthropods comprise a diverse assemblage. Many were accommodated in the Trilobitoidea (120), but this is an artificial group (158). The limbs of *Marrella splendens, Burgessia bella,* and *Waptia fielden-sis* may show some trilobite-like features (57, 159), but the majority of genera cannot be placed in any higher taxon. They demonstrate an early facet of arthropod radiation, but they throw little light on the question of arthropod polyphyly (73).

The most abundant arthropod is *Marrella* (91, 108, 113, 117, 137, 148). A wedge-shaped cephalic shield bore two pairs of elongate spines. Uniramous first and second antennae preceded biramous appendages composed of a jointed walking leg and filamentous gill branch (Figures 1H, 2E) (153, 154). *Burgessia* (91, 116, 117, 137, 148) had a circular carapace. The cephalon carried anterior antennae and three pairs of biramous appendages consisting of a jointed walking leg and a slender flagellum. With the exception of the last pair, all of the trunk appendages were biramous with a walking leg and gill branch (56). The cephalic region of *Waptia* (91, 117, 137, 148) was equipped with two pairs of antennae, four pairs of walking legs, and pedunculate eyes. The trunk had six pairs of gills. The legs and gills, therefore, show alternate reduction from a more primitive biramous condition (57). The notion that *Waptia* is closely comparable to the decapodan protozoea larva (52) appears to be unfounded (C. P. Hughes, personal communication).

The cephalon of *Yohoia tenuis* (91, 116, 117, 137) bore a shield and appendages; the latter included three pairs of walking legs and a remarkable pair of large appendages with distal articulating spines that could have been used to grasp food (155). *Leanchoilia superlata* (94, 116, 117, 137, 148) had a cephalic shield with upturned rostrum and a pair of enormous anterior appendages. The trunk appendages were biramous and each included a filamentous branch (17). The anterior part of *Branchiocaris pretiosa* (96, 117) was enclosed in a bivalved carapace. The cephalon possessed antennae and chela-like appendages. The trunk seems to have consisted of a large number of segments bearing lamelliform appendages (12).

Sidneyia inexpectans (91, 114, 117, 134, 137, 140) and *Emeraldella brocki* (91, 115, 117, 137) both had gnathobasic limbs. *Sidneyia* had certain similarities to the modern merostome *Limulus* (17). The triramous appendages of *Emeraldella* with an inner walking leg and two large foliaceous lobes

(Bruton in 53) are crustacean in aspect (17). Arthropod remains comparable to *Sidneyia* have also been noted from Indochina (72), but the supposed specimen from Greenland (25) is inorganic (44).

Tuzoia [supposedly represented by four or five species, type *T. retifera:* (13, 96, 117, 137)], *Carnarvonia venosa, Hurdia victoria* and *H. triangulata, Isoxys acutangulus* and *I. longissimus* (117, 137), and *Proboscicaris agnosta* and *P. ingens* (106, 117) are only known by their carapaces. In the absence of appendages, identification as phyllocarid crustaceans cannot be positive. *Tuzoia* and *Isoxys* have been recorded from other Cambrian strata (48, 84, 96, 99, 101, 132), but the supposed *Hurdia* from Australia (19) is inorganic (3). *Tuzoia* and *Isoxys* may owe their wide geological and geographical ranges to comparatively heavy sclerotization.

Phyllocarids with soft parts preserved include *Canadaspis perfecta* (14, 117, 137), *Perspicaris dictynna* and *P. recondita* (13, 117), and more tentatively *Plenocaris plena* (117, 137, 155). In *Canadaspis* a bivalved carapace covered the cephalon and thorax. Anteriorly there were two pairs of antennae and a pair of pedunculate eyes. Chewing mandibles were succeeded by first and second maxillae. The latter appendages were primitive and similar to the eight pairs of thoracic appendages, each having an inner jointed walking leg and a filamentous outer branch (14). *Anomalocaris gigantea,* generally regarded as the abdomen of a phyllocarid (96, 117, 137), has recently been reinterpreted as the appendage of a large arthropod (13, 15). *Anomalocaris* is found in much greater abundance in the nearby *Ogygopsis* Shale (133, 152) and has been recorded from other Cambrian strata (15, 96, 99).

Aysheaia pedunculata (18, 58, 59, 136, 148, 159) has excited the greatest interest on account of its remarkable resemblance to the modern onychophores (e.g. *Peripatus*), which resemble the hypothetical ancestral uniramian arthropod (73). Whittington (160) concluded that *Aysheaia* differs in certain respects from modern onychophores and tardigrades, but conceded that the uniramian ancestor probably closely resembled *Aysheaia.*

The placement of *Skania fragilis* with trilobite-like forms (120, 148) has been criticized (18), but whether comparison with the peculiar *Parvancorina minchami* from the latest Precambrian (47) can be upheld is uncertain (C. P. Hughes, in preparation). Additional arthropods awaiting detailed redescription are: *Mollisonia gracilis* and *M.* (?) *rara* (93, 115, 117, 137, 148), *Odaraia alata* (117, 137), *Alalcomenaeus cambricus* (116, 117), *Molaria spinifera, Habelia optata* (91, 115, 117, 137), *Helmetia expansa* (117, 142, 148), ostracode-like forms (137, 154), and several other genera (117).

Echinoderms

The Burgess Shale has yielded the earliest known crinoid, *Echmatocrinus brachiatus.* A large conical calyx and tapering holdfast was covered with irregularly arranged plates. Plated uniserial arms with probable tube feet arose from the calyx. An eocrinoid, *Gogia*(?) *radiata,* and the "arms" of an unknown ?eocrinoid have also been described (119). The edrioasteroid *Walcottidiscus* (6, 7) awaits restudy (J. Sprinkle, personal communication). The notion that the soft-bodied medusiform *Eldonia ludwigi* was a holothurian (22, 23, 135, 140; but see 24) was reaffirmed by Durham (43), who demonstrated that the interpretation of *Eldonia* as a siphonophore (69–71) is untenable. The supposed specimens of *Eldonia* from Eire (110) are inorganic (41). *Laggania cambria* has been interpreted both as a holothurian (23, 135) and a polychaete (70), but the only known specimen is a composite fossil formed by the superposition of the medusoid *Peytoia nathorsti* and a sponge (35).

Molluscs

A single specimen was identified as *Helcionella* (46), but it is undeterminable. A low orthoconic shell attributed [perhaps incorrectly (E. L. Yochelson, personal communication)] to the monoplacophoran *Scenella* (90, 137) is abundant. Hyolithids, which however may not be molluscs (109), are common. *Hyolithes carinatus* sometimes has the operculum and associated curved appendages in place (136, 162). *Wiwaxia corrugata* (75, 136, 148) was covered with ribbed scales and bore elongate dorso-lateral spines. It has been regarded as a polychaete (55, 136), but the radula-like feeding apparatus suggests an affinity with primitive molluscs (S. Conway Morris, in preparation).

Lophophorates

The brachiopods are typical Cambrian forms. Two species, however, are exceptional: In some specimens the mantle setae and pedicle have been preserved (8, 97, 139, 146). *Odontogriphus omalus* had a dorso-ventrally compressed body with a ventral feeding apparatus identified as a lophophore (27). Criticism of this interpretation (65) appears unfounded. Minute conical objects associated with the lophophore have been interpreted tentatively as conodonts, and *Odontogriphus* is regarded as a conodont animal (27; see also 80).

Chordates and Hemichordates

"Ottoia" tenuis (136), which is unrelated to the type species *Ottoia prolifica,* is similar to an enteropneust worm. Another undescribed animal resembles an enormous rhabdopleuroid pterobranch (S. Conway Morris, in preparation). Although Cambrian graptolites are known (102), those described as *Chaunograptus scandens* from the Burgess Shale (108) require restudy.

The presence of a longitudinal bar (notochord) and sigmoidally deflected segments (myotomes) in *Pikaia gracilens* (Figure 1D) indicates that it is a

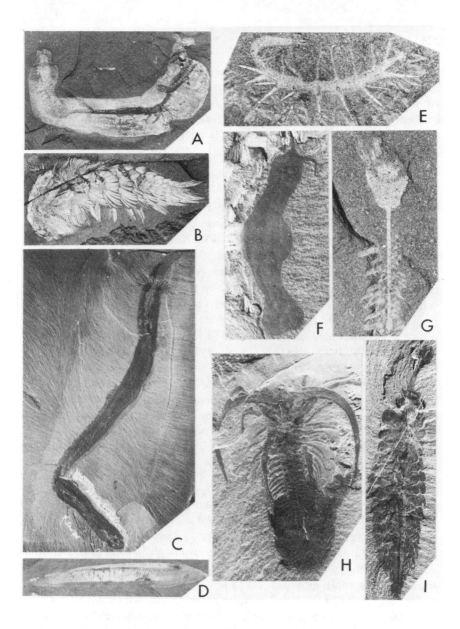

Figure 1 Representative Burgess Shale specimens photographed in ultraviolet light—A: *Ottoia prolifica* Walcott (G.S.C. 40972), priapulid, X 0.7 (30); B: *Canadia spinosa* Walcott (U.S.N.M. 198724), polychaete, X 1.4 (37); C: *Louisella pedunculata* Walcott (U.S.N.M. 198648), priapulid, X 0.5 (30); D: *Pikaia gracilens* Walcott (U.S.N.M. 198684), chordate, X 1.2.; E: *Hallucigenia sparsa* (Walcott) (U.S.N.M. 198658), phylum uncertain, X 3.7 (29); F: *Amiskwia sagittiformis* Walcott (U.S.N.M. 57644), phylum uncertain, X 2.3 (32); G: *Dinomischus isolatus* Conway Morris (M.C.Z. 1083), phylum uncertain X 2.1 (31); H: *Marrella splendens* Walcott (G.S.C. 26592), arthropod, X 2.5 (154); I: *Opabinia regalis* Walcott (U.S.N.M. 57684), phylum uncertain, X 0.9 (156).

primitive chordate (34) rather than a polychaete (55, 136, 148). The earliest fish scales are Upper Cambrian (95), and *Pikaia* may not be far removed from the ancestral fish. Whether study of this chordate and another undescribed form will lend credence to the suggestion that the vertebrates are derived from mitrate echinoderms (60) is uncertain (S. Conway Morris, in preparation).

Polychaetes

The only annelids represented are the Polychaeta. None of them has a mouth armature (37); polychaetes acquired jaws in the Ordovician (62). The parapodia of *Canadia spinosa* (136, 148) had broad notosetae extending across the dorsal surface, large fascicles of narrower neurosetae, and interramal lobate gills (Figure 1B). The parapodia bear a striking resemblance to those of the modern Palmyridae, but this may be due to convergence. The other genera show no close affinities with modern families (37). The parapodia of *Burgessochaeta setigera* carried identical notosetae and neurosetae, whereas *Peronochaeta dubia* (136, 148) had uniramous parapodia with acicular and capillary setae (37). *Insolicorypha psygma* had biramous parapodia with elongate neuropodia bearing cirri and long slender setae (37). The leaf-shaped *Pollingeria grandis* has been regarded as the detached scales, supposedly furrowed by a commensal worm, of a polychaete (55, 136). This interpretation appears very doubtful; *Pollingeria* may not even be an animal (S. Conway Morris, in preparation).

Priapulids

There is a rich assemblage of priapulids (30) which morphologically, at least, are more diverse than modern forms (64, 88). *Ottoia prolifica* (136) shows magnificent preservation (Figure 1A): The intestine, retractor muscles, and nerve cord are identifiable. *Ottoia* has the closest relationship with modern priapulids, especially *Halicryptus spinulosus* (30). *Selkirkia columbia,* which was mistaken for a polychaete (55, 136, 148), is a tubicolous form. Possible examples of *Selkirkia* have been noted from other Cambrian rocks (30, 98, 99). *Louisella pedunculata,* which has been misinterpreted both as a holothurian (23, 136) and a polychaete (55, 70), is the largest (up to 20 cm) of the priapulids. It was unusual in having papillate proboscis scalids and a trunk armed with concentric zones of spines and two longitudinal rows of papillae (Figures 1C, 2C) that probably functioned as gills (30). *Fieldia lanceolata,* previously mistaken for an arthropod (137), had a small proboscis and a very spiny trunk (30). *Ancalagon minor* (136, 148) is of special interest because although it is a priapulid it has a marked resemblance to a hypothetical free-living ancestor (50) of the endoparasitic Acanthocephala (30).

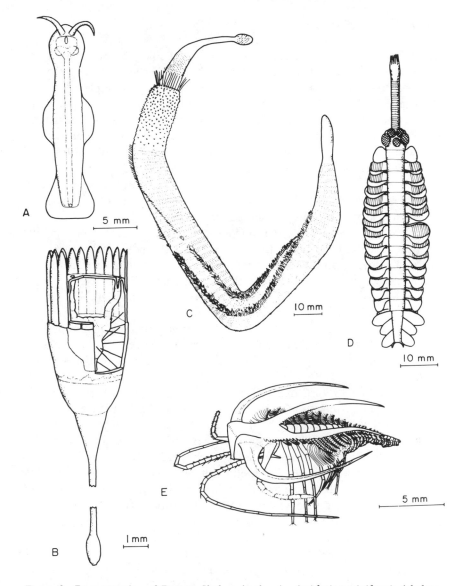

Figure 2 Reconstruction of Burgess Shale animals—A: *Amiskwia sagittiformis* (phylum uncertain). Ventral view showing cerebral ganglia and gut with subterminal openings (32). B: *Dinomischus isolatus* (phylum uncertain). Portion of the upper calyx and pointed bracts cut away to reveal gut supported by fibers. Most of the stem is omitted (31). C: *Louisella pedunculata* (priapulid). Proboscis fully everted and trunk twisted along its axis to show spinose zones and papillate gills (30). D: *Opabinia regalis* (phylum uncertain). Dorsal view with frontal process extending forward. Right lateral lobe and pleated gill of segment 7 cut away to reveal lobe and gill of segment 8 (156). E: *Marrella splendens* (arthropod). Oblique lateral view. Left gill branches 1–4 and 10–26 cut away to reveal walking legs (154).

Miscellaneous Worms

This heterogenous and artificial group is united solely by our inability to accommodate its members in any known phylum. Some of these specimens probably represent new phyla. *Opabinia regalis* has been placed in the arthropods (58, 91, 94, 116, 117, 120, 137), but it has only a most distant connection with that group. The cephalon bore five eyes and a frontal process that was flexible and armed with distal teeth. The trunk carried lateral lobes and, with the exception of the first segment, gills with dorsal lamellae (Figures 1I, 2D) (156). *Nectocaris pteryx* had an elongate and streamlined body. The head carried anterior appendages, large eyes, and more posteriorly a shield-like structure. Dorsal and ventral fins supported by fin rays arose from the segmented trunk (28).

Hallucigenia sparsa (136, 148) had a most unusual appearance. A globular head and elongate trunk that curved upwards posteriorly were supported by seven pairs of elongate spines. Seven tentacles terminating in cuticularized bifid tips arose from the dorsal trunk (Figure 2E). Its zoological affinities and mode of life remain problematical (29). In *Banffia constricta* (136) an anterior annulated section was separated from a sac-like posterior by a constriction. This curious anatomy may represent a primitive adaptation that isolated a more passive posterior from pressure fluctuations set up by the locomotory movements of the anterior (33).

Amiskwia sagittiformis had a prominent pair of tentacles arising from an oval head and a trunk supporting lateral and caudal fins (Figures 1F, 2A). This worm has been regarded both as a chaetognath (55, 136) and a bathypelagic nemertean (63, 83), but since neither assignment can be supported the animal's phyletic position remains obscure (32). The calyx of *Dinomischus isolatus* was supported by an elongate and slender stem with swollen terminal holdfast. Plate-like bracts arose from the calyx edge. A recurved gut with an enlarged stomach was supported in the body cavity by fibers (Figures 1G, 2B). *Dinomischus* has certain similarities, which may be only superficial, to the Entoprocta (31).

Oesia disjuncta had a swollen anterior section, which contrary to Walcott's (136) observation appears to be unarmed, and an elongate trunk. *Oesia* has been identified as both an annelid (55, 125, 136) and an appendicularian tunicate (68), but neither suggestion seems likely (S. Conway Morris, in preparation). *Redoubtia polypodia* and *Portalia mira* were interpreted as holothurians (142, 148), although Madsen (70) considered them to be a polychaete and sponge, respectively. None of these proposals is convincing; together with *Worthenella cambria* (136) and other miscellaneous worms, these animals await restudy.

Coelenterates

Mackenzia costalis (135) is generally regarded as an actinian (23, 24, 142, 151). *Peytoia nathorsti* had a peculiar medusiform body composed of thirty-two lobes around a central cavity. Each lobe was armed with a proximal pair of prongs (35, 135). Despite the claim that *Peytoia* is a scyphozoan (135), it remains of uncertain systematic position (51). An undescribed form has a strong resemblance to a pennatulacean or sea pen (S. Conway Morris, in preparation).

Sponges

The prolific fauna is represented by demosponges [e.g. *Hazelia* (type species *H. palmata*)], hexactinellids (e.g. *Protospongia hicksi*), and heteractinids [e.g. *Eiffelia globosa, Chancelloria eros* (45, 103, 144)]. Some of the genera are known from other lower Paleozoic strata, but the Burgess Shale specimens are exceptionally well-preserved (103). Phylogenetic discussions have often referred to Burgess Shale species, and *Hazelia* has been given a key position in demosponge phylogeny (45, 161). However, a comprehensive restudy of the sponge fauna is now underway (J. K. Rigby, personal communication).

Trace Fossils

Evidence for trace fossils, especially in the Phyllopod bed, is almost completely lacking (86, 153). Rare structures filled with pyrite from the Phyllopod bed may represent infilled burrows. Narrow burrows associated with arthropod carapaces (*Leanchoilia, Canadaspis*) have also been recorded, although these rare specimens apparently originate from above the Phyllopod bed (D. M. Rudkin, personal communication). Irregularly shaped clumps of hyolithids, sometimes with associated brachiopods, may represent coprolites of an unidentified (?arthropod) predator. They are especially common from the Raymond quarry.

PALEOECOLOGY OF THE BURGESS SHALE FAUNA

Most (87.5%) of the fauna was benthonic; the pelagic component was less well-represented (Figure 3D). The benthonic fauna evidently inhabited the muds of the pre-slide environment adjacent to the base of the algal reef (36). The presence of burrowing worms and an extensive epifauna suggests the sediment was fairly well consolidated. Nevertheless the muds were unstable, perhaps due to rapid rates of deposition on submarine slopes, and periodically they slumped into the poisonous post-slide environment.

   The persistence of abundant genera [e.g. *Marrella* (153, 154), *Ottoia, Selkirkia* (30), and *Canadaspis* (14)] throughout the Phyllopod bed suggests that only one association or community was present (36). The fluctuations in abundance of these common forms, together with the more restricted vertical distribution of other species (29, 30, 37, 56, 137, 155, 157), probably reflect a patchy distribution of the fauna over the seafloor (37).

The presence of an infauna demonstrates that it, and probably the rest of the benthos, was present from the inception of the mud flows (37). As the flows moved downslope they may have eroded additional epifauna. The movement of the sediment destroyed most evidence of the life-positions of the fauna. Life-positions within the pre-slide environment can be established only by comparing the morphology of each species with that of living relatives and analogous forms.

Infaunal, Epifaunal, and Pelagic Components

The vagrant infauna was dominated by burrowing priapulids (Figure 3D). *Ottoia* (30) and *Louisella* (Figure 1A, C) may also have occupied temporary burrows, while the latter genus probably aerated its gills by dorsoventral undulations of the compressed body. Other members of the vagrant infauna probably include the polychaete *Peronochaeta* (37) and the probable enteropneust worm. The polychaete *Burgessochaeta* possibly lived in a semi-permanent burrow (37), but in general the sessile infauna appears to have been restricted.

In contrast the sessile epifauna was extensive and included the sponges (Figure 3D) (144), brachiopods (139, 146), echinoderms [(119); except *Eldonia* (43, 135)], the enigmatic *Dinomischus* (Figure 1G) (31), and, among the coelenterates, *Mackenzia* (135) and the probable sea pen. The vagrant epifauna was dominated by arthropods (Figure 3D) that walked across the seafloor or perhaps on occasion swam close to it; some of them (e.g. *Burgessia*) may have burrowed (56). The vagrant epifauna also included the molluscs and enigmatic beasts such as *Opabinia* (Figures 1I, 2D) (156), and *Hallucigenia* (Figure 1E) (29). The polychaete *Canadia* (37) and chordate *Pikaia* (Figure 1B, D) (34, 136, 148) appear to have been well adapted for swimming; they were probably at least partially nektobenthonic. The pelagic species (Figure 3D) may have been derived from different depths and may include representatives swept in from the open ocean. The agnostid and eodiscid trilobites are generally regarded as pelagic (105). The other pelagic animals are identified on the dual basis of morphology and rarity. The carapace of *Isoxys* shows convergence with the modern pelagic ostracode *Conchoecia daphnoides;* this may indicate a similar mode of life. The abundance of gelatinous tissue in *Amiskwia* (Figures 1F, 2A) (32), *Odontogriphus* (27), and *Eldonia* (135) or marked adaptations to swimming [e.g. *Nectocaris* (28), *Insolicorypha* (37)] are taken as key features. Their rarity is a reflection of the low probability of becoming involved in benthonic turbidites. Unlike the other pelagic genera, *Eldonia* is abundant (43); while it may not have been pelagic, its restriction to a limited horizon and area (27, 135, 137, 153) suggests the specimens were trapped as a shoal.

The more impoverished Raymond quarry fauna differs somewhat in character from that of the Phyllopod bed (36). The lower diversity could be due in part to more selective fossilization; but the presence of creatures such as the infaunal worm *Banffia* (33, 136) and the epifaunal arthropod *Leanchoilia* (94, 137, 148), which are very rare in the Phyllopod bed, suggests that a different association is represented (36).

Feeding Methods

In only a few cases are identifiable gut contents known. Feeding methods are usually established by determining the likely function of the food-collecting organs. In many cases (33%) the method of feeding remains either unknown or very uncertain. Three feeding types are recognized: suspension feeders (35% of the fauna), deposit feeders (13.5%, includes swallowers and detritus collectors), and carnivores/scavengers (18.5%).

Many of the members of the sessile epifauna were suspension feeders. These animals probably exploited different water levels for food (Figure 3A), as has been demonstrated in other paleocommunities (11, 66, 67). Vagrant filter feeders included *Leanchoilia* (17); the identification of a polychaete worm in the gut of this arthropod (116) is incorrect. Pelagic suspension feeders are represented by *Eldonia, Odontogriphus* (27), and perhaps *Amiskwia* (Figures 1F, 2A) (32).

Deposit feeders include the priapulid *Fieldia,* which usually had sediment in its midgut (30), the hyolithids (74), and probably the monoplacophoran *Scenella* (E. L. Yochelson, personal communication). This feeding type has been identified with varying degrees of certainty in the arthropods *Branchiocaris* (12), *Canadaspis* (14), *Plenocaris, Yohoia* (155), *Marrella* (Figures 1H, 2E) (73, 154), and possibly *Burgessia* (56) and *Naraoia* (158). The polychaete *Burgessochaeta* may have picked food off the sediment surface with its elongate tentacles (37). A poorly known polychaete [Type A in (37)] has its gut packed with sediment.

Predators and scavengers form a significant proportion of the fauna. Gut contents of the priapulid *Ottoia* include hyolithids, often with opercula in situ, and more rarely brachiopods. The hyolithids were presumably eaten alive and their usual orientation within the gut suggests that they were hunted (30). Since hyolithids are regarded as epifaunal (74), *Ottoia* may have sought them along the sediment-water interface. That one specimen of *Ottoia* contains another individual of *Ottoia* in its gut is taken as evidence

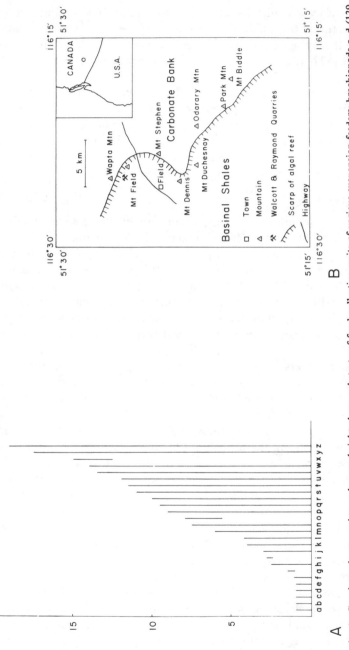

Figure 3 A: Graph to show maximum known height above substrate of food collecting units of various suspension feeders: brachiopods a–d (139, 146), sponges e,f,h,j–n,p–w,y,z (144), echinoderms g,o,x (119), phylum uncertain i (31). The sponges are assumed to have filtered water over entire length of body. Brachiopods attached to algae (143) or sponges (30) are not taken into account. Depth of insertion into sediment assumed to be minimal and species assumed to have lived vertically. a–d, various brachiopods; e, *Hazelia (H.) mammillata*; f, *Diagoniella hindei*; g, *Walcottidiscus magister*; h, *Pirania muricata*; i, *Dinomischus isolatus*; j, *Eiffelia globosa*; k, *Vauxia (V.) dignata*; l, *Takakkawia lineata*; m, *H. palmata*; n, *Corralio undulata*;

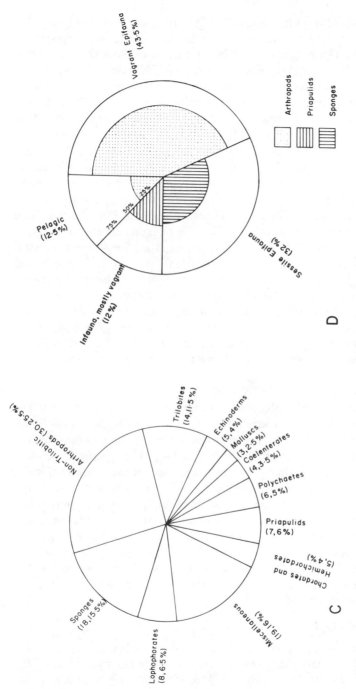

o, *Echmatocrinus brachiatus*; p, *H. delicatula*; q, *Chancelloria eros*; r, *H. obscura*; s, *V. gracilenta*; t, *V. bellula*, u, *H. nodulifera*; v, *H. conferta*; w, *V. densa*; x, *Gogia? radiata*; y, *Halichondrites elissa*; z, *Wapkia grandis*. B: locality map of the area around Field, British Columbia. Inset shows location of area with respect to western North America. C: pie diagram of composition of fauna on a generic basis. The number of genera and its approximate percentage of the fauna are given after each group. The figures given for the sponges are estimates and may be conservative. A few additional genera of nontrilobitic arthropods have been described but their validity is not certain. D: pie diagram of living habits of the fauna on a generic basis. The phylum that dominates each habitat is also shown.

for cannibalism (30). Other priapulids [*Selkirkia, Louisella* (Figures 1C, 2C), *Ancalagon*] had prominent mouth armatures and were apparently predatory (30). The prominent gnathobasic limbs of the arthropods *Olenoides* (157), *Naraoia* (158), *Emeraldella,* and *Sidneyia* (17) [which has prominent gut contents (117, 134)], were probably used for predation. There is circumstantial evidence for scavenging by ostracodes (30, 154). The frequent association of *Aysheaia* with sponges suggests that they formed the diet of this primitive arthropod (160). The armed frontal process of the enigmatic *Opabinia* (Figures 1I, 2D) probably captured food and passed it back to the mouth (156). The dorsal tentacles of *Hallucigenia* (Figure 1E) may have been used for feeding. The clustering of about twenty specimens of this bizarre beast on another worm suggests that they congregated to scavenge the corpse (29). The radula-like mouthparts of *Wiwaxia* may have been used for scavenging or grazing. That the elongate dorso-lateral spines of some specimens of *Wiwaxia* are broken off is ascribed to unsuccessful predation (S. Conway Morris, in preparation). *Nectocaris* is considered to be predatory and may have occupied an ecological niche similar to that of modern chaetognaths (28). The coelenterate *Mackenzia* may have been carnivorous, but contrary to the case for other probable Cambrian coelenterates (1) no direct evidence for predation has been noted. The body of the medusoid *Peytoia* may have been able to contract radially and bring together its prongs to grasp prey (S. Conway Morris, in preparation).

Other Aspects of Paleoecology

Of interactions other than interspecies feeding little is presently known. The crinoid *Echmatocrinus* is invariably attached to a worm tube, a hyolithid, or another hard object (119). The sponges, however, are only occasionally attached to worm tubes or brachiopods (30, 144). The specific association between the brachiopod *Dictyonina,* which settled on the elongate spicules of the sponge *Pirania,* may represent a case of commensalism (30), though brachiopods of other species are occasionally found attached. Inarticulate brachiopods are sometimes found attached to algae (143). That some species are found crowded together over limited areas may indicate gregarious habits. Examples have been noted among sponges (144), the crustacean *Canadaspis* (14), the priapulid *Ottoia,* and the possible enteropneust worm. Study of the association of species on the collected slabs might throw more light on species interaction.

Rotting and Decay

Despite the exquisite preservation of the fossils there is evidence of decay. In some species a dark stain (Figure 1H) is associated with some of the specimens (14, 29, 30, 36, 37, 56, 153–156, 158, 160). The stain comprises decay products that oozed from the body into the surrounding mud (29, 30).

Its frequent location around the anterior or posterior regions suggests that its usual points of egress were the mouth and anus. In rare cases decay has resulted in detachment of external (e.g. setae) or internal organs (e.g. intestine) (37, 148, 153, 154). Additional evidence for decay was documented in the priapulid *Ottoia,* where a gradation was demonstrated from perfectly preserved specimens, to those where the body wall has folded away from the cuticle and the internal organs are visible, to specimens that consist of collapsed and folded cuticle (30).

THE WIDER-SCALE IMPORTANCE OF THE BURGESS SHALE FAUNA

1. About 16% of the fauna cannot be placed in known phyla. Some of these creatures [e.g. *Hallucigenia* (29), *Opabinia* (156)] represent "experimental" groundplans (Figures 1E,I, 2D) that were ultimately unsuccessful. These forms may represent relicts of an earlier metazoan radiation, but the persistence of novel animals during the Paleozoic (61, 81, 82) demonstrates how scanty our knowledge is of metazoan diversification.
2. Some 80% of the fauna is soft-bodied. The low diversity of normal Cambrian faunas (see 2, 126) may be due partly to the scarcity of groups possessing readily fossilizable hard parts.
3. The pervasive idea that certain groups must have had hard parts to function effectively is negated by the existence of relatives [e.g. *Naraoia* among trilobites (158)] with fragile cuticles. The existence of "soft-bodied" trilobites (unfossilizable in normal circumstances) is postulated by Crimes (38) from study of early Cambrian trace fossils.
4. Banks (4) ascribed late Precambrian and Cambrian trace fossils to the activities of arthropods, annelids, and molluscs, rather than to Metazoa as a whole. The diversity of the Burgess Shale fauna illustrates that many other potential trace producers were present in the Cambrian and that in the absence of diagnostic features noncommital identifications are preferable.
5. As is not the case in modern marine environments, priapulids are more abundant than polychaetes in the Burgess Shale (34). The decline of predatory priapulids in favor of polychaetes may have begun in the Ordovician when jaws evolved among eunicid-like polychaetes (30). Further research should show how the Burgess Shale species occupying various ecological niches were replaced by different forms. For instance, certain Burgess Shale arthropods may have occupied niches presently taken by various crustaceans, while other larger arthropods might have approximated the role of modern benthic fish (33). Sponges were evidently the dominant suspension feeders, and other groups (e.g. brachiopods, echinoderms) were less important.

6. Descendants of some groups represented in the Burgess Shale avoided extinction by migrating into marginal niches. In contrast with modern priapulids, those of the Cambrian evidently dominated at least some infaunal assemblages (Figure 3D). The modern Priapulidae have narrow distribution and often occupy habitats unattractive to many metazoans (64). The other two families (Tubiluchidae, Maccabeidae) are meiofaunal (64, 88) and may be minaturized descendants of Cambrian priapulids. As other examples of minaturization, the entoprocts may derive from a *Dinomischus*-like form (Figures 1G, 2B) (31) and the tardigrades from an *Aysheaia*-like animal (18, 73, 160). The endoparasitic Acanthocephala, possibly derived from an *Ancalagon*-like priapulid (30), might illustrate another method of "escape" from competition.

ACKNOWLEDGMENTS

I thank H. B. Whittington, F. R. S., and C. P. Hughes for critically reading the manuscript. They and D. E. G. Briggs, D. L. Bruton, F. J. Collier, D. H. Collins, R. M. Finks, W. H. Fritz, I. A. McIlreath, W. D. I. Rolfe, D. M. Rudkin, J. Sprinkle, and E. L. Yochelson provided valuable information. I am grateful to the Palaeontological Association (Figures 1A, C, E, G; 2B, C), the Royal Society (Figures 1B, I; 2D), E. Schweizerbart'sche (Figures 1F; 2A), and the Geological Survey of Canada (Figure 1H; 2E) for permission to reproduce figures. My wife Zoë typed several drafts of the manuscript. My work on the fauna was supported by the Natural Environment Research Council and St. John's College, Cambridge.

Literature Cited

1. Alpert, S. P., Moore, J. N. 1975. Lower Cambrian trace fossil evidence for predation on trilobites. *Lethaia* 8:223–30
2. Bambach, R. K. 1977. Species richness in marine benthic habitats through the Phanerozoic. *Paleobiology* 3:152–67
3. Banks, M. R. 1962. On *Hurdia? davidi* Chapman from the Cambrian of Tasmania. *Aust. J. Sci.* 25:222
4. Banks, N. L. 1970. Trace fossils from the late Precambrian and Lower Cambrian of Finnmark, Norway. *Geol. J. Spec. Iss.* 3:19–34
5. Barthel, K. W. 1970. On the deposition of the Solnhofen lithographic limestone (Lower Tithonian, Bavaria, Germany). *Neues Jahrb. Geol. Palaeontol. Abh.* 135:1–18
6. Bassler, R. S. 1935. The classification of the Edrioasteroidea. *Smithson. Misc. Collect.* 93(8):1–11
7. Bassler, R. S. 1936. New species of

American Edrioasteroidea. *Smithson. Misc. Collect.* 95(6):1–33
8. Bell, W. C. 1941. Cambrian Brachiopoda from Montana. *J. Paleontol.* 15:193–255
9. Bergström, J. 1969. Remarks on the appendages of trilobites. *Lethaia* 2:395–414
10. Breder, C. M. 1957. A note on preliminary stages in the fossilization of fishes. *Copeia* 1957:132–35
11. Breimer, A. 1969. A contribution to the paleoecology of Palaeozoic stalked crinoids. *Proc. K. Ned. Akad. Wet.* 72(B):139–50
12. Briggs, D. E. G. 1976. The arthropod *Branchiocaris* n. gen., Middle Cambrian, Burgess Shale, British Columbia. *Bull. Geol. Surv. Can.* 264:1–29
13. Briggs, D. E. G. 1977. Bivalved arthropods from the Cambrian Burgess Shale of British Columbia. *Palaeontology* 20:595–621

14. Briggs, D. E. G. 1978. The morphology, mode of life, and affinities of *Canadaspis perfecta* (Crustacea: Phyllocarida), Middle Cambrian, Burgess Shale, British Columbia. *Philos. Trans. R. Soc. London, Ser. B* 281:439–87

15. Briggs, D. E. G. 1978. A new trilobite-like arthropod from the Lower Cambrian Kinzers Formation, Pennsylvanian. *J. Paleontol.* 52:132–40

16. Brooks, H. K., Caster, K. E. 1956. *Pseudoarctolepis sharpi,* n. gen., n. sp. (Phyllocarida), from the Wheeler Shale (Middle Cambrian) of Utah. *J. Paleontol.* 30:9–14

17. Bruton, D. L. 1977. Appendages of *Sidneyia, Emeraldella* and *Leanchoilia* and their bearing on trilobitoid classification. *J. Paleontol.* 51:Suppl. to 2, Pt. 3, 4–5 (Abstr.)

18. Cave, L. D., Simonetta, A. M. 1975. Notes on the morphology and taxonomic position of *Aysheaia* (Onychophora?) and of *Skania* (undetermined phylum). *Monit. Zool. Ital.* 9(N.S.): 67–81

19. Chapman, F. 1926. On a supposed phyllocarid from the older Palaeozoic of Tasmania. *Pap. Proc. R. Soc. Tasmania,* pp. 79–80

20. Cisne, J. L. 1973. Beecher's Trilobite Bed revisited: ecology of an Ordovician deepwater fauna. *Postilla* 160:1–25

21. Cisne, J. L. 1975. Anatomy of *Triarthrus* and the relationships of the Trilobita. *Fossils & Strata* 4:45–63

22. Clark, A. H. 1912. Restoration of the genus *Eldonia,* a genus of free swimming holothurians from the Middle Cambrian. *Zool. Anz.* 39:723–25

23. Clark, A. H. 1913. Cambrian holothurians. *Am. Nat.* 47:488–507

24. Clark, H. L. 1912. Fossil holothurians. *Science* 35:274–78

25. Cleaves, A. B., Fox, E. F. 1935. Geology of the west end of Ymer Island, East Greenland. *Geol. Soc. Am. Bull.* 46:463–88

26. Cloud, P. 1976. Beginnings of biospheric evolution and their biogeochemical consequences. *Paleobiology* 2: 351–87

27. Conway Morris, S. 1976. A new Cambrian lophophorate from the Burgess Shale of British Columbia. *Palaeontology* 19:199–222

28. Conway Morris, S. 1976. *Nectocaris pteryx,* a new organism from the Middle Cambrian Burgess Shale of British Columbia. *Neues Jahrb. Geol. Palaeontol. Monatsh. H.* 12:705–13

29. Conway Morris, S. 1977. A new metazoan from the Cambrian Burgess Shale of British Columbia. *Palaeontology* 20:623–40

30. Conway Morris, S. 1977. Fossil priapulid worms. *Spec. Pap. Palaeontol.* 20:iv,1–95

31. Conway Morris, S. 1977. A new entoproct-like organism from the Burgess Shale of British Columbia. *Palaeontology* 20:833–45

32. Conway Morris, S. 1977. A redescription of the Middle Cambrian worm *Amiskwia sagittiformis* Walcott from the Burgess Shale of British Columbia. *Palaeontol. Z.* 51:271–87

33. Conway Morris, S. 1977. Burgess Shale metazoan association. In *Patterns of Evolution,* ed. A. Hallam, pp. 31–33. Amsterdam: Elsevier. 591 pp.

34. Conway Morris, S. 1977. Aspects of the Burgess Shale fauna, with particular reference to the non-arthropod component. *J. Paleontol.* 51: Suppl. to No. 2, Pt 3, pp. 7–8 (Abstr.)

35. Conway Morris, S. 1978. *Laggania cambria:* A composite fossil. *J. Paleontol.* 52:126–31

36. Conway Morris, S. 1979. The Burgess Shale. In *Encyclopedia of Paleontology,* ed. R. W. Fairbridge, D. Jablonski, Stroudsburg, Pa: Dowden, Hutchinson & Ross. In press

37. Conway Morris, S. 1979. Middle Cambrian polychaetes from the Burgess Shale of British Columbia. *Philos. Trans. R. Soc. London, Ser. B.* 285:227–74

38. Crimes, T. P. 1975. The stratigraphical significance of trace fossils. In *The Study of Trace Fossils,* ed. R. W. Frey, pp. 109–30. NY: Springer. 562 pp.

39. Dean, B. 1902. The preservation of muscle fibres in sharks of the Cleveland Shale. *Am. Geol.* 30:273–78

40. Degens, E. T., Mopper, K. 1976. Factors controlling the distribution and early diagenesis of organic material in marine sediments. In *Chemical Oceanography,* ed. J. P. Riley, R. Chester, 6:59–113. London: Academic. 414 pp. 2nd ed.

41. Dhonau, N. B., Holland, C. H. 1974. The Cambrian of Ireland. In *Cambrian of the British Isles, Norden, and Spitzbergen,* ed. C. H. Holland, 2:157–76. London: Wiley. 300 pp.

42. Dunbar, C. O. 1925. Antennae in *Olenellus getzi,* n. sp. *Am. J. Sci.* 9:303–8

43. Durham, J. W. 1974. Systematic position of *Eldonia ludwigi* Walcott. *J. Paleontol.* 48:750–55

44. Eha, S. 1953. The pre-Devonian sediments on Ymers φ, Suess Land, and Ella φ (East Greenland) and their tectonics. *Medd. Groenl.* 111(2):1–105

45. Finks, R. M. 1970. The evolution and ecologic history of sponges during Palaeozoic times. *Symp. Zool. Soc. London* 25:3–22

46. Fritz, W. H. 1971. Geological setting of the Burgess Shale. *Proc. 1st N. Am. Paleontol. Conv., Chicago, 1969* I:1155–70

47. Glaessner, M. F. 1958. New fossils from the base of the Cambrian in South Australia. *Trans. R. Soc. South Aust.* 81:185–88

48. Glaessner, M. F. 1976. Early Phanerozoic annelid worms and their geological and biological significance. *J. Geol. Soc. London* 132:259–75

49. Golubic, S. 1976. Organisms that build stromatolites. *Dev. Sedimentol.* 20:113–26

50. Golvan, Y. J. 1958. Le phylum des Acanthocephala. Première note sa place dans l'échelle zoologique. *Ann. Parasitol. Hum. Comp.* 33:538–602

51. Harrington, H. J., Moore, R. C. 1956. Medusae incertae sedis and unrecognizable forms. In *Treatise on Invertebrate Paleontology,* ed. R. C. Moore, F:F153–61. Lawrence, Kans: Univ. Kansas. 498 pp.

52. Heldt, J. H. 1954. *Waptia fieldensis* Walcott et les stades larvaires des Pénéides. *Bull. Soc. Sci. Nat. Tunis* 6:177–80

53. Hessler, R. R., Newman, W. A. 1975. A trilobitomorph origin for the Crustacea. *Fossils & Strata* 4:437–59

54. Hofmann, H. J., Parsley, R. L. 1966. Antennae of *Ogygopsis. J. Paleontol.* 40:209–11

55. Howell, B. F. 1962. Worms. See Ref. 51, W:W144–77

56. Hughes, C. P. 1975. Redescription of *Burgessia bella* from the Middle Cambrian Burgess Shale, British Columbia. *Fossils & Strata* 4:415–35

57. Hughes, C. P. 1977. The early arthropod *Waptia fieldensis. J. Paleontol.* 51:suppl. to 2, Pt. 3, p. 15(Abstr.)

58. Hutchinson, G. E. 1930. Restudy of some Burgess Shale fossils. *Proc. US Natl. Mus.* 78(11):1–24

59. Hutchinson, G. E. 1969. *Aysheaia* and the general morphology of the Onychophora. *Am. J. Sci.* 267:1062–66

60. Jefferies, R. P. S., Lewis, D. N. 1978. The English Silurian fossil *Placocystites forbesianus* and the ancestry of the vertebrates. *Phil. Trans. R. Soc. London, Ser. B.* 282:205–323

61. Johnson, R. G., Richardson, E. S. 1969. Pennsylvanian invertebrates of the Mazon Creek area, Illinois: The morphology and affinities of *Tullimonstrum. Fieldiana Geol.* 12(8):119–49

62. Kielan-Jaworowska, Z. 1968. Scolecodonts versus jaw apparatuses. *Lethaia* 1:39–49

63. Korotkevich, V. S. 1967. Systematic position of *Amiskwia sagittiformis* from Middle Cambrian of Canada. *Paleontol. J.* 1967 (4):115–18 (transl. from *Paleontol. Zh.*)

64. Land, J. 1970. Systematics, zoogeography, and ecology of the Priapulida. *Zool. Verh. Rijksmus. Nat. Hist. Leiden* 112:1–118

65. Landing, E. 1977. *"Prooneotodus" tenuis* (Müller, 1959) apparatuses from the Taconic allochthon, eastern New York: Construction, taphonomy and the protoconodont "supertooth" model. *J. Paleontol.* 51:1072–84

66. Lane, N. G. 1963. The Berkeley crinoid collection from Crawfordsville, Indiana. *J. Paleontol.* 37:1001–8

67. Lane, N. G. 1973. Paleontology and paleoecology of the Crawfordsville fossil site (Upper Osagian; Indiana). *Univ. Calif. Publ. Geol. Sci.* 99:1–141

68. Lohmann, H. 1922. *Oesia disjuncta* Walcott, eine Appendicularie aus dem Kambrium. *Mitt. Zool. Mus. Hamb.* 38:69–75

69. Madsen, F. J. 1956. *Eldonia,* a cambrian siphonophore—formerly interpreted as a holoturian [sic]. *Vidensk. Medd. Dan. Naturhist. Foren. Khobenhavn* 118:7–14

70. Madsen, F. J. 1957. On Walcott's supposed Cambrian holothurians. *J. Paleontol.* 31:281–82

71. Madsen, F. J. 1962. The systematic position of the Middle Cambrian fossil *Eldonia. Medd. Dan. Geol. Foren.* 15:87–89

72. Mansuy, H. 1912. Paléontologie. Etude geologique du Yun-Nan Oriental. *Mem. service Geol. Indochine* 1(2):1–146

73. Manton, S. M. 1977. *The Arthropoda. Habits, Functional Morphology, and Evolution.* Oxford: Clarendon, 527 pp.

74. Marek, L., Yochelson, E. L. 1976. Aspects of the biology of Hyolitha (Mollusca). *Lethaia* 9:65–82

75. Matthew, G. F. 1899. Studies on Cambrian faunas, No. 3. Upper Cambrian fauna of Mount Stephen, British Columbia. The trilobites and worms. *Trans. R. Soc. Can. 2nd Ser.* 5(4):39–66

76. Matthew, G. F. 1902. Notes on Cambrian faunas. No. 8. Cambrian Brachi-

opoda and Mollusca of Mt. Stephen, B.C., with the description of a new species of *Metoptoma. Trans. R. Soc. Can. 2nd Ser.* 8(4):107–12

77. McIlreath, I. A. 1974. Stratigraphic relationships at the western edge of the Middle Cambrian carbonate facies belt, Field, British Columbia. *Geol. Surv. Can. Pap.* 74–1 (Part A):333–34

78. McIlreath, I. A. 1975. Stratigraphic relationships at the western edge of the Middle Cambrian carbonate facies belt, Field, British Columbia. *Geol. Surv. Can. Pap.* 75–1 (Part A):557–58

79. McIlreath, I. A. 1977. Accumulation of a Middle Cambrian, deep-water limestone debris apron adjacent to a vertical, submarine carbonate escarpment, southern Rocky Mountains, Canada. *Soc. Econ. Paleontol. Mineral. Spec. Publ.* 25:113–24

80. Nicoll, R. S. 1977. Conodont apparatuses in an Upper Devonian palaeoniscoid fish from the Canning Basin, Western Australia. *Aust. Bur. Miner. Resour. Geol. Geophys. J.* 2:217–28

81. Nitecki, M. H., Schram, F. R. 1976. *Etacystis communis,* a fossil of uncertain affinities from the Mazon Creek fauna (Pennsylvanian of Illinois). *J. Paleontol.* 50:1157–61

82. Nitecki, M. H., Solem, A. 1973. A problematic organism from the Mazon Creek (Pennsylvanian) of Illinois. *J. Paleontol.* 47:903–7

83. Owre, H. B., Bayer, F. M. 1962. The systematic position of the Middle Cambrian fossil *Amiskwia* Walcott. *J. Paleontol.* 36:1361–63

84. P'an, K. 1957. On the discovery of Homopoda from South China. *Acta Palaeontol. Sinica* 5:523–26

85. Parker, B. C., Dawson, E. Y. 1965. Non-calcareous marine algae from California Miocene deposits. *Nova Hedwigia Z. Kryptogamenkd.* 10:273–95

86. Piper, D. J. W. 1972. Sediments of the Middle Cambrian Burgess Shale, Canada. *Lethaia* 5:169–75

87. Piper, D. J. W. 1972. Turbidite origin of some laminated mudstones. *Geol. Mag.* 109:115–26

88. Por, F. D., Bromley, H. J. 1974. Morphology and anatomy of *Maccabeus tentaculatus* (Priapulida: Seticoronaria). *J. Zool.* 173:173–97

89. Rasetti, F. 1951. Middle Cambrian stratigraphy and faunas of the Canadian Rocky Mountains. *Smithson. Misc. Collect.* 116(5):1–277

90. Rasetti, F. 1954. Internal shell structures in the Middle Cambrian gastropod *Scenella* and the problematic genus *Stenothecoides. J. Paleontol.* 28:59–66

91. Raymond, P. E. 1920. The appendages, anatomy, and relationships of trilobites. *Mem. Conn. Acad. Arts Sci.* 7:1–169

92. Raymond, P. E. 1930. Report on invertebrate paleontology. *Mus. Comp. Zool. (Harv. Univ.) Annu. Rep.* 1929–1930:31–33

93. Raymond, P. E. 1931. Notes on invertebrate fossils, with descriptions of new species. *Bull. Mus. Comp. Zool. Harv. Univ.* 55:165–213

94. Raymond, P. E. 1935. *Leanchoilia* and other Mid-Cambrian Arthropoda. *Bull. Mus. Comp. Zool. Harv. Univ.* 76:205–30

95. Repetski, J. E. 1978. A fish from the Upper Cambrian of North America. *Science* 200:529–31

96. Resser, C. E. 1929. New Lower and Middle Cambrian Crustacea. *Proc. US Natl. Mus.* 76(9):1–18

97. Resser, C. E. 1938. Fourth contribution to nomenclature of Cambrian fossils. *Smithson. Misc. Collect.* 97(10):1–43

98. Resser, C. E. 1939. The Spence Shale and its fauna. *Smithson. Misc. Collect.* 97(12):1–29

99. Resser, C. E., Howell, B. F. 1938. Lower Cambrian *Olenellus* zone of the Appalachians. *Geol. Soc. Am. Bull.* 49:195–248

100. Richardson, E. S., Johnson, R. G. 1971. The Mazon Creek faunas. *Proc. 1st N. Am. Paleontol. Conv., Chicago, 1969* I:1222–35

101. Richter, R., Richter, E. 1927. Eine Crustacee (*Isoxys carbonelli* n. sp.) in den *Archaeocyathus*—Bildungen der Sierra Morena und ihre stratigraphische Beurteilung. *Senckenbergiana* 9:188–95

102. Rickards, R. B. 1977. Patterns of evolution in the graptolites. See Ref. 33, pp. 333–58

103. Rigby, J. K. 1976. Some observations on occurrences of Cambrian Porifera in western North America and their evolution. *Brigham Young Univ. Geol. Stud.* 23:51–60

104. Robison, R. A. 1969. Annelids from the Middle Cambrian Spence Shale of Utah. *J. Paleontol.* 43:1169–73

105. Robison, R. A. 1972. Mode of life of agnostid trilobites. *Proc. 24th Int. Geol. Congr.* 7:33–40

106. Rolfe, W. D. I. 1962. Two new arthropod carapaces from the Burgess Shale (Middle Cambrian) of Canada. *Breviora* 160:1–9

107. Rominger, C. 1887. Description of primordial fossils from Mount Stephens,

N. W. Territory of Canada. *Proc. Acad. Nat. Sci. Philadelphia* 1887 (part 1): 12–19

108. Ruedemann, R. 1931. Some new Middle Cambrian fossils from British Columbia. *Proc. US Natl. Mus.* 79(27): 1–18

109. Runnegar, B., Pojeta, J., Morris, N. J., Taylor, J. D., Taylor, M. E., McClung, G. 1975. Biology of the Hyolitha. *Lethaia* 8:181–91

110. Ryan, W. J., Hallissy, T. 1912. Preliminary notice of some new fossils from Bray Head, County Wicklow. *Proc. R. Ir. Acad. Sect. B.* 29:246–51

111. Satterthwaite, D. F. 1976. *Paleobiology and paleoecology of Middle Cambrian algae from western North America.* PhD thesis. Univ. California, Los Angeles. 121 pp.

112. Simonetta, A. M. 1961. Osservazioni su *Marria Walcotti* Ruedemann: un graptolite e non un artropodo. *Boll. Zool.* 28:569–72

113. Simonetta, A. M. 1962. Note sugli artropodi non triloboti della Burgess Shale, Cambriano medio della Columbia Britannica (Canada). I° Contributo: Il genere *Marrella* Walcott 1912. *Monit. Zool. Ital.* 69:172–85

114. Simonetta, A. M. 1963. Osservazioni sugli artropodi non triloboti della "Burgess Shale" (Cambriano Medio). II Contributo: I Generi *Sidneya* [sic] ed *Amiella* Walcott 1911. *Monit. Zool. Ital.* 70–71:97–100

115. Simonetta, A. M. 1964. Osservazioni sugli artropodi non triloboti della "Burgess Shale" (Cambriano medio). III Contributo: I generi *Molaria, Habelia, Emeraldella, Parahabelia* (Nov.), *Emeraldoides* (Nov.). *Monit. Zool. Ital.* 72:215–31

116. Simonetta, A. M. 1970. Studies on non-trilobite arthropods of the Burgess Shale (Middle Cambrian). The genera *Leanchoilia, Alalcomenaeus, Opabinia, Burgessia, Yohoia* and *Actaeus. Palaeontogr. Ital.* 66(N.S. 36):35–45

117. Simonetta, A. M., Cave, L. D. 1975. The Cambrian non-trilobite arthropods from the Burgess Shale of British Columbia. A study of their comparative morphology taxinomy [sic] and evolutionary significance. *Palaeontogr. Ital.* 69(N.S. 39):1–37

118. Sondheimer, E., Dence, W. A., Mattick, L. R., Silverman, S. R. 1966. Composition of combustible concretions of the alewife, *Alosa pseudoharengus. Science* 152:221–23

119. Sprinkle, J. 1973. Morphology and evolution of blastozoan echinoderms. *Mus. Comp. Zool. (Harv. Univ.) Spec. Publ. 1–283*

120. Størmer, L. 1959. Trilobitoidea. See Ref. 51, O:O23–37

121. Stürmer, W. 1970. Soft parts of cephalopods and trilobites: some surprising results of X-ray examinations of Devonian slates. *Science* 170:1300–2

122. Stürmer, W., Bergström, J. 1973. New discoveries on trilobites by X-rays. *Palaeontol. Z.* 47:104–41

123. Stürmer, W., Bergström, J. 1976. The arthropods *Mimetaster* and *Vachonisia* from the Devonian Hunsrück Shale. *Palaeontol. Z.* 50:78–111

124. Stürmer, W., Bergström, J. 1978. The arthropod *Cheloniellon* from the Devonian Hunsrück Shale. *Palaeontol. Z.* 52:57–81

125. Tarlo, L. B. 1960. The invertebrate origins of the vertebrates. *Proc. 21st Int. Geol. Congr.* 22:113–23

126. Valentine, J. W., Foin, T. C., Peart, D. 1978. A provincial model of Phanerozoic diversity. *Paleobiology* 4:55–66

127. Voigt, E. 1939. Fossil red blood corpuscles found in a lizard from the Middle Eocene lignite of the Geiseltal near Halle. *Res. Prog.* 5:53–56

128. Wade, M. 1968. Preservation of soft-bodied animals in Precambrian sandstones at Ediacara, South Australia. *Lethaia* 1:238–67

129. Walcott, C. D. 1884. On a new genus and species of Phyllopoda from the Middle Cambrian. *US Geol. Surv. Bull.* 10:330–31

130. Walcott, C. D. 1886. Second contribution to the studies on the Cambrian faunas of North America. *US Geol. Surv. Bull.* 30:731–195

131. Walcott, C. D. 1888. Cambrian fossils from Mount Stephens, Northwest Territory of Canada. *Am. J. Sci.* 36 (3rd Ser.):161–66

132. Walcott, C. D. 1890. The fauna of the Lower Cambrian or *Olenellus* zone. *Annu. Rep. US Geol. Surv. 1888–1889* 10(1):509–760

133. Walcott, C. D. 1908. Mount Stephen rocks and fossils. *Can. Alp. J.* 1:232–48

134. Walcott, C. D. 1911. Middle Cambrian Merostomata. *Smithson. Misc. Collect.* 57:17–40

135. Walcott, C. D. 1911. Middle Cambrian holothurians and medusae. *Smithson. Misc. Collect.* 57:41–68

136. Walcott, C. D. 1911. Middle Cambrian annelids. *Smithson. Misc. Collect.* 57: 109–44

137. Walcott, C. D. 1912. Middle Cambrian Branchiopoda, Malacostraca, Trilobita, and Merostomata. *Smithson. Misc. Collect.* 57:145–228

138. Walcott, C. D. 1912. Cambrian of the Kicking Horse Valley, B. C. *Geol. Surv. Can. Rep.* 26:188–91

139. Walcott, C. D. 1912. Cambrian Brachiopoda. *Geol. Surv. US Monogr.* 51(1):1–872; (2):1–363

140. Walcott, C. D. 1916. Evidences of primitive life. *Smithson. Inst. Annu. Rep. 1915*, pp. 235–55

141. Walcott, C. D. 1918. Appendages of trilobites. *Smithson. Misc. Collect.* 67:115–216

142. Walcott, C. D. 1918. Geological explorations in the Canadian Rockies. *Smithson. Misc. Collect.* 68:4–20

143. Walcott, C. D. 1919. Middle Cambrian Algae. *Smithson. Misc. Collect.* 67:217–60

144. Walcott, C. D. 1920. Middle Cambrian Spongiae. *Smithson. Misc. Collect.* 67:261–364

145. Walcott, C. D. 1921. Notes on structure of *Neolenus. Smithson. Misc. Collect.* 67:365–456

146. Walcott, C. D. 1924. Cambrian and Ozarkian Brachiopoda, Ozarkian Cephalopoda and Notostraca. *Smithson. Misc. Collect.* 67:477–554

147. Walcott, C. D. 1928. Pre-Devonian Paleozoic formations of the Cordilleran province of Canada. *Smithson. Misc. Collect.* 75:175–368

148. Walcott, C. D. 1931. Addenda to descriptions of Burgess Shale fossils (with explanatory notes by Charles E. Resser). *Smithson. Misc. Collect.* 85:1–46

149. Walton, J. 1923. On the structure of a Middle Cambrian alga from British Columbia (*Marpolia spissa* Walcott). *Biol. Rev. Cambridge Philos. Soc.* 1:59–62

150. Weigelt, J. 1935. Some remarks on the excavations in the Geisel Valley. *Res. Prog.* 1:155–59

151. Wells, J. W., Hill, D. 1956. Zoanthiniaria, Corallimorpharia, and Actiniaria. See Ref. 51, F:F232–33

152. Whiteaves, J. F. 1892. Description of a new genus and species of phyllocarid Crustacea from the Middle Cambrian of Mount Stephen, B.C. *Can. Rec. Sci.* 5:205–8

153. Whittington, H. B. 1971. The Burgess Shale: History of research and preservation of fossils. *Proc. 1st N. Am. Paleontol. Conv., Chicago* I:1170–1201

154. Whittington, H. B. 1971. Redescription of *Marrella splendens* (Trilobitoidea) from the Burgess Shale, Middle Cambrian, British Columbia. *Bull. Geol. Surv. Can.* 209:1–24

155. Whittington, H. B. 1974. *Yohoia* Walcott and *Plenocaris* n. gen., arthropods from the Burgess Shale, Middle Cambrian, British Columbia. *Bull. Geol. Surv. Can.* 231:1–27

156. Whittington, H. B. 1975. The enigmatic animal *Opabinia regalis,* Middle Cambrian, Burgess Shale, British Columbia. *Philos. Trans. R. Soc. London Ser. B.* 271:1–43

157. Whittington, H. B. 1975. Trilobites with appendages from the Middle Cambrian, Burgess Shale, British Columbia. *Fossils & Strata* 4:97–136

158. Whittington, H. B. 1977. The Middle Cambrian trilobite *Naraoia,* Burgess Shale, British Columbia. *Philos. Trans. R. Soc. London Ser. B.* 280:409–43

159. Whittington, H. B. 1977. *Aysheaia* and arthropod relations. *J. Paleontol.* 51: Suppl. to 2, Pt. 3, p. 31(Abstr.)

160. Whittington, H. B. 1978. The lobopod animal *Aysheaia pedunculata,* Middle Cambrian, Burgess Shale, British Columbia. *Philos. Trans. R. Soc. London, Ser. B.* 284:165–97

161. Wiedenmayer, F. 1977. The Nepheliospongiidae Clarke 1900 (Demospongea, Upper Devonian to Recent), an ultraconservative, chiefly shallow-marine sponge family. *Eclogae Geol. Helv.* 70:885–918

162. Yochelson, E. L. 1961. The operculum and mode of life of *Hyolithes. J. Paleontol.* 35:152–61

Ann. Rev. Ecol. Syst. 1979. 10:351–71
Copyright © 1979 by Annual Reviews Inc. All rights reserved

THE PHYSIOLOGICAL ❖4166
ECOLOGY OF PLANT SUCCESSION

F. A. Bazzaz

Department of Botany, University of Illinois, Urbana, Illinois 61801

INTRODUCTION

Succession is a process of continuous colonization of and extinction on a site by species populations. The process has long been central in ecological thinking; much theory and many data about succession have accumulated over the years.

Since nearly all species in all communities participate in successional interactions, and because physiological ecology encompasses everything that a plant does during its life cycle, a complete review of physiological ecology of all species in all successions is not possible. Thus in this review I discuss the physiological adaptations of species of one successional gradient—from open field to broad-leaved deciduous forest. I concentrate on the physiological adaptations of early successional plants to environmental variability and collate the literature on tree physiology to make comparisons with early successional plants. My discussion may not be applicable to seres where there is little difference in physiognomy between early and late successional plants or where the designation of species as early or late successional is unjustified (e.g. for certain desert and tundra habitats). I discuss the nature of successional environments, seed germination, seedling and mature plant development, plant growth, photosynthesis, water use, and the physiological ecology of competition and interference.

THE NATURE OF SUCCESSIONAL ENVIRONMENTS

The environment of a plant may vary daily, seasonally, vertically, and horizontally. The level of variability is determined by many factors including climate, geographical location, geomorphological features, the nature of site disturbances, and the number and kind of species present. The influence of the environment on the plant depends not only upon the level of environmental variability and the predictability of that variation, but also on the

351

0066-4162/79/1120-0351$01.00

change in plant size and physiology through time. It is generally thought that environmental variability in open, early successional habitats is higher than in closed, late successional ones. The variability of the physical environment is related mainly to the amount of energy that reaches the soil surface and the way in which it is dispersed from the surface. In an open field, energy exchange occurs at or near the soil surface, light energy reaches the surface unaltered and maximum temperature fluctuations occur there. In a later successional forest the surface of energy exchange is the upper layers of the canopy. Temperature fluctuations below the canopy are buffered by the vegetation itself, and progressively less energy penetrates toward the forest floor; light at the floor is markedly depleted of photosynthetically active wavelengths and is high in far-red wavelengths. Thus seedlings of the late successional species, except in large light gaps, experience a less variable and less extreme environment in the forest with respect to temperature, humidity, and wind. However, sunflecks under a canopy result in extremely variable light intensity and perhaps rapid fluctuation of leaf temperature. The extent, frequency, and magnitude of these events, and the physiological response of plants to them, have not been investigated.

Maximum fluctuation in temperature and soil moisture occurs at or near the soil surface in open early successional habitats (70, 73). Thus germinating seeds and young seedlings may experience a wider range of fluctuations in these two variables than do mature plants. In open early successional habitats CO_2 concentrations are higher than ambient just above the soil surface, increase with soil depth, and reach maximum values just above the water table (80). In forests CO_2 concentrations may rise above ambient levels within the forest canopy, especially early in the day and late at night (29, 100).

Although there is general agreement about the relative levels of environmental variability in early and late successional habitats, quantification and interpretation of this variability are still rather difficult and may be frustrated by phenomena such as sunflecks. A more serious problem, however, is the fact that the plant itself and not variation in the physical factors per se determines the effect of variability. It is likely that similar levels of variation of an environmental parameter cause quite different responses in different species: For some a certain level may be of no consequence to their function; for others it may be detrimental or stimulatory.

ECOPHYSIOLOGICAL CHARACTERISTICS OF SUCCESSIONAL PLANTS

Seed Germination

Seeds of many early successional plants live for years in the soil (45, 87). Seeds of early successional trees long dormant in the soil may germinate in large numbers when the canopy opens (54). In contrast, seeds of late successional trees lose viability quickly (e.g. 2, 83).

The relationship between seed germination and various parameters of the physical environment has been reviewed, with emphasis on its adaptive significance (e.g. 46, 90). Early and late successional environments differ primarily with respect to light intensity and spectral quality. Seeds of early successional plants are sensitive to light (37, 38, 79, 94) and their germination is strongly inhibited by vegetation-filtered (high far-red/red) light (30, 44, 81, 85). In contrast, seeds of later successional plants, especially those found in climax forests, do not require light for germination—e.g. *Fagus grandifolia* (77) and *Acer saccharum* (59). Furthermore, seeds of species from open habitats require more light for germination than do those of woodland species (34). Fluctuating temperatures also enhance the germination of many species and may be the most important factor in seed germination of annuals (22, 86, 91).

Seeds of early successional plants germinate at or near the soil surface. Here the seeds experience unfiltered light, high daytime temperatures early in the growing season, much variation in daily temperature, and low CO_2 concentrations. Thus seed germination of early successional plants is related to disturbance that brings some seeds from deep in the soil closer to the soil surface. Furthermore, both unfiltered high light (rich in red wavelengths) and fluctuating temperatures are associated with disturbance in forests, and the germination of some successional trees is also keyed to this disturbance (e.g. *Prunus pensylvanica* (54) and *Betula alleghaniensis* (25). Seed germination of early successional plants may be linked to disturbance in other ways. For example, KNO_3 and other nitrate salts enhance seed germination of several species, including some early successional herbs (36, 46, 66). In devegetated areas a flush of nitrates may occur early in the spring (88) and act as a cue for germination as well as a resource for the young seedlings.

Another aspect of germination in early successional plants is the development of induced (secondary) dormancy—e.g. in *Ambrosia trifida* (22), *A. artemisiifolia* (5), and *Amaranthus* sp. (26). This strategy should protect the seed bank in the event that the site is disturbed again when environmental conditions may be unsuitable for seedling growth.

Ambrosia artemisiifolia, perhaps the most common annual of oldfield succession, possesses a complex germination strategy combining several of the features common to colonists. Germination of the species is closely linked to disturbance, which ensures the availability of resources and reduces the probability of competition with later-successional species (Figure 1). The seeds are dormant when shed. After winter stratification a shift in germinator/inhibitor ratio takes place (97) and the seeds become ready to germinate. If the seeds are brought up to or near the surface by disturbance they experience unfiltered light, fluctuating temperatures, and reduced CO_2 concentrations, all factors that have been shown to increase seed germination in this species (4, 5, 67). If there is no disturbance some seeds germinate; most do not, and after some time these develop an induced dormancy. Later disturbance causes little germination; the remainder of the seeds require another stratification before they germinate. Seed germination of this species and of several other early successional plants is not an all-or-none phenomenon. Some germination occurs regardless of environmental conditions; additional germination occurs after restratification. This suggests polymorphisms for germination in this and several other species (3, 62, 98).

Germination in early successional plants is epigeal. The cotyledons green up quickly, are photosynthetic (7, 66, 78), and enlarge their surface quickly. Forest trees with large seeds—e.g. *Aesculus* and *Quercus*—have hypogeal germination, while those with small seeds—e.g. *Fagus* and *Acer*—have epigeal germination. The photosynthetic advantages of epigeal germination have associated costs: expenditures for mechanical support and the apparency of cotyledons to above-ground predators.

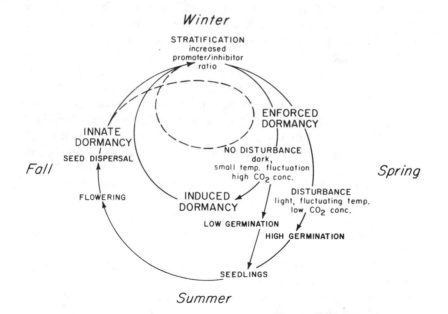

Figure 1 Schematic representation of seed germination in *Ambrosia artemisiifolia* L., a common colonizer in oldfield succession. Dashed line represents seed morphs that require more than one stratification cycle to germinate.

Emergence and Physiology of Seedlings of Herbaceous Colonizers

Winter annuals (mostly composites) have small seeds adapted for long-distance dispersal. They have little or no innate dormancy (37, 43) and are ready to germinate as soon as they land in a suitable site.They germinate under a wide range of temperatures in the field in late summer and fall. Their emergence may be spread over weeks or months, however, and flushes of germination seem to follow periods of rain (73). The seedlings develop into rosettes that overwinter. Most rosettes die in winter owing to frost heaving; the probability of survival is higher for larger ones (73). However, these rosettes are capable of photosynthesizing at relatively high rates and accumulate reserves during winter. They photosynthesize over a wide range of temperatures, utilize a wide range of light intensities, lower their light compensation points at low leaf temperatures in winter, and begin photosynthesis within minutes after exposure to light. In summer their photosynthetic optima shift to higher temperatures (72). Summer annuals produce relatively large seeds that are well protected by heavy seed coats. They are dormant when shed, have complex dormancy mechanisms (Figure 1), and remain inactive for extended periods until the environment becomes conducive to their germination. They then germinate quickly. In the presence of winter annuals, the summer annuals are much suppressed and contribute little to community production. The dominance of the winter annuals is achieved by the preemption of environmental resources—e.g. light, water, space, and especially nutrients (71).

Time of seed maturation, dormancy characteristics, and response to temperature interact to determine the time of seedling emergence and therefore the period of growth of the winter annuals and their time of prominence in the field. Such differences in phenology may contribute to coexistence of these species. Seedlings of the summer annuals emerge in the spring somewhat according to their temperature requirements for germination. Recruitment of new seedlings into the population usually stops in early summer.

In open habitats moisture fluctuates widely near the soil surface but much less below (73), and maximum daily temperature fluctuations occur at the soil surface (10, 70). The growing seedlings should be selected for ecophysiological adaptations to these variable environmental factors: They must be broad-niched. They must be able to grow rapidly, both above and below ground, to escape this zone of maximum variability. Evidence for broad response may be deduced from germination and emergence of some early successional plants over a wide range of temperature (e.g. 37, 67, 89) and moisture (68, 70). Although the physiological ecology of seedlings has not been studied in many species, our data (7, 66) suggest that early successional plants have high rates of photosynthesis and respiration over a wide range of temperatures and soil moisture. Comparative studies of seedling physiology of early and late successional plants are required for a better understanding of the modes of adaptation to the contrasting physical and biological environments of these plants. Young seedlings of *A. artemisiifolia* exhibit another adaptation to the variable conditions near the surface. They may become desiccated in the field, but upon wetting they develop new roots and the seedlings resume growth. Since soil disturbance could occur at any time during the germination period, and because plowing may expose already established seedlings to the drying conditions at the soil surface, this behavior may be adaptive. Whether this is common in early successional plants is unknown.

Growth and Development

PHOTOSYNTHESIS   Light has been recognized as a major factor in species replacement, especially in forest succession (14, 47, 48, 52, 53). The degree of shade tolerance and the arrangement of the foliage and branching patterns are important in determining successional sequences in deciduous forest (40, 95). Most ecophysiological research has emphasized photosynthesis because of its direct relationship to plant survival and growth (56). The light saturation curve (Figure 2), has proven useful in comparing plants from contrasting habitats. It shows several physiological properties of the plants—i.e. rate of dark respiration, the initial slope of photosynthesis (an index of quantum efficiency), the light compensation point, the light saturation point, and the maximum rate of photosynthesis. If the environment around the leaf is closely monitored and transpiration is simultaneously measured with photosynthesis, a variety of other parameters of ecological interest (e.g. leaf resistance and water use efficiency) can be obtained.

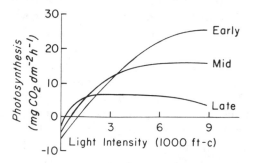

Figure 2   Idealized light saturation curves for early-, mid-, and late-successional plants.

Rates of photosynthesis are often higher in sun-adapted than in shade-adapted species (13). Herbaceous species have higher photosynthetic rates than woody species (49, 50, 56). Data from our work and various other sources (Table 1) show that the rate of photosynthesis per unit of leaf area generally declines with succession (Figure 3). Under optimal conditions the rate of photosynthesis of early successional plants may be as high as 50 mg CO_2 dm^{-2} h^{-1}. This is not merely a reflection of their herbaceous growth habit, since late successional herbs of deciduous forest have lower rates. Early successional trees have intermediate rates and late successional plants have low rates. Furthermore, the rate of photosynthesis in the late successional group tends to decline with increased shade tolerance (Table 1).

The light saturation curves of photosynthesis of early and late successional plants show how they differ in a number of ways (Figure 2). First, light saturation occurs at higher light intensities in early successional plants (7, 66, 73, 96). Early successional trees also saturate at high light intensity (1, 11, 47, 48, 61, 74). Light saturation in late successional plants occurs at much lower light intensities (~10–15% of full sunlight). This has been shown for several important, late successional trees—e.g. *Fagus grandifolia*, *Quercus rubra* (51), *Acer saccharum*, and *Quercus coccinea* (101). In early

Table 1 Some representative photosynthetic rates (mg CO_2 dm^{-2} h^{-1}) of plants in a successional sequence[a]

Plant	Rate	Plant	Rate
Summer annuals		Early successional trees	
Abutilon theophrasti	24	*Diospyros virginiana*	17
Amaranthus retroflexus	26	*Juniperus virginiana*	10
Ambrosia artemisiifolia	35	*Populus deltoides*	26
Ambrosia trifida	28	*Sassafras albidum*	11
Chenopodium album	18	*Ulmus alata*	15
Polygonum pensylvanicum	18		
Setaria faberii	38	Late successional trees	
		Liriodendron tulipifera	18
Winter annuals		*Quercus velutina*	12
		Fraxinus americana	9
Capsella bursa-pastoris	22	*Quercus alba*	4
Erigeron annuus	22	*Quercus rubra*	7
Erigeron canadensis	20	*Aesculus glabra*	8
Lactuca scariola	20	*Fagus grandifolia*	7
		Acer saccharum	6
Herbaceous perennials			
Aster pilosus	20		

[a]Late successional trees are arranged according to their relative successional position. Data from sources cited in text.

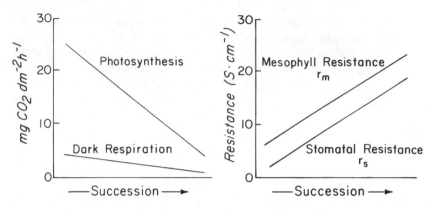

Figure 3 General trends of photosynthesis, dark respiration, mesophyll, and stomatal resistances in relation to successional position.

successional herbs and trees (the sun-adapted plants) the rate of photosynthesis remains unchanged if the leaves are exposed to intensities above saturation. However in late successional plants the photosynthetic rates may decline under similar conditions (solarization) (e.g. 14, 19, 47). Furthermore, solarization occurs faster in the more shade-adapted tree species.

The initial slope of the light response curve has been analyzed for a number of species and has been shown to be steeper for shade-adapted than for sun-adapted species (15, 51) and for climax than for successional trees (48). Thus late successional trees are photosynthetically more efficient at low light intensities than early successional herbs. Early successional trees have values between those of early successional herbs and those of late successional trees (1).

RESPIRATION   The rate of dark respiration is high in early successional plants and generally decreases with succession. It ranges from ~5 mg CO_2 dm^{-2} h^{-1} for *Ambrosia artemisiifolia* (7) to less than 1 mg CO_2 dm^{-2} h^{-1} in *Fagus grandifolia* and several other late successional trees (51). These trends are consistent with the fact that sun-adapted species have higher respiration rates and generally faster metabolism than do shade-adapted species (31, 49, 50, 92). Early successional trees have intermediate dark respiration rates (1, 11, 61, 74). The rate of decline of dark respiration with succession seems to be less than that of photosynthesis (Figure 3). Thus the photosynthesis/respiration ratio decreases with succession. The rate of dark respiration and the initial slope of photosynthetic response to light determine the light compensation point, the intensity at which photosynthesis balances light respiration. The light compensation point is predictably lower for shade-adapted, late successional plants than for sun-adapted, early successional ones (Figure 2). Supportive evidence for this is found in many of the studies on photosynthesis and respiration discussed above.

In climax forests photosynthetic production in the understory is limited by light intensity. Most species of the mature forest are capable of photosynthesizing at rates higher than those usually achieved under the canopy. Whereas dark respiration rates are similar for *Acer saccharum, Quercus rubra,* and *Q. alba* whether in a gap or under the shade of a closed canopy, their photosynthetic rates are significantly lower in the shade (29). But respiration rates of *Fagus grandifolia* and *Liriodendron tulipifera* increase significantly when grown under low light (51). Whether dark respiration rates are the same in a gap and under closed canopy or significantly decline in the shade, the decline of photosynthesis in the shade results in a less favorable carbon balance. The photosynthesis to respiration ratios (P/R) of a species in the shade and in a gap reflect the level of its shade adaptation and successional position. The P/R of *A. saccharum* is highest under shade and lowest in the gap. The reverse is true for *Q. alba,* while *Q. rubra* is intermediate under both light conditions. Thus the climax *A. saccharum* has a more favorable carbon balance in the shade, while the successional *Q. alba* has a more favorable carbon balance in the gap.

When the CO_2 fluxes into and out of the leaf in a closed illuminated container become equal, CO_2 concentration in the surrounding air ceases to change and the CO_2 compensation point is reached. This point varies among plants and is an indicator of photorespiration (42). Since photorespiration is known to increase with light intensity, the sun-adapted, early successional plants may have higher rates than shade-adapted late successional plants. Measurements of CO_2 compensation points for 5 early successional herbs and 5 early successional trees revealed no differences between the two groups ($\bar{x} = 48.1 \pm 3.7$ and 50.5 ± 2.2 ppm CO_2, respectively). Since no data are available for late successional species comparisons are not possible.

TRANSPIRATION, PLANT RESISTANCE, AND WATER USE   Data from numerous sources indicate that transpiration rates are generally high in early successional plants. In this respect early successional plants behave much like other herbaceous sun-adapted plants as reported by Larcher (50). Late successional plants generally have low rates of transpiration (about one quarter that of early successional plants). The differences in rates of transpiration and photosynthesis between early and late successional plants reflect differences in stomatal and mesophyll resistances, both of which generally increase with succession [Figure 3, based on data from (9, 24, 39, 101–103)]. Energy balance considerations of open and shaded environments indicate that low leaf resistances in early successional species may be effective in preventing harmfully high leaf temperatures. In contrast, the high leaf resistances of shaded, late successional species may be necessary to prevent excessive cooling. In shade-adapted herbs, transpiration rates seem to be higher than those of the trees.

Water use efficiency (WUE) is a measure of the amount of water expended in transpiration to obtain a unit of carbon dioxide from the surrounding air. It is influenced by the magnitudes of the leaf resistances (boundary layer, stomatal, and mesophyll) to gas exchange. Generally mesophyll resistance is highest, stomatal resistance is intermediate, and boundary layer resistances are lowest (39). The general water status of the plant is further influenced by the total plant resistance to water transport from the soil to the leaf. There are few measurements of resistance to water flow in plants. However, herbaceous plants have lower resistance to water transport than do deciduous trees (50) because herbs have a more efficient water transport system than do woody species (18). In 4 early successional annuals [(9) and F. A. Bazzaz, unpublished data] we found very low plant water resistance values ranging from 0.5×10^6 to 1.6×10^6 s cm^{-1}. In contrast the value for *Acer saccharum* was 2.3×10^6 s cm^{-1}. The measurement of this and the other three resistances requires accurate control and measurement of the plant's environment and are not common in ecological literature.

The ratio rate-of-photosynthesis/rate-of-transpiration per unit area per unit time for several early successional plants [(69), F. A. Bazzaz, R. W. Carlson, unpublished] suggests that, under optimal conditions, WUE is high ($\sim$7.2 mg g^{-1}). Using a different method of calculation [resistance to water vapor diffusion (R_{H_2O})/resistance to CO_2 assimilation (R_{CO_2}) (39)] Wuenscher & Kozlowski (101, 102) concluded that WUE was higher for the xeric, early successional *Quercus velutina* than for the climax *Acer saccharum*. Generally, however, resistance to H_2O and CO_2 diffusion and water use efficiencies found in the literature are difficult to compare and relate confidently to successional position because of (a) the different ways they are measured and expressed and (b) their dependence on leaf age (23, 24) and soil moisture levels during growth (69).

The speed and the degree of change of stomatal resistance to changes in light intensities may be adaptive and may be related to the successional position of the species. Stomatal opening in response to increased light is consistently faster in shade-tolerant than shade-intolerant trees (99). *Acer saccharum* seedlings respond faster than do *Fraxinus americana* seedlings. Transpiration declines faster in *Acer* and at less negative water potentials than in *Fraxinus* (21). Thus in the shade *Acer* can function better than *Fraxinus* because of its ability to use sunflecks but is less capable of surviving if subjected to water limitation.

RESPONSE TO WATER LIMITATION   When water potential of the leaves declines, stomata begin to close; CO_2 and water vapor exchange also decline. Stomatal sensitivity to changes in leaf water potential differs widely among plants, especially with regard to the water potentials at which reduction in photosynthesis and respiration begin, the steepness of their decline, and the potential at which they become negligible (16). In contrast to plants with low photosynthetic rates, plants with high photosynthetic rates do not extract water from dry soil. Their photosynthetic rates decline quickly with decreased potential of the soil (60). This may suggest that photosynthesis declines at less negative water potentials in early successional plants (high photosynthetic rates) than in late successional plants (low photosynthetic rates). There are a few data on the relationship between photosynthetic rates and water potential in successional species, but generalizations may be difficult to draw without a thorough study of this relationship for plants in a sere. It appears, however, that late successional plants are perhaps more sensitive to declining moisture levels (Figure 4). For example, photosynthesis in *Acer* begins to decline at about −200 kPa and becomes quite low at −1000 kPa. In *Ambrosia* it begins to decline at about −800 kPa and continues at 20% of maximum at −2200 kPa (7). In some early successional trees

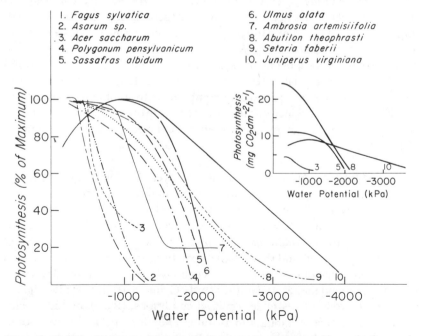

Figure 4   Relationship between photosynthetic rate and water potential in species from early- (4, 7, 8, 9), mid- (5, 6, 10), and late- (1, 2, 3) successional habitats.

photosynthetic rates may not decline until water potentials are as low as −1500 kPa (Figure 4). In *Ulmus alata* and *Juniperus virginiana* photosynthesis even rises initially with decrease in leaf moisture level, and maximum photosynthetic rates are reached at moisture levels that these plants experience in the field (1, 61).

Differences in photosynthetic responses to water potential may be found within successional groups. *Setaria faberii* and *Polygonum pensylvanicum* are early successional annuals that exhibit quite different daytime patterns in leaf water potential. Whereas *Polygonum* maintains its potential at about −400 kPa, *Setaria* potentials vary from −200 to −2200 kPa. The differences are related to their root locations in the soil (63, 96). However *Setaria* maintains near-maximum photosynthesis to as low as −1200 kPa and photosynthesizes at 30% of maximum at −2400 kPa. Photosynthesis in *Polygonum* begins to decline at ∼−400 kPa and becomes negligible at −1800 kPa. Both species maintain active photosynthesis in the field environment by using compensatory strategies and therefore coexist.

In early successional plants the recovery of leaf water potential, transpiration, and photosynthesis after watering moisture-stressed plants is rapid but not uniform among species. Maximum rates are restored within several hours after watering several annuals (1, 7, 17, 61, 96). Recovery in late successional plants has not been examined except in *Acer saccharum*. Seedlings were found to recover fully 4 days after rewatering (J. W. Geis, R. L. Tortorelli, unpublished).

Since early successional plants generally key in on disturbance they stand astride pulses of available resources. Rapidly obtaining and using them before they subside and preventing competitors from obtaining them may be adaptive. The rapid increase in the growth rate of early successional trees and herbs when the canopy above them is removed (54, 65) is an example of this strategy.

GROWTH RATES   No systematic study has yet used formal growth analysis techniques to compare the growth, under uniform conditions, of species of a sere. However, Grime (31) found that the shade-adapted climax species *Acer saccharum* and *Quercus rubra* had lower rates than the sun-adapted successional trees *Ailanthus altissima*, *Rhus glabra*, and *Fraxinus americana*. He also found that relative growth rates were highest for several arable weeds (including some early successional herbs). Additional supportive data are found in (35). In another study Grime & Hunt (34) investigated the relative growth rates under optimal conditions, (R_{max}), of 132 species from a wide range of habitats in Britain. They found that woody species generally exhibited a bias toward low R_{max}, that annual plants were most frequent in the high R_{max} category, and that in disturbed habitats fast-growing species were predominant.

Early successional plants are fugitives. Their continued survival depends on their dispersal to open sites. Being competitively inferior they must grow and consume available resources rapidly. High relative growth rate is therefore an integral component of their fugitive strategy.

Compression of Environmental Extremes

In addition to their broad physiological response to environmental gradients, early successional species may be capable of compressing environmental variability around them. In the field on a clear day in July we found that leaf temperature of *Polygonum* and *Abutilon* remained rather constant between 25–28°C, near optimum for photosynthesis (96), from 900–1600 hr while air temperature was 33–35°C. The reduction of leaf temperature variation relative to air temperature in *Abutilon* has been attributed to a more or less passive energy balance phenomenon (103) and is related to the air-temperature–leaf-temperature crossover described by Gates (28). More active control, however, may be possible through alteration of the angle of the leaf relative to the sun. *Abutilon* normally displays variation in leaf angle from 0°–60° off horizontal. Such shifts of leaf angle can result in differences in leaf temperatures as high as 8°C when the sun is directly overhead. The generality of compression of environmental variability in early successional plants and the difference from late successional plants in ability to ameliorate environmental stress should be investigated.

Acclimation

Another aspect of the physiological ecology of early successional plants that has adaptive value in their variable environments is their ability to acclimate rapidly. The rapid and efficient acclimation of several winter annuals—e.g. *Erigeron canadensis, E. annuus,* and *Lactuca scariola*—to seasonally changing temperature (72) is an example of this adaptation (Figure 5). Acclimation to the light environment that the plants experience during their growth seems to be more pronounced in the early (6, 12, 20, 64) than in the several late successional plants we examined (Figure 6). The seeds of *Aster pilosus* usually germinate under the shade of other plants. However the rosettes develop under less shade and the plants bolt and flower in full sun after the opening of the canopy by the demise of annuals above them. The different photosynthetic response to light intensity of the three stages reflects their respective light environments (66). The curves are respectively similar to those of shade-adapted late successional, mid-successional, and sun-adapted early successional plants.

Forest Herbs

Herbaceous plants of temperate deciduous forests may experience an environment that has features of early and late succession. Growth in several

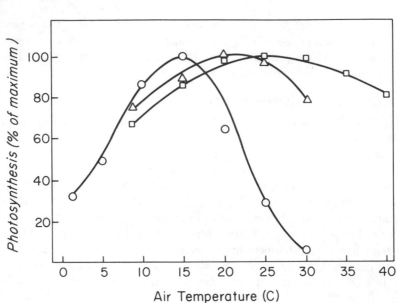

Figure 5   Shifts in photosynthetic response to temperature in *Erigeron canadensis.* Measurements made in February (○), July (□), and October (△). Data from (72).

of these species begins soon after temperatures rise above freezing and continues after canopy closure. The development of the canopy results in reductions in light intensity, temperature fluctuation, vapor pressure deficit, and windspeed (57). Sparling (82) studied dark respiration rates, light compensation points, light saturation, and maximum photosynthetic rates before, during, and after canopy closure in 25 species of woodland herbs. Based on their responses he divided the plants into 3 groups. First are those that are light saturated at 1000–3000 fc and have a compensation point above 40 fc. They behave as sun species and usually become dormant after canopy closure. Second are those that are light saturated between 250 and 1000 fc and have lower compensation points than plants of the first group. Their metabolic activities decline with canopy closure (84). Finally, a number of species are shade-tolerant. They light-saturate at <250 fc and have compensation points <25 fc. They also have a steeper initial slope of photosynthesis, indicating that they are more efficient at low light intensities than the sun-adapted group. Photosynthetic efficiency at low light intensity is aided by the general increase in chlorophyll content of leaves after canopy closure (8). Further examination of Sparling's data reveals that members of the sun-adapted group have, on the average, the highest mean photosynthetic rates (8.8 mg CO_2 g h^{-1}), members of the second group have interme-

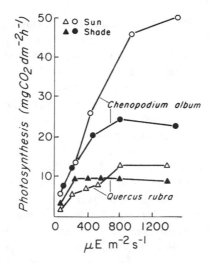

Figure 6   Light saturation curves for *Chenopodium album,* an early successional annual, and *Quercus rubra,* a late successional tree, grown in sun and in shade (F. A. Bazzaz, R. W. Carlson, unpublished data).

diate rates (6.1 mg CO_2 g h^{-1}) and the shade-adapted plants have the lowest rates (3.5 mg CO_2 g h^{-1}).

A remarkable adaptation to the changing light environment of a deciduous climax forest is that found in *Hydrophyllum appendiculatum* (57). The species produces two types of leaves with two different morphologies. The first, produced under open canopy (sun leaves), light-saturate at 2500 fc., light-compensate at 100 fc, photosynthesize at 8.5 mg CO_2 dm^{-2} h^{-1}, and respire at 1.8 mg CO_2 dm^{-2} h^{-1}. The second, produced under closed canopy (shade leaves), have corresponding values of 700 fc, 60 fc, 2.4, and 0.5 mg CO_2 dm^{-2} h^{-1}. The initial rise in photosynthetic rate with increased light is steeper in the shade than in the sun leaves. These leaf types behave like early successional and late successional leaves, respectively.

Members of the sun-adapted group, like early successional plants, are expected to respond quickly to changes in the environment, including rapid uptake of nutrient flushes. This behavior has been observed in *Erythronium americanum,* which begins its growth with a quick increase in plant weight and fast uptake of spring nutrient flushes (58).

Competition and Interference

Competition may be absent only early in the initial stages of colonization; as the young seedlings grow, competition between them begins. Later in succession the situation becomes more complex, and seedlings may compete with individuals of different ages and sizes belonging to many species. Competition between early successional plants is preemptive (93). Individuals and species that arrive first use up resources and starve later arrivals. Competitive losers become winners later if their life history characteristics allow them to grow when their competitors become inactive (66, 72). Competitive interactions as causal factors in succession have been discussed recently (32). The replacement of the pioneer *Erigeron* by *Aster,* and that in turn by *Andropogon,* involves increased competitive ability (43). Suppression of summer annuals is caused by preemption of resources by the already established winter annuals (71, 73). In the Piedmont the failure of pine seedlings to compete with the established trees leads to their replacement by the climax oaks, which compete for water better than pines (47). In late succession individuals and species that have survived at a low level of activity (having been suppressed by preemption and contest competition from neighbors) may replace their competitors when these weaken or die. Thus species replacement involves the interaction of competition with stress tolerance.

Detailed studies of the role of allelopathy in succession in Oklahoma (76) have shown that several members of the first "stage" of succession produce toxins, mostly phenolic compounds, that are inhibitory to several other species from that stage and sometimes to themselves. For example, the rapid disappearance of certain pioneer weeds and their replacement by *Aristida oligantha* are caused by allelopathic interference. Most of the pioneer plants do not inhibit *A. oligantha;* but extracts from several of the species and from *A. oligantha* inhibit nitrogen-fixation by soil algae and bacteria. *Aristida oligantha* stands are maintained for a long time because they occupy nutrient-poor habitats that are kept poor by allelopathy and are not suitable for the "climax" *Andropogon scoparius.* Thus, allelopathy speeds succession to *A. oligantha* and slows its replacement by the climax grasses. Chemicals released from several plants at various positions along this successional gradient control nutrient availability by differentially inhibiting nutrifying organisms (76), thereby influencing species replacement and community composition.

Allelochemics are also produced by some midsuccessional herbs and early successional trees. The perennial *Andropogon virginicus* produces chemicals that inhibit invasion by shrubs and therefore maintains itself in large populations (75). Allelopathic effects of *Solidago, Aster,* and grasses, prevent tree regeneration in glades (41). The presence of patches of early successional trees with little understory may be caused by their allelopathic action—e.g. in *Ailanthus altissima* (55) and *Rhus copallina* (76). *Sassafras albidum* maintains itself in pure patches by producing inhibitory chemicals (27), but there may be alternate explanations for this clumping (4).

The role of allelopathy in community composition is controversial. Several contradicting reports exist on whether a species is strongly, mildly, or nonallelopathic. Furthermore, the relative contributions of allelopathy and competition for environmental resources in determining both community composition and the rate of species replacement are still unknown.

CONCLUSIONS

Early and late successional plants in seres from open field to forest have many contrasting physiological attributes (Table 2). Evidence for some of these characterizations is strong; for others, additional research is required. Research on the physiological ecology of successional plants in tropical habitats is needed for a more accurate comparison with temperate succession. Studies should be conducted on the physiological ecology of plants in seres in which there is little change in growth form with succession—e.g. in grasslands. Finally, because of the increasing interest in species turnover

Table 2   Physiological characteristics of early and late successional plants

Attribute	Early successional plants	Late successional plants
Seeds		
dispersal in time	long	short
secondary (induced) dormancy	common	uncommon ?
Seed germination		
enhanced by		
light	yes	no
fluctuating temperatures	yes	no
high NO_3^- concentrations	yes	no ?
inhibited by		
far-red light	yes	no
high CO_2 concentrations	yes	no ?
Light saturation intensity	high	low
Light compensation point	high	low
Efficiency at low light	low	high
Photosynthetic rates	high	low
Respiration rates	high	low
Transpiration rates	high	low
Stomatal and mesophyll resistances	low	high
Resistance to water transport	low	high
Acclimation potential	high	low
Recovery from resource limitation	fast	slow
Ability to compress environmental extremes	high	low ?
Physiological response breadth	broad	narrow
Resource acquisition rates	fast	slow ?
Material allocation flexibility	high	low ?

in forest gaps and their role in regulating species diversity and coexistence, research on the physiological ecology of gap species in temperate and tropical forests would increase our understanding of species replacement in nature.

ACKNOWLEDGMENTS

Barbara Benner, Roger Carlson, Toni Hartgerink, David Hartnett, Tom Lee, David Peterson, Merle Schmierbach, Art Zangerl, and especially Judy Parrish and Steward Pickett, read drafts of this paper and made several helpful suggestions. I am indebted to them and to several former members of our laboratory, especially Dudley Raynal, David Regehr, and Nancy Wieland, for their significant contributions to some of the research discussed in this review.

Literature Cited

1. Bacone, J., Bazzaz, F. A., Boggess, W. R. 1976. Correlated photosynthetic responses and habitat factors of two successional tree species *Oecologia* 23: 63–74
2. Barton, L. V. 1961. *Seed Preservation and Longevity*. NY: Interstate. 216 pp.
3. Baskin, J. M., Baskin, C. C. 1976. Germination dimorphism in *Heterotheca subaxillaris* var. *subaxillaris*. *Bull. Torrey Bot. Club* 103:201–6
4. Bazzaz, F. A. 1968. Succession on abandoned fields in the Shawnee Hills, southern Illinois. *Ecology* 49:924–36
5. Bazzaz, F. A. 1970. Secondary dormancy in the seeds of the common ragweed *Ambrosia artemisiifolia*. *Bull. Torrey Bot. Club* 97:302–5
6. Bazzaz, F. A. 1973. Photosynthesis of *Ambrosia artemisiifolia* L. plants grown in greenhouse and in the field. *Am. Midl. Nat.* 90:186–90
7. Bazzaz, F. A. 1974. Ecophysiology of *Ambrosia artemisiifolia:* A successional dominant. *Ecology* 55:112–19
8. Bazzaz, F. A., Bliss, L. C. 1971. Net primary production of herbs in a Central Illinois deciduous forest. *Bull. Torrey Bot. Club* 98:90–94
9. Bazzaz, F. A., Boyer, J. S. 1972. A compensating method for measuring carbon dioxide exchange, transpiration, and diffusive resistances of plants under controlled environmental conditions. *Ecology* 53:343–49
10. Bazzaz, F. A., Mezga, D. M. 1973. Primary production and microenvironment in an *Ambrosia*-dominated old field. *Am. Midl. Nat.* 90:70–78

11. Bazzaz, F. A., Paape, V., Boggess, W. R. 1972. Photosynthetic and respiratory rates of *Sassafras albidum. For. Sci.* 18:218–22
12. Berry, J. A. 1975. Adaptation of photosynthetic process to stress. *Science* 188:644–50
13. Boardman, N. K. 1977. Comparative photosynthesis of sun and shade plants. *Ann. Rev. Plant Physiol.* 28:355–77
14. Bormann, F. H. 1953. Factors determining the role of loblolly pine and sweetgum in early old-field succession in the Piedmont of North Carolina. *Ecol. Monogr.* 23:339–58
15. Bourdeau, P. F., Laverick, M. L. 1958. Tolerance and photosynthetic adaptability to light intensity in White Pine, Red Pine, Hemlock and *Ailanthus* seedlings. *For. Sci.* 4:196–207
16. Boyer, J. S. 1976. Water deficits and photosynthesis. In *Water Deficits and Plant Growth,* 4:153–90. NY: Academic
17. Brix, H. 1962. The effect of water stress on the rates of photosynthesis and respiration in tomato plants and Loblolly Pine seedlings. *Physiol. Plant.* 15:10–20
18. Camacho-B., S. E., Hall, A. E., Kaufmann, M. R. 1974. Efficiency and regulation of water transport in some woody and herbaceous species. *Plant Physiol.* 54:169–72
19. Carpenter, S. B., Hanover, J. W. 1974. Comparative growth and photosynthesis of Black Walnut and Honeylocust seedlings. *For. Sci.* 20:317–24
20. Chabot, B. F., Chabot, J. F. 1977. Effects of light and temperature on leaf

anatomy and photosynthesis in *Fragaria vesca. Oecologia* 26:363–77

21. Davies, W. J., Kozlowski, T. T. 1975. Stomatal responses to changes in light intensity as influenced by plant water stress. *For. Sci.* 21:129–33

22. Davis, W. E. 1930. Primary dormancy, after-ripening, and the development of secondary dormancy in embryos of *Ambrosia trifida. Contrib. Boyce Thompson Inst.* 2:285–303

23. Drew, A. P., Bazzaz, F. A. 1979. Response of stomatal resistance and photosynthesis to night temperature in *Populus deltoides. Oecologia* 36: In press

24. Federer, C. A. 1976. Differing diffusive resistance and leaf development may cause differing transpiration among hardwoods in spring. *For. Sci.* 22:359–64

25. Forcier, L. K. 1975. Reproductive strategies and co-occurrence of climax tree species. *Science* 189:808–9

26. Frost, R. A., Cavers, P. B. 1975. The ecology of pigweeds (*Amaranthus*) in Ontario. I. Interspecific and intraspecific variation in seed germination among local collections of *A. powellii* and *A. retroflexus. Can. J. Bot.* 53:1276–84

27. Gant, R. E., Clebsch, E. E. C. 1975. The allelopathic influences of *Sassafras albidum* in old-field succession in Tennessee. *Ecology* 56:604–15

28. Gates, D. M. 1968. Transpiration and leaf temperature. *Ann. Rev. Plant Physiol.* 19:211–33

29. Geis, J. W., Tortorelli, R. L., Boggess, W. R. 1971. Carbon dioxide assimilation of hardwood seedlings in relation to community dynamics in Central Illinois. *Oecologia* 7:276–89

30. Górski, T. 1975. Germination of seeds in the shadow of plants. *Physiol. Plant.* 34:342–46

31. Grime, J. P. 1966. Shade avoidance and shade tolerance in flowering plants. In *Light as an Ecological Factor,* ed. R. Bainbridge, G. C. Evans, O. Rackham, pp. 187–207. Oxford: Blackwell

32. Grime, J. P. 1977. Evidence for the existence of three primary strategies in plants and its relevance to ecological and evolutionary theory. *Am. Nat.* 111:1169–94

33. Grime, J. P., Hunt, R. 1975. Relative growth rate: Its range and adaptive significance. *J. Ecol.* 63:393–422

34. Grime, J. P., Jarvis, B. C. 1974. Shade avoidance and shade tolerance in flowering plants. In *Light as an Ecological Factor,* ed. G. C. Evans, R. Bainbridge, O. Rackham, 2:525–32. Oxford: Blackwell Scientific

35. Grime, J. P., Jeffery, D. W. 1965. Seedling establishment in vertical gradients of sunlight. *J. Ecol.* 53:621–42

36. Harper, J. L. 1977. *Population Biology of Plants.* London: Academic. 892 pp.

37. Hayashi, I., Numata, M. 1967. Ecology of pioneer species of early stages in secondary succession. *Bot. Mag. Tokyo* 80:11–22

38. Holm, R. E., Miller, M. R. 1972. Hormonal control of weed seed germination. *Weed Sci.* 20:209–12

39. Holmgren, P., Jarvis, P. G., Jarvis, M. S. 1965. Resistance to carbon dioxide and water vapor transfer in leaves of different plant species. *Physiol. Plant.* 18:557

40. Horn, H. S. 1971. *The Adaptive Geometry of Trees.* Princeton, NJ: Princeton Univ. Press. 144 pp.

41. Horsley, S. B. 1977. Allelopathic inhibition of black cherry by fern, grass, goldenrod, and aster. *Can. J. For. Res.* 7:205–16

42. Jackson, W. A., Volk, R. J. 1970. Photorespiration. *Ann. Rev. Plant Physiol.* 21:385–432

43. Keever, C. 1950. Causes of succession on old field of the Piedmont, North Carolina. *Ecol. Monogr.* 20:229–50

44. King, T. J. 1975. Inhibition of seed germination under leaf canopies in *Arenaria serpyllifolia, Veronica arvensis* and *Cerastrum holosteoides. New Phytol.* 75:87–90

45. Kivilaan, A., Bandurski, R. S. 1973. The ninety-year period for Dr. Beal's seed viability experiment. *Am. J. Bot.* 60:140–45

46. Koller, D. 1972. Environmental control of seed germination. In *Seed Biology,* ed. T. T. Kozlowski, 2:1–102. NY: Academic

47. Kozlowski, T. T. 1949. Light and water in relation to growth and competition of Piedmont forest tree species. *Ecol. Monogr.* 19:207–31

48. Kramer, P. J., Decker, J. P. 1944. Relation between light intensity and rate of photosynthesis of loblolly pine and certain hardwoods. *Plant Physiol.* 19:350–58

49. Larcher, W. 1969. The effect of environmental and physiological variables on the carbon dioxide gas exchange of trees. *Photosynthetica* 3:167–98

50. Larcher, W. 1975. *Physiological Plant Ecology.* Berlin: Springer. 252 pp.

51. Loach, K. 1967. Shade tolerance in tree seedlings. I. Leaf photosynthesis and respiration in plants raised under artificial shade. *New Phytol.* 66:607–21

52. Loucks, O. L. 1970. Evolution of diversity, efficiency and community stability. *Am. Zool.* 10:17–25

53. MacArthur, R. H., Connell, J. H. 1966. *The Biology of Populations.* NY: Wiley. 200 pp.

54. Marks, P. L. 1974. The role of pin cherry (*Prunus pensylvanica* L.) in the maintenance of stability in northern hardwood ecosystems. *Ecol. Monogr.* 44:73–88

55. Mergen, F. 1959. A toxic principle in the leaves of *Ailanthus*. *Bot. Gaz.* 121:32–36

56. Mooney, H. A. 1972. The carbon balance of plants. *Ann. Rev. Ecol. Syst.* 3:315–46

57. Morgan, M. D. 1971. Life history and energy relationships of *Hydrophyllum appendiculatum*. *Ecol. Monogr.* 41:329–49

58. Muller, R. N. 1978. The phenology, growth and ecosystem dynamics of *Erythronium americanum* in the northern hardwood forest. *Ecol. Monogr.* 48:1–20

59. Olson, D. F. Jr., Gabriel, W. J. 1974. Acer L. In *Seeds of Woody Plants in the United States. For. Serv. USDA Handbook No. 450*, ed. C. Schopmeyer, pp. 187–94. Washington, DC: GPO

60. Orians, G. H., Solbrig, O. T. 1977. A cost income model of leaves and roots with special reference to arid and semiarid areas. *Am. Nat.* 111:677–90

61. Ormsbee, P., Bazzaz, F. A., Boggess, W. R. 1976. Physiological ecology of *Juniperus virginiana* in oldfields. *Oecologia* 23:75–82

62. Palmblad, I. G. 1969. Population variation in germination of weedy species. *Ecology* 50:746–48

63. Parrish, J. A. D., Bazzaz, F. A. 1976. Underground niche separation in successional plants. *Ecology* 57:1281–88

64. Patterson, D. T., Duke, S. O., Hoagland, R. E. 1978. The effects of irradiance during growth on adaptive photosynthetic characters of velvet leaf and cotton. *Plant Physiol.* 61:402–5

65. Perozzi, R. E. 1976. *Structure and dynamics of early successional communities dominated by Setaria feberii.* PhD thesis. Univ. Illinois, Urbana. 75 pp.

66. Peterson, D. L., Bazzaz, F. A. 1978. Life cycle characteristics of *Aster pilosus* in early successional habitats. *Ecology* 59:1005–13

67. Pickett, S. T. A., Baskin, J. M. 1973. The role of temperature and light in the germination behavior of *Ambrosia artemisiifolia*. *Bull. Torrey Bot. Club* 100:165–70

68. Pickett, S. T. A., Bazzaz, F. A. 1978. Germination of co-occurring annual species on a soil moisture gradient. *Bull. Torrey Bot. Club* 105:312–16

69. Pickett, S. T. A., Bazzaz, F. A. 1979. Physiological basis for competitive ability in co-occurring annual plants. *Ecology.* Submitted

70. Raynal, D. J., Bazzaz, F. A. 1973. Establishment of early successional plant populations on forest and prairie soil. *Ecology* 54:1335–41

71. Raynal, D. J., Bazzaz, F. A. 1975. Interference of winter annuals with *Ambrosia artemisiifolia* in early successional fields. *Ecology* 56:35–49

72. Regehr, D. L., Bazzaz, F. A. 1976. Low temperature photosynthesis in successional winter annuals. *Ecology* 57: 1297–1303

73. Regehr, D. L., Bazzaz, F. A. 1979. On the population dynamics of *Erigeron canadensis*, a successional winter annual. *J. Ecol.* In press

74. Regehr, D. L., Bazzaz, F. A., Boggess, W. R. 1975. Photosynthesis, transpiration and leaf conductance of *Populus deltoides* in relation to flooding and drought. *Photosynthetica* 9:52–61

75. Rice, E. L. 1972. Allelopathic effects of *Andropogon virginicus* and its persistance in old fields. *Am. J. Bot.* 59:752–55

76. Rice, E. L. 1974. *Allelopathy.* NY: Academic. 353 pp.

77. Rudolf, P. O., Leach, W. B. 1974. *Fagus* L. See Ref. 59, pp. 401–5

78. Sasaki, S., Kozlowski, T. T. 1968. The role of cotyledons in early development of pine seedlings. *Can. J. Bot.* 46: 1173–83

79. Sauer, J., Struik, G. 1964. A possible ecological relation between soil disturbance, light flash and seed germination. *Ecology* 45:884–86

80. Schwartz, D. M., Bazzaz, F. A. 1973. In situ measurements of carbon dioxide gradients in a soil-plant-atmosphere system. *Oecologia* 12:161–67

81. Smith, H. 1973. Light quality and germination: ecological implications. In *Seed Ecology, Proc. 19th Easter School Agric. Sci., Univ. Nottingham*, ed. W. Heydecker, pp. 219–31. University Park, Pa: Penn. State Univ. Press

82. Sparling, J. H. 1967. Assimilation rates of some woodland herbs in Ontario. *Bot. Gaz.* 128:160–68

83. Stein, W. I., Slabaugh, P. E., Plummer, A. P. 1974. Harvesting, processing and storage of fruits and seeds. See Ref. 59

84. Taylor, R. J., Pearcy, R. W. 1976. Seasonal patterns of the CO_2 exchange characteristics of understory plants from a deciduous forest. *Can. J. Bot.* 54:1094–103

85. Taylorson, R. B., Borthwick, H. A. 1969. Light filtration by foliar canopies: significance for light-controlled weed seed germination. *Weed Sci.* 17:48–51

86. Thompson, K., Grime, J. P., Mason, G. 1977. Seed germination in response to diurnal fluctuations of temperature. *Nature* 267:147–49

87. Toole, E. H., Brown, K. 1946. The final results of the Duvel buried seed experiment. *J. Agric. Res.* 72:201–10

88. Vitousek, P. M., Reiners, W. A. 1975. Ecosystem succession and nutrient retention: a hypothesis. *BioScience* 25:376–81

89. Wagner, R. H. 1967. Application of a thermal gradient bar to the study of germination patterns in successional plants. *Am. Midl. Nat.* 77:86–92

90. Wareing, P. F. 1966. Ecological aspects of seed dormancy and germination. In *Reproductive Biology and Taxonomy of Vascular Plants,* ed. J. G. Hawos, pp. 103–21. Oxford: Pergamon

91. Warington, K. 1936. The effect of constant and fluctuating temperatures on the germination of the weed-seeds in arable soil. *J. Ecol.* 24:185–204

92. Went, F. W. 1957. *The Experimental Control of Plant Growth.* Waltham, Mass: Chronica Botanica. 343 pp.

93. Werner, P. A. 1976. Ecology of plant populations in successional environments. *Syst. Bot.* 1:246–68

94. Wesson, G., Wareing, P. F. 1969. The role of light in germination of naturally occurring populations of buried weed seeds. *J. Exp. Bot.* 20:402–13

95. Whitney, G. G. 1976. The bifurcation ratio as an indicator of adaptive strategy in woody plant species. *Bull. Torrey Bot. Club* 103:67–72

96. Wieland, N. K., Bazzaz, F. A. 1975. Physiological ecology of three codominant successional annuals. *Ecology* 56:681–88

97. Willemsen, R. W., Rice, E. L. 1972. Mechanism of seed dormancy in *Ambrosia artemisiifolia. Am. J. Bot.* 59:248–57

98. Williams, J. T., Harper, J. L. 1965. Seed polymorphism and germination. I. The influence of nitrates and low temperatures on the germination of *Chenopodium album. Weed Res.* 5:141–50

99. Wood, D. B., Turner, W. C. 1971. Stomatal response to changing light of four tree species of varying shade tolerance. *New Phytol.* 70:77–84

100. Woodwell, G. M., Dykeman, W. R. 1966. Respiration of a forest measured by CO_2 accumulation during temperature inversions. *Science* 154:1031–34

101. Wuenscher, J. E., Kozlowski, T. T. 1970. Carbon dioxide transfer resistance as a factor in shade tolerance of tree seedlings. *Can. J. Bot.* 48:453–56

102. Wuenscher, J. E., Kozlowski, T. T. 1971. Relationship of gas-exchange resistance to tree-seedling ecology. *Ecology* 52:1016–23

103. Zangerl, A. R. 1978. Energy exchange phenomena, physiological rates and leaf size variation. *Oecologia* 34:107–12

Ann. Rev. Ecol. Syst. 1979. 10:373–98

COMMUNICATION IN THE ULTRAVIOLET

❖4167

Robert E. Silberglied

Department of Biology, Harvard University, Cambridge, Massachusetts 02138;
and Smithsonian Tropical Research Institute, P. O. Box 2072, Balboa, Panama

INTRODUCTION

Biologists are physically handicapped: They view the world through limited senses that overlap, but are not congruent with, those of the organisms they study. By technologically extending their senses, biologists have only recently begun to explore other sensory worlds, and in so doing have revealed a diverse array of unexpected phenomena. Among the most interesting and least explored of these sensory realms is that part of the electromagnetic spectrum which is normally invisible to humans, but to which most other organisms are visually sensitive—the ultraviolet.

Ultraviolet (UV) light is electromagnetic radiation of wavelengths between about 40 and 400 nm (110). While approximately 7% of solar radiation is emitted in this region, only about 3% of the sunlight reaching the ground is of wavelengths shorter than 400 nm. This difference is due to reflection and absorption of UV in the atmosphere (92). The spectral range of terrestrial UV extends from a short-wave limit of 286 nm to the violet end (400 nm) of the so-called "visible spectrum."

Is UV light qualitatively different from "visible" light? No. There is nothing 'special' about UV: It is the vertebrate eye that is unusual. In most vertebrate species, wavelengths shorter than about 400 nm are prevented from reaching the UV-sensitive retina by "shielding" pigments in the lens and/or cornea. Its existence was not suspected until 1801, when Johann Ritter demonstrated the chemical effect of such radiation.

That organisms respond behaviorally to UV light was discovered by Sir John Lubbock (Lord Avebury) in 1876 (131). By taking advantage of the fact that ants ordinarily keep their larvae and pupae shielded from light, and when exposed carry them swiftly to darkness, he tested their responses to light that had been dispersed into a spectrum by a prism. The ants carried

373

0066-4162/79/1120-0373$01.00

their pupae away from the violet and blue parts of the spectrum, and deposited them in the red. But they *first* removed pupae that lay in the "dark," beyond the violet! Using a fluorescent compound, Lubbock was able to demonstrate the presence of UV light in that region. The ants' response was abolished when the UV rays were blocked by a seemingly transparent, but UV-absorbing, solution. Lubbock concluded that "it would appear that the colours of objects and the general aspect of nature must present to them a very different appearance from what it does to us" (131).

How do "the colour of objects and the general aspect of nature" appear to other organisms? Four factors are involved:

1. *The light environment.* The spectral quality and intensity of light vary over time and space.
2. *The optical properties of natural objects.* Light is reflected, absorbed, and transmitted by matter. The spectral quality and intensity of reflected and transmitted light carry information about matter to the eye.
3. *The visual system.* Light is refracted, filtered, and absorbed in the eye. The optical properties of the eye and the spectral absorption of receptor pigments determine what information will be transduced to electrical signals.
4. *The brain.* The appearance of the world to the organism depends upon how the brain interprets the spatial and chromatic information sent to it by the visual receptors.

Since the sensitivity of most organisms extends over a broad spectral range that usually includes, but is not limited to, the ultraviolet, it makes no sense at all to consider 'vision in the ultraviolet' as a subject apart. Any attempt to understand the visual world of an organism that sees UV light must take into account all the information available about that species' whole color vision system, and should include information from all spectral regions in which the animal is sensitive. And yet ". . . to know what the world looks like to an insect . . . is really impossible because so much of the story remains untold, namely the integrative capacities of visual interneurons" (25).

The Light Environment

UV light has more energy per quantum than other solar radiation reaching the earth's surface. It is strongly absorbed by various organic molecules (190), including proteins and nucleic acids, and it is responsible for many biological effects (cf 21–24, 32, 69, 92, 106, 214). The spectral composition of sunlight in the region responsible for most physiological effects (286–315 nm, or UV-B) has been carefully studied, both for medical reasons and because certain human activities detrimentally affect the stratospheric ozone filter that normally protects us (2, 5, 69, 92).

In contrast, UV-A (315–400 nm) has received much less attention. This is unfortunate for two reasons: (*a*) UV-A radiation contains over 90% of solar ultraviolet reaching the ground (180), and (*b*) ultraviolet-sensitive visual receptors are most responsive to UV-A radiation (see 71, 92). The spectral quality and quantity of the ultraviolet component of direct sunlight and skylight vary with time of day, season, latitude, altitude, and with meteorological and other conditions (49, 70, 73, 94, 107, 108). Reflection from and transmission through water, leaves, and other natural objects also affect the light environment. While I cannot in the space available describe photic environments in detail (cf 77) certain generalizations should be made.

UV radiation is strongest at high altitudes, where the air mass is thinnest, and in the tropics, where the ozone concentration is lowest (9, 136). In polar and temperate regions UV radiation is greatest during summer; it varies less with season in the tropics (107, 108). Its daily variation is similar to that of "visible" light; its nocturnal intensity depends upon the phase of the moon. Since lunar spectral reflectance is somewhat lower in the UV than in the "visible spectrum" (11) nocturnal illumination contains less UV, relative to "visible" light, than does diurnal illumination. Little is known about the relative levels of near UV and "visible" light during twilight hours, when many organisms are active (cf 137). No organisms are known to produce substantial UV light by bioluminescent reactions (J. W. Hastings, personal communication).

Optical Properties of Natural Objects in the Ultraviolet

The blue of the sky is due to scattering. Since (Rayleigh) scattering is inversely proportional to the fourth power of wavelength, UV light is scattered more than blue, so skylight is very rich in UV radiation (Figures 1–2). Polarization is also due to scattering, and skylight is most strongly polarized in the UV. Shadows are lit primarily by skylight (although to a greater or lesser extent by light reflected from or transmitted through objects as well); they are therefore not as "dark" in UV as in "visible" light. Snow, ice (123–126), sand (65), and some other materials reflect UV strongly, but water and most biological materials transmit and absorb it, except at high (grazing) angles of incidence. Clean water, whether fresh or salt, transmits UV well, though not so well as blue; UV transmittance is adversely affected by turbidity and by the presence of dissolved organic material ("gelbstoff") (7, 17, 86, 93). Most vegetation absorbs 80–90% of incident UV radiation (22, 28, 66; see 106), as does vertebrate hair (123–125, 215).

Since humans cannot ordinarily see UV light, we study the optical properties of natural objects with technological assistance. Reflectance spectrophotometry is used for detailed description of UV reflection, but is not practical for rapid surveys nor usable under natural illumination or with moving subjects. Fluoroscopy, photography, and "video-viewing" through appropriate filters have been used to transduce UV reflectance to mono-chrome images we can see and quantify (3, 6, 31, 35, 38, 40, 51–53, 57, 62, 64, 81, 102, 123, 124, 126, 132, 196). Chemical (115; see 14) and fluorescence (73) techniques have been devised to reveal certain UV-absorbing pigments.

One cannot assume that reflection of a particular color of light implies a communicative role for the color—one would hardly draw such a conclusion from the green of chlorophyll or the yellow of cytochrome. The optical properties of biological tissues are products of many selective forces, including those that determine the structural components of the tissues and the physiological functions of those substances (cf 77). One need not "explain" all reflectance patterns in adaptive terms. However, if we are familiar with the usual range of UV reflectances characteristic of plant and animal tissues, unexpected patterns may alert us to those that *might* be used in communication. The distribution and diversity of such patterns may suggest testable hypotheses about their function in communication.

FLOWER PATTERNS   As seen through a UV-restricted viewing device, or in a UV photograph, many flowers stand out distinctly from the surrounding vegetation or landscape. Frequently the flowers reflect UV light more strongly than does the background, but the reverse may occur (65). While there is often little correlation between "visible" color and UV reflection (38, 100, 157, 174), UV-reflecting flowers are most often yellow or violet (76). Less frequently, flowers of other "visible" colors, such as green, blue, or red, reflect UV.

Many flowers have detailed UV patterns composed of reflecting and absorbing regions. Often the centers of UV-reflecting flowers or inflorescences have conspicuous UV-absorbing marks, producing a UV "bull's-eye." These markings may correspond with "visible" spots or lines located at or near nectaries, or at the opening to the floral tube within which nectaries are found. Such markings are usually called "nectar-guides" or "honey guides" (38, 112, 146, 218), regardless of their UV and "visible" spectral properties. While usually UV-absorbing in a UV-reflecting field, on occasion nectar-guides are reflecting on a dark field (115, 213). Sometimes one petal may contrast with others (36, 122–123), or all petals may be

→

Figures 1–4   Panchromatic "visible" (400–700 nm, *left*) and ultraviolet (305–700 nm, *right*) landscapes of flowering meadows. Skylight is very intense in UV (Figure 2), while vegetation tends to absorb UV strongly (Figure 4). Note the differences between the "visible" and UV reflection patterns of the various Compositae represented; see also Figures 9, 10. (Photographs courtesy of T. Eisner.)

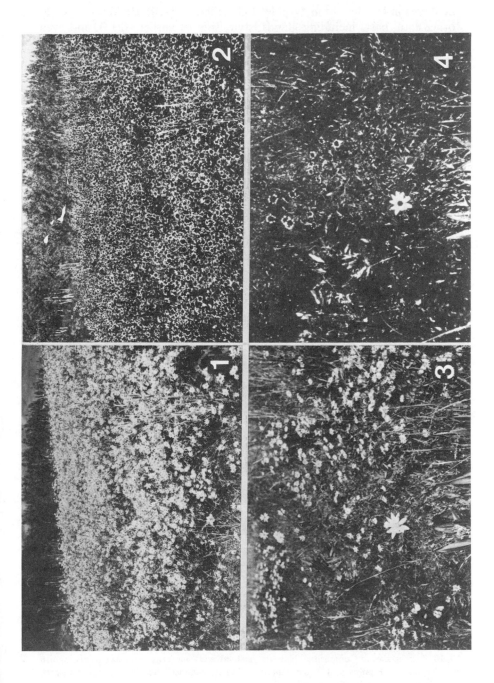

bicolored in UV but not in "visible" light (Figures 5–10). Besides the petals, other floral parts including glands, style, stigma, anthers (213), and even pollen or nectar (211, 212, cf 101) may contrast with the rest of the flower in UV-reflectance. The surfaces of floral parts may be modified physically to increase or reduce reflectance, and UV-absorbing flavenoids may contribute to the darker parts of the pattern (14, 76, 186–189, 210).

Diverse floral patterns of UV reflection are found in many families (see Table 1). Floral UV patterns are often species- or population-specific (26,

Figures 5–10   Floral patterns: panchromatic "visible" (*left*) and UV (*right*) photographs of (Figures 5, 6) *Jussiaea* sp. (Onagraceae); (Figures 7, 8) an unidentified crucifer; and (Figures 9, 10) three species of Compositae. Note the distinctive UV patterns that serve to distinguish similar species (Figures 9, 10) and function as "nectar-guides" for flower-visiting insects. (Photographs 9 and 10 courtesy of T. Eisner.)

Table 1 Some descriptions and discussions of ultraviolet optical properties of various organisms[a]

Subject	Reference numbers
Plants	
Vegetation	2, 22, 28, 66, 106, 178
Flowers	
General	14, 38, 56, 57, 65, 76, 83, 100, 112, 115–117, 129, 133, 135, 139, 141, 147, 149, 150, 153, 155, 157, 174, 175, 177, 193, 218, 227
Agavaceae	213
Boraginaceae	65
Caryophyllaceae	65
Compositae	1, 14, 57, 58, 65, 82, 102, 105, 186, 188, 189, 210
Cruiferae	87
Fumariaceae	144
Guttiferae	56, 82
Hydrophyllaceae	36
Leguminosae	26, 74, 88, 95–97
Liliaceae	216
Loganiaceae	58
Menyanthaceae	169
Oleaceae	56, 58
Onagraceae	41
Orchidaceae	118, 119, 209
Oxalidaceae	83
Papaveraceae	83, 114
Polemoniaceae	128
Ranunculaceae	57, 82, 83, 143
Scrophulariaceae	102, 138, 140, 142, 187
Buds	56
Nectar	101, 211, 212
Color production	14, 41, 58, 76, 101, 186–189, 210, 216
Animals	
Arthropods	
Araneae	57, 84, 100, 133
Collembola	100, 103
Odonata	20, 179
Coleoptera	82, 83, 172
Diptera	43, 82, 83, 202
Hymenoptera	147
Lepidoptera	
General	18, 33, 82, 83, 135, 147, 148, 150, 176, 182, 192, 194, 195, 197–199
Moths	82, 135, 148, 176
Butterflies	
Nymphalidae (s. lat.)	45, 59, 60, 67, 135, 147, 148, 208
Papilionidae	62, 90, 91, 135
Pieridae	4, 57, 62, 63, 67, 68, 75, 80, 83, 145, 147, 148, 160–165, 167, 168, 171, 181–185, 194, 195, 198, 199, 204
Larvae	20, 201
Color production	4, 30, 44, 67, 68, 80, 90, 91, 145, 161, 204,
Vertebrates	123–126, 215

[a]Most references dealing with several families are listed under "general."

36, 87, 88, 97, 169) and are usefully employed as genetic markers or taxonomic characters. Ultraviolet reflection appears to be commoner among large flowers than among small ones (76) and among insect-pollinated species than among wind- or bird-pollinated species (57, 116, 175). Geographic, ecological, and seasonal distributions of floral colors, including UV, are not well-defined and are subject to conflicting interpretations (100, 157).

BUTTERFLY PATTERNS   The wings of butterflies bear a wide variety of UV reflection patterns, including the most intense such reflection found in nature. In many species, these patterns show little or no correspondence with "visible" features (Figures 11–15). In the absence of UV-absorbing pigments, some butterfly wings are strongly UV-reflecting owing to scattering from the intricate ultrastructure of wing scales. The females of many Pierini [sensu (109); e.g. Pieris spp.] are ultraviolet-reflecting for this reason (167, 168, 171), while male wings are strongly UV-absorbing owing to the presence of pigments in the wing scales (44, 145; cf. 80). On the other hand, males of many "sulfur" [Rhodocerini (109)] and "orange-tip" [Euchloini (109)] butterflies have brilliant UV reflection caused by interference of light reflected from multiple thin films in the wing scales (4, 67, 68, 82, 160–165, 194, 195). Fluorescent compounds (30), which absorb strongly in the ultraviolet, may supress UV-reflection while enhancing colors in the "visible" spectrum (91).

Crane's study (33), the first to investigate the colors of many species using a quantitative colorimetric technique, reached the erroneous conclusion that UV reflection is rare in butterflies. Her sample consisted of a collection of Trinidad species peculiarly low in ultraviolet reflection. Recent studies (e.g. 147, 148, 164, 194, 195) have revealed numerous examples of spectacular patterns entirely restricted to UV, especially among the Pieridae. In many nymphalids (e.g. Morpho, Agrias, Prepona) and lycaenids, ultraviolet reflection patterns closely correspond to regions of "visible" iridescence (45, 135).

Sexual dimorphism is more common, and often better expressed, in UV than in "visible" reflection patterns (194). "Iridescent" (interference-produced) reflection is found on the wings of many male Pieridae, Nymphalidae [sensu (55)] and Lycaenidae; when found in the female it is always better-developed in the conspecific male. Large areas with this kind of reflection (Figures 14, 15) are usually confined to the dorsal wing surfaces.

As with floral patterns, UV reflection can be employed as a taxonomic tool; it may also be so employed by the butterflies. Among the Pieridae it has been used to separate "problem" species in Gonepteryx (160–165), Colias (63, 194, 195, 198, 199), and Phoebis (4, 194, 195); and

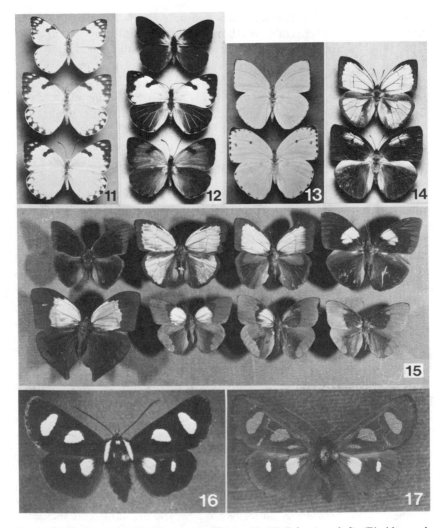

Figures 11–17 Patterns of lepidopterans: (Figures 11, 12) *Belenois zochalia* (Pieridae; male above, two famales below), a species that is monomorphic in (Figure 11) "visible" light but which has sexual dimorphism and female polymorphism in (Figure 12) ultraviolet (see 194). (Figures 13, 14) *Phoebis argante* (Pieridae; male above, female below), a species with high-intensity, "iridescent" (Figure 14) UV reflection patterns due to interference in the male (poorly developed in the female) that are entirely hidden in (Figure 13) "visible" light (see 4). (Figure 15) Males of eight species of *Phoebis* (Pieridae, UV photograph), illustrating intrageneric diversity of male patterns characteristic of many "sulfur" butterflies [Rhodocerini (109)]; these patterns are entirely hidden in "visible" light (not shown) (see 4). (Figures 16, 17) *Alypia octomaculata* (Agaristidae), a diurnal moth with strong (Figure 17) UV reflection from hindwing spots that appear similar to those on the forewings in (Figure 16) "visible" light; the hindwing spots may be revealed or concealed by appropriate posturing by the moth [R. Silberglied, unpublished data; see also (176)].

in the Nymphalidae to characterize a distinctive new *Prepona* (45). The "iridescent" UV patterns of males show considerable interspecific variation [(164, 194); Figure 15]. Geographic, seasonal (192), and genetic polymorphism in UV reflection characters have been documented among *Colias* spp. (63, 181, 194, 195, 198, 199), *Belenois zochalia* [(194); Figures 11, 12], *Pieris rapae* (194), and *Gonepteryx* spp. (164, 165). Forms of "visibly" polymorphic species, on the other hand, do not necessarily differ in ultraviolet reflection [e.g. *Papilio glaucus:* (18, 62, 135, 176)]. The genetics of UV reflection has been studied in *Colias* by means of interspecific hybridization and was found to be inherited as a sex-limited, sex-linked character (75, 198).

The intense UV reflection caused by interference has unusual properties. Since it is produced in exposed ridges of the outer layer of wing scales, spatially separated from the pigment-bearing regions (67), it may be combined with "visible" color as an overlay. Thus, while in most organisms the spectral properties of "visible" pigments constrain the possible range and intensity of UV reflection, in these butterflies any "visible" color may be combined with intense UV reflection. In an extreme example, an extensive UV-reflecting area covers both the white spots and part of the surrounding black areas on the wings of *Hypolimnas misippus:* The butterfly has spots of "visible" white + intense UV, surrounded by areas of "visible" black + intense UV (195).

These interference colors also have unusual physical characteristics. With most colored surfaces, the intensity of reflected light approximates a linear trigonometric function of the angle between light source, reflecting surface, and observer (e.g. 204); with interference colors, reflectance at a given wavelength "cuts on" and "cuts off" much more sharply. Furthermore, the peak wavelength (hence the color) shifts with changes of angle (67). At certain angles the reflected light should be strongly polarized. As seen through an ultraviolet-viewing device, flying male *Colias eurytheme, Phoebis argante,* and other species resemble flashing beacons. "[W]ith every wingbeat, a flying *Morpho* butterfly changes the angle of light incidence through the entire possible range. To the human eye, a *Morpho* in flight is simply a flickering flash of varying tints of blue. However, to another *Morpho,* in sunlight, there should be a brilliant shift from blue-green or blue to ultraviolet, then momentary extinction and back again through the spectral arc; conceivably this may be an exceptionally potent stimulus" (33; cf 200).

PATTERNS OF OTHER ARTHROPODS   Little is known about UV reflection of arthropods other than butterflies, and even less about how they contrast with their natural surroundings. Most arthropod integuments strongly absorb UV light, but occasional exceptions occur. While many spiders are UV-absorbing (84, 100, 133), other closely related species are UV-reflecting (57). Some Collembola (100) are UV-absorbing. Strong UV reflection occurs among libellulid dragonflies, especially on the abdomens of mature males [(179); Figure 19]; it is probably a Tyndall effect, due to the "pruinosity" that gives rise to blue or white in many species (20). UV reflection occurs on the iridescent body regions of some male and female damselflies (e.g. *Argia fumipennis;* R. Silberglied, unpublished observations). Among beetles, UV-reflecting areas are often found in association with pubescence, dried secretions or subcuticular layers (82, 83, 172). "Pruinose" areas of flattened projections produce UV reflection in ephydrid flies (43, 202); the silvery-white bands on mosquitoes also reflect UV (82, 83). Bumblebees living at high altitudes reflect ultraviolet light more strongly than those from lowlands (147). Certain moths have strong UV reflection [(82, 135, 148, 176); Figures 16–17], with sexual dimorphism well-developed in some species (135). The larvae of tent caterpillars (*Malacosoma* spp.) have UV-reflecting spots and stripes, which produce Tyndall scattering among cuticular filamentous processes (20, 201).

Visual Systems

Most animals show behavioral responses to UV light: Sea anemones retract their tentacles (27), planarians turn away (16), arthropods exhibit various responses, etc. Even vertebrates (see below) show behavioral responses, though it is not certain to what extent they visualize UV images.

ARTHROPODS   It is difficult to generalize about insect vision, owing to the great diversity among the visual systems of over a million species. One of the most constant features in the few insects that have been studied is the presence of UV-sensitive receptors (see 46, 71, 92, 150). Both ocelli and compound eyes are UV-sensitive (54, 170, 224; see 71). Besides insects, other arthropods respond to UV radiation, including spiders (47), xiphosurans (121, 122, 166, 220, 222), and crustaceans (130, 223).

Owing to the ease with which honeybees can be conditioned to respond to visual stimuli, more has been learned about their color vision than about that of any other animal except ourselves (see 38, 61, 71, 151, 152, 218). Worker honeybees have a trichromatic color vision system with ultraviolet, blue, and yellow receptor types (8, 37–39, 120, 218). Honeybees discriminate not only among these three primaries, but also among combinations of them: blue + yellow ("blue-green"), UV + blue ("bee-violet"), and UV + yellow ("bee-purple"). "Bee-white" is a mixture of ultraviolet, blue, and yellow (37, 39). The UV receptors are more sensitive than the other receptor types. Honeybees can discriminate as distinct from pure yellow a mixture

Figures 18–19 (Figure 18) Male of *Colias eurytheme* (Pieridae) courting a sitting female (UV photograph). In this species the intense UV reflection of the male's wings (visible only on the right forewing in this photograph), is a signal used by females in conspecific mate-selection, and by males as a sexual recognition cue (see 195, 198, 199). (Figure 19) UV photograph of mature male (*left*) and female (*right*) *Plathemis lydia* and mature male *Libellula pulchella* (*below*) (Odonata: Libellulidae), illustrating sexual dimorphism and interspecific differences commonly found among dragonflies. The areas that reflect UV also reflect "visible" light, but not as intensely; the reflection is a "Tyndall effect" (see 20, 179).

of 99% yellow + 1% UV (37–39). In addition to hue, saturation and intensity are important parameters of optical stimuli for bees (38, 97, 100).

Many other diurnal insects have trichromatic color-vision systems, usually with one of the receptors maximally sensitive in the UV (e.g. 12, 85; see 71, 150). There are, however, also numerous exceptions to the "general" pattern of trichromatic color vision. The mantis *Tenodera sinensis,* for example, has only a single visual pigment, which absorbs strongly at 515 nm and has a secondary absorption peak at 370 nm. At least two different receptor types are required for wavelength discrimination, so this mantis is probably color-blind. Parts of some insect eyes are composed of only one receptor type, often UV. Dichromatic color-vision systems are known in a cockroach (*Periplaneta*), a neuropteran (*Ascalaphus*), and a beetle (*Dineutes*). Extended red-sensitivity is known in some Lepidoptera (10, 34, 173; see 71, 203).

The limited behavioral flexibility of most insects precludes easy bioassays of color-discrimination ability, so far more is known about receptor physiology than about integration of color information. While there may be similarities at the receptor level between many insects' visual systems (e.g. 12), integration may differ considerably between groups. In the literature on floral colors and other visual stimuli, the honeybee color-vision system is often generalized to other insects. One should bear in mind that this is done out of ignorance, not knowledge.

VERTEBRATES   The lack of UV-sensitivity in humans and many other vertebrates has been attributed to the presence of UV-absorbing pigments in the ocular media (13, 78, 79, 98, 158, 219). Humans (19, 127, 219) and frogs (98) whose lenses have been removed (aphakics) can see ultraviolet light. Owing to the widespread occurrence of such UV-absorbing pigments in vertebrate eyes, several investigators came to the conclusion that, under natural conditions, vertebrates as a group are blind to UV (e.g. 205, 221).

However, some vertebrates have UV-transparent ocular media—e.g. many nocturnal species and those that live under low light conditions in the ocean (42, 98, 156; cf 159). Since many vertebrate lenses fluoresce "visibly" when stimulated by UV (15, 226), and since sensitivity to UV in cats *decreases* when the lens is removed (50), it is conceivable that some vertebrates perceive UV only as an out-of-focus haze caused by lens fluorescence.

UV-sensitivity has been demonstrated in a frog (72), a toad (48), a newt (104), several species of lizards (154), and two species of birds [hummingbird: (89); homing pigeon: (111, 225)]. None of these animals has been tested for the ability to discriminate among different UV patterns under natural conditions, but recent laboratory experiments (M. Kreithen, T.

Eisner, unpublished) reveal high acuity and UV image recognition in the homing pigeon. The extent to which vertebrates with UV-transparent ocular media see and use ultraviolet information in behavior and communication remains unknown.

COMMUNICATION SYSTEMS

Pollination

Visual communication between flowers and anthophilous organisms occurs in one direction only: Flowers transmit information and animals receive it. A floral pattern (*a*) advertises the presence of the "receptive" flower, (*b*) provides characters for identification of the flowering species, and (*c*) directs the visitor's behavior in a manner that assures pollen transfer. UV patterns contribute to all three.

ADVERTISEMENT   The presence of a "receptive" flower is communicated to potential visitors by a visual signal that contrasts with the surrounding vegetation (Figures 1–4). Leaves generally absorb strongly in the ultraviolet, blue and yellow-red regions of the spectrum, providing a desaturated, rather neutral ("bee-gray") background against which the colored floral signals are deployed (38). Flowers are rarely pure UV to bees [an exception is the poppy *Papaver rhoeas* (38, 114, 129)]; most commonly they reflect "bee-purple" (yellow + UV), yellow, "bee-violet" (blue + UV), or blue. The floral colors of many species have been analyzed with respect to honeybee color vision and attractiveness (38, 97, 100, 157).

Where there is high ambient UV reflection, the situation may be reversed: UV-*absorption* may be important as advertising (65). UV advertisement is far from universal, however: Many insect-visited flowers do not contrast with vegetation in the UV, and many insect visitors do not require UV reflection patterns to "release" feeding behavior (cf 153).

The timing of the advertisement is critical to the plant, since visits at inappropriate times might damage the flower or condition unrewarded visitors to avoid such flowers in the future. The deployment of the floral signal is controlled by various means: During early development the petals may be hidden between sheathing sepals or bracts, the petals may be folded, and they may be colored differently on their exposed surfaces, etc. Buds of *Hypericum peltatum, Jasminium primulinum,* and other species, for example differ from flowers in UV reflection but not in "visible" color (56). Only those outer portions of the folded petals that are exposed in the bud absorb UV. It appears that exposure to sunlight mediates the UV darkening of these outer parts (56). Flowers that have been pollinated usually reduce their signals by dropping petals, changing color, or by other means (96, 113, 213, 217).

The attractiveness of the floral signal is due to both innate and learned behavior on the part of the visitor. Bees are especially attracted to figures having complex outlines: Stars and crosses are preferred over circles of the same total area, etc (see 218). UV patterns enrich the complexity of floral signals, and in so doing may increase overall attractiveness.

IDENTIFICATION   Flower constancy is the tendency of individual foragers to restrict visitation to single species of flowers at a time. It is characteristic of all groups of anthophilous organisms, though it is better developed in some than in others (29, 174). Flower constancy is mediated in part by visual signals, which for insects involve a UV component. Considerable diversity of UV and "visible" patterns has been demonstrated among sympatric flowering plants with overlapping blooming periods (38, 100, 157). It is commonly assumed that insects can discriminate between flower species that exhibit strong differences in pattern; however, except in the honeybee this ability has rarely been tested (cf 134, 153).

Daumer (38) demonstrated that several species of flowers that differ in UV but not "visible" light could be discriminated by honeybees. He showed that this ability could be abolished by blocking the UV component with a filter (38). Jones (95) utilized a "reciprocal bouquet" technique (29) to show that between two sympatric species of *Cercidium* (Leguminosae), flower constancy is probably mediated by visual responses of pollinators to UV-reflection differences between species.

GUIDANCE   As indicated above, many flowers bear special markings that serve to orient visitors. For example, the UV-absorbing "banner" petal of caesalpinoid legumes is used by bees for orientation to the sagittal plane of the flower, either before or after landing (96). "Nectar-guide" markings indicate the way to nectaries or to the entrances of floral tubes containing nectaries (112, 115). Visiting insects, ordinarily attracted to the edges of petals, are directed instead by these high-contrast markings to their goal (146). Nectar-guides may contrast in "visible" and/or UV light. Daumer (38) elegantly demonstrated the effectiveness of UV nectar-guides in leading honeybees about on "dummy" flowers with artificial markings. As might be expected, flowers pollinated by birds generally have "visible" nectar-guides with weak UV contrast (57, 116). The UV markings of orchids having "pseudocopulatory pollination" may also be used as recognition cues or orientation guides by male visitors (118, 119).

Butterfly Communication

Communication among butterflies is primarily sexual. Males generally respond to visual stimuli of the female and begin courtship, presenting visual and chemical signals to which the female responds. Most studies of butterfly courtship have concentrated upon the stimuli "releasing" male courtship behavior; far less is understood about the basis for mate acceptance and rejection by females (see 191, 197).

Since many butterfly species are sexually dimorphic in UV reflectance, one potential function of these patterns might be sexual recognition. The extreme differences among the patterns of males of many closely related species suggest that they might be used for species discrimination by females. The "iridescent" UV-reflection produced by interference, with its high intensity, spectral purity, and abrupt flashing with the wingbeat, also seems ideal as a long-range signal. UV-absorption may also be a signal; what is important is the contrast with the surrounding environment in which the wings are displayed (148).

INTRASPECIFIC COMMUNICATION   Petersen et al (171) were the first to demonstrate that the UV component of an insect color pattern might serve a communicative function. They found that zinc-white (UV-reflecting) butterfly models were far more attractive to male *Pieris napi* and *P. bryoniae* than were lead-white (UV-absorbing) ones. Female *P. napi* and *P. bryoniae* (195), as well as other species of *Pieris* (172, 174), reflect UV more strongly than do conspecific males. Investigators (167, 168) using zinc-white and other "model" butterflies, showed that in *Pieris rapae crucivora* UV reflection of the female wings is a necessary component for the release of male courtship behavior.

Many "sulfur" butterflies [Rhodocerini (109)] differ from most "whites" [Pierini (109)] in that UV signals releasing male courtship behavior are reversed with respect to the sexes: The females are more UV-absorbing than are the males (194). The intensely reflecting male patterns of many species are produced by interference. In *Eurema lisa* and *Colias eurytheme,* males are attracted to the UV-absorbing females and are inhibited by UV-reflecting models (182, 184, 185, 199). In *C. eurytheme,* males are also repelled from mating pairs by a special UV flash signal given by the male *in copulo* (195, 199).

The nymphalid butterfly *Anartia fatima* is unusual in that both its UV reflection and visible color change with age: Its yellow, UV-absorbing bands become white and UV-reflecting in both sexes. Older, UV-reflecting females are more attractive to mate-seeking males [(59, 60, 208); R. E. Silberglied, A. Aiello, D. M. Windsor, unpublished data].

INTERSPECIFIC COMMUNICATION   Ultraviolet patterns are also involved in interspecific isolation among "sulfur" butterflies. In *Eurema, Phoebis, Colias,* and other genera, there are strong interspecific differences in male UV reflection patterns among congeneric sympatric species. Female *C. eurytheme* accept conspecific males that have strong UV reflection;

experimental obliteration of the male UV pattern (without changing the visible color) results in a decreased mating frequency relative to control males (195, 199). On the other hand, the females of *C. philodice* (the males of which are UV-absorbing), do not discriminate against conspecific males that were experimentally adorned with *C. eurytheme* UV-reflecting wing patches (195, 199) but there is no reason to expect that *all* "sulfurs" will use UV cues. Females of *E. lisa* (182–184) and *C. eurytheme* (185, 195, 199, 206, 207) discriminate using both UV reflection and olfactory cues, but the female of *C. philodice* relies entirely on olfaction (75, 195, 199, 207). In *C. eurytheme* and *C. philodice,* the genetic loci determining the UV reflection pattern are linked on the sex chromosomes with loci that determine species-specific male pheromones and female mate-selection behavior (75).

Communication between Prey and Predator

CAMOUFLAGE Several authors have remarked upon the close resemblance between certain insects and their backgrounds in UV. Certain high arctic Collembola closely resemble the flowers upon which they sit (100, 103). The distinctive UV patterns of some lepidopterous larvae (20) may camouflage them from visually orienting predators, such as predatoid and parasitoid wasps. Some caterpillars (e.g. *Synchlora* sp., Geometridae) clothe themselves in materials such as leaves or petals; whether or not these protect them depends upon the match between garb and substrate (T. Eisner, R. E. Silberglied, D. Aneshansley, et al, unpublished observations).

Some arthropod predators rely upon camouflage as they wait in ambush for prey. Kevan (100) and Hinton (84) suggest that UV-absorbing thomisid spiders might be well-concealed from potential insect prey in their UV-absorbing floral lairs. The red marks on *Misumena vatia* may be warning colors aimed at vertebrate predators but invisible to insects (84). On the other hand, one UV-reflecting species sits conspicuously on the UV-absorbing ray bases of *Rudbeckia* (57), and a UV-absorbing spider sits on a UV-reflecting flower (133)! Much remains to be learned about the degree to which thomisids, phymatids, mantids, and other flower-haunting predators contrast with their chosen resting-places, and about the extent to which flower-visitors avoid inflorescences with conspicuous predators. The same can be said for predators that lurk concealed on other substrates, such as bark or leaves.

MIMICRY   If we assume that terrestrial vertebrates do not see ultraviolet light, UV reflection patterns should be free from the selection pressures of vertebrate predators. For example, since visually oriented birds and lizards account for much insect predation, we might expect mimicry to be less precise in a spectral region where they are blind. The few data bearing on mimicry are equivocal. The UV patterns of models and mimics studied by Lutz (135) were "as good—or better—matches, one for another, as the matches we see." Brues (18) found some models and mimics to be similar, others less so, in UV reflectance. Remington (176) found that African butterfly models and mimics resembled one another in UV reflection pattern more than did New World models and mimics. He suggested that more African than New World predators might see in the ultraviolet, or that mimicry complexes in Africa have had more time to develop. Hinton (82, 83) reports a case of mimicry among diurnal scarab beetles that "breaks down" in the ultraviolet, and suggests that the beetles might use their patterns for conspecific recognition.

Several factors may influence UV similarity between model and mimic. First, in the presence of strong selection in the adjacent "visible" spectrum, aposematic patterns evolve that usually employ a limited array of bio-chromes and structural colors in a particular spatial distribution. The in-trinsic spectral properties of the available materials may, in some cases, preclude differentiation in the nearby ultraviolet. Second, unless there is selection *for* differentiation in the ultraviolet, we needn't expect it to occur. Since communication within and between many mimetic species involves olfactory stimuli (197), selection for distinctive visual signals may be weak. Third, some mimetic species have arthropod predators that orient to prey by means of vision—e.g. salticid spiders, mantids, odonates, asilid flies, etc. If any of these predators avoids warning colors or patterns, there would be selection for resemblance in UV as well as in "visible" light. Fourth, the book is not closed on whether all vertebrate predators are functionally UV-blind.

SUMMARY

Ultraviolet light is probably invisible to most vertebrates but is an important spectral component of the color vision systems of insects. Flowers possess UV reflection patterns which, in conjunction with other visual cues, stimulate identification and orientation behavior in anthophilous insects. Wing-borne UV patterns function in sex-recognition and as reproductive isolating mechanisms in butterflies. UV patterns are probably important in other contexts. The greatest obstacle to understanding the function and importance of reflection patterns is our ignorance of the integrative mechanisms of insect color vision.

ACKNOWLEDGMENTS

I thank T. Eisner, M. Kreithen, and D. Udovic for permission to use unpublished information, and A. Aiello for technical assistance.

Literature Cited

1. Abrahamson, W., McCrea, K. D. 1977. Ultraviolet light reflection and absorption patterns in populations of *Rudbeckia* (Compositae). *Rhodora* 79:269–277
2. Allen, L. H. Jr., Gausman, H. W., Allen, W. A. 1975. Solar ultraviolet radiation in terrestrial plant communities. *J. Environ. Qual.* 4:285–294
3. Allman, M. 1973. Photomacrography using direct ultraviolet radiation and the problem of sharp focus. *J. Photogr. Sci.* 21:265–270
4. Allyn, A. C. Jr., Downey, J. C. 1977. Observations on male UV reflectance and scale ultrastructure in *Phoebis* (Pieridae). *Bull. Allyn Mus.* 42:1–20
5. Alyea, F. N., Cunnold, D. M., Prinn, R. G. 1975. Stratospheric ozone destruction by aircraft-induced nitrogen oxides. *Science* 188:117–121
6. Aneshansley, D., Eisner, T. 1975. Ultraviolet viewer. *Science* 188:82
7. Armstrong, F. A. J., Boalch, G. T. 1961. The ultraviolet absorption of sea water. *J. Mar. Biol. Ass. UK* 41:591–597
8. Autrum, H., Von Zwehl, V. 1964. Die spektrale Empfindlichkeit einzelner Sehzellen des Bienenauges. *Z. Vergl. Physiol.* 48:357–384
9. Bener, P. 1972. Approximate values of intensity of natural ultraviolet radiation for different amounts of atmospheric ozone. Final Report, European Research Office, U.S. Army, London
10. Bernard, G. D. 1979. Red-absorbing visual pigment of butterflies. *Science,* in press
11. Biberman, L. M., Dunkelman, L., Fickett, M. I., Finke, R. G. 1966. Levels of nocturnal illumination. *Inst. Defense Anal. Res. Pap.* p-232 viii + 130 pp. + 68 pp. appendix. Washington DC: GPO
12. Bishop, L. G., Chung, D. W. 1972. Convergence of visual sensory capabilities in a pair of Batesian mimics. *J. Insect Physiol.* 18:1501–1508
13. Boettner, E. A., Wolter, J. R. 1962. Transmission of the ocular media. *Invest. Ophthalmol.* 1:776–783
14. Brehm, B. G., Krell, D. 1975. Flavonoid localization in epidermal papillae of flower petals: a specialized adaptation for ultraviolet absorption. *Science* 190:1221–1223
15. Brolin, S. E., Cederlund, C. 1958. The fluorescence of the lens of the eyes of different species. *Acta Ophthalmol. (Kbh.)* 36:324–328
16. Brown, H. M. 1967. Effects of ultraviolet and photorestorative light on the phototaxic behavior of *Planaria.* In: *Chemistry of Learning: Invertebrate Research. Proceedings of a Symposium. 7–10 September, 1966. East Lansing, Mich.* pp. 295–309. New York: Plenum Press
17. Brown, M. 1977. Transmission spectroscopy examinations of natural waters: C. Ultraviolet spectral characteristics of the transition from terrestrial humus to marine yellow substance. *Estuar. Coastal Mar. Sci.* 5:309–317
18. Brues, C. T. 1941. Photographic evidence on the visibility of color patterns in butterflies to the human and insect eye. *Proc. Amer. Acad. Arts Sci.* 74:281–285
19. Burian, H. M., Ziv, B. 1959. Electric response of the phakic and aphakic human eye to stimulation with near ultraviolet. *AMA Arch. Ophthalmol.* 61:347–350
20. Byers, J. R. 1975. Tyndall blue and surface white of tent caterpillars, *Malacosoma* spp. *J. Insect Physiol.* 21:401–415
21. Caldwell, M. M. 1968. Solar ultraviolet radiation as an ecological factor for alpine plants. *Ecol. Monogr.* 38:243–268
22. Caldwell, M. M. 1971. Solar UV irradiation and the growth and development of higher plants. *Photophysiology* 4:131–177. New York: Academic Press
23. Caldwell, M. M. 1972. Biologically effective solar ultraviolet irradiation in the Arctic. *Arct. Alp. Res.* 4:39–43
24. Caldwell, M. M. 1977. The effects of solar UV-B radiation (280–315 nm) on higher plants: implications of stratospheric ozone reduction. In *Research in Photobiology,* ed. A. Castellani, pp. 597–607. New York: Plenum Publ. Corp.
25. Carlson, S. D., Chi, C. 1979. The functional morphology of the insect photoreceptor. *Ann. Rev. Entomol.* 24:379–416
26. Carter, A. M. 1974. Evidence for the hybrid origin of *Cercidium sonorae* (Leguminosae: Caesalpinioideae) of northwestern Mexico. *Madroño* 22:266–272
27. Clark, E. D., Kimeldorf, D. J. 1971. Behavioral reactions of the sea anemone, *Anthopleura xanthogrammica,* to ultraviolet and visible radiations. *Radiat. Res.* 45:166–175
28. Clark, J. B., Lister, G. R. 1975. Photosynthetic action spectra of trees. II. The

relationship of cuticle structure to the visible and ultraviolet spectral properties of needles from four coniferous species. *Plant Physiol.* 55:407–413

29. Clements, F. E., Long, F. L. 1923. *Experimental Pollination—an Outline of the Ecology of Flowers and Insects.* Washington, DC: Carnegie Inst. Wash. Publ. 336. vii + 274 pp. + 17 pl.

30. Cockayne, E. A. 1924. The distribution of fluorescent pigments in the Lepidoptera. *Trans. Roy. Entomol. Soc. London* 1924:1–19

31. Conlon, V. M. 1971. Some standard photographic procedures suitable for archaeological documentation. *Med. Biol. Illus.* 21:60–65

32. Coohill, T. P., Fingerman, M. 1975. Relative effectiveness of ultraviolet and visible light in eliciting pigment dispersion in melanophores of the fiddler crab, *Uca pugilator,* through the secondary response. *Physiol. Zool.* 48:57–63

33. Crane, J. 1954. Spectral reflectance characteristics of butterflies (Lepidoptera) from Trinidad, B.W.I. *Zoologica (N.Y.)* 39:85–115

34. Crane, J. 1955. Imaginal behavior of a Trinidad butterfly, *Heliconius erato hydara* Hewitson, with special reference to the social use of color. *Zoologica (N.Y.)* 40:167–196

35. Cronin, J. F., Rooney, T. P., Williams, R. S. Jr., Molineux, C. E., Bliamptis, E. E. 1973. Ultraviolet radiation and the terrestrial surface. In: *The Surveillant Science, Remote Sensing of the Environment,* ed. R. K. Holz. Boston: Houghton Mifflin Co.

36. Cruden, R. W. 1972. Pollination biology of *Nemophila menziesii* (Hydrophyllaceae) with comments on the evolution of oligolectic bees. *Evolution* 26:373–389

37. Daumer, K. 1956. Reizmetrische Untersuchung des Farbensehens der Bienen. *Z. Vergl. Physiol.* 38:413–478

38. Daumer, K. 1958. Blumenfarben, wie sie die Bienen sehen. *Z. Vergl. Physiol.* 41:49–110

39. Daumer, K. 1963. Kontrastempfindlichkeit der Bienen für "weiss" verschiedenen UV-Gehalts. Ein Beitrag zur Frage, woran die Bienen den Sonnenstand hinter einer Wolkendecke erkennen. *Z. Verg. Physiol.* 46:336–350

40. De Bruin, J. P. 1961. Principles of ultraviolet light and some of its applications in photography. *J. Biol. Photogr. Ass.* 29(2):53–63

41. Dement, W. A., Raven, P. H. 1974. Pigments responsible for ultraviolet pat-terns in flowers of *Oenothera* (Onagraceae). *Nature (London)* 252:705–706

42. Denton, E. J. 1956. Recherches sur l'absorption de la lumière par le cristallin des poissons. *Bull. Inst. Océanogr. Monaco* 1071:1–10

43. Deonier, D. L. 1974. Ultraviolet-reflective surfaces on *Ochthera mantis mantis* (DeGeer) (Diptera: Ephydridae). Preliminary Report. *Entomol. News* 85: 193–201

44. Descimon, H. 1976. Biology of pigmentation in Pieridae butterflies. In *Chemistry and Biology of Pteridines,* ed. H. Descimon, pp. 805–840. Berlin: Walter de Gruyter

45. Descimon, H., Mast de Maeght, J., Stoffel, J. R. 1973–4. Contribution à l'étude des nymphalides néotropicales. Description de trois nouveaux *Prepona* Mexicains. *Alexanor* 8:101–5, 155–59, 235–40

46. Dethier, V. G. 1963. *The Physiology of Insect Senses.* London: Methuen. ix + 266 pp.

47. De Voe, D. 1972. Dual sensitivities of cells in wolf spider eyes at ultraviolet and visible wavelengths of light. *J. Gen. Physiol.* 59:247–269

48. Dietz, M. 1972. Erdkröten können UV-Licht sehen. *Naturwissenschaften* 59:316

49. Diffey, B. L. 1977. The calculation of the spectral distribution of natural ultraviolet radiation under clear day conditions. *Phys. Med. Biol.* 22:309–316

50. Dodt, E., Walther, J. B. 1958. Netzhautsensitivität, Linsenabsorption und physikalische Lichtstreuung. *Pflügers Arch. Gesamte Physiol.* 266:167–174

51. Eastman Kodak Company. Publication M-3. *Infrared and Ultraviolet Photography.* Rochester, N.Y.: Eastman Kodak Co.

52. Eastman Kodak Company. Publication M-13. *Ultraviolet Sensitizing of Kodak Plates.* Rochester, N.Y.: Eastman Kodak Co.

53. Eastman Kodak Company. Publication M-27. *Ultraviolet and Fluorescence Photography.* Rochester, N.Y.: Eastman Kodak Co.

54. Eaton, J. L. 1976. Spectral sensitivity of the ocelli of the adult cabbage looper moth, *Trichoplusia ni. J. Comp. Physiol. A Sens. Neural Behav. Physiol.* 109: 17–24

55. Ehrlich, P. R. 1958. The comparative morphology, phylogeny and higher classification of the butterflies (Lepidoptera: Papilionoidea). *Univ. Kansas Sci. Bull.* 39:305–370

56. Eisner, T., Eisner, M., Aneshansley, D. 1973. Ultraviolet patterns on rear of flowers: basis of disparity of buds and blossoms. *Proc. Nat. Acad. Sci. USA* 70:1002–1004

57. Eisner, T., Silberglied, R. E., Aneshansley, D., Carrell, J. E., Howland, H. C. 1969. Ultraviolet video-viewing: the television camera as an insect eye. *Science* 166:1172–1174

58. Eisner, T., Eisner, M., Hyypio, P. A., Aneshansley, D., Silberglied, R. E. 1973. Plant taxonomy: ultraviolet patterns of flowers visible as fluorescent patterns in pressed herbarium specimens. *Science* 179:486–487

59. Emmel, T. C. 1972. Mate selection and balanced polymorphism in the tropical nymphalid butterfly, *Anartia fatima*. *Evolution* 26:96–107

60. Emmel, T. C. 1973. On the nature of the polymorphism and mate selection phenomena in *Anartia fatima* (Lepidoptera: Nymphalidae). *Evolution* 27:164–165

61. Erber, J. 1975. The dynamics of learning in the honeybee. *J. Comp. Physiol.* 99:231–255

62. Ferris, C. D. 1972. Ultraviolet photography as an adjunct to taxonomy. *J. Lepidopt. Soc.* 26:210–215

63. Ferris, C. D. 1973. A revision of the *Colias alexandra* complex (Pieridae) aided by ultraviolet reflectance photography with designation of a new subspecies. *J. Lepidopt. Soc.* 27:57–73

64. Ferris, C. D. 1975. A note on films and ultraviolet photography. *News Lepidopt. Soc.* 6:6–7

65. Frohlich, M. W. 1976. Appearance of vegetation in ultraviolet light: absorbing flowers, reflecting backgrounds. *Science* 194:839–841

66. Gausman, H. W., Rodriguez, R. R., Escobar, D. E. 1975. UV radiation reflectance, transmittance and absorptance by plant leaf epidermises. *Agron. J.* 67:720–724

67. Ghiradella, H., Aneshansley, D., Eisner, T., Silberglied, R. E., Hinton, H. E. 1972. Ultraviolet reflection of a male butterfly: interference color caused by thin-layer elaboration of wing scales. *Science* 178:1214–1217; 179:415

68. Ghiradella, H. 1974. Development of ultraviolet reflecting butterfly scales: how to make an interference filter. *J. Morphol.* 142:395–409

69. Giese, A. C. 1977. *Living with Our Sun's Ultraviolet Rays*. New York: Plenum Press. xii + 185 pp.

70. Goldberg, B., Klein, W. H. 1977. Variations in the spectral distribution of daylight at various geographical locations on the earth's surface. *Solar Energy* 19:3–13

71. Goldsmith, T. H., Bernard, G. D. 1974. The visual system of insects. In *The Physiology of Insecta*, Vol. 2, ed. M. Rockstein, pp. 165–272. New York: Academic Press. 2nd ed.

72. Govardovskii, V. I., Zueva, L. V. 1974. Spectral sensitivity of the frog eye in the ultraviolet and visible region. *Vision Res.* 14:131–132

73. Green, A. E. S., Sawada, T., Shettle, E. P. 1974. The middle ultraviolet reaching the ground. *Photochem. Photobiol.* 19:251–259

74. Green, T. W., Bohart, G. E. 1975. The pollination ecology of *Astragalus cibarius* and *Astragalus utahensis* (Leguminosae). *Amer. J. Bot.* 62:379–386

75. Grula, J. 1978. *The inheritance of traits maintaining ethological isolation between two species of* Colias *butterflies*. Ph.D. Thesis, University of Kansas, Lawrence

76. Guldberg, L. D., Atsatt, P. R. 1975. Frequency of reflection and absorption of ultraviolet light in flowering plants. *Amer. Midl. Natur.* 93:35–43

77. Hailman, J. 1977. *Optical Signals*. Bloomington & London: Indiana Univer. Press. xix + 362 pp.

78. Hemmingsen, E. A., Douglas, E. 1967. Snow blindness in animals. *Antarctic J. U.S.* 2:99–100

79. Hemmingsen, E. A., Douglas, E. L. 1970. Ultraviolet radiation thresholds for corneal injury in antarctic and temperate-zone animals. *Comp. Biochem. Physiol.* 32:593–600

80. Hidaka, T., Okada, M. 1970. Sexual differences in wing scales of the white cabbage butterfly, *Pieris rapae crucivora*, as observed with a scanning electron microscope. *Zool. Mag. (Dobutsugaku Zasshi)* 79:181–184

81. Hill, R. J. 1977. Technical note: ultraviolet reflectance-absorbance photography: an easy, inexpensive research tool. *Brittonia* 29:382–390

82. Hinton, H. E. 1973. Some recent work on the colours of insects and their likely significance. *Proc. Brit. Entomol. Natur. Hist. Soc.* 6:43–54

83. Hinton, H. E. 1973. Natural deception. In *Illusion in Nature and Art*, ed. R. L. Gregory, E. H. Gombrich, pp. 97–160. London: Duckworth

84. Hinton, H. E. 1976. Possible significance of the red patches of the female crab spider *Misumena vatia*. *J. Zool.* 180:35–39

85. Hoeglund, G., Hamdorf, K., Rosner, G. 1973. Trichromatic visual system in an insect and its sensitivity control by blue light. *J. Comp. Physiol.* 86:265–279

86. Højerslev, N. K. 1978. Solar middle ultraviolet (UV-B) measurement in coastal waters rich in yellow substance. *Limnol. Oceanogr.* 23:1076–1079

87. Horovitz, A., Cohen, Y. 1972. Ultraviolet reflectance characteristics in flowers of crucifers. *Amer. J. Bot.* 59:706–713

88. Horovitz, A. Harding, J. 1972. Genetics of *Lupinus*. V. Intraspecific variability for reproductive traits in *Lipinus nanus*. *Bot. Gaz.* 133:155–165

89. Huth, H. H. 1972. Der spektrale Sehbereich eines Violettohr-Kolibris. *Naturwissenschaften* 59:650

90. Huxley, J. 1975. The basis of structural color variation in two species of *Papilio*. *J. Entomol.* (*A*) 50:9–22

91. Huxley, J. 1976. The coloration of *Papilio zalmoxis* and *P. antimachus,* and the discovery of Tyndall blue in butterflies. *Proc. Roy. Soc. London B Biol. Sci.* 193:441–453

92. [Institute for Defense Analysis] 1975. *Impacts of Climatic Change on the Biosphere. C.I.A.P. Monograph 5, Ultraviolet Radiation Effects.* Washington, DC: US Dep. Trans. (NTIS: DOT-TST-75-55)

93. Jerlov, N. G. 1966. Aspects of light measurement in the sea. In *Light as an Ecological Factor,* ed. R. Bainbridge, G. C. Evans, O. Rackham, pp. 91–98. Oxford and Edinburgh: Blackwell Scientific Publications

94. Johnson, F. S., Mo, T., Green, A. E. S. 1976. Average latitudinal variation in ultraviolet radiation at the earth's surface. *Photochem. Photobiol.* 23:179–188

95. Jones, C. E. 1978. Pollinator constancy as a prepollination isolating mechanism between sympatric species of *Cercidium*. *Evolution* 32:189–198

96. Jones, C. E., Buchmann, S. L. 1974. Ultraviolet floral patterns as functional orientation cues in hymenopterous pollination systems. *Anim. Behav.* 22:481–485

97. Kauffeld, N. M., Sorenson, E. L. 1971. Interrelations of honeybee preference of alfalfa clones and flower color, aroma, nectar volume, and sugar concentration. *Kansas State Univ. Agr. Appl. Sci.* (*Manhattan*), *Kansas Agr. Exp. Sta. Publ.* No. 163. ii + 14 pp.

98. Kennedy, D., Milkman, R. D. 1956. Selective light absorption by the lenses of lower vertebrates, and its influence on spectral sensitivity. *Biol. Bull. Mar. Biol. Lab. Woods Hole* 111:375–386

99. Kevan, P. G. 1972. Collembola on flowers on Banks Island, N.W.T. *Quaest. Entomol.* 8:121

100. Kevan, P. G. 1972. Floral colors in the high arctic with reference to insect-flower relations and pollination. *Can. J. Bot.* 50:2289–2316

101. Kevan, P. G. 1976. Fluorescent nectar. *Science* 194:341–342

102. Kevan, P. G., Grainger, N. D., Mulligan, G. A., Robertson, A. R. 1973. A gray scale for mₑ uring reflectance and color in the insect and human visual spectra. *Ecology* 54:924–926

103. Kevan, P. G., Kevan, D. K., McE. 1970. Collembola as pollen feeders and flower visitors with observation from the high arctic. *Quaest. Entomol.* 6:311–326

104. Kimeldorf, D. J., Fontanini, D. F. 1974. Avoidance of near-ultraviolet radiation exposures by an amphibious vertebrate. *Environ. Physiol. Biochem.* 4:40–44

105. King, R. M., Krantz, V. E. 1975. Ultraviolet reflectance patterns in the Asteraceae: I. Local and cultivated species. *Phytologia* 31:66–114

106. Klein, R. M. 1978. Plants and near-ultraviolet radiation. *Bot. Rev.* 44:1–127

107. Klein, W. H., Goldberg, B. 1974. *Solar Radiation Measurements, 1968–1973.* Rockville, Maryland: Smithsonian Radiation Biology Laboratory

108. Klein, W. H., Goldberg, B. 1976. *Solar Radiation Measurements, 1974–1975.* Rockville, Maryland: Smithsonian Radiation Biology Laboratory

109. Klots, A. B. 1931. A generic revision of the Pieridae (Lepidoptera), together with a study of the male genitalia. *Entomol. Amer.* (*n.s.*) 12:139–254

110. Koller, L. R. 1965. *Ultraviolet Radiation.* New York: John Wiley and Sons. 2nd ed.

111. Kreithen, M., Eisner, T. 1978. Ultraviolet light detection by the homing pigeon. *Nature* (*London*) 272:347–348

112. Kugler, H. 1930. Blütenökologische Untersuchungen mit Hummeln. Der Farbensinn der Tiere—die optische Bindung in der Natur—das Saftmalproblem. *Planta* 10:229–280

113. Kugler, H. 1936. Die Ausnutzung der Saftmalsumfärbung bei den Rosskastanienblüten durch Bienen und Hummeln. *Ber. Dtsch. Bot. Ges.* 54:394–400

114. Kugler, H. 1947. Hummeln und die UV-Reflexion an Kronblättern. *Naturwissenschaften,* 34:315–316
115. Kugler, H. 1963. UV-Musterungen auf Blüten und ihr Zustandekommen. *Ber. Dtsch. Bot. Ges.* 75(Sondernummer): 49–54
116. Kugler, H. 1966. UV-Male auf Blüten. *Ber. Dtsch. Bot. Ges.* 79:57–70
117. Kugler, H. 1971. UV-Musterung bei Alpenblumen. *Jahrb. Ver. Schultze Alpenfl. Tiere* 36:61–65
118. Kullenberg, B. 1956. On the scents and colours of *Ophrys* flowers and their specific pollinators among the aculeate Hymenoptera. *Svensk Bot. Tidskr.* 50:25–46
119. Kullenberg, B. 1961. Studies in *Ophrys* pollination. *Zool. Bidr. Uppsala* 34:1–340 + 51 pl.
120. Labhart, T. 1974. Behavioral analysis of light intensity discrimination and spectral sensitivity in the honey bee, *Apis mellifera. J. Comp. Physiol. A Sens. Neural. Behav. Physiol.* 95:203–216
121. Lall, A. B. 1970. Spectral sensitivity of intracellular responses from visual cells in median ocellus of *Limulus polyphemus. Vision Res.* 10:905–909
122. Lall, A. B., Chapman, R. M. 1973. Phototaxis in *Limulus* under natural conditions: Evidence for reception of near-ultraviolet light in the median dorsal ocellus. *J. Exp. Biol.* 58:213–224.
123. Lavigne, D. M. 1976. Counting harp seals with ultraviolet photography. *Polar Rec.* 18:269–277
124. Lavigne, D. M., Øritsland, N. A. 1974. Ultraviolet photography: a new application for remote sensing of mammals. *Can. J. Zool.* 52:939–941
125. Lavigne, D. M., Øritsland, N. A. 1974. Black polar bears. *Nature* 251:218–219
126. Lavigne, D. M., Øritsland, N. A., Falconer, A. 1977. Remote sensing and ecosystem management. *Norsk Polarinst. Skrift.* No. 166. 52 pp.
127. Lerman, S. 1976. Lens fluorescence in aging and cataract formation. *Documenta Ophth.* 8:241–260
128. Levin, D. A. 1972. The adaptiveness of corolla-color variants in experimental and natural populations of *Phlox drummondii. Amer. Natur.* 106:57–70
129. Lotmar, R. 1933. Neue Untersuchungen über den Farbensinn der Bienen, mit besonderer Berücksichtigung des Ultravioletts. *Z. Vergl. Physiol.* 19:637–723
130. Lubbock, J. 1882. On the sense of color among some of the lower animals. *J. Linn. Soc. (Zool.)* 16[1883]:121–127
131. Lubbock, J. 1882. *Ants, Bees and Wasps: a Record of Observations on the Habits of the Social Hymenoptera.* New York: D. Appleton & Co. xix + 448 pp. + 5 pl.
132. Luner, S. J. 1968. Ultraviolet photography for visualization of separation patterns of zone electrophoresis and other methods. *Anal. Biochem.* 23:357–358
133. Lutz, F. E. 1924. Apparently non selective characters and combinations of characters, including a study of ultraviolet in relation to the flower-visiting habits of insects. *Ann. N. Y. Acad. Sci.* 29:181–283
134. Lutz, F. E. 1933. Experiments with "stingless bees" (*Trigona cressoni parastigma*) concerning their ability to distinguish ultraviolet patterns. *Amer. Mus. Novit.* No. 641:26 pp.
135. Lutz, F. E. 1933. "Invisible" colors of flowers and butterflies; attempting to get a better idea of how things in this world look to its principal inhabitants, the insects. *Natur. Hist.* 33:565–576
136. McClatchey, R. A., Fenn, R. W., Selby, J. E. A., Volz, F. E., Garing, J. S. 1972. *Optical Properties of the Atmosphere.* Cambridge, Mass: Air Force Cambridge Research Laboratories Environmental Research Paper No. 411 (NTIS: AFCRL-72-0497). 3rd ed.
137. McFarland, W. N., Munz, F. W. 1975. The visible spectrum during twilight and its implications to vision. In *Light as an Ecological Factor: II,* eds. G. C. Evans, R. Bainbridge, O. Rackham, pp. 249–270. Oxford: Blackwell Scientific Publications
138. Macior, L. W. 1968. Pollination adaptation in *Pedicularis groenlandica. Amer. J. Bot.* 55:927–932
139. Macior, L. W. 1971. Co-evolution of plants and animals—systematic insights from plant-insect interactions. *Taxon* 20:17–28
140. Macior, L. W. 1973. The pollination ecology of *Pedicularis* on Mount Rainier. *Amer. J. Bot.* 60:863–871
141. Macior, L. W. 1974. Pollination ecology of the front range of the Colorado Rocky Mountains. *Melanderia* [*Washington State Entomol. Soc.*]. 15: ii + 59 pp.
142. Macior, L. W. 1975. The pollination ecology of *Pedicularis* (Scrophulariaceae) in the Yukon Territory. *Amer. J. Bot.* 62:1065–1072
143. Macior, L. W. 1975. The pollination ecology of *Delphinium tricorne* (Ranunculaceae). *Amer. J. Bot.* 62:1009–1016

144. Macior, L. W. 1978. Pollination interactions in sympatric *Dicentra* species. *Amer. J. Bot.* 65:57–62
145. Makino, K., Kiyoo, S., Masahiko, K., Ueno, N. 1952. Sex in *Pieris rapae* L. and the pteridin content of their wings. *Nature* (*London*) 170:933–934
146. Manning, A. 1956. The effect of honeyguides. *Behaviour* 9:114–139
147. Mazokhin-Porshnyakov, G. A. 1954. [Ultraviolet radiation of the sun as a factor in insect habitats.] *Zh. Obsh.; Biol.* 15:362–367 (In Russian) [English translation available from British Library Lending Division]
148. Mazokhin-Porshnyakov, G. A. 1957. [Reflecting properties of butterfly wings and the role of ultra-violet rays in the vision of insects.] *Biofizika* 2:358–368 (In Russian) [English translation: *Biophysics* 2:352–362]
149. Mazokhin-Porshnyakov, G. A. 1959. Reflection of ultraviolet rays by flowers, and insect vision. *Entomol. Obozrenie* 38:285–296
150. Mazokhin-Porshnyakov, G. A. 1969. *Insect Vision.* New York: Plenum. xiv + 306 pp.
151. Menzel, R. 1963. Das Gedächtnis der Honigbiene für Spektralfarben, I: Kurzzeitiges und langzeitiges Behalten. *Z. Vergl. Physiol.* 63:290–309
152. Menzel, R., Erber, J. 1978. Learning and memory in bees. *Sci. Amer.* 239(1):102–110
153. Miyakawa, M. 1976. Flower-visiting behavior of small white butterfly, *Pieris rapae crucivora. Annot. Zool. Jpn.* 49:261–273
154. Moehn, L. D. 1974. The effect of quality of light on agonistic behavior of iguanid and agamid lizards. *J. Herpetol.* 8:175–183.
155. Mosquin, T. 1969. The spectral qualities of flowers in relation to photoreception in pollinating insects. *11th Int. Bot. Congr., Seattle, Washington,* p. 153 (Abstr.)
156. Motais, R. 1957. Sur l'absorption de la lumière par le cristallin de quelques poissons de grand profundeur. *Bull. Inst. Océanogr. Monaco* 1094:1–4
157. Mulligan, G. A., Kevan, P. G. 1973. Color, brightness, and other floral characteristics attracting insects to the blossoms of some Canadian weeds. *Can. J. Bot.* 51:1939–1952
158. Muntz, W. R. A. 1972. Inert absorbing and reflecting pigments. In *Handbook of Sensory Physiology, volume 7-1: Photochemistry of Vision,* ed. H. J. A. Dartnall, pp. 529–565. Berlin: Springer-Verlag
159. Muntz, W. R. A. 1975. The visual consequences of yellow filtering in the eyes of fishes occupying different habitats. In *Light as an Ecological Factor: II.,* eds. G. C. Evans, R. Bainbridge, O. Rackham, pp. 281–287. Oxford: Blackwell Scientific Publ.
160. Nekrutenko, Y. P. 1964. The hidden wing-pattern of some Palaearctic species of *Gonepteryx* and its taxonomic value. *J. Res. Lepidoptera* 3:65–68
161. Nekrutenko, Y. P. 1965. 'Gynandromorphic effect' and the optical nature of hidden wing-pattern in *Gonepteryx rhamni* L. (Lepidoptera, Pieridae). *Nature* (*London*) 205:417–418
162. Nekrutenko, Y. P. 1965. Three cases of gynandromorphism in *Gonepteryx;* an observation with ultraviolet rays. *J. Res. Lepidoptera* 4:103–108
163. Nekrutenko, Y. P. 1967. Sproba klyucha dlya vyznachennya vydiv rodu *Gonepteryx* Leach (1815) (Lepidoptera, Pieridae) iz vykorystannyam prykhovanoho krylovoho malyunka. *Dopov. Acad. Nauk. Ukr. Res. Ser. B Geol. Geofiz, Khim. Biol.* 29:263–265 (In Russian)
164. Nekrutenko, Y. P. 1968. *Phylogeny and geographical distribution of the genus* Gonepteryx (*Lepidoptera, Pieridae): an attemapt* [sic] *of study in historical zoogeography.* Kiev: Naukova Dumka. 128 pp. (In Russian)
165. Nekrutenko, Y. P., Didmanidze, E. A. 1975. [New data on geographic variation of ultraviolet reflectance pattern in *Gonepteryx rhamni* L. (Lepidoptera, Pieridae) in the Caucasus.] *Dopov. Akad. Nauk. Ukr. Res. Ser. B Geol. Geofiz. Khim. Biol.* 4:370–373 (In Ukranian)
166. Nolte, J., Brown, J. E. 1972. Ultraviolet-induced sensitivity to visible light in ultraviolet receptors of *Limulus. J. Gen. Physiol.* 59:186–200
167. Obara, Y. 1970. Studies on the mating behavior of the white cabbage butterfly, *Pieris rapae crucivora* Boisduval: III. Near-ultraviolet reflection as the signal of intraspecific communication. *Z. Vergl. Physiol.* 69:99–116
168. Obara, Y., Hidaka, T. 1968. Recognition of the female by the male, on the basis of ultra-violet reflection, in the white cabbage butterfly, *Pieris rapae crucivora* Boisduval. *Proc. Jpn. Acad.* 44:829–832
169. Ornduff, R., Mosquin, T. 1970. Variation in the spectral qualities of flowers in

the *Nymphoides indica* complex (Menyanthaceae) and its possible adaptive significance. *Can. J. Bot.* 48:603–605

170. Pappas, L. G., Eaton, J. L. 1977. The internal ocellus of *Manduca sexta:* electroretinogram and spectral sensitivity. *J. Insect Physiol.* 23:1355–1358

171. Petersen, B., Törnblom, O., Bodin, N. O. 1952. Verhaltenstudien am Rapsweisslung und Bergweissling (*Pieris napi* L. und *Pieris bryoniae* Ochs.). *Behavior* 4:67–84

172. Pope, R. D., Hinton, H. E. 1977. A preliminary survey of ultraviolet reflectance in beetles. *Biol. J. Linn. Soc.* 9:331–348

173. Post, C. T. Jr., Goldsmith, T. H. 1969. Physiological evidence for color receptors in the eye of a butterfly. *Ann. Entomol. Soc. Amer.* 62:1497–1498

174. Proctor, M., Yeo, P. 1973. *The Pollination of Flowers*. London: Collins, New Naturalist. 418 pp.

175. Raven, P. H. 1972. Why are bird-visited flowers predominately red? *Evolution* 26:674

176. Remington, C. L. 1973. Ultraviolet reflectance in mimicry and sexual signals in the Lepidoptera. *J. N. Y. Entomol. Soc.* 81:124

177. Richtmeyer, F. K. 1923. The reflection of ultraviolet by flowers. *J. Opt. Soc. Amer.* 7:151–168

178. Robberecht, R., Caldwell, M. M. 1978. Leaf epidermal transmittance of ultraviolet radiation and its implications for plant sensitivity to ultraviolet-radiation induced injury. *Oecologia* (*Berlin*) 32:277–288

179. Robey, C. W. 1975. Observations on the breeding behavior of *Pachydiplax longipennis* (Odonata: Libellulidae). *Psyche* 82:89–96

180. Robinson, N. 1966. *Solar Radiation*. Amsterdam: Elsevier

181. Roland, J. 1978. Variation in spectral reflectance of alpine and arctic *Colias* (Lepidoptera: Pieridae). *Can. J. Zool.* 56:1447–1453

182. Rutowski, R. L. 1977. The use of visual cues in sexual and species discrimination by males of the small sulphur butterfly *Eurema lisa* (Lepidoptera, Pieridae). *J. Comp. Physiol.* 115:61–74

183. Rutowski, R. L. 1977. Chemical communication in the courtship of the small sulphur butterfly *Eurema lisa* (Lepidoptera, Pieridae). *J. Comp. Physiol.* 115:75–85

184. Rutowski, R. L. 1978. The courtship behavior of the small sulphur butterfly,

Eurema lisa (Lepidoptera, Pieridae). *Anim. Behav.* 26:892–903

185. Rutowski, R. L. 1978. The form and function of ascending flights in *Colias* butterflies. *Behav. Ecol. Sociobiol.* 3:163–172

186. Scogin, R. 1976. Floral UV patterns and anthochlor pigments in the genus *Coreopsis* (Asteraceae). *Aliso* 8:425–427

187. Scogin, R. 1976. Anthochlor pigments and pollination biology: I. The UV absorption of *Antirrhinum majus* flowers. *Aliso* 8:429–431

188. Scogin, R., Young, D. A., Jones, C. E. Jr. 1977. Anthochlor pigments and pollination biology: II. The ultraviolet floral pattern of *Coreopsis gigantea* (Asteraceae). *Bull. Torrey Bot. Club* 104:155–159

189. Scogin, R., Zakar, K. 1976. Anthochlor pigments and floral UV patterns in the genus *Bidens*. *Biochem. Syst. Ecol.* 4:165–167

190. Scott, A. I. 1964. *Interpretation of the Ultraviolet Spectra of Natural Products*. Oxford: Pergamon Press. 443 pp.

191. Scott, J. A. 1973. Mating of butterflies. *J. Res. Lepidoptera* 11[1973]:99–127

192. Scott, J. A. 1974. Survey of ultraviolet reflectance of Nearctic butterflies. *J. Res. Lepidoptera* 12:151–160

193. Seybold, A., Weissweiler, A. 1945. Spectrophotometrische Messungen an Blumenblättern. *Bot. Arch. Z. Ges. Bot. Grenzgeb.* 45:358–386

194. Silberglied, R. E. 1969. *Ultraviolet reflection of pierid butterflies. Phylogenetic implications and biological significance.* M.S. thesis. Cornell University

195. Silberglied, R. E. 1973. *Ultraviolet reflection of butterflies and its behavioral role in the genus* Colias (*Lepidoptera-Pieridae*). Ph.D. thesis, Harvard University

196. Silberglied, R. E. 1976. Visualization and recording of longwave ultraviolet reflection from natural objects. *Func. Photogr.* (*Photogr. Appl. Sci. Tech. Med.*) 11(2):20–29; (3):30–33

197. Silberglied, R. E. 1977. Communication in the Lepidoptera. In *How Animals Communicate*, ed. T. A. Sebeok, pp. 362–402. Bloomington: Indiana Univ.

198. Silberglied, R. E., Taylor, O. R. 1973. Ultraviolet differences between the sulphur butterflies, *Colias eurytheme* and *C. philodice*, and a possible isolating mechanism. *Nature* (*London*) 241:406–408

199. Silberglied, R. E., Taylor, O. R. Jr. 1978. Ultraviolet reflection and its be-

havioral role in the courtship of the sulfur butterflies, *Colias eurytheme* and *C. philodice* (Lepidoptera, Pieridae). *Behav. Ecol. Sociobiol.* 3:203–243

200. Simon, H. 1971. *The Splendor of Iridescence; Structural Colors in the Animal World.* New York: Dodd, Mead and Company, 268 pp.

201. Stehr, F. W., Cook, E. F. 1968. A revision of the genus *Malacosoma* Hübner in North America (Lepidoptera: Lasiocampidae): systematics, biology, immatures, and parasites. *Bull. US Nat. Mus.* 276:1–321

202. Steinly, B. A., Deonier, D. L., Regensburg, J. T. 1978. Scanning electron microscopy of ultraviolet-reflective pruinosity in species of *Ochthera* (Diptera: Ephydridae). *Entomol. News* 89:117–124

203. Swihart, S. L., 1967. Neural adaptations in the visual pathway of certain heliconiine butterflies, and related forms, to variations in wing coloration. *Zoologica* (*N. Y.*) 52:1–14

204. Takizawa, T., Koyama, N. 1974. Reflection of ultraviolet light from the wing surface of the cabbage butterfly, *Pieris rapae crucivora* Boisduval (Lepidoptera: Pieridae). *J. Fac. Text Sci. Tech. Shinshu Univ. Ser. A Biol.* 17:1–12

205. Tansley, K. 1965. *Vision in Vertebrates.* London: Chapman & Hall

206. Taylor, O. R. Jr. 1972. Random vs. non-random mating in the sulfur butterflies, *Colias eurytheme* and *Colias philodice* (Lepidoptera: Pieridae). *Evolution* 26:344–356

207. Taylor, O. R. Jr. 1973. Reproductive isolation in *Colias eurytheme* and *C. philodice* (Lepidoptera: Pieridae): use of olfaction in mate selection. *Ann. Entomol. Soc. Amer.* 66:621–626

208. Taylor, O. R. 1973. A non-genetic "polymorphism" in *Anartia fatima* (Lepidoptera: Nymphalidae). *Evolution* 27:161–164

209. Thien, L. B., Marcks, B. G. 1972. The floral biology of *Arethusa bulbosa, Calopogon tuberosus,* and *Pogonia ophioglossoides* (Orchidaceae). *Can. J. Bot.* 50:2219–2235

210. Thompson, W. R., Meinwald, J., Aneshansley, D., Eisner, T. 1972. Flavonols: Pigments responsible for ultraviolet absorption in nectar guide of flower. *Science* 177:528–530

211. Thorp, R. W., Briggs, D. L., Estes, J. R., Erickson, E. H. 1975. Nectar fluorescence under UV irradiation. *Science* 189:476–478

212. Thorp, R. W., Briggs, D. L., Estes, J. R., Erickson, E. H. 1976. [Reply to Kevan (101).] *Science* 194:342

213. Udovic, D. 1978. Ultraviolet floral patterns in *Yucca whipplei* and their potential adaptive significance. Ms., submitted to *Amer. J. Bot.*

214. Urbach, F., ed. 1969. *The Biologic Effects of Ultraviolet Radiation (with Emphasis on the Skin).* London: Pergamon Press. xv + 704 pp.

215. Ustinov, D. A., Kozunin, I. I. 1963. K metodike izucheniya propuskaniya ul'-trafioletovykh luchei sherstnymi pokrovami zhivotnykh s pomoshch'yu spektrofotometra SF-4. *Byul. Eksptl. Biol. Med.* 56:124–126

216. Utech, F. H., Kowano, S. 1975. Biosystematic studies in *Erythronium* (Liliaceae-Tulipeae): I. Floral biology of *E. japonicum* Decne. *Bot. Mag. (Tokyo)* 88:163–176

217. Vogel, S. 1950. Farbwechsel und Zeichnungsmuster bei Blüten. *Oesterr. Bot. Z.* 97:45–100

218. Von Frisch, K. 1967. *The Dance Language and Orientation of Bees.* Cambridge: Belknap Press of Harvard University Press. xiv + 566 pp.

219. Wald, G. 1952. Alleged effects of the near ultraviolet on human vision. *J. Opt. Soc. Amer.* 42:171–177

220. Wald, G., Krainin, J. M. 1963. The median eye of *Limulus:* An ultraviolet receptor. *Proc. Nat. Acad. Sci. USA* 50:1011–1017

221. Walls, G. L. 1942. *The Vertebrate Eye and its Adaptive Radiation.* Bloomfield Hills, Michigan: Cranbrook Inst. of Science

222. Wasserman, G. S. 1976. *Limulus* psychophysics: spectral sensitivity of the ventral eye. *J. Exp. Psychol.* 105:240–253

223. Waterman, T. H. 1961. Light sensitivity and vision. In *The Physiology of Crustacea, vol. 2, Sense Organs, Integration, and Behavior,* ed. T. H. Waterman, pp. 1–64. N.Y.: Academic Press

224. Wilson, M. 1978. The functional organisation of locust ocelli. *J. Comp. Physiol. A Sens. Neural. Behav. Physiol.* 124:297–316

225. Wright, A. A. 1972. The influence of ultraviolet-radiation on the pigeon's color discrimination. *J. Exp. Anal. Behav.* 17:325–337

226. Wyszecki, G., Stiles, W. L. 1967. *Color Science.* New York: John Wiley & Sons. xiv + 628 pp.

227. Ziegenspeck, H. 1955. Die Farben- und UV-Photographie und ihre Bedeutung für die Blütenbiologie. *Mikroskopie,* 10:323–328

Ann. Rev. Ecol. Syst. 1979. 10:399–422

BIOGEOGRAPHICAL ASPECTS ◆4168
OF SPECIATION IN THE
SOUTHWEST AUSTRALIAN FLORA

Stephen D. Hopper

Western Australian Wildlife Research Centre, Department of Fisheries and
Wildlife, P.O. Box 51, Wanneroo, W.A., 6065 Australia.

INTRODUCTION

The southwest of the Australian continent is a region of biogeographic
interest, having an angiosperm flora remarkably rich in endemic species. It
was estimated that 2450 (68%) of the 3600 species known in the region
by the mid 1960s were endemics (12, 13, 107). Recent taxonomic research
suggests that the actual level of endemism may approach 75–80%, since
southwestern genera are being enlarged on average by 10–30% in current
monographs (19, 28–30, 36, 38, 61, 62, 67, 72–74, 76, 112, 113, 122, 126,
154, 155).

As pointed out by Hooker (70) and subsequent authors (34, 37, 38, 76,
123), the rich flora of the southwest is all the more remarkable in view of
its low relief and subdued topography. The Great Plateau of Western
Australia (89), an extensive deeply weathered landform that rarely exceeds
500 m in height, dominates much of the southwestern landscape (Figure 1).
Areas of relatively high relief are confined to a series of small ranges along
its southern margin and uplifted or resistant remnants of Tertiary plateau
levels along its western margin. All of these small mountains and plateau
remnants are less than 1000 m high, with the exception of a few peaks in
the Stirling Range that rise to 1109 m. In contrast, other regions of the
world rich in endemic species have much higher and more extensive mountain systems, e.g. South Africa (66, 149), California (5, 96, 132, 144, 150),
Turkey and Greece (52, 148, 158), and some oceanic islands (26, 37).

The broader aspects of southwestern biogeography, including floristic relationships with other areas of Australia and the nature of the barriers that isolated the flora since the mid-Tertiary, have been adequately treated by previous reviewers (34, 37, 51, 53, 57, 68, 76, 86, 87, 107, 121, 123, 128). However, the central question of the factors responsible for the origin of so many endemic species within the southwest has not yet been addressed in detail. This is perhaps understandable, since sufficient floristic, biosystematic, paleoclimatic, and geomorphic data for an evolutionary overview

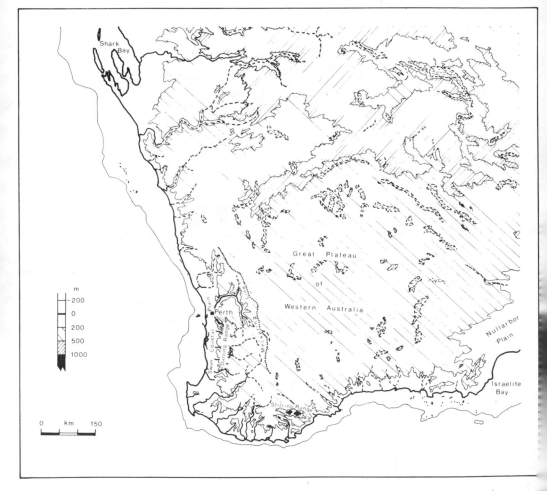

Figure 1 Topography and major drainage systems in southwestern Australia. Seasonal water bodies are demarcated by dashed lines. The Meckering Line indicates a major drainage divide separating river systems of the west coast from the salt lakes inland (117, 118).

of speciation in the southwest have become available only recently. It is the purpose of the present review to present a synthesis of these data. Particular emphasis is placed on the role of climatic fluctuations and landscape evolution in facilitating speciation. The reader is referred to two cogent reviews (83, 84) for discussions of cytoevolutionary aspects of speciation in the southwest.

RAINFALL ZONES AND THEIR VEGETATION

The area under review agrees approximately with that included in the South West Botanical Province and South West Interzone of Western Australia [(14, 15, 34, 53, 57, 123); Figure 2]. From the viewpoint of regional differentiation in species richness, three main rainfall-vegetation zones appear to provide a useful subdivision of this area (Figure 2). These are (a) the high-rainfall zone (800–1400 mm annual rainfall)—forests and woodlands dominated by jarrah (*Eucalyptus marginata*), marri (*E. calophylla*), and karri *(E. diversicolor)*—occupies the lower southwest and extends northwards along the Swan Coastal Plain and Darling Range to the vicinity of Perth; (b) the transitional-rainfall zone (300–800 mm annual rainfall)—woodlands, mallees and heathlands—reaches the west coast from Perth to Shark Bay, extends southeast inland of the forests through an area known locally as the Wheatbelt, and occurs as an elongate strip along the south coast eastward to Israelite Bay; (c) the arid zone (less than 300 mm annual rainfall)—*Eucalyptus* woodlands, *Acacia* shrublands, and *Triodia* hummock grassland—occupies much of central Australia and is marginal to the area under review, extends along the western coast northwards of Shark Bay to the tropics and along the south coast from Israelite Bay eastward across the low open shrubland of the Nullarbor Plain to temperate South Australia.

Although there is an approximate parallel between vegetation structure and annual rainfall across these three zones, soil characteristics are of primary importance in determining local distribution patterns of vegetation (14, 15, 53, 57, 69, 95, 97, 115, 120, 139, 152, 153). For example, in the transitional-rainfall zone, heathlands and mallees are best developed on highly leached nutrient-deficient sands and laterites, particularly in the two main near-coastal regions (Figure 2), while woodlands and shrublands predominate in inland areas of the Wheatbelt where soils have higher nutrient status and clay content. Similarly, in the high-rainfall zone, eucalypt forests are dominant on the extensive lateritic soils of the Darling Range and Great Plateau while eucalypt and *Banksia* woodlands occur on the leached sands of the Swan Coastal Plain.

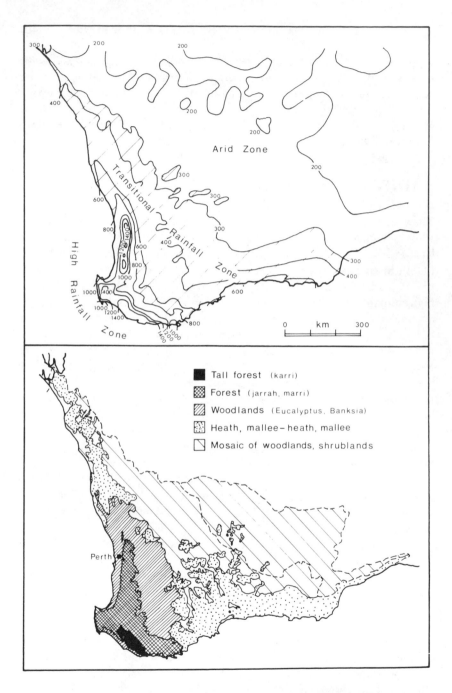

Figure 2 Distribution of annual rainfall isoheyts (mm), rainfall zones, and dominant vegetation formations in southwestern Australia. After the *Western Australian Year Book 1974* (rainfall), and modified from Beard [(14, 15), unpublished map (vegetation)]. Dashed lines indicate boundaries of the South West Botanical Province and South West Interzone.

GEOGRAPHICAL PATTERNS OF SPECIES RICHNESS

Since the pioneering research of Diels (53), biogeographers have noted that the flora of the southwest is richest in the coastal and inland sandheaths of the transitional rainfall zone and relatively species poor in the high rainfall forests and arid zone communities (34, 57, 63, 90, 107, 123, 139). These descriptive observations have been confirmed in several quantitative studies where the region has been subdivided into even-sized grid cells and the number of species per cell determined for a given genus or family (Figure 3). A recurrent pattern of up to three times as many species in the transitional-rainfall zone grid cells as in high-rainfall and arid zone cells has emerged for most genera of five or more species for which quantitative data are available. Examples include *Anigozanthos, Conostylis* [Haemodoraceae (73), S. D. Hopper unpublished data], *Adenanthos, Banksia, Conospermum, Dryandra, Grevillea, Hakea, Isopogon, Lambertia, Petrophile* [Proteaceae (123, 139)], *Eriostemon, Phebalium* [Rutaceae (P. G. Wilson, S. D. Hopper, unpublished data)], *Acacia* [Mimosaceae (76)], *Astartea, Baeckea, Beaufortia, Calothamnus, Calytrix, Chamelaucium, Darwinia, Eremaea, Eucalyptus, Kunzea, Leptospermum, Lhotskya, Melaleuca, Micromyrtus, Regelia, Scholtzia, Thryptomene, Verticordia* and *Wehlia* [Myrtaceae (B. L. Rye, unpublished data; S. D. Hopper, unpublished data; 63)]. In contrast, a few genera are known to be as rich in species in the high-rainfall zone as in the transitional-rainfall zone, e.g. *Persoonia, Synaphaea* [Proteaceae (139)], *Boronia* [Rutaeae (P. G. Wilson, S. D. Hopper, unpublished data)], *Agonis,* and *Hypocalymma* [Myrtaceae (B. L. Rye unpublished data)].

Several authors have emphasized the species richness of the western and southern near-coastal heathlands (Figure 2) of the transitional-rainfall zone and suggested that the intervening communities of the Wheatbelt are relatively species poor (53, 57, 107, 139). A number of genera indeed show such a bimodal heathland pattern of species richness (e.g. *Beaufortia, Calothamnus, Calytrix, Conospermum, Conostylis, Darwinia, Dryandra, Grevillea, Hakea, Isopogon, Leptospermum, Petrophile, Phebalium, Regelia, Verticordia*) or else are concentrated in the southern heathlands (e.g. *Adenanthos, Astartea, Banksia, Eucalyptus, Kunzea, Lambertia*) or in the western heathlands (e.g. *Eremaea, Scholtzia*) but not in the adjacent Wheatbelt. However, some genera have nodes of species richness throughout the transitional-rainfall zone, including both near-coastal heathland areas and the Wheatbelt (e.g. *Acacia, Baeckea, Melaleuca*), or else have nodes extending into the Wheatbelt from the southern heathlands (e.g. *Chamelaucium*) or from the western heathlands (e.g. *Micromyrtus, Thryptomene*). Moreover, a few genera are richest in species in the Wheatbelt itself, and are relatively species-poor in the two heathland areas (e.g. *Lhotskya, Eriostemon, Wehlia*). A recent ecological survey (115) suggested that some woodland communities of the Wheatbelt in fact contain more species than adjacent heathland communities. Thus the entire transitional-rainfall zone, including both its heathlands and its woodlands, may be regarded as species-rich relative to the high-rainfall and arid zones.

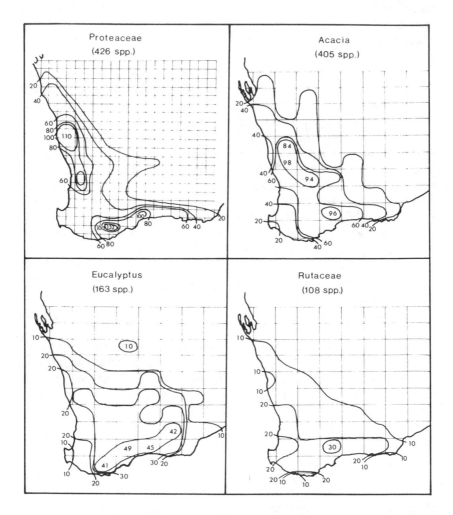

Figure 3 Maps illustrating the relative number of species of Proteaceae, *Acacia, Eucalyptus* and Rutaceae in regions of southwestern Australia. Isoflors (species richness isopleths) encompass even-sized grids that have comparable numbers of species (45 km^2 grids for the Proteaceae analysis, 1° latitude X 1.5° longitude grids for the other three). From [(139), Proteaceae; (76), *Acacia;* (41) and S. D. Hopper, unpublished data, *Eucalyptus;* P. G. Wilson, S. D. Hopper, unpublished data, Rutaceae].

A feature of the transitional-rainfall zone is the high number of local endemics characterizing each of the main heathland centers and, to a lesser extent, the Wheatbelt (53, 139). Species occurring throughout the transitional-rainfall zone are relatively uncommon. For example, data (139) indicate that only 16% of the 303 species of Proteaceae occurring in the two heathland centers (cf Figure 3) are common to both.

DISTRIBUTION OF RELICT AND RECENTLY EVOLVED SPECIES

If environmental circumstances have provided a stimulus to the development of the striking differences in species richness in rainfall zones of the southwest, then common trends in the distribution of taxonomically unrelated relict and recently evolved species should occur (144). Thus, the species richness of the transitional-rainfall zone may have arisen because it contained habitats either more favorable as refugia for evolutionary relicts or more favorable for recent speciation than those of the high-rainfall and arid zones. Accordingly, one would expect to find either relict species or recently evolved species in greater numbers in the transitional-rainfall zone than in the other two zones.

Ideally, the identification of relict and recently evolved species should rest on paleontological evidence. However, the known fossil record for present-day Western Australian species is limited to Holocene deposits no older than 8000 years B. P. (43, 44). Relative ages of closely related living species can be inferred, however, from comparative systematic data, on the assumptions that morphological and genetic divergence generally increases between species through time and certain changes in chromosome number tend to be unidirectional (84, 141). Thus, morphologically distinct taxa (e.g. species of monotypic genera) may be regarded as evolutionary relicts (144), while subspecies, semispecies, and sibling species may be regarded as recently evolved; diploids may be regarded as older than their polyploid derivatives; and species with high chromosome numbers in a dysploid series may be regarded as older than those with low numbers.

Table 1 summarizes distribution data for cytogenetically studied species or chromosome races regarded as relict or recent on the basis of the above criteria; a list of these taxa together with chromosome numbers and authorities consulted may be found in (73). Examples have been drawn from a wide sample of genera and families in the flora, including *Agrostocrinum, Dianella, Hodgsoniola, Laxmania, Murchisonia, Wurmbea* (Liliaceae), *Baxteria, Kingia, Lomandra* (Xanthorrhoeaceae), *Anigozanthos, Blancoa, Conostylis, Macropidia* (Haemodoraceae), *Cypselocarpus, Tersonia* (Gyrostemonaceae), *Drosera* (Droseraceae), *Chorizema, Cupulanthus, Euchilopis, Eutaxia, Isotropis* (Fabaceae), *Boronia, Chorilaena, Geleznowia* (Rutaceae),

Table 1 Number of relict and recently evolved species or chromosome races of cytogenetically studied genera occuring in the high-rainfall and transitional-rainfall zones of southwestern Australia[a]

| | Number of taxa in | | | |
	High-rainfall zone	Transitional-rainfall zone	Both or borderline	Total
Relict or parental taxa	11 (9)[b]	40 (42)	18	69
Recent or derived taxa				
diploids	0 (2)	9 (7)	6	15
low dysploids	2 (3)	13 (12)	8	23
polyploids	6 (5)	24 (25)	7	37
Total	8 (9)	46 (45)	21	75

[a]Species, chromosome races, and authorities consulted are listed in (73); genera are listed in the present text.

[b]Numbers in parentheses indicate those expected on the basis of the relative areas of the two zones (the transitional-rainfall zone occupies an area 5 times larger than that of the high-rainfall zone).

Hybanthus (Violaceae), *Calytrix, Darwinia, Leptospermum, Thryptomene* (Myrtaceae), *Dampiera, Goodenia, Pentaptilon* (Goodeniaceae), and *Stylidium* (Stylidiaceae).

Given that the transitional-rainfall zone occupies an area approximately five times greater than that encompassed by the high-rainfall zone, it would appear that relict and recently evolved taxa occur in each zone at close to the frequencies expected on the basis of relative areas (Table 1). On the cytogenetic data at hand, there is not a disproportionately high number of either relict or recently evolved taxa in the transitional-rainfall zone. However, there are suggestions in the data that (*a*) recent diploid and possibly dysploid speciation has occurred more frequently in the transitional-rainfall zone than in the high-rainfall zone, while recent polyploids have arisen as frequently or more frequently in the high-rainfall zone; and that (*b*) slightly more relict species occur in the high-rainfall zone than expected.

Although the data on species for which chromosome counts are available provide only weak support for these trends, greater confidence in their validity emerges from a consideration of the distribution of relict and recently evolved species in other groups for which critical taxonomic evidence has been acquired. Thus, in *Acacia,* a genus that appears to have speciated in Australia largely at the diploid level, species presumed to be recently evolved on the basis of their taxonomic relationships are up to three times more frequent in 1° latitude × 1.5° longitude areas in the transitional-rainfall zone than in comparable areas of the high-rainfall zone (76). A similar concentration of recently evolved species in the transitional zone is found in *Eucalyptus* (28–30, 34, 41, 75) and, indeed, appears to hold for a number of other southwestern genera (7, 19, 36, 38, 55, 61, 62, 67, 72–74, 122, 126, 154). Moreover, recently evolved subspecies of several genera are also concentrated in the transitional-rainfall zone and relatively sparse in the high-rainfall zone [e.g. *Conostylis* (67, 72), *Anigozanthos* (73, 74), *Acacia* (112, 113), *Eucalyptus* (28, 41), *Stylidium* (7, 36, 38, 48, 83, 85), and *Kennedya* (138)].

A study of monotypic genera (N. G. Marchant, S. D. Hopper unpublished data) produced results that concur with the above data as to areas where relictual taxa are concentrated. Of the 46 southwestern monotypics, only 6 extend into or are restricted endemics of the arid zone, namely *Didymanthus* (Chenopodiaceae); *Emblingia* (Emblingiaceae), *Erichsenia* (Fabaceae); *Hemiphora* (Dicrastylidaceae); *Neogoodenia* (Goodeniaceae); and *Neotysonia* (Asteraceae). In contrast, 12 are endemic in the high-rainfall zone—*Baxteria* (Xanthorrhoeaceae); *Chaetanthus* and *Meeboldina* (Restionaceae); *Diplopogon* (Poaceae); *Reedia* (Cyperaceae); *Hodgsoniola* (Liliaceae); *Cephalotus* (Cephalotaceae); *Chorilaena* (Rutaceae); *Diaspasis* (Goodeniaceae); *Eremosyne* (Saxifragaceae); *Jansonia* (Fabaceae); and *Meziella* (Haloragaceae)—and 18 are endemic in the transitional-rainfall zone, being particularly concentrated in the western and southern near-coastal heathlands—*Blancoa* and *Macropidia* (Haemodoraceae); *Harperia* and *Onychosepalum* (Restionaceae); *Murchisonia* (Liliaceae); *Rhizanthella* (Orchidaceae); *Spartochloa* (Poaceae); *Cupulanthus* (Fabaceae); *Cypselocarpus* (Gyrostemonaceae); *Geleznowia, Muiriantha, Nematolepis* and *Rhadinothamnus* (Rutaceae); *Nigromnia* and *Pentaptilon* (Goodeniaceae); *Oligarrhena* (Epacridaceae); *Siegfriedia* (Rhamnaceae); and *Spirogardnera* (Santalaceae). The remaining 10 monotypics are more widespread, occurring in both the high-rainfall and transitional-rainfall zones, i.e. *Agrostocrinum* (Liliaceae); *Epiblema* (Orchidaceae); *Kingia* (Xanthorrhoeaceae); *Cosmelia* and *Needhamiella* (Epacridaceae); *Euchilopsis* (Fabaceae); *Nuytsia* (Loranthaceae); *Tersonia* (Gyrostemonaceae); *Philydrella* (Philydraceae); and *Pronaya* (Pittosporaceae). Thus, on the basis of the relative areas of the three rainfall zones, these data indicate that the high-rainfall zone has a disproportionately high number of monotypic relict genera. In various animal groups, also, it is noticeable that the high-rainfall zone has more monotypic genera than would be expected on the basis of relative areas, while fewer than expected occur in the transitional and arid zones (2, 9, 31, 35, 49, 56, 65, 78, 90, 94, 103, 104, 133, 137).

Studies of large arid-zone genera, such as *Triodia* [Poaceae (32)], *Eremophila* [Myoporaceae (8, 54)], *Ptilotus* [Amaranthaceae (145)], *Cassia* [Caesalpiniaceae (129, 130)], *Solanum* [Solanaceae (130)], and *Acacia* [Mimosaceae (76)], suggest that the flora is generally depauperate in terms of the area of the zone and that polyploidy has been involved in much of the recent speciation. In these respects the arid zone and high-rainfall zone are similar.

These recurrent trends in the distribution of relict and recently evolved species in diverse groups of plants strongly suggest that environmental circumstances of regional impact have elicited common evolutionary responses in many components of the flora. Thus, it would seem that conditions favorable for prolific speciation were present in the transitional-rainfall zone in the recent geological past, whereas conditions in the high-rainfall and arid zones have been less favorable for recent speciation. On the other hand, conditions allowing for the persistence of relict species must have been effective for a considerable geological period in the high-rainfall zone, while they have been less effective in this regard in the transitional and arid zones.

If climatic and edaphic controls of the distribution of species have been as stringent throughout the evolutionary history of southwestern angiosperms as they are today (14, 15, 34, 43, 44, 53, 57, 69, 73, 95, 97, 115, 120, 139, 152, 153), then rates at which populations persisted, became extinct, were isolated, migrated, hybridized, and ultimately speciated would have been under the influence of past climates and patterns of landscape evolution. The following section reviews patterns of landscape evolution and climatic history that have indeed concurred with those expected to produce the existing regional zonation of recently evolved and relict species.

LANDSCAPE DEVELOPMENT, CLIMATIC HISTORY, AND SPECIATION IN THE TERTIARY AND QUATERNARY

The Tertiary

Churchill (43, 45), Marchant (107), and Beard (14) noted that at the time of maximum marine intrusion onto the Australian continent during the Eocene, a mosaic of numerous islands and peninsulas would have flanked the southwestern coastline. The resultant fragmentation of the flora, as in present-day oceanic archipelagos, may well have precipitated speciation (26, 37). The origin of genera and some bradytelic (slowly evolving) species endemic to the Stirling Range [e.g. the monotypic genera *Cupulanthus* (Fabaceae) and *Muiriantha* (Rutaceae)] and other monadnocks dotted along the southern coast may have occurred during the Eocene period of insular evolution. However, this period seems too remote to account for the genesis of most endemic species in these areas, since many have close morphological relatives in adjacent regions and must have evolved in the more recent geological past (86).

Two post-Eocene geomorphological developments in the southwest may have had a profound influence on regional habitat diversity and opportunities for speciation. First, the entire Great Plateau and flanking sedimentary basins appear to have been uplifted, since old beach levels from an Eocene cycle of sedimentation at about 300 m above present sea level have been found at several locations on the southern, eastern, and northern margins of the plateau (14, 15, 88). While there is no evidence that this uplift significantly altered the low relief and sluggish uncoordinated drainage of inland areas of the Great Plateau, coastal rivers were rejuvenated and dissected the early Tertiary plateau into numerous insular-peninsular remnants bordered by erosional and depositional landforms (14, 15, 118). Thus the plateau margins today have relatively rugged and more diverse topographical features than inland areas (Figure 1). This applies in particular to those plateau margins in areas of the transitional-rainfall zone where heathlands now predominate (cf Figure 2).

The second important geomorphological development was the widespread formation of lateritic soils on the Great Plateau during the Oligocene and/or Miocene (88) and their continued formation in climatically suitable areas (i.e. with a good supply of seasonal rainfall) to the present day (116–118). Weathering of these soils gave rise to nutrient-deficient sandplains (20, 27, 120), thereby expanding the habitat available to xeromorphic heathland species and apparently stimulating the evolution of sclerophylly in several groups that were to become increasingly dominant during the progressive aridity of late Tertiary-Quaternary time (11, 16, 34, 43, 64, 86, 87, 111, 135). Moreover, the intensity of weathering and consequent distribution of lateritic gravels and sands varied regionally, presumably with significant effects on population structures and rates of speciation. This regional variation in weathering has been well documented for plateau landforms near to and inland from Perth (20, 47, 101, 117–120). A geomorphological boundary [named the Meckering Line by Mulcahy (117); cf Figure 1] that runs approximately parallel to but 50 km inland of the boundary between the high-rainfall and transitional-rainfall zones demarcates major changes in the drainage and erosional patterns. Inland of the Meckering Line landforms are subdued, consisting of low lateritic-capped divides separated by broad flat-floored valleys that contain chains of salt lakes and occasional emergent granite monadnocks. For much of the Tertiary and Quaternary, streams in this region have been ephemeral, with low erosive power. West of the Meckering Line an effective westward-flowing river system now operates (Figure 1), with a change from shallow flat-floored valleys in headwater areas of the Wheatbelt to deep V-shaped valleys where streams pass through the Darling Range and issue from the Darling Scarp (101, 117, 118). The distribution of lateritic, erosional, and depositional landforms shows a striking change from the high-rainfall areas in the Darling Range to the transitional-rainfall areas near the Meckering Line (Figure 4). Where rainfall is high, lateritic soils (duricrusts, gravels, and sands) form a continuous mantle on the gently undulating hills, being absent only adjacent to sharply incised streams on the steeper valley sides. To the east where rainfall is lower, lateritic soils are much dissected, occurring on a multitude of insular steep-sided plateau remnants (mesas bounded by "breakaways") similar to those in heathland areas on the western and southern plateau margins. Younger erosional and depositional soils occupy in mosaic fashion a greater proportion of the landscape in this transitional-rainfall region than they do in the Darling Range. The greater erosional dynamism in the transitional-rainfall zone has been attributed to shallower (therefore less resistant) lateritic duricrusts and to vegetation changes resulting in landscape instability during late Tertiary-Quaternary climatic fluctuations (20, 47, 101, 118, 120).

Thus, as a result of coastal drainage rejuvenation consequent upon uplift and differential erosion of lateritic landforms, topography and soils are considerably more diverse throughout present-day transitional-rainfall areas near the western and southern coasts and inland through the western Wheatbelt than in the high-rainfall and arid zones. The development of these regional differences in landform dissection and edaphic complexity must have had a significant influence on rates of speciation, particularly when the extreme sensitivity of many extant species to edaphic change is borne in mind. The preservation and continued formation of lateritic soils throughout the mid-late Tertiary and Quaternary in the high-rainfall zone would have allowed for the existence of large continuous populations and favored close adaptation to laterite. Such a population structure and habitat continuity favors slow evolution and speciation (143, 144). Hence, conditions favorable for the preservation of relict species have prevailed in the high-rainfall zone. In contrast, the more active erosional history of transitional rainfall landscapes led to recurrent fragmentation, isolation, extinction, migration, confluence, and possibly hybridization of populations, and elicited adaptation of more flexible components of the flora to derived erosional and depositional soils. Such flux in habitat and population structure is conducive to rapid speciation (143, 144), and must have been a primary cause of the present-day concentration of recently evolved species in the transitional-rainfall zone (76). Additionally, the isolated lateritic residuals in the transitional zone would have provided habitat continuity allowing for the persistence of some relict species (cf 43, 95, 120).

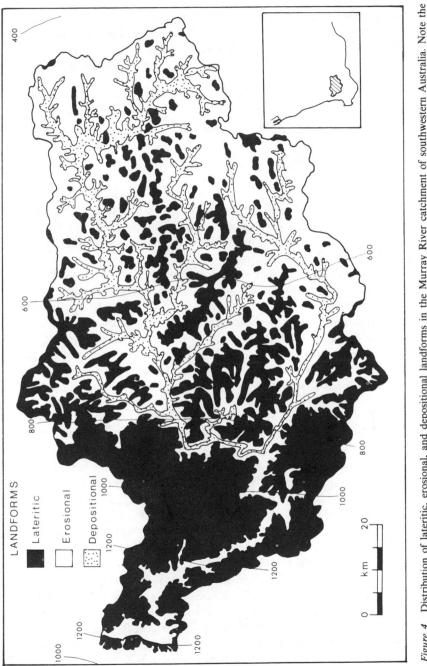

Figure 4 Distribution of lateritic, erosional, and depositional landforms in the Murray River catchment of southwestern Australia. Note the continuity of lateritic landforms in the western sector which experiences more than 800 mm annual rainfall, and the much-dissected lateritic landforms in the eastern sector of transitional rainfall. After (101).

A final development during the mid-late Tertiary that appears to have had a profound influence on the composition of the southwestern flora was a climatic change throughout southern Australia from subtropical conditions through progressively more arid phases to a climatic pattern resembling that of today by the beginning of the Quaternary (23, 43, 46, 59, 64, 109, 111). This climatic change led to the extinction of many of the subtropical rainforest elements in southwestern Australia (6, 43, 45, 50, 77, 110, 111). Rates of extinction were no doubt accentuated because arid conditions would have developed earliest on the northwestern coast of Western Australia (16) and thereby cut off the only migration route available to southern species capable of keeping pace with the northward retreat of tropical-subtropical climatic zones. The only refuge open to subtropical mesophytes was in high-rainfall areas of the southwest. Indeed, many of the relict species presently concentrated in the high-rainfall zone may be remnants of the early Tertiary subtropical flora that avoided extinction because of the prevailing humid climate in the area.

While mesophytic subtropical elements were affected adversely by the progressive aridity, groups that had become sclerophyllous in response to the poor nutrient status of lateritic soils were preadapted to the climate change and proliferated in the growing areas of semi-aridity and mediterranean climate (11, 34, 43, 86, 87, 111, 135). Moreover, in these groups the fluctuating but progressively more arid climate would have provided an additional stimulus to speciation in transitional-rainfall areas in a manner (outlined below) that became most effective in the climatically turbulent Quaternary.

The Quaternary

The Pliocene-Pleistocene boundary marks the last accurately dated fossil occurrence of subtropical rainforest species in areas of southern Australia that are presently semi-arid (46, 64, 109, 111). Hence it was during the short two million years of the Quaternary that many of the southern Australian sclerophyllous genera achieved dominance (64, 111). Moreover, it seems that the majority of recently evolved species may owe their origins to evolutionary events stimulated by the climatic fluctuations, landscape erosion, and sea level changes of the Quaternary.

The concept of a generally pluvial Pleistocene followed by a mid-Holocene "Great Arid" period achieved prominence in the Australian geomorphological and biogeographical literature for several decades, largely under the influence of a hypothesis developed by Crocker & Wood (51). It was proposed that during this arid period "the desiccation was so severe and sudden that it resulted in a considerable portion of the pre-arid flora being entirely wiped out. The surviving remnants were isolated in numerous refuges where habitat diversity, especially climatic diversity, was greatest. The present-day plant communities are the result of re-colonisation of vast, virtually bare, areas, especially in the arid regions." Adherence to this view of Quaternary climates led to claims of remarkably rapid speciation for groups inhabiting transitional and arid zones. For example, Burbidge (33, 34) proposed that many of the mallee eucalypts evolved their distinctive habit (multiple stems emerging from an underground lignotuber) and speciated during the Holocene as a consequence of the "Great Arid."

While subsequent research has supported the hypothesis that arid conditions have had a profound evolutionary effect on the Australian flora, the concept of a single "Great Arid" period in the Holocene is no longer tenable. Rather, there is now strong palynological and geomorphological evidence that several arid phases of varying severity occurred throughout the Quaternary and in the late Tertiary (23–25, 50, 108, 109, 111, 156, 157). The most recent Quaternary period of major aridity has been reliably dated as prior to the Holocene at 18,000–16,000 years B. P., the time of the last glacial maximum (23, 108). Bowler (23) suggested that the causes of major aridity were to be sought in greatly intensified atmospheric circulation aided by increased continental extent corresponding to glacial low sea levels. Thus each of the main periods of glaciation during the Pleistocene would have produced arid conditions in southern Australia, while the interglacials would have been more pluvial.

Evidence of climatic change and active landscape evolution in the southwest during the Quaternary comes from a number of sources, including coastal and lacustrine geomorphology (14, 15, 23, 98, 100, 135, 156, 157), the fossil occurrence of extant species in areas drier or wetter than those encompassed by present distributions (10, 42–44, 93, 94, 99), the occurrence of aboriginal stone artifacts that must have been obtained from deposits presently under the sea (127), and the occurrence of outlying populations of many species in regions wetter or drier than those occupied by the main species population (43, 73, 95). The climatic changes and fluctuating sea levels of the Quaternary would have accentuated established patterns of Tertiary landscape evolution on the Great Plateau and hence promoted differential rates of speciation in rainfall zones, as outlined previously. Additional stimulants to Quaternary speciation may well have been the deposition of new coastal landforms and the occurrence of regional differences in the severity of climatic stresses on plant populations. With regard to the latter, Gentilli (58, 60) and Burbidge (33, 34; see also 24) noted that the vegetation of present-day transitional-rainfall areas would have been much more susceptible to stress during minor climatic changes than would the vegetation of high-rainfall and arid zones. This was inferred from the distribution of present rainfall gradients, which are steepest in the high-rainfall zone and more gradual towards the arid zone (Figure 2). A slight decrease in annual rainfall could advance the desert margin tens or even hundreds of kilometres into the transitional zone, while having a negligible effect on the boundary of the high-rainfall zone. Even during a severe arid period, a belt of high rainfall would persist along the south coast and in the Darling Range as a result of onshore moisture-laden winds. Conversely, a period of exceptionally wet weather would extend the high-rainfall zone into transitional areas, but would be unlikely to exert a strong influence on the arid core of the interior. Hence during the recurrent climatic fluctuations of the Quaternary, plant populations in the transitional-rainfall zone would have experienced soil moisture stresses more than their counterparts in high-rainfall and arid areas. Consequently, population flush-crash cycles (39) would have occurred more often and rates of isolation, migration, extinction, confluence, and possible hybridization would have been highest in the transitional-rainfall zone. Thus climatic stresses, as well as the erosional consequences of climatic fluctuations, would have produced labile and fragmented population systems ideal for rapid speciation in the transitional zone.

Coastal landforms were in a state of considerable flux throughout the Quaternary as a result of changes in sea level (40). Alternate phases of marine invasion and regression opened and closed coastal migration routes, particularly across the southern margin of the Nullarbor Plain (68, 91, 102, 104–106, 108, 121, 123–125, 136, 137), and led to the formation of coastal plains and limestone islands (14, 15, 100, 135). These, too, have been active sites of population isolation, extinction, migration, and confluence (1, 42, 43) which, in some cases, have led to speciation—e.g. *Conostylis pauciflora* (71, 72), tetraploid races of *Dampiera linearis* (21, 22), and *Conostylis setigera* (S. D. Hopper, unpublished data) on the Swan Coastal Plain.

CONCLUSIONS

The fossil record indicates that much that is familiar and unique in the southwestern vegetation today—the dominance of sclerophyll hardwoods, the species richness of the heathlands, the high specific endemism—developed only in the recent geological past as a consequence of progressive aridity in the late Tertiary and the climatic fluctuations of the Quaternary. Throughout most of its evolutionary history, the flora enjoyed subtropical conditions and sclerophyllous genera formed only a minor component of the vegetation, presumably on infertile soils and in communities marginal to the luxuriant subtropical rainforests.

It would appear that the species richness and high endemism of the modern flora developed despite the absence of high mountain systems, as a consequence of the following specific geohistorical circumstances:

1. The preservation until today of some early Tertiary landscapes on the Great Plateau due to tectonic stability, low relief, and slow, local patterns of erosion. Even in the transitional-rainfall zone where erosion has been most active, remnants of the Tertiary plateau surface still exist. This feature of the landscape, combined with the persistence of humid climatic conditions in the high-rainfall zone, has provided habitat continuity over a long geological period, thereby saving many relict species from extinction.

2. The existence of marine, edaphic, or climatic barriers to migration on all sides of the southwest since the Eocene. These barriers have effectively isolated most components of the flora from related groups in eastern Australia, and have been responsible primarily for the maintenance of high specific endemism in the region.

3. The extensive formation of lateritic soils on the Great Plateau in the Oligocene and/or Miocene, and their continued genesis in areas of seasonal rainfall to the present day. Weathering of laterites gave rise to nutrient-deficient sands and gravels favoring a sclerophyllous heathland flora preadapted to the progressive aridity of the late Tertiary and Quaternary. The unique character of much of the modern southwestern flora is due to the prolific speciation and adaptive radiation of heathland genera.

4. Erosional dynamism and recurrent climatic stresses in the transitional rainfall zone during the late Tertiary and Quaternary. These factors produced a mosaic of landforms and soils together with labile population structures in the transitional-zone flora. Populations on the older plateau surfaces were fragmented; rates of migration, isolation, extinction, confluence, and hybridization were increased among species adapted to younger soils; and accelerated rates of speciation resulted. Habitat and climatic stability were greater in the high-rainfall zone during these periods and hence favored the persistence of a more conservative relict flora. Thus evolutionary studies of southwestern angiosperms provide support for the emerging concept that semi-arid transitional climatic zones are particularly favorable sites for speciation, while permanently humid and permanently arid zones favor evolutionary stability (3–5, 131, 132, 140, 142–144, 151).

At least two of the four factors suggested by Raven (131) as being responsible for the evolution of the rich endemic floras in mediterranean climatic regions seem applicable in southwestern Australia. These are the existence of complex edaphic mosaics (cf Figure 4) and the susceptibility of plant populations to major climatic changes and year-to-year fluctuations. As indicated above, both these factors appear to have produced population structures particularly conducive to rapid genetic differentiation and speciation. On the other hand, Raven's suggestion that evolutionary divergence in mediterranean climates has been promoted by the environment's favoring rapidly evolving annuals seems unlikely to apply to the southwestern flora, since annual herbs form a relatively small component of the total species (13, 57). Raven's fourth suggestion—that speciation in mediterranean regions has been facilitated by the occurrence of large numbers of species-specific hymenopteran pollinators—warrants further investigation in the Australian southwest. Present evidence is equivocal, since some hymenopteran groups in the region appear to be catholic pollen and nectar gatherers [(71), S. D. Hopper, unpublished data; G. J. Keighery, personal communication], while other groups are known to be species-specific, e.g. the thynnid wasps that pollinate pseudocopulatory orchids (146, 147). Apart from differential adaptation to pollinators, the flora appears to have responded to several other environmental parameters whose evolutionary significance remains to be systematically examined (e.g. fire, soil nutrients, soil water, predators, competitors).

While present knowledge allows for an understanding of the major climatic and geomorphological contingencies that appear to have influenced speciation in the southwestern flora, critical information on particular evolutionary developments remains fragmentary. Moreover, it should be stressed that geohistorical circumstances alone have not been wholly responsible for patterns of speciation in the Australian southwest. It is clear from the work of S. H. James and his students (7, 17, 18, 21, 22, 48, 55, 71, 73, 75, 79–85, 92, 134) that properties of the genetic systems of transitional-rainfall zone taxa have exerted a profound influence over evolutionary potentialities. The future promises rich rewards for integrative studies, such as those on *Dampiera linearis* (21, 22) or *Stylidium crossocephalum* (48), where evolutionary interpretations are based on a thorough understanding of dynamic geohistorical contingencies and the genetic systems of southwestern species.

ACKNOWLEDGMENTS

I wish to thank A. A. Burbidge, A. H. Burbidge, D. J. Coates, A. J. M. Hopkins, R. J. Hnatiuk, S. H. James, G. J. Keighery, A. R. Main, N. G. Marchant, and B. L. Rye for constructive comments on the manuscript. I am grateful to J. S. Beard, A. H. Burbidge, K. Clarke, D. J. Coates, A. J. M. Hopkins, P. G. Farrell, S. H. James, G. J. Keighery, B. J. Keighery, T. D. Macfarlane, N. G. Marchant, B. L. Rye, G. Stone, and P. G. Wilson for permission to include details of their unpublished research in this publication. I am also indebted to Dr. A. A. Burbidge (Chief Research Officer, Western Australian Wildlife Research Centre), Professor J. S. Pate (Botany Department, University of Western Australia) and Dr. J. W. Green (Curator, Western Australian Herbarium) for the provision of research facilities at their respective institutions while work for the review was in progress.

My research on the phytogeography and evolutionary biology of the southwest Australian flora has been supported by grants from the Australian Biological Resources Study, the Australian Research Grants Committee, the Western Australian Department of Fisheries and Wildlife, and a Commonwealth Postgraduate Research Award.

Literature Cited[1]

1. Abbott, I. 1977. Species richness, turnover and equilibrium in insular floras near Perth, Western Australia. *Aust. J. Bot.* 25:193–208
2. Archer, M. 1977. Revision of the dasyurid marsupial genus *Antechinomys* Krefft. *Mem. Queensl. Mus.* 18:17–29
3. Axelrod, D. I. 1967. Drought, diastrophism and quantum evolution. *Evolution* 21:201–9
4. Axelrod, D. I. 1972. Edaphic aridity as a factor in angiosperm evolution. *Am. Nat.* 106:311–20
5. Axelrod, D. I. 1973. History of the Mediterranean Ecosystem in California. In *Ecological Studies 7. Mediterranean Type Ecosystems,* ed F. di Castri, H. A. Mooney, pp. 225–77. Berlin: Springer. 405 pp.
6. Balme, B. E., Churchill, D. M. 1959. Tertiary sediments at Coolgardie, Western Australia. *J. R. Soc. West. Aust.* 42:37–43
7. Banyard, B. J., James, S. H. 1979. Biosystematic studies in the *Stylidium crassifolium* species complex (Stylidiaceae). *Aust. J. Bot.* 27:27–37
8. Barlow, B. A. 1971. Cytogeography of the genus *Eremophila. Aust. J. Bot.* 19:295–310
9. Baynes, A. 1975. The distributions of the native mammals of south-western Australia. *Aust. Mammal.* 1:404–5
10. Baynes, A., Merrilees, D., Porter, J. K. 1975. Mammal remains from the upper levels of a late Pleistocene deposit in Devil's Lair, Western Australia. *J. R. Soc. West. Aust.* 58:97–126
11. Beadle, N. C. W. 1979. Origins of the Australian angiosperm flora. In *Biogeography and Ecology of Australia,* Vol. 1, ed. A. Keast. The Hague: Junk. In press
12. Beard, J. S. 1969. Endemism in the Western Australian flora at the species level. *J. R. Soc. West. Aust.* 52:18–20
13. Beard, J. S. 1970. *A Descriptive Catalogue of West Australian Plants.* Sydney: Soc. Growing Aust. Plants. 142 pp. 2nd ed.
14. Beard, J. S. 1975. *Nullarbor. Explanatory Notes to Sheet 4, 1:1000 000 Series, Vegetation Survey of Western Australia.* Nedlands: Univ. West. Aust. Press. 104 pp.
15. Beard, J. S. 1976. *Murchison. Explanatory Notes to Sheet 6, 1:1 000 000 Series, Vegetation Survey of Western Australia.* Nedlands: Univ. West. Aust. Press. 141 pp.
16. Beard, J. S. 1977. Tertiary evolution of the Australian flora in the light of latitudinal movements of the continent. *J. Biogeog.* 4:111–18
17. Beltran, I. C., James, S. H. 1970. Complex hybridity in *Isotoma petraea* III. Lethal system in $\odot_{12}$ Bencubbin. *Aust. J. Bot.* 18:223–32
18. Beltran, I. C., James, S. H. 1974. Complex hybridity in *Isotoma petraea* IV. Heterosis in interpopulational hybrids. *Aust. J. Bot.* 22:251–64
19. Bennett, E. M. 1972. A revision of the Australian species of *Hybanthus* Jacquin (Violaceae). *Nuytsia* 1:218–41
20. Bettenay, E., Hingston, F. J. 1964. Development and distribution of soils in the Merredin area, Western Australia. *Aust. J. Soil. Res.* 2:173–86
21. Bousfield, L. R. 1970. *Chromosome races in* Dampiera linearis R. Br. PhD thesis. Univ. West Aust. Perth
22. Bousfield, L. R., James, S. H. 1976. The behaviour and possible cytoevolutionary significance of B chromosomes in *Dampiera linearis* (Angiospermae: Goodeniaceae). *Chromosoma* 55:309–23
23. Bowler, J. M. 1976. Aridity in Australia: age, origins and expression in aeolian landforms and sediments. *Earth Sci. Rev.* 12:279–310
24. Bowler, J. M. 1976. Recent developments in reconstructing Late Quaternary environments in Australia. In *The Biological Origin of the Australians,* ed. R. L. Kirk, A. G. Thorne, pp. 55–77. Canberra: Aust. Inst. Aborig. Stud. 449 pp.

[1]Literature search ended December 1978

25. Bowler, J. M., Hope, G. S., Jennings, J. N., Singh, G., Walker, D. 1976. Late Quaternary climates of Australia and New Guinea. *Quat. Res.* 6:359–94
26. Bramwell, D. 1972. Endemism in the flora of the Canary Islands. In *Taxonomy, Phytogeography and Evolution*, ed. D. H. Valentine, pp. 141–59. London: Academic. 431 pp.
27. Brewer, R., Bettenay, E. 1973. Further evidence concerning the origin of the Western Australian sand plains. *J. Geol. Soc. Aust.* 19:541–53
28. Brooker, M. I. H. 1972. Four new taxa of *Eucalyptus* from Western Australia. *Nuytsia* 1:242–53
29. Brooker, M. I. H. 1974. Six new species of *Eucalyptus* from Western Australia. *Nuytsia* 1:297–314
30. Brooker, M. I. H., Blaxell, D. F. 1978. Five new species of *Eucalyptus* from Western Australia. *Nuytsia* 2:220–31
31. Burbidge, A. A., Kirsch, J. A. W., Main, A. R. 1974. Relationships within the Chelidae (Testudines: Pleurodira) of Australia and New Guinea. *Copeia* 1974:392–409
32. Burbidge, N. T. 1953. The genus *Triodia* R.Br. (Graminae). *Aust. J. Bot.* 1:121–84
33. Burbidge, N. T. 1953. The significance of the mallee habit in Eucalyptus. *Proc. R. Soc. Queensl.* 62:73–78
34. Burbidge, N. T. 1960. The phytogeography of the Australian Region. *Aust. J. Bot.* 8:75–209
35. Calaby, J. H., Corbett, L. K., Sharman, G. B., Jonnston, P. G. 1974. The chromosomes and systematic position of the Marsupial Mole, *Notoryctes typhlops*. *Aust. J. Biol. Sci.* 27:529–32
36. Carlquist, S. 1969. Studies in Stylidiaceae: new taxa, field observations, evolutionary tendencies. *Aliso* 7:13–64
37. Carlquist, S. 1974. *Island Biology*. NY: Columbia Univ. Press. 660 pp.
38. Carlquist, S. 1976. New species of *Stylidium*, and notes on Stylidiaceae from south western Australia. *Aliso* 8:447–63
39. Carson, H. L. 1975. The genetics of speciation at the diploid level. *Am. Nat.* 109:83–92
40. Chappell, J. 1976. Aspects of Late Quaternary palaeogeography of the Australian–East Indonesian Region. See Ref. 24, pp. 11–22
41. Chippendale, G. M. 1973. *Eucalypts of the Western Australian Goldfields (and the Adjacent Wheatbelt)*. Canberra: Aust. Govt. Publ. Serv. 218 pp.
42. Churchill, D. M. 1960. Late Quaternary changes in the vegetation on Rottnest Island. *West. Aust. Nat.* 7:160–66
43. Churchill, D. M. 1961. *The Tertiary and Quaternary vegetation and climate in relation to the living flora in South Western Australia*. PhD thesis. Univ. West. Aust. Perth. 241 pp.
44. Churchill, D. M. 1968. The distribution and prehistory of *Eucalyptus diversicolor* F. Muell., *E. marginata* Donn. ex Sm., and *E. calophylla* R.Br. in relation to rainfall. *Aust. J. Bot.* 16:125–51
45. Churchill, D. M. 1973. The ecological significance of tropical mangroves in the early Tertiary floras of southern Australia. *Spec. Publ. Geol. Soc. Aust.* 4:79–86
46. Churchill, D. M. 1973. Fossil pollen from Dulcurna Station, Cal Lal, S. W. New South Wales, Australia. *Mem. Natl. Mus. Victoria* 34:183–84
47. Churchward, H. M. 1970. Erosional modification of a lateritized landscape over sedimentary rocks. Its effect on soil distribution. *Aust. J. Soil Res.* 8:1–19
48. Coates, D. J., James, S. H. 1979. Chromosome variation in *Stylidium crossocephalum* (Angiospermae: Stylidiaceae) and the dynamic coadaptation of its lethal system. *Chromosoma* 72:357–76
49. Cogger, H. G. 1975. *Reptiles and Amphibians of Australia*. Sydney: Reed. 584 pp.
50. Cookson, I. C. 1954. The occurrence of an older Tertiary microflora in Western Australia. *Aust. J. Sci.* 17:37–38
51. Crocker, R. L., Wood, J. G. 1947. Some historical influences on the development of the South Australian vegetation communities and their bearing on concepts and classifications in ecology. *Trans. R. Soc. S. Aust.* 71:91–136
52. Davis, P. H. 1971. Distribution patterns in Anatolia with particular reference to endemism. In *Plant Life of South-West Asia*, ed. P. H. Davis, P. C. Harper, I. C. Hedge, pp. 15–27. Aberdeen: University Press. 335 pp.
53. Diels, L. 1906. Die Pflanzenwelt von West-Australien südlich des Wendekreises. In *Die Vegetation der Erde*, ed. A. Engler, O. Drude, Vol. VII. Leipzig: Engelmann. 413 pp.
54. Ey, T. M., Barlow, B. A. 1972. Distribution of chromosome races in the *Eremophila glabra* complex. *Search* 3:337–38
55. Farrell, P. G., James, S. H. 1979. *Stylidium ecorne* (F. Muell. ex Erickson

and Willis) comb. et stat. nov. (Stylidiaceae). *Aust. J. Bot.* 27:39–45

56. Forshaw, J. M. 1969. *Australian Parrots.* Melbourne: Landsdowne Press. 306 pp.

57. Gardner, C. A. 1944. The vegetation of Western Australia with special reference to the climate and soils. *J. R. Soc. West. Aust.* 28:11–87

58. Gentilli, J. 1951. Bioclimatic changes in Western Australia. *West. Aust. Nat.* 2:175–84

59. Gentilli, J. 1961. Quaternary climates of the Australian Region. *Ann. NY Acad. Sci.* 95:465–501

60. Gentilli, J. 1971. Climatic fluctuations. In *Climates of Australia and New Zealand,* ed. J. Gentilli, Amsterdam: Elsevier. 405 pp.

61. George, A. S. 1974. Seven new species of *Grevillea* (Proteaceae) from Western Australia. *Nuytsia* 1:370–74

62. George, A. S. 1974. Five new species of *Adenanthos* (Proteaceae) from Western Australia. *Nuytsia* 1:381–86

63. George, A. S., Hopkins, A. J. M., Marchant, N. G. 1979. The heathlands of Western Australia. In *Ecosystems of the World. Heathlands and Related Shrublands,* ed. R. L. Specht. Amsterdam: Elsevier. In press

64. Gill, E. D. 1975. Evolution of Australia's unique flora and fauna in relation to the plate tectonics theory. *Proc. R. Soc. Victoria* 87:215–34

65. Glauert, L. 1967. *A Handbook of the Snakes of Western Australia.* Perth: West Aust. Nat. Club. 3rd ed. 62 pp.

66. Goldblatt, D. 1978. An analysis of the flora of Southern Africa: its characteristics, relationships, and origins. *Ann. Mo. Bot. Gard.* 65:369–436

67. Green, J. W. 1960. The genus *Conostylis* R.Br. II. Taxonomy *Proc. Linn. Soc. New South Wales* 85:334–73

68. Green, J. W. 1964. Discontinuous and presumed vicarious plant species in southern Australia. *J. R. Soc. West. Aust.* 47:25–32

69. Havel, J. J. 1975. Site-vegetation mapping in the northern jarrah forest (Darling Range). 1. Definition of site-vegetation types. *For. Dept. West. Aust. Bull.* 86. 115 pp.

70. Hooker, J. D. 1860. *The Botany The Antarctic Voyage of H. M. Discovery Ships Erebus and Terror in the years 1839–1843. Part III. Flora Tasmaniae Vol. I. Dicotyledones.* London: Lovell Reeve. 359 pp.

71. Hopper, S. D. 1977. Variation and natural hybridization in the *Conostylis*

aculeata R.Br. species group near Dawesville, Western Australia. *Aust. J. Bot.* 25:395–411

72. Hopper, S. D. 1978. Nomenclatural notes and new taxa in the *Conostylis aculeata* group (Haemodoraceae). *Nuytsia* 2:254–64

73. Hopper, S. D. 1978. *Speciation in the kangaroo paws of southwestern Australia (Anigozanthos and Macropidia: Haemodoraceae).* PhD thesis. Univ. West. Aust., Perth. 418 pp.

74. Hopper, S. D., Campbell, N. A. 1977. A multivariate morphometric study of species relationships in kangaroo paws (*Anigozanthos* Labill. and *Macropidia* Drumm. ex Harv.: Haemodoraceae). *Aust. J. Bot.* 25:523–44

75. Hopper, S. D., Coates, D. J., Burbidge, A. H. 1978. Natural hybridization and morphometric relationships between three mallee eucalypts in the Fitzgerald River National Park, W. A. *Aust. J. Bot.* 26:319–33

76. Hopper, S. D., Maslin, B. R. 1978. Phytogeography of *Acacia* in Western Australia. *Aust. J. Bot.* 26:63–78

77. Hos, D. 1975. Preliminary investigation of the palynology of the Upper Eocene Werillup Formation, Western Australia. *J. R. Soc. West. Aust.* 58:1–14

78. Iredale, T. 1939. A review of the land Mollusca of Western Australia. *J. R. Soc. West. Aust.* 25:1–88

79. James, S. H. 1963. *Cytological Studies in the Lobeliaceae.* PhD thesis. Univ. Sydney, Australia. 176 pp.

80. James, S. H. 1965. Complex hybridity in *Isotoma petraea.* I. The occurrence of interchange heterozygosity, autogamy and a balanced lethal system. *Heredity* 20:341–53

81. James, S. H. 1970. A demonstration of a possible mechanism of sympatric divergence using simulation techniques. *Heredity* 25:241–52

82. James, S. H. 1970. Complex hybridity in *Isotoma petraea.* II. Components and operation of a possible evolutionary mechanism. *Heredity* 25:53–78

83. James, S. H. 1973. Cytogenetic aspects of the speciation process in plants. *J. R. Soc. West. Aust.* 56:36–43

84. James, S. H. 1979. Cytoevolutionary patterns, genetic systems and the phytogeography of Australia. See Ref. 11. In press

85. James, S. H. 1979. Chromosome numbers and genetic systems in the trigger plants of Western Australia (*Stylidium:* Stylidiaceae). *Aust. J. Bot.* 27:17–25

86. Johnson, L. A. S., Briggs, B. G. 1975. On the Proteaceae—the evolution and classification of a southern family. *Bot. J. Linn. Soc.* 70:83–182

87. Johnson, L. A. S., Briggs, B. G. 1979. Three old southern families—Myrtaceae, Proteaceae and Restionaceae. See Ref. 11. In press

88. Johnstone, M. H., Lowry, D. C., Quilty, P. G. 1973. The geology of south western Australia—a review. *J. R. Soc. West. Aust.* 56:5–15

89. Jutson, J. T. 1934. *The Physiography (Geomorphology) of Western Australia. Bull. Geol. Surv. West. Aust. No. 95.* 366 pp.

90. Keast, A. 1960. The unique plants and animals of south western Australia. *Aust. Mus. Mag.* 13:152–57

91. Keast, A. 1961. Bird speciation on the Australian continent. *Bull. Mus. Comp. Zool.* 123:305–495

92. Keighery, G. J. 1975. Chromosome numbers in the Gyrostemonaceae Endl. and the Phytolaccaceae Lindl.: a comparison. *Aust. J. Bot.* 23:335–38

93. Kendrick, G. W. 1977. Middle Holocene marine molluscs from near Guildford, Western Australia, and evidence for climatic change. *J. R. Soc. West. Aust.* 59:97–104

94. Kendrick, G. W. 1978. New species of fossil nonmarine molluscs from Western Australia and evidence of late Quaternary climatic change in the Shark Bay district. *J. R. Soc. West. Aust.* 60:49–60

95. Lange, R. T. 1960. Rainfall and soil control of tree species distribution around Narrogin, Western Australia. *J. R. Soc. West. Aust.* 43:104–10

96. Lewis, H. 1972. The origin of endemics in the California flora. See Ref. 26, pp. 179–89

97. Loneragan, W. A. 1975. The ecology of a graveyard. *Aust. J. Bot.* 23:803–14

98. Lowry, D. C. 1970. Geology of the Western Australian part of the Eucla Basin. *Bull. Geol. Surv. West. Aust. No. 122.* 201 pp.

99. Lundelius, E. L. 1960. Post Pleistocene faunal succession in Western Australia and its climatic interpretation. *Rep. Int. Geol. Congr. 21st Session, Pt. IV,* 142–53

100. McArthur, W. M., Bettenay, E. 1960. *The Development and Distribution of the Soils of the Swan Coastal Plain, Western Australia.* Melbourne: CSIRO Soil Publ. No. 16. 55 pp.

101. McArthur, W. M., Churchward, H. M., Hick, P. T. 1977. *Landforms and Soils of the Murray River Catchment Area of Western Australia.* Melbourne: CSIRO Aust. Div. Land Resour. Manag. Ser. No. 3, 1–23

102. Mackerras, I. M. 1962. Speciation in Australian Tabanidae. In *The Evolution of Living Organisms,* ed. G. W. Leeper, pp. 328–58. Melbourne: Melbourne Univ. Press. 459 pp.

103. Main, A. R. 1965. *Frogs of Southern Western Australia.* Perth: West. Aust. Nat. Club. 73 pp.

104. Main, A. R. 1968. Ecology, systematics and evolution of Australian frogs. *Adv. Ecol. Res.* 5:37–86

105. Main, A. R., Lee, A. K., Littlejohn, M. J. 1958. Evolution in three genera of Australian frogs. *Evolution* 12:224–33

106. Main, B. Y. 1962. Adaptive responses and speciation in the spider genus *Aganippe* Cambridge. See Ref. 102, pp. 359–69

107. Marchant, N. G. 1973. Species diversity in the south-western flora. *J. R. Soc. West. Aust.* 56:23–30

108. Martin, H. A. 1973. Palynology and historical ecology of some cave excavations in the Australian Nullarbor. *Aust. J. Bot.* 21:283–316

109. Martin, H. A. 1973. Upper Tertiary palynology in southern New South Wales. *Spec. Publ. Geol. Soc. Aust.* 4:35–54

110. Martin, H. A. 1977. The history of *Ilex* (Aquifoliaceae) with special reference to Australia: Evidence from pollen. *Aust. J. Bot.* 25:655–73

111. Martin, H. A. 1979. The Tertiary flora. See Ref. 11. In press

112. Maslin, B. R. 1975. Studies in the genus *Acacia* (Mimosaceae). 4. A revision of the series Pulchellae. *Nuytsia* 1:388–494

113. Maslin, B. R. 1978. Studies in the genus *Acacia* (Mimosaceae). 7. The taxonomy of some diaphyllodinous species. *Nuytsia* 2:200–19

114. Monzu, N. 1977. *Coexistence of carrion breeding Caliphoridae (Diptera) in Western Australia.* PhD thesis. Univ. West. Aust., Perth. 221 pp.

115. Muir, B. G. 1977. Biological survey of the Western Australian wheatbelt. Part 2.: Vegetation and habitat of Bendering Reserve. *Rec. West. Aust. Mus. Suppl. No. 3.* 142 pp.

116. Mulcahy, M. J. 1960. Laterites and lateritic soils in south-western Australia. *J. Soil Sci.* 11: 206–25

117. Mulcahy, M. J. 1967. Landscapes, laterites and soils in south western Australia. In *Landform Studies from Australia*

and New Guinea, ed. J. N. Jennings, J. A. Mabbutt, pp. 211–30. Canberra: Aust. Natl. Univ. Press. 434 pp.

118. Mulcahy, M. J. 1973. Landforms and soils of south western Australia. *J. R. Soc. West. Aust.* 56:16–22

119. Mulcahy, M. J., Churchward, H. M., Dimmock, G. M. 1972. Landforms and soils on an uplifted peneplain in the Darling Range, Western Australia. *Aust. J. Soil Res.* 10:1–14

120. Mulcahy, M. J., Hingston, F. J. 1961. *The Development and Distribution of the Soils of the York-Quairading Area, Western Australia*. Melbourne: CSIRO Soil Publ. No. 17. 43 pp.

121. Nelson, E. C. 1974. Disjunct plant distributions on the south-western Nullarbor Plain, Western Australia. *J. R. Soc. West. Aust.* 57:105–7

122. Nelson, E. C. 1978. A taxonomic revision of the genus *Adenanthos* (Proteaceae). *Brunonia* 1:303–406

123. Nelson, E. C. 1979. Phytogeography of southern Australia. See Ref. 11. In press.

124. Parsons, R. F. 1969. Distribution and palaeogeography of two mallee species of *Eucalyptus* in southern Australia. *Aust. J. Bot.* 17. 30

125. Parsons, R. F. 1970. Mallee vegetation of the southern Nullarbor and Roe Plains, Australia. *Trans. R. Soc. S. Aust.* 94:227–42

126. Paust, S. 1974. Taxonomic studies in *Thomasia* and *Lasiopetalum* (Sterculiaceae). *Nuytsia* 1:348–66

127. Pearce, R. H. 1977. Relationship of chert artefacts at Walyunga in southwest Australia, to Holocene sea levels. *Search* 10:375–77

128. Pryor, L. D., Johnson, L. A. S. 1979. *Eucalyptus*, the universal Australian. See Ref. 11. In press

129. Randell, B. R. 1970. Adaptations in the genetic system of Australian arid zone *Cassia* species (Leguminosae, Caesalpinioideae). *Aust. J. Bot.* 18:77–97

130. Randell, B. R., Symon, D. E. 1977. Distribution of *Cassia* and *Solanum* species in arid regions of Australia. *Search* 8:206–7

131. Raven, P. H. 1973. The evolution of Mediterranean floras. See Ref. 5, pp. 213–24

132. Raven, P. H., Axelrod, D. I. 1978. Origin and relationships of the California flora. *Univ. Calif. Publ. Bot.* 72:1–134

133. Ride, W. D. L. 1970. *A guide to the Native Mammals of Australia*. London: Oxford Univ. Press. 249 pp.

134. Rye, B. L. 1979. Chromosome number variation in the Myrtaceae and its taxonomic implications. *Aust. J. Bot.* 27. In press

135. Seddon, G. 1972. *Sense of Place. A Response to an Environment—the Swan Coastal Plain, Western Australia*. Nedlands: Univ. West. Aust. Press. 274 pp.

136. Serventy, D. L. 1953. Some speciation problems in Australian birds: with particular reference to the relations between Bassian and Eyrean 'Species-Pairs'. *Emu* 53:131–45

137. Serventy, D. L., Whittell, H. M. 1976. *Birds of Western Australia*. Perth: Univ. West. Aust. Press. 481 pp. 5th ed.

138. Silsbury, J. H., Brittan, N. H. 1955. Distribution and ecology of the genus *Kennedya* Vent. in Western Australia. *Aust. J. Bot.* 3:113–35

139. Speck, N. H. 1958. *The vegetation of the Darling-Irwin botanical districts and an investigation of the distribution patterns of the family Proteaceae in South Western Australia*. PhD thesis. Univ. West. Aust., Perth. 637 pp.

140. Stebbins, G. L. 1952. Aridity as a stimulus to evolution. *Am. Nat.* 86:33–44

141. Stebbins, G. L. 1971. *Chromosomal Evolution in Higher Plants*. London: Edward Arnold. 216 pp.

142. Stebbins, G. L. 1972. Ecological distribution of centers of major adaptive radiation in angiosperms. See Ref. 26, pp. 7–34

143. Stebbins, G. L. 1974. *Flowering Plants: Evolution Above the Species Level*. Cambridge, Mass: Harvard Univ. Press. 399 pp.

144. Stebbins, G. L., Major, J. 1965. Endemism and speciation in the California flora. *Ecol. Monogr.* 35:1–35

145. Stewart, D. A., Barlow, B. A. 1976. Infraspecific polyploidy and gynodioecism in *Ptilotus obovatus* (Amaranthaceae). *Aust. J. Bot.* 24:237–48

146. Stoutamire, W. P. 1974. Australian terrestrial orchids, thynnid wasps, and pseudocopulation. *Bull. Am. Orchid Soc.* 43:13–18

147. Stoutamire, W. P. 1975. Pseudocopulation in Australian terrestrial orchids. *Bull. Am. Orchid Soc.* 44:226–33

148. Turrill, W. B. 1929. *The Plant-life of the Balkan Peninsula—A Phytogeographical Study*. Oxford: Clarendon Press.

149. Weimarck, H. 1941. Phytogeographical groups, centres and intervals within the Cape flora. *Lunds Univ. Arsskr. Avd. 2*, 37(5): 143 pp.

150. Whittaker, R. H. 1961. Vegetation history of the Pacific Coast States and the

"central" significance of the Klamath Region. *Madroño* 16:5–23

151. Whittaker, R. H. 1977. Evolution of species diversity in land communities. *Evol. Biol.* 10:1–67

152. Williams, R. F. 1932. An ecological analysis of the plant communities of the jarrah region on a small area near Darlington. *J. R. Soc. West. Aust.* 18:105–24

153. Williams, R. F. 1945. An ecological study near Beraking forest station. *J. R. Soc. West. Aust.* 31:19–31

154. Wilson, P. G. 1970. A taxonomic revision of the genera *Crowea, Eriostemon* and *Phebalium* (Rutaceae). *Nuytsia* 1:3–155

155. Wilson, P. G. 1975. A taxonomic revision of the genus *Maireana* (Chenopodiaceae). *Nuytsia* 2:2–83

156. Wyrwoll, K. H. 1977. Late Quaternary events in Western Australia. *Search* 8:32–34

157. Wyrwoll, K. H., Milton, D. 1976. Widespread late Quaternary aridity in Western Australia. *Nature* 264:429–30

158. Zohary, M. 1973. *Geobotanical Foundations of the Middle East,* Vol. 1. Stuttgart: Gustav Fischer. 340 pp.

AUTHOR INDEX

423

SUBJECT INDEX

CUMULATIVE INDEXES

CONTRIBUTING AUTHORS, VOLUMES 6–10

450

CHAPTER TITLES, VOLUMES 6–10